Springer Series on Biofilms

Series Editors:

Mark Shirtliff, Baltimore, USA
Paul Stoodley, Southhampton, United Kingdom
Thomas Bjarnsholt, Copenhagen Ø, Denmark

For further volumes:
http://www.springer.com/series/7142

Kendra P. Rumbaugh • Iqbal Ahmad
Editors

Antibiofilm Agents

From Diagnosis to Treatment and Prevention

Volume 8

Springer

Editors
Kendra P. Rumbaugh
Department of Surgery
Texas Tech University
  Health Sciences Center
Lubbock
Texas
USA

Iqbal Ahmad
Department of Ag. Microbiology
Aligarh Muslim University
Aligarh
India

ISSN 1863-9607   ISSN 1863-9615 (electronic)
ISBN 978-3-642-53832-2   ISBN 978-3-642-53833-9 (eBook)
DOI 10.1007/978-3-642-53833-9
Springer Heidelberg New York Dordrecht London

Library of Congress Control Number: 2014936011

© Springer-Verlag Berlin Heidelberg 2014
This work is subject to copyright. All rights are reserved by the Publisher, whether the whole or part of the material is concerned, specifically the rights of translation, reprinting, reuse of illustrations, recitation, broadcasting, reproduction on microfilms or in any other physical way, and transmission or information storage and retrieval, electronic adaptation, computer software, or by similar or dissimilar methodology now known or hereafter developed. Exempted from this legal reservation are brief excerpts in connection with reviews or scholarly analysis or material supplied specifically for the purpose of being entered and executed on a computer system, for exclusive use by the purchaser of the work. Duplication of this publication or parts thereof is permitted only under the provisions of the Copyright Law of the Publisher's location, in its current version, and permission for use must always be obtained from Springer. Permissions for use may be obtained through RightsLink at the Copyright Clearance Center. Violations are liable to prosecution under the respective Copyright Law.
The use of general descriptive names, registered names, trademarks, service marks, etc. in this publication does not imply, even in the absence of a specific statement, that such names are exempt from the relevant protective laws and regulations and therefore free for general use.
While the advice and information in this book are believed to be true and accurate at the date of publication, neither the authors nor the editors nor the publisher can accept any legal responsibility for any errors or omissions that may be made. The publisher makes no warranty, express or implied, with respect to the material contained herein.

Printed on acid-free paper

Springer is part of Springer Science+Business Media (www.springer.com)

Preface

The editors would like to dedicate this text to the late Dr. Bill Costerton, who is regarded as "The Father of Biofilm." Bill spent the good part of his career working tirelessly to alert and convince the medical community about the existence and importance of biofilms. The fact that many medical specialties are now addressing the "biofilm problem" is in no small degree because of his contributions and those of the scientists he trained and mentored.

Biofilms comprise microbial microcolonies adhered to a surface and surrounded by a sticky exopolysaccharide matrix. Once adherent, microbes multiply and anchor themselves in quite intricate structures, which appear to allow for communication and transfer of nutrients, waste, and signaling compounds. Microbial biofilms constitute a major cause of chronic infections, especially in association with medical devices. Biofilms are extremely difficult to eradicate with conventional antibiotics and therefore represent an enormous healthcare burden.

While the "biofilm concept" has, for the most part, become accepted by the medical community, clinicians are left with the dilemma of how to diagnose and treat these infections. While there are a number of books highlighting research progress on understanding mechanisms of biofilm establishment and their roles in disease, there are currently no existing resources which provide a comprehensive review of the available antibiofilm options.

The purpose of this book is to provide a survey of the recent progress that has been made on the development of antibiofilm agents. Biofilm experts from across the globe have contributed and related their expertise on topics ranging from diagnosing and characterizing biofilm infections to treatment options and finally regulatory challenges to the commercial development of antibiofilm drugs. We intend for this book to serve as a valuable resource for medical professionals seeking to treat biofilm-related disease, academic and industry researchers interested in drug discovery, and instructors who teach microbial pathogenesis and medical microbiology.

Lubbock, TX Kendra P. Rumbaugh
Aligarh, Uttar Pradesh, India Iqbal Ahmad
November 2013

Contents

Part I Medical Biofilms

Biofilms in Disease 3
Michael Otto

The Use of DNA Methods to Characterize Biofilm Infection 15
Randall Wolcott and Stephen B. Cox

Imaging Biofilms in Tissue Specimens 31
Garth James and Alessandra Marçal Agostinho Hunt

Mechanisms of Drug Resistance in Fungi and Their Significance
in Biofilms 45
Rajendra Prasad, Abdul Haseeb Shah, and Sanjiveeni Dhamgaye

Horizontal Gene Transfer in Planktonic and Biofilm Modes 67
Melanie Broszat and Elisabeth Grohmann

The Role of Quorum Sensing in Biofilm Development 97
Kendra P. Rumbaugh and Andrew Armstrong

Part II Strategies for Biofilm Control

Current and Emergent Control Strategies for Medical Biofilms 117
Mohd Sajjad Ahmad Khan, Iqbal Ahmad, Mohammad Sajid,
and Swaranjit Singh Cameotra

The Effect of Plasmids and Other Biomolecules
on the Effectiveness of Antibiofilm Agents 161
L.C. Gomes, P.A. Araújo, J.S. Teodósio, M. Simões,
and F.J. Mergulhão

Antimicrobial Coatings to Prevent Biofilm Formation on Medical Devices 175
Phat L. Tran, Abdul N. Hamood, and Ted W. Reid

Medicinal Plants and Phytocompounds: A Potential Source of Novel Antibiofilm Agents 205
Iqbal Ahmad, Fohad Mabood Husain, Meenu Maheshwari, and Maryam Zahin

***Staphylococcus aureus* Biofilm Formation and Inhibition** 233
Carolyn B. Rosenthal, Joe M. Mootz, and Alexander R. Horswill

Novel Targets for Treatment of *Pseudomonas aeruginosa* Biofilms 257
Morten Alhede, Maria Alhede, and Thomas Bjarnsholt

Inhibition of Fungal Biofilms 273
Christopher G. Pierce, Anand Srinivasan, Priya Uppuluri, Anand K. Ramasubramanian, and José L. López-Ribot

Biofilm Control Strategies in Dental Health 291
Jorge Frias-Lopez

Control of Polymicrobial Biofilms: Recent Trends 327
Derek S. Samarian, Kyung Rok Min, Nicholas S. Jakubovics, and Alexander H. Rickard

Antibiofilm Strategies in the Food Industry 359
Pilar Teixeira and Diana Rodrigues

Part III The Future of Antibiofilm Agents

Biofilm Inhibition by Nanoparticles 385
D. Bakkiyaraj and S.K. Pandian

Drug Delivery Systems That Eradicate and/or Prevent Biofilm Formation 407
Mohammad Sajid, Mohd Sajjad Ahmad Khan, Swaranjit Singh Cameotra, and Iqbal Ahmad

Eradication of Wound Biofilms by Electrical Stimulation 425
Chase Watters and Matt Kay

The Effects of Photodynamic Therapy in Oral Biofilms 449
Michelle Peneluppi Silva, Juliana Campos Junqueira, and Antonio Olavo Cardoso Jorge

Clinical and Regulatory Development of Antibiofilm Drugs: The Need, the Potential, and the Challenges 469
Brett Baker, Patricia A. McKernan, and Fred Marsik

Index 487

About the Editors

Iqbal Ahmad is a senior faculty and Chairman of the Department of Agricultural Microbiology. Trained at the Central Drug Research Institute, Lucknow and The Himalaya Drug Company, India, Dr. Ahmad Joined AMU Aligarh as lecturer in 1995. His research works encompasses various disciplines such as drug resistance and virulence in *E. coli* and plasmid biology, microbial ecology, impact of wastewater on soil health and crop productivity, drug and metal resistance in microbes of clinical and environmental origin, and biological activities of Indian Medicinal plants. His present research interest is in the fields of molecular basis of drug resistance and virulence factors linkage in pathogenic bacteria, bacterial quorum sensing, and modulation of quorum sensing-linked bacterial traits and biofilms by natural products. Dr. Ahmad has guided many Ph.D. and several M.Sc. theses, completed four research projects, published seven Books. His research works have received fair citations by scientific community.

Kendra Rumbaugh is an Associate Professor of Surgery, with joint appointments in the departments of Immunology and Molecular Microbiology and Cell Biology and Biochemistry at the Texas Tech University Health Sciences Center in Lubbock, Texas. Dr. Rumbaugh's research focuses on understanding and treating wound infections, and she is especially interested in how biofilms, polymicrobial interactions and quorum sensing contribute to bacterial pathogenesis.

Part I
Medical Biofilms

Biofilms in Disease

Michael Otto

Abstract Biofilms contribute to a majority of infectious diseases caused by bacterial and fungal pathogens. These range from chronic infections of indwelling medical devices and wounds to frequently fatal, serious infections like endocarditis. Biofilm research was initially focused on "environmental" biofilms, such as those present in wastewater tubing. More recently, "medical" biofilms as present during human infection have gained increased attention, and several animal models to mimic biofilm-associated infection in vivo have been established. Furthermore, biofilm research has shifted from the use of laboratory to clinical strains and is being complemented by the genetic analysis of isolates originating from biofilm infection. Often these investigations showed that in vitro results only have limited relevance for the in vivo situation, revealing the necessity of more intensive in vivo biofilm research. This introductory chapter will present an overview of biofilm infections, resistance, and the general model of biofilm development. It will also introduce important biofilm molecules and principles of regulation in premier biofilm-forming pathogens and finish with a general outline of possible routes of anti-biofilm drug development.

1 Introduction

According to the World Health Organization, infectious diseases are the second most frequent cause of death worldwide, responsible for more than 13 millions of deaths per year, which is second only to diseases of the heart. Many of these deaths are due to bacteria. Acute respiratory infections are the most frequent causes of deaths among infectious diseases; they are often directly due to, or exacerbated by,

M. Otto (✉)
Pathogen Molecular Genetics Section, Laboratory of Human Bacterial Pathogenesis, National Institute of Allergy and Infectious Diseases, The National Institutes of Health, Bethesda, MD, USA
e-mail: motto@niaid.nih.gov

bacterial pathogens. In addition, bacteria cause a wide array of nonfatal but nevertheless severe infections, such as infections of the skin and soft tissues, the lung, the intestine, and the urinary tract, to name but a few.

Many bacterial infections occur in the hospital in patients with a weakened immune system, which is due to underlying genetic or infection-related immune deficiencies, or the generally weakened status of the patient. Widespread antibiotic resistance often makes these infections extremely difficult to treat.

Another important reason for the problems associated with treating bacterial infections is the formation of biofilms. The National Institutes of Health estimated that more than 60 % of microbial infections proceed with the involvement of biofilms. Biofilms are sticky agglomerations of bacteria or other microorganisms. They significantly decrease the efficacy of antibiotics and the patient's immune defenses.

In nature, bacteria commonly form biofilms. However, for more than a century, microbiological research was limited to growing bacteria under artificial conditions which we now know barely reflect their natural biofilm mode of growth. William J. Costerton, a pioneer of biofilm research, introduced the biofilm concept and the term "biofilms" to microbiology in the 1970s (Costerton et al. 1978). Initially focused on in vitro research and "environmental" bacteria, biofilm research over time increasingly included the investigation of "medical" biofilms formed by bacterial pathogens during infection.

Medical biofilm research comes with significant challenges that biofilm researchers are still struggling to cope with. This is due to the fact that in vitro biofilm models, despite the fact that they revealed many molecular determinants and principles of biofilm formation, barely reflect the situation that the bacteria encounter in the human host. The more recent focus on establishing animal models of biofilm infection and the capacity to directly investigate infectious isolates by modern genetic methods has taken biofilm research to a new level. Notably, concepts developed based on in vitro biofilm research often were not confirmed on the in vivo level, demonstrating the necessity to complement in vitro biofilm research by appropriate methods to ascertain their in vivo relevance (Joo and Otto 2012).

2 Biofilm Infections

Among the many types of infection in which biofilms are involved, a few have gained particular attention from researchers, owing to their frequency, severity, or potential model character for other biofilm-associated infections. Infections on indwelling medical devices, such as catheters or joint prostheses, are virtually always biofilm related. Owing to the high number of surgical interventions being performed nowadays, they are very common. By far the most important pathogens causing infections of indwelling medical devices are *Staphylococcus aureus* and coagulase-negative staphylococci, such as *Staphylococcus epidermidis* (Otto 2008). As these bacteria are commensals on the human skin and mucosal surfaces, device-related infections commonly are caused by contamination of the devices during insertion, with the infectious isolates originating either from healthcare personnel or the patient. Of note, infected devices can be a source for life-threatening secondary infections, such as septicemia.

Biofilms on contact lenses are a common cause of keratitis (Elder et al. 1995). Similar to device-associated infections, they develop by contamination with commensal bacteria, often involving coagulase-negative staphylococci, corynebacteria, bacilli, *Streptococcus pneumoniae*, *Pseudomonas aeruginosa*, or *Serratia marcescens*. The fungi *Candida albicans* and *Fusarium* ssp. also are frequent causes of biofilms on contact lenses.

Probably the most widespread biofilm infection is dental plaque, the source of several dental infections such as caries or periodontitis (Pihlstrom et al. 2005). In contrast to infections of indwelling medical devices, which are normally due to one single infectious isolate, dental plaque is a multi-species bacterial biofilm community (Hojo et al. 2009). Group B streptococci and lactobacilli are especially frequent among dental plaque-causing bacteria. We are only beginning to understand the many interactions between the members of the dental plaque biofilm community.

Urinary tract infections often involve biofilms. Most frequently the infecting bacterium is *Escherichia coli* (Marcus et al. 2008). Middle-ear infection (*Otitis media*) also is a common biofilm-associated disease, especially in children (Bakaletz 2007). The infecting bacteria include predominantly *S. pneumoniae*, *Haemophilus influenzae*, and *Moraxella catarrhalis*. Moreover, biofilms may contribute to streptococcal pharyngitis ("Strep throat") (Murphy et al. 2009) and chronic wound infections (Percival et al. 2012). The latter often contain polymicrobial biofilms with skin-related and other bacteria, including anaerobes.

Unless complications occur, the biofilm-associated infections discussed so far are not life-threatening. However, there are also examples of extremely severe and frequently fatal diseases that involve biofilms. Infective endocarditis has a particularly high fatality rate and involves bacterial biofilms forming on the valves of the heart (Que and Moreillon 2011). *S. aureus*, Viridans group streptococci, and coagulase-negative staphylococci are the most common causes.

Cystic fibrosis (CF, mucoviscidosis) is an autosomal recessive genetic disorder, caused by a mutation in the gene coding for the cystic fibrosis transmembrane conductance regulator (CFTR), which is involved in regulating sodium and chloride transfer across membranes (Riordan et al. 1989). Patients suffering from CF are particularly prone to chronic bacterial infection (Cohen and Prince 2012). *S. aureus* and *H. influenzae* dominate at early age, while *P. aeruginosa* is isolated in 80 % of cases from patients older than 18 years (Rajan and Saiman 2002). *P. aeruginosa* bacteria infecting the lungs of CF patients very likely grow in biofilms (Singh et al. 2000). Accordingly, *P. aeruginosa* CF isolates often show a characteristic "mucoid" phenotype associated with biofilm formation (May et al. 1991).

Owing to its involvement in CF infection as an especially severe form of biofilm-associated infection, in addition to the fact that molecular tools are more readily available for this bacterium compared to many other biofilm pathogens, biofilm formation in *P. aeruginosa* has been, and still is, the most intensely investigated biofilm-forming bacterium. Much of what we know about biofilms and biofilm development stems from investigation using *P. aeruginosa*. However, we have also increasingly become aware of the fact that many mechanisms of biofilm formation discovered in *P. aeruginosa* have less of a model character than previously assumed, as the molecular mechanisms of biofilm formation may significantly differ between different biofilm-forming pathogens. Of note, an important problem associated with the model character of *P. aeruginosa* for biofilm infection is the fact that *P. aeruginosa* CF infections are difficult to mimic in animal infection models (Hoffmann 2007).

3 Biofilm Resistance

It has often been stressed that biofilms provide resistance to mechanisms of host defense, in particular, leukocyte phagocytosis. However, there have been conflicting results as to whether biofilm cells are inherently resistant to phagocytosis (Gunther et al. 2009). As investigation performed in staphylococcal biofilms suggests, protection is likely due mainly to the production of the extracellular biofilm matrix, which may inhibit the engulfment of biofilm cell clusters by phagocytes (Guenther et al. 2009; Vuong et al. 2004a). Furthermore, the matrix, which consists of polymers with low immunogenicity, shields biofilm cells from recognition of bacterial cell surface-exposed epitopes by the immune system (Thurlow et al. 2011).

Many antibiotics have significantly lower efficacy against biofilm as compared to planktonic (i.e., free-floating) cells (Stewart and Costerton 2001). The difference can reach factors of around 1,000 (Davies 2003). Biofilm resistance (or strictly speaking, tolerance, as opposed to specific mechanisms of resistance) is due to different reasons. The extracellular biofilm matrix provides a mechanical shield, preventing at least some antibiotics from reaching their target, often the bacterial peptidoglycan, the cytoplasmic membrane, or intracellular targets such as protein or DNA biosynthesis molecules. Furthermore, biofilm tolerance is due to the physiological status of biofilm cells, which is characterized by low activity of cell processes such as cell wall, protein, or DNA biosynthesis. Thus, the many antibiotics that target those processes are barely active against cells in biofilms (Davies 2003).

4 General Model of Biofilm Formation

Research initially performed in *P. aeruginosa*, but in the meantime also in many other bacteria, revealed a general model of how biofilms develop (O'Toole et al. 2000). For bacterial pathogens, the first step is attachment to tissue surfaces. Rarely, attachment may proceed directly on abiotic surfaces, such as on catheters, but because human matrix proteins soon cover any foreign device in the human body, this form of attachment likely only plays a minor role even in device-associated biofilm infections. In the case of motile bacteria, such as *P. aeruginosa*, attachment may be preceded by active motion toward the surface, whereas nonmotile bacteria have to rely on passive modes of motion in that first step of biofilm development.

After attachment is accomplished, the bacteria proliferate and surround themselves with the characteristic biofilm matrix. This matrix is composed of many different molecules. Some are specific to the given bacterium, such as the exopolysaccharides and secreted proteins produced by many biofilm bacteria. Others may be produced by a large subset of bacteria, such as teichoic acids found in Gram-positive bacteria. As biofilms are in a stationary mode of growth, the biofilm matrix also comprises molecules that are released from dying cells. In particular, extracellular DNA (eDNA) was found to contribute to the biofilm matrix in many bacteria (Whitchurch et al. 2002). Electrostatic interactions between oppositely charged matrix polymers are believed to play a key role in matrix formation. It needs to be stressed that for some of these molecules, evidence for a participation in the biofilm matrix is only derived from in vitro investigation, such as in the case of eDNA. The environment in the human host contains factors, such as nucleases and proteases, which have the potential to interfere strongly with the composition of the biofilm matrix. Especially eDNA may be degraded by the efficient human serum DNaseI (Whitchurch et al. 2002). It may be because the human host cannot degrade them that biofilm bacteria produce specific biofilm exopolysaccharides, several of which have a proven function in in vivo biofilm formation (Rupp et al. 1999; Conway et al. 2004; Hoffmann et al. 2005).

Were it only for the biofilm matrix components, biofilms would be unstructured "clumps" of cells, and expansion of a biofilm would hardly be possible without leaving cells in deeper layers prone to death due to limited nutrient availability. However, we know from microscopic analysis that biofilms have a characteristic three-dimensional structure with cellular agglomerations in "mushroom" shape and channels that provide nutrients to those deeper layers. The molecular factors that facilitate channel formation have recently gained much attention. Several biofilm-forming bacteria were found to produce surfactant molecules to structure biofilms in that fashion (Otto 2013). Notably, the same forces that underlie channel formation are responsible for the detachment of cell clusters from a biofilm, a mechanism that leads to dissemination of the pathogenic bacteria to the bloodstream, and thus may cause second-site infections.

Biofilm formation is under the control of a series of regulatory systems, which often differ considerably between different biofilm-forming bacteria. However, there are also generally applicable concepts in biofilm regulation. In several bacteria, such as *E. coli*, sensory and regulatory systems trigger biofilm development upon contact with a surface (Otto and Silhavy 2002). Furthermore, the general switch from the planktonic to the biofilm mode of growth is often under control of the second messenger cyclic di-GMP (Romling et al. 2013). Finally, cell density-dependent regulation ("quorum sensing," QS) controls biofilm differentiation in many microorganisms (Irie and Parsek 2008).

5 Biofilm Pathogens

While the general model of biofilm formation gives a good overall outline that is applicable to many biofilm-forming bacteria, most biofilm microorganisms produce highly specific biofilm factors. Some of those that were thoroughly investigated shall briefly be introduced in the following.

Biofilm formation in *P. aeruginosa* is best understood, at least in vitro. This species produces three main biofilm exopolysaccharides, the negatively charged alginate, the mannose-rich neutral "Psl," and the glucose-rich "Pel" exopolysaccharides (Ryder et al. 2007). Production of alginate in particular is associated with the "mucoid" phenotype of *P. aeruginosa* strains isolated from cystic fibrosis infection (May et al. 1991). The impact of QS on biofilms was first described in *P. aeruginosa*, where as in many other bacterial pathogens, it has a strong impact on the production of biofilm factors and biofilm development in general (Davies et al. 1998). QS regulation in *P. aeruginosa* involves at least three systems (Rhl, Las, and Qsc) forming a QS network (Jimenez et al. 2012). Early experiments performed in *P. aeruginosa* indicated that QS is a positive regulator of biofilm expansion (Davies et al. 1998), but we know now that the impact of QS on biofilm development is more complicated, affecting a series of factors involved in biofilm growth and structuring (Joo and Otto 2012). Rhamnolipids, for example, are QS-controlled surfactants that facilitate *P. aeruginosa* biofilm structuring (Boles et al. 2005). Furthermore, pili (or fimbriae) in *P. aeruginosa* provide motility and are not only important for reaching a surface, but also in QS-regulated detachment processes (Gibiansky et al. 2010), where cells regain pili-mediated motility starting in the center of biofilm "mushrooms" (Purevdorj-Gage et al. 2005).

S. aureus and coagulase-negative staphylococci contribute to a number of biofilm infections and dominate among pathogens causing infections of indwelling medical devices. Much of our knowledge on staphylococcal biofilm formation stems from research on the human commensal *S. epidermidis* (Otto 2009). *S. epidermidis*—as most other staphylococci—produces an exopolysacharide termed polysaccharide intercellular adhesin (PIA) or poly-N-acetyl glucosamine (PNAG). PIA/PNAG is a linear homopolymer of N-acetyl glucosamine with partial de-acetylation that introduces positive changes in the otherwise neutral molecule (Mack et al. 1996; Vuong et al. 2004b). It has a demonstrated significant function in in vitro and in vivo biofilm formation, although not all staphylococcal biofilm-forming strains (especially *S. aureus*) appear to rely on PIA/PNAG to form biofilms (Rohde et al. 2007). A large number of proteins also contribute to the formation of the staphylococcal biofilm matrix, such as the accumulation-associated protein Aap (Conrady et al. 2008). The biofilm-structuring surfactant phenol-soluble modulin (PSM) peptides of staphylococci are controlled by the accessory gene regulator

(Agr) QS system (Periasamy et al. 2012; Wang et al. 2007) and the exopolysaccharide PIA/PNAG by the LuxS QS system (Xu et al. 2006).

Group B Streptococci (GBS) such as *Streptococcus mutans* participate to a significant extent in dental plaque formation. *S. mutans* secretes glucosyl transferases and glucan binding proteins, which produce water-soluble and -insoluble glucans that facilitate biofilm formation (Banas and Vickerman 2003). Many GBS produce a polysaccharide capsule that contains moieties with similarity to host saccharides, which thus—in addition to their role in biofilm matrix formation—may provide protection from host defenses (Wyle et al. 1972). Biofilm formation in streptococci is regulated by a series of global regulators, including competence systems, which regulate the uptake of DNA (Suntharalingam and Cvitkovitch 2005). The competence/QS signal peptide CSP (competence-stimulating peptide) has a major role in controlling these phenotypes (Li et al. 2001).

In *E. coli*, a pathogen frequently involved in urinary tract infection, different forms of pili (type I fimbriae, curli fimbriae, and conjugative pili) participate in attachment and biofilm formation (Beloin et al. 2008). The Cpx system senses the surface and neighboring bacteria, affecting production of flagellae and biofilm maturation (Otto and Silhavy 2002). Interestingly, despite the fact that *E. coli* is not closely related to staphylococci, it produces the same matrix exopolysaccharide PIA/PNAG (called PGA in *E. coli*) (Wang et al. 2004), indicating that specific biofilm-related genes have been distributed far beyond species and genus barriers.

Acinetobacter baumannii is a biofilm-forming pathogen often involved with hospital-acquired pneumonia that has recently received much attention (Cerqueira and Peleg 2011). *A. baumannii* can form biofilms on abiotic surfaces that survive for several days, in which pili produced by the *csu* operon play a preeminent role (Tomaras et al. 2008). However, these pili are not important for attachment to mammalian cells (de Breij et al. 2009), exemplifying that in vitro results regarding biofilm factors may have limited relevance for the in vivo situation. *A. baumannii* produces two biofilm molecules that have previously been described in staphylococci: PIA/PNAG (Choi et al. 2009) and the biofilm-associated protein (Bap) (Loehfelm et al. 2008), again showing that key biofilm factors were distributed across genus barriers even between Gram-negative and Gram-positive bacteria.

C. albicans is the most frequent fungal human pathogen. While *C. albicans* biofilm development follows the same general model as do bacteria, the participating molecules are not related, owing to the fact that this pathogen is a eukaryotic organism (Cuellar-Cruz et al. 2012). Attachment occurs via cell wall proteins and is followed by the production of hyphae and a matrix that consists of several different polymers. Similar to bacteria, QS regulation has a strong impact on *Candida* biofilm development, with tyrosol and farnesol being the most important QS signals (Singh and Del Poeta 2011).

6 In Vitro and In Vivo Analysis of Biofilm Development

Analyzing biofilm formation in in vitro models ranges from simple microtiter plate assays to sophisticated flow reactors. Flow constantly provides fresh media to the biofilm cells and is often applied to mimic environmental biofilms, such as those formed in wastewater tubing. Which in vitro model best mimics "medical" biofilms as present during infection is debatable. Many observations and findings indicate that results achieved using in vitro biofilm models are difficult to transfer to the in vivo situation (Joo and Otto 2012). Nevertheless, modeling biofilm formation in vitro has the advantage that the biofilms can be analyzed using state-of-the-art microscopic techniques, such as confocal laser scanning microscopy (CLSM). By taking regular interval pictures of a biofilm forming in a flow cell, movies can be produced using CLSM that give detailed insight into biofilm development.

In addition to the genetic analysis of infectious isolates, the analysis of biofilms during infection relies primarily on animal models of biofilm-associated infection. Some biofilm infections, such as indwelling device-related infection, are easier to mimic in animal models than others, such as lung infection during cystic fibrosis or dental plaque formation. For that reason, we have a better understanding of in vivo biofilm factors in bacteria that cause device-related infections than many other biofilm-related diseases. Clearly, the development of better models of biofilm-associated infection is a premier task of current and future biofilm research.

7 Targeting Medical Biofilms

Biofilm formation is still a problem for drug development that has not been satisfactorily addressed. With the development of novel antibiotics almost having come to a halt (Cooper and Shlaes 2011), companies are often not focusing on biofilm-associated infections, as those are regarded as even more complicated to tackle. At least it is now common practice to monitor the efficacy of a drug in development against in vitro biofilms.

Generally, one can envision two different approaches to combat medical biofilms. First, novel antibiotics may be developed that have increased efficacy against biofilms. These should be antibiotics that penetrate the biofilm matrix and have a bactericidal rather than bacteriostatic mode of action. Second, drugs specifically inhibiting attachment, proliferation, or even biofilm structuring may target biofilm formation itself. It is also conceivable to develop drugs that promote biofilm dispersal, leaving biofilm cells more prone to attack by conventional antibiotics. However, biofilm molecules that are conserved in different biofilm pathogens are rare. This approach thus has the disadvantage of limited applicability and marketability. Some regulatory factors may be more widespread, but inhibiting regulators in antibacterial drug development requires much caution. Unfortunately, the outlook regarding the timeframe for the availability of drugs that are active against biofilms is rather bleak, necessitating more extensive efforts both in general biofilm research and in the development of biofilm-active antibiotics.

Acknowledgments This work was supported by the Intramural Research Program of the National Institute of Allergy and Infectious Diseases, U.S. National Institutes of Health.

References

Bakaletz LO (2007) Bacterial biofilms in otitis media: evidence and relevance. Pediatr Infect Dis J 26(10 Suppl):S17–S19
Banas JA, Vickerman MM (2003) Glucan-binding proteins of the oral streptococci. Crit Rev Oral Biol Med 14(2):89–99
Beloin C, Roux A, Ghigo JM (2008) *Escherichia coli* biofilms. Curr Top Microbiol Immunol 322:249–289
Boles BR, Thoendel M, Singh PK (2005) Rhamnolipids mediate detachment of *Pseudomonas aeruginosa* from biofilms. Mol Microbiol 57(5):1210–1223
Cerqueira GM, Peleg AY (2011) Insights into *Acinetobacter baumannii* pathogenicity. IUBMB Life 63(12):1055–1060
Choi AH, Slamti L, Avci FY, Pier GB, Maira-Litran T (2009) The *pgaABCD* locus of *Acinetobacter baumannii* encodes the production of poly-beta-1-6-N-acetylglucosamine, which is critical for biofilm formation. J Bacteriol 191(19):5953–5963
Cohen TS, Prince A (2012) Cystic fibrosis: a mucosal immunodeficiency syndrome. Nat Med 18 (4):509–519
Conrady DG, Brescia CC, Horii K, Weiss AA, Hassett DJ, Herr AB (2008) A zinc-dependent adhesion module is responsible for intercellular adhesion in staphylococcal biofilms. Proc Natl Acad Sci USA 105(49):19456–19461
Conway BA, Chu KK, Bylund J, Altman E, Speert DP (2004) Production of exopolysaccharide by *Burkholderia cenocepacia* results in altered cell-surface interactions and altered bacterial clearance in mice. J Infect Dis 190(5):957–966
Cooper MA, Shlaes D (2011) Fix the antibiotics pipeline. Nature 472(7341):32
Costerton JW, Geesey GG, Cheng KJ (1978) How bacteria stick. Sci Am 238(1):86–95
Cuellar-Cruz M, Lopez-Romero E, Villagomez-Castro JC, Ruiz-Baca E (2012) Candida species: new insights into biofilm formation. Future Microbiol 7(6):755–771
Davies D (2003) Understanding biofilm resistance to antibacterial agents. Nat Rev Drug Discov 2 (2):114–122
Davies DG, Parsek MR, Pearson JP, Iglewski BH, Costerton JW, Greenberg EP (1998) The involvement of cell-to-cell signals in the development of a bacterial biofilm. Science 280 (5361):295–298
de Breij A, Gaddy J, van der Meer J, Koning R, Koster A, van den Broek P, Actis L, Nibbering P, Dijkshoorn L (2009) CsuA/BABCDE-dependent pili are not involved in the adherence of *Acinetobacter baumannii* ATCC19606(T) to human airway epithelial cells and their inflammatory response. Res Microbiol 160(3):213–218
Elder MJ, Stapleton F, Evans E, Dart JK (1995) Biofilm-related infections in ophthalmology. Eye (Lond) 9(Pt 1):102–109
Gibiansky ML, Conrad JC, Jin F, Gordon VD, Motto DA, Mathewson MA, Stopka WG, Zelasko DC, Shrout JD, Wong GC (2010) Bacteria use type IV pili to walk upright and detach from surfaces. Science 330(6001):197
Guenther F, Stroh P, Wagner C, Obst U, Hansch GM (2009) Phagocytosis of staphylococci biofilms by polymorphonuclear neutrophils: *S. aureus* and *S. epidermidis* differ with regard to their susceptibility towards the host defense. Int J Artif Organs 32(9):565–573

Gunther F, Wabnitz GH, Stroh P, Prior B, Obst U, Samstag Y, Wagner C, Hansch GM (2009) Host defence against *Staphylococcus aureus* biofilms infection: phagocytosis of biofilms by polymorphonuclear neutrophils (PMN). Mol Immunol 46(8–9):1805–1813

Hoffmann N (2007) Animal models of chronic *Pseudomonas aeruginosa* lung infection in cystic fibrosis. Drug Discov Today 4(3):99–104

Hoffmann N, Rasmussen TB, Jensen PO, Stub C, Hentzer M, Molin S, Ciofu O, Givskov M, Johansen HK, Hoiby N (2005) Novel mouse model of chronic *Pseudomonas aeruginosa* lung infection mimicking cystic fibrosis. Infect Immun 73(4):2504–2514

Hojo K, Nagaoka S, Ohshima T, Maeda N (2009) Bacterial interactions in dental biofilm development. J Dent Res 88(11):982–990

Irie Y, Parsek MR (2008) Quorum sensing and microbial biofilms. Curr Top Microbiol Immunol 322:67–84

Jimenez PN, Koch G, Thompson JA, Xavier KB, Cool RH, Quax WJ (2012) The multiple signaling systems regulating virulence in *Pseudomonas aeruginosa*. Microbiol Mol Biol Rev 76(1):46–65

Joo HS, Otto M (2012) Molecular basis of in vivo biofilm formation by bacterial pathogens. Chem Biol 19(12):1503–1513

Li YH, Lau PC, Lee JH, Ellen RP, Cvitkovitch DG (2001) Natural genetic transformation of *Streptococcus mutans* growing in biofilms. J Bacteriol 183(3):897–908

Loehfelm TW, Luke NR, Campagnari AA (2008) Identification and characterization of an *Acinetobacter baumannii* biofilm-associated protein. J Bacteriol 190(3):1036–1044

Mack D, Fischer W, Krokotsch A, Leopold K, Hartmann R, Egge H, Laufs R (1996) The intercellular adhesin involved in biofilm accumulation of *Staphylococcus epidermidis* is a linear beta-1,6-linked glucosaminoglycan: purification and structural analysis. J Bacteriol 178 (1):175–183

Marcus RJ, Post JC, Stoodley P, Hall-Stoodley L, McGill RL, Sureshkumar KK, Gahlot V (2008) Biofilms in nephrology. Expert Opin Biol Ther 8(8):1159–1166

May TB, Shinabarger D, Maharaj R, Kato J, Chu L, DeVault JD, Roychoudhury S, Zielinski NA, Berry A, Rothmel RK et al (1991) Alginate synthesis by *Pseudomonas aeruginosa*: a key pathogenic factor in chronic pulmonary infections of cystic fibrosis patients. Clin Microbiol Rev 4(2):191–206

Murphy TF, Bakaletz LO, Smeesters PR (2009) Microbial interactions in the respiratory tract. Pediatr Infect Dis J 28(10 Suppl):S121–S126

O'Toole G, Kaplan HB, Kolter R (2000) Biofilm formation as microbial development. Annu Rev Microbiol 54:49–79

Otto M (2008) Staphylococcal biofilms. Curr Top Microbiol Immunol 322:207–228

Otto M (2009) *Staphylococcus epidermidis* – the 'accidental' pathogen. Nat Rev Microbiol 7:555–567

Otto M (2013) Staphylococcal infections: mechanisms of biofilm maturation and detachment as critical determinants of pathogenicity. Annu Rev Med 64:175–188

Otto K, Silhavy TJ (2002) Surface sensing and adhesion of *Escherichia coli* controlled by the Cpx-signaling pathway. Proc Natl Acad Sci USA 99(4):2287–2292

Percival SL, Hill KE, Williams DW, Hooper SJ, Thomas DW, Costerton JW (2012) A review of the scientific evidence for biofilms in wounds. Wound Repair Regen 20(5):647–657

Periasamy S, Joo HS, Duong AC, Bach TH, Tan VY, Chatterjee SS, Cheung GY, Otto M (2012) How *Staphylococcus aureus* biofilms develop their characteristic structure. Proc Natl Acad Sci USA 109(4):1281–1286

Pihlstrom BL, Michalowicz BS, Johnson NW (2005) Periodontal diseases. Lancet 366 (9499):1809–1820

Purevdorj-Gage B, Costerton WJ, Stoodley P (2005) Phenotypic differentiation and seeding dispersal in non-mucoid and mucoid *Pseudomonas aeruginosa* biofilms. Microbiology 151 (Pt 5):1569–1576

Que YA, Moreillon P (2011) Infective endocarditis. Nat Rev Cardiol 8(6):322–336

Rajan S, Saiman L (2002) Pulmonary infections in patients with cystic fibrosis. Semin Respir Infect 17(1):47–56

Riordan JR, Rommens JM, Kerem B, Alon N, Rozmahel R, Grzelczak Z, Zielenski J, Lok S, Plavsic N, Chou JL et al (1989) Identification of the cystic fibrosis gene: cloning and characterization of complementary DNA. Science 245(4922):1066–1073

Rohde H, Burandt EC, Siemssen N, Frommelt L, Burdelski C, Wurster S, Scherpe S, Davies AP, Harris LG, Horstkotte MA, Knobloch JK, Ragunath C, Kaplan JB, Mack D (2007) Polysaccharide intercellular adhesin or protein factors in biofilm accumulation of *Staphylococcus epidermidis* and *Staphylococcus aureus* isolated from prosthetic hip and knee joint infections. Biomaterials 28(9):1711–1720

Romling U, Galperin MY, Gomelsky M (2013) Cyclic di-GMP: the first 25 years of a universal bacterial second messenger. Microbiol Mol Biol Rev 77(1):1–52

Rupp ME, Ulphani JS, Fey PD, Bartscht K, Mack D (1999) Characterization of the importance of polysaccharide intercellular adhesin/hemagglutinin of *Staphylococcus epidermidis* in the pathogenesis of biomaterial-based infection in a mouse foreign body infection model. Infect Immun 67(5):2627–2632

Ryder C, Byrd M, Wozniak DJ (2007) Role of polysaccharides in *Pseudomonas aeruginosa* biofilm development. Curr Opin Microbiol 10(6):644–648

Singh A, Del Poeta M (2011) Lipid signalling in pathogenic fungi. Cell Microbiol 13(2):177–185

Singh PK, Schaefer AL, Parsek MR, Moninger TO, Welsh MJ, Greenberg EP (2000) Quorum-sensing signals indicate that cystic fibrosis lungs are infected with bacterial biofilms. Nature 407(6805):762–764

Stewart PS, Costerton JW (2001) Antibiotic resistance of bacteria in biofilms. Lancet 358 (9276):135–138

Suntharalingam P, Cvitkovitch DG (2005) Quorum sensing in streptococcal biofilm formation. Trends Microbiol 13(1):3–6

Thurlow LR, Hanke ML, Fritz T, Angle A, Aldrich A, Williams SH, Engebretsen IL, Bayles KW, Horswill AR, Kielian T (2011) *Staphylococcus aureus* biofilms prevent macrophage phagocytosis and attenuate inflammation in vivo. J Immunol 186(11):6585–6596

Tomaras AP, Flagler MJ, Dorsey CW, Gaddy JA, Actis LA (2008) Characterization of a two-component regulatory system from *Acinetobacter baumannii* that controls biofilm formation and cellular morphology. Microbiology 154(Pt 11):3398–3409

Vuong C, Voyich JM, Fischer ER, Braughton KR, Whitney AR, DeLeo FR, Otto M (2004a) Polysaccharide intercellular adhesin (PIA) protects *Staphylococcus epidermidis* against major components of the human innate immune system. Cell Microbiol 6(3):269–275

Vuong C, Kocianova S, Voyich JM, Yao Y, Fischer ER, DeLeo FR, Otto M (2004b) A crucial role for exopolysaccharide modification in bacterial biofilm formation, immune evasion, and virulence. J Biol Chem 279(52):54881–54886

Wang X, Preston JF 3rd, Romeo T (2004) The *pgaABCD* locus of *Escherichia coli* promotes the synthesis of a polysaccharide adhesin required for biofilm formation. J Bacteriol 186(9):2724–2734

Wang R, Braughton KR, Kretschmer D, Bach TH, Queck SY, Li M, Kennedy AD, Dorward DW, Klebanoff SJ, Peschel A, DeLeo FR, Otto M (2007) Identification of novel cytolytic peptides as key virulence determinants for community-associated MRSA. Nat Med 13(12):1510–1514

Whitchurch CB, Tolker-Nielsen T, Ragas PC, Mattick JS (2002) Extracellular DNA required for bacterial biofilm formation. Science 295(5559):1487

Wyle FA, Artenstein MS, Brandt BL, Tramont EC, Kasper DL, Altieri PL, Berman SL, Lowenthal JP (1972) Immunologic response of man to group B meningococcal polysaccharide vaccines. J Infect Dis 126(5):514–521

Xu L, Li H, Vuong C, Vadyvaloo V, Wang J, Yao Y, Otto M, Gao Q (2006) Role of the *luxS* quorum-sensing system in biofilm formation and virulence of *Staphylococcus epidermidis*. Infect Immun 74(1):488–496

The Use of DNA Methods to Characterize Biofilm Infection

Randall Wolcott and Stephen B. Cox

Abstract Because of biofilm's fundamental properties—its polymicrobial nature (genetic diversity) and "viable but not culturable" microbial constituents—clinical cultures are wholly unsuited for evaluating chronic infections associated with biofilm. DNA-based technologies (molecular methods) have a number of advantages for evaluating human infections. Real-time PCR and sequencing technologies are particularly robust for identifying microorganisms in human environments because of development of their methods by the human microbiome project. DNA methods enjoy much higher sensitivity and specificity than cultivation methods for identifying microorganisms regardless of their phenotype. Moreover, real-time PCR can be quantitative in an absolute sense, while sequencing methods yield accurate relative quantification of all constituents of the sampled infection. All methods for microbial identification have biases, yet molecular methods suffer the least from these biases. Although DNA-based identification of microorganisms has the limitation that sensitivities to antibiotics cannot be determined in a Petri dish and must be determined by identifying mobile genetic resistance elements within the microbes, molecular methods are a significant improvement in the identification of microorganisms for human infections and are currently the only reliable technology for diagnosing biofilm infection.

1 DNA-Based Testing

In the land of the blind, the one-eyed man is king—Erasmus.

R. Wolcott (✉)
Southwest Regional Wound Care Center, 2002 Oxford Avenue, Lubbock, TX, 79410, USA

Research and Testing Laboratory, 4321 Marsha Sharp Freeway, Lubbock, TX, 79407, USA
e-mail: randy@randallwolcott.com

S.B. Cox
Research and Testing Laboratory, 4321 Marsha Sharp Freeway, Lubbock, TX, 79407, USA

The human microbiome project (HMP) has forever changed how microorganisms will be identified (Chain et al. 2009). The HMP was established to identify and to quantitate bacteria living in normal human environments such as the gut, oral cavity, skin, urogenital, etc. Several challenges for the project were that the microbes in these host environments are polymicrobial, they are not quantifiable by cultivation methods, and they generally exist in a biofilm phenotype. In fact, the vast majority of the species known to inhabit normal host environments are not routinely culturable (Petrosino et al. 2009), which is characteristic of the biofilm phenotype (Fux et al. 2005). These facts led investigators to employ molecular methods.

Molecular methods are based on the idea of direct examination of the bacterial DNA existing in the sample to allow for identification of the bacteria that are present. There has been a very rapid and fluid progression of molecular technologies that can analyze microbial DNA. However, to get any of these molecular technologies to give a meaningful analysis, high-quality DNA first must be obtained. Therefore, one of the most important obstacles to using molecular methods for identifying and quantitating microorganisms in human infections is obtaining good microbial DNA from the sample (i.e., the process of DNA extraction). There are a number of excellent kits and laboratory methods for obtaining microbial DNA from mixed samples (samples that contain both microbial and human DNA). However, each method has different extraction efficiencies, and these efficiencies may vary for the different species within the sample. Yet even with these challenges, many extraction methods can approach 96 % efficiency (Fitzpatrick et al. 2010).

The process of DNA extraction, especially from samples that contain some of the host products, also can extract substances that inhibit later analysis of the microbial DNA. For example, polymerase chain reaction (PCR) is a common method used to amplify microbial DNA, yet the process can be inhibited by substances found in the sample. These PCR inhibitors include complex polysaccharides, bile salts, hemoglobin degradation products, polyphenolic compounds, heavy metals, and, most frequently, large amounts of human DNA (Stauffer et al. 2008). Many of the more common PCR inhibitors can be effectively mitigated, but if the inhibitors cannot be identified and controlled, resampling may be necessary. Once good DNA is obtained from the sample, most current molecular instrumentation can obtain reliable clinical results.

PCR is a widely used method of processing DNA that has a relatively long history of use in the clinic (Krishna and Cunnion 2012; Reddington et al. 2013). PCR utilizes primers that attach to complementary regions of bases in the microbial DNA and, through a polymerase reaction, create copies of this area. This copying process doubles the amount of target sequence with every cycle of the PCR. Real-time PCR has the ability to quantitate, in an absolute sense, how much microbial DNA is in the original sample. The number of cycles required before the real-time signal reaches a detection threshold (cycle threshold number or ct number) can be correlated to an absolute number of microbes present in the original sample. This is an extremely powerful feature of real-time PCR that can used to quantify "bacterial load" (Verhoeven et al. 2012). However, PCR has several important limitations. Most limiting is the fact that real-time PCR requires a primer sequence to be developed for each species of microorganism present in the sample. With the literally thousands of different microorganisms that can be in human chronic infections, constructing thousands of primers for each analysis is inefficient, costly, and currently not feasible.

There also are a number of different parameters involved in the process of performing PCR, such as chemistries (e.g., Syber Green, TaqMan), platforms (e.g., Roche v. Abbott), factors in plate preparation, etc., that can impact results. Incomplete optimization of these parameters can lead to amplification inefficiencies, inconsistent reproducibility, random PCR products, and other problems. The optimization of chemistries, primers, and instrument variables is focused on improving sensitivity of the primer to the target microbe without sacrificing specificity for the organism (prevention of cross-reactivity with other species). Optimization must also take into account dynamic range so that minor species are detected and quantitated as accurately as the dominant species in the sample.

Diagnostic laboratories painstakingly optimize all of the PCR variables by choosing appropriate instruments, chemistries, and primers to mitigate the potential negative impacts of these variables. However, there are still limits to quantitative PCR methods. For example, even though the reported results will be extremely specific for the microbial species present, due to DNA extraction efficiencies for different species, different amplification efficiencies for different species, and other variables, quantification of the microbes in the sample remains mildly inconsistent. Calculating bacterial load by real-time PCR often yields up to an order of magnitude variation for known quantities (usually lower), yet this seems to be an acceptable level of variability for clinical decision-making.

Although real-time PCR can rapidly yield usable information on bacterial load and identify a limited number of microbial species, it is impractical for PCR to be used alone for the identification and quantification of microbes in most human infections. Investigators in the HMP encountered the same limitations and quickly turned to sequencing (Aagaard et al. 2012). One of the technologies used early on in the HMP was a whole metagenome survey of the microbes present. This methodology looks at all genes present in a sample, which is an excellent way to determine species of fungi, bacteria, and even viruses present in an infection. The problem with determining all the genes present was that it required a massive number of sequencing base pairs (bp-ATCG) for a sample, which only allowed a small number of samples to be evaluated per sequencing run. These surveys also lost some of their quantitative ability (Fodor et al. 2012).

An alternative methodology was developed in which a very specific gene, the ribosomal 16S rDNA gene for bacteria or 18S rDNA gene for fungi, can be amplified through a PCR step and then sequenced. The use of the 16S rDNA gene as an indicator of bacterial taxonomic relationships traces back to the pioneering work of Woese and Fox (1977). This method provides two important pieces of information. Once the 16S or 18S region has been sequenced, it can be compared to a database of known sequences, thus yielding the genus and species with a high level of confidence. In addition, this method can allow for relative quantification of the microbes present within each sample. The number of "copies" of the gene for each species in the sample can be totaled, allowing for each species to be expressed as a percent of that total number. Although it does not provide for absolute quantification, this method does allow investigators to determine the dominant, major, and minor species within a sample (Rhoads et al. 2012a). Because this approach focused on sequencing only a single gene from each microbe, it allowed for several hundred samples to be analyzed on the same plate in a single run, greatly reducing the cost and increasing the speed of analysis. It was mainly through the development of sequencing technologies and methods that allowed investigators to elucidate fully the microorganisms present in the human microbiome (Morgan et al. 2013).

Sequencing is the molecular method for determining the exact order of nucleotides (i.e., adenine, thymine, cytosine, guanine) of a specific fragment of DNA or an entire genome. Sequencing instruments, such as the Roche 454, the PacBio (Pacific Biosciences), and Ion Torrent (Life Technologies), use different methods, but they all accurately determine the sequences of long segments of specific regions of microbial DNA, such as the 16S rDNA gene for bacteria and the 18S rDNA gene for fungi. These technologies can give a 99 % accurate code for the targeted gene, which is easily translated into taxonomic identification.

The microbial gene that codes for the 16S ribosomal subunit is conserved in all prokaryotic organisms except for a small subgroup of Archaea. The 16S ribosomal DNA has about 1,500 nucleotides, which contain nine hypervariable regions (v1–v9), and allows for the ability to identify bacteria at the species level. Fortunately, v1 can differentiate *Staphylococcus* to a genus level, and if the first three regions (v1–v3) can be sequenced, then the majority of other bacteria can be resolved to a genus level with a high degree of certainty. The 16S ribosomal DNA has been called the genomic fingerprint, and a 400+ nucleotide sequence of the 16S ribosomal DNA region is capable of reliably reading this genomic fingerprint.

Often, sequencing is carried out at multiple points along the 16S gene. It has been demonstrated that sequencing two fragments of the 16S gene consisting of 762 based pairs and 598 base pairs is more accurate in identifying bacteria than a single fragment of 1,343 base pairs (Jenkins et al. 2012). Therefore, sequencing methods often use primer sets consisting of two or more primers that cover different regions of the 16S gene. These primer sets can have some bias in how efficiently they sequence specific bacterial species.

Once sequencing has been completed, a data analysis pipeline is needed to begin processing the data. The data analysis process consists of two major stages: quality checking and diversity analysis. During the quality checking stage, denoising (Quince et al. 2009, 2011) and chimera checking (Haas et al. 2011) are performed on all the reads within the data. Each read is quality scanned and deficient reads are removed from the sample. The primary output of this stage is high-quality sequences. During the diversity analysis, sequences from each sample are run through an algorithm (typically involving a match to a database of known sequences) to determine the taxonomic information for each sequence. Reference databases exist for sequences from the 16S, 18S, 23S, ITS, and/or SSU regions.

Bioinformatics, the post-analysis processing of the massive data, therefore becomes the overseer of the quality of the reported results to the clinician. It is very difficult for clinicians to abandon the visible, tangible, and familiar microorganism growing in a Petri dish for the very complex "black box" type of results produced by bioinformatics. However, current laboratory regulations requiring strict validation and reproducibility coupled with proficiency testing of unknown samples can allow the clinician to feel very comfortable with these new molecular methods. Also, a closer examination of clinical cultures demonstrates that clinicians may have placed their faith in an insufficient method all along.

2 Clinical Cultures: The Land of the Blind

Medical microbiology has clung to cultivation methods even while environmental microbiology migrated to DNA methods for microbial identification decades ago. This failure to take advantage of new technologies to improve microbial identification has left clinicians "blind" to the microbial reality of most infections. Many deficiencies in traditional cultivation methods make routine clinical cultures unacceptable for medical microbiology.

Only a handful of media, such as tryptic soy agar, blood agar, nutrient agar, brain–heart infusion agar, and a few others are used to plate routine samples and they are grown at only one temperature (usually 37 °C) for 24–48 h. These experimental conditions have been worked out to be adequate for *Staphylococcus* species, *Streptococcus* species, *Pseudomonas aeruginosa*, and several other bacteria that can grow under these limitations. However, the vast majority of bacterial and fungal species do not grow under these laboratory conditions. Therefore, hundreds to even thousands of specialty media have been developed along with various algorithms for microbes that require different atmospheres, nutrients, length of time, temperature, etc., to be grown. No other single fact could be more convincing for making the argument that routine clinical cultures are inadequate for diagnosing human infection.

Also, bacteria in the biofilm phenotype are notoriously difficult to grow in routine clinical cultures because they are "viable but not culturable." Biofilm infections also tend to be polymicrobial. Early investigators at the time of Koch found, "No matter how ingenious the machinery, how careful the researchers, they kept ending up with beakers of mixed bacteria. The inability to get anything but mixed cultures led many scientists to believe that the bacteria had to be in mixed groups in order to thrive, that they could never be separated..." (Hager 2006). To solve this problem, Koch developed the methodology of pure culture very similar to that of our current clinical culture.

Koch found on the semiliquid surface of agar infused with necessary nutrients that only one species of bacteria in his clinical sample would propagate and the rest of the bacteria, "the contaminants," would not grow or would be outcompeted. What we now know is that the experimental design of the nutrient-enriched agar plate encourages planktonic phenotype propagation of the bacterial species in an exponential growth phase pattern. We also know that the experimental design has significant bias for the bacterial species that propagate well under the experimental conditions of temperature, nutrient, time, etc. This creates a huge selection bias to grow the microorganisms which the medical microbiologists have decided in advance are the pathogens. With molecular methods, we have discovered even more shortcomings of clinical cultures.

Many clinicians continue to hold Koch's view of one microorganism producing one clinical infection. While this generally may be true for acute infections that are commonly produced by bacteria in the planktonic phenotype, it does not hold true for biofilm infection. Chronic infections are associated with biofilm phenotype bacteria (Del Pozo and Patel 2007) and are often polymicrobial, which confounds the methods of clinical cultures. When molecular methods are compared with clinical culture to identify the microbes, we start to understand why clinical cultures provide little help in managing most chronic infections.

In pleural effusion samples, which tend to be culture negative even when the patient shows clear signs of infection, the use of universal 16S PCR, "bacterial load," demonstrated bacteria in 82 % of the clinically infected samples, whereas clinical cultures grew bacteria only 55 % of the time. Utilizing a single molecular test improved bacterial identification by 27 %. It should also be noted that this individual PCR test had only 0.9 % false positives whereas clinical cultures had a 2.6 % false positive rate (Insa et al. 2012).

Also, it has been found to be more advantageous to first identify the microorganisms utilizing molecular methods and then select media and growth conditions to cultivate the microorganisms present. Up to 20 different growth conditions were necessary to cultivate microorganisms in a single cystic fibrosis study (Sibley et al. 2011). This demonstrates that the "one size fits all" routine clinical culture is inadequate to handle the diversity of chronic infections.

A retrospective study that evaluated 168 chronic wounds with both clinical culture and molecular diagnostics (PCR and pyrosequencing) revealed the comprehensiveness of molecular methods (Rhoads et al. 2012a). Evaluating chronic wounds at a genus level for bacterial taxa only, cultures identified 17 different bacterial genera, whereas the DNA methods identified 338 bacterial taxa. Cultures underreported the diversity of the wound microbiota, but even more importantly, they failed to identify the most abundant bacteria in the wound over half the time (Rhoads et al. 2012b). Cultures obtained from polymicrobial biofilm infections fail to identify the diversity by a factor of 20-fold and fail to identify the cornerstone genus over half the time.

To improve on the design of the previous study, a prospective study was conducted in which 51 consecutive chronic wounds had a single sample taken from their surface (Rhoads et al. 2012b). The sample was homogenized and a portion was sent for clinical culture, a portion sent for PCR and pyrosequencing, and the remaining saved for further analysis if necessary. Once the clinical culture was complete and all the sub-plates identified by phenotypic methods (biochemistries) the sub-plates were submitted for sequencing. The results showed that 5 wounds (10 %) were culture negative and 9 of the 46 remaining wounds (19 %) had discrepant results between the bacterial isolate identified by culture versus sequencing. For example, culturing methods identified *P. aeruginosa*, whereas sequencing evaluating the same sub-plate identified *Salmonella enterica*. Once again, culture failed to demonstrate the most abundant species over 50 % of the time (Rhoads et al. 2012a). It may be that one main reason clinicians struggle to manage chronic infections is because traditional culturing methods consistently report minor constituents of the infections rather than the dominant culprits.

Over 68 % of patients receive at least one course of antibiotics for the management of their chronic wounds (Howell-Jones et al. 2005). Unfortunately, multiple studies have demonstrated that treating wounds based on culture results does not improve the outcomes of the healing of the wound (Lipsky et al. 2004, 2011; Siami et al. 2001). This information has led some investigators to conclude that even though pathogens such as *P. aeruginosa* may be present in the wound, the pathogen is not doing any harm. That conclusion is made because when chronic infections are treated with anti-pseudomonal antibiotics specifically for *P. aeruginosa* identified by culture, there is no improvement in wound healing outcomes (Joseph 2013). The confusing results from clinical culture, which leads clinicians and scientists alike to conclude that pathogens may not behave pathogenically or that bacteria don't matter in certain chronic infections (O'Meara et al. 2010), may be due to the inadequacies of the cultivation methods.

Although routine clinical cultures are inadequate for evaluating chronic infections, we must first determine if the proposed replacement (i.e., molecular methods) is any better. That is, will adopting molecular methods improve clinical outcomes for chronic infections produced by biofilm phenotype microorganisms? After all, by growing bacteria, medical microbiologists can apply antibiotic discs and determine the "real-world" sensitivity of the isolated bacteria. Also, even though it has been demonstrated that DNA degrades quite quickly (2–3 days) once the bacteria dies within the host infection (Post et al. 1996), there is no clear determination that the microbial DNA identified by molecular methods is associated with a living bacterial cell. However, in a chronic wound infection model, when wound biofilm was comprehensibly diagnosed utilizing molecular methods and the microorganisms identified specifically treated, healing outcomes did improve (Dowd et al. 2011). Regardless, the primary tenant of medicine is for the clinician to fully diagnose the disease, and as demonstrated above, clinical cultures are mostly blind to the microbial reality of polymicrobial biofilm infection.

3 Advantages of DNA Diagnosis: The One Eye

An Oslerian (Sir William Osler) model of medicine mandates that the clinician diagnose a malady as fully as possible to formulate the most appropriate treatment available. Evidence-based medicine often requires not only diagnosis before the treatment regimen but also frequent intervals of reevaluation during the treatment to show efficacy. So, no matter the generation of the clinician or which model of medicine to which the clinician ascribes, diagnosis of the condition is fundamental. Diagnosis prior to treatment is especially important in the management of chronic infections.

However, most clinicians treating chronic infections have abandoned the fundamental principle of initial diagnosis. The problem seems to lie not in the clinicians but in the diagnostic tools available. Many different culturing methods have been tried, yet they do not improve outcomes in the treatment of chronic infections. The inadequacy of cultivation methods has led to a de facto management of chronic infections by an educated guess, trial and error method.

The transition toward adopting molecular methods for medical microbiology need not be difficult. For virology there are no other reliable methods other than nucleic acid-based analysis. Almost a decade ago it was established that not only was DNA-based testing more accurate and reliable than clinical culture, but it also had the advantage of reduced time to diagnosis and high throughput (Mothershed and Whitney 2006). New methods have also been developed to identify various different antibiotic resistance determinants while at the same time providing genetic surveillance for new and existing pathogens (Weile and Knabbe 2009). Indeed from 2001 to 2007, 215 novel bacterial species were identified in human infections by sequencing methods with 100 of these new species identified in four or more individual patients (Woo et al. 2008). Molecular methods offer faster and higher throughputs while staying true to the original purpose of identifying and quantifying microbes. Recent studies demonstrate that close to 100 % sensitivity and specificity can be achieved for evaluating clinical infections (Hansen et al. 2010). One issue is that molecular methods may be identifying too many microorganisms, leading the clinician to over treat a specific infection.

DGGE and imaging methods showed that there was much more diversity present in wounds than clinical cultures were reporting (Davies et al. 2004; James et al. 2008). Clinicians managing other chronic infections such as chronic rhinosinusitis (Stephenson et al. 2010), cystic fibrosis (Goddard et al. 2012), middle ear infections (Laufer et al. 2011), and burns utilized molecular methods to show similar findings. It has been generally agreed that these and other chronic infections are associated with bacteria propagating in biofilm phenotype (Del Pozo and Patel 2007). Although molecular methods can identify microbes regardless of their mode of growth, the same is not true for clinical cultures. Molecular technology provides the clinician a more robust understanding of the infection, but also forces the clinician to consider multiple microbial species. At the same time, molecular methods do not provide any clear information on which species are producing the infection and which species are merely contaminants.

New methods are rapidly developing where microRNA (Martens-Uzunova et al. 2013) and messenger RNA (Mutz et al. 2013) can be sequenced and identified. This will provide critical information as to the inner workings of microbial cells which should provide insight as to strategies being used to cause infection. This may shed light on which microorganisms within the community are behaving as pathogens.

Before a bacterial species can be deemed a pathogen, or more importantly before that species can be dismissed as a contaminant, the clinician must take into account the synergies which arise within a polymicrobial infection. By including multiple bacterial and/or fungal species into a single community, the biofilm achieves numerous advantages such as passive resistance (Elias and Banin 2012), metabolic cooperation (Fischbach and Sonnenburg 2011), by-product influence (Elias and Banin 2012), quorum-sensing systems, an enlarged gene pool with more efficient DNA sharing (Madsen et al. 2012), and many other synergies that give the polymicrobial infection a competitive advantage. It is best to view a biofilm as a single entity possessing multiple genetic resources to allow it to adapt and thrive regardless of the stresses it encounters. In general, a more diverse population (i.e., greater the gene pool) will make the biofilm more robust in terms of its survivability (Tuttle et al. 2011).

Metabolic cross feeding has been well established between genetically distinct species. It has been shown that *Streptococcus gordonii* produces peroxide that can cause *Aggregatibacter actinomycetemcomitans* (Aa) to produce a factor H binding protein which limits the host's ability to kill Aa through a complement mediated lysis (Ramsey et al. 2011). This metabolic cooperation has been identified in numerous polymicrobial models (Dalton et al. 2011; Mikx and van der Hoeven 1975; Kuboniwa et al. 2006).

Waste products, molecules that bacteria produce that are end products and are of no benefit to the metabolizing member, are released into the local biofilm environment. Many of these metabolites such as ammonia, lactic acid, and carbon dioxide can have significant influence on the surrounding microorganisms (Elias and Banin 2012). Studies have demonstrated that *Fusobacterium nucleatum* and *Prevotella intermedia* generate ammonia which raises the pH suitable for *Porphyromonas gingivalis* (Takahashi 2003) and that *F. nucleatum* also provides an increased carbon dioxide environment which increases the pathogenicity of *P. gingivalis* (Diaz et al. 2002).

Passive resistance is when one of the members in the biofilm possesses a resistance factor that can protect other members of the biofilm which do not have the factor. There are numerous biofilm defenses which limit the effectiveness of antibiotics. For example, a beta-lactamase producing strain of *Haemophilus influenza* was cocultured with *Streptococcus pneumoniae* deficient in any resistance factors. *Haemophilus influenza* increased the MIC/MBC of *S. pneumoniae* by amoxicillin (Weimer et al. 2011).

The clinical concern relative to the synergies of polymicrobial biofilm is that the infection will be more severe and recalcitrant to treatment. There are many examples which show that this is indeed the case. Low levels of *P. aeruginosa* mixed with *Staphylococcus aureus* increased infection rates in a rat model (Hendricks et al. 2001). In the mouse model, *Prevotella* increases the pathogenicity of *S. aureus* (Mikamo et al. 1998). *Escherichia coli* produced marked increase in the size of abscess formation with *Bacteroides fragilis* in a diabetic mouse model (Mastropaolo et al. 2005). There also is clinical evidence to suggest that polymicrobial infections are more severe (Tuttle et al. 2011).

The synergies and general recalcitrance produced by polymicrobial infections argue for the full evaluation of every infection. This means not only identification of all species present but also their quantification. However, there is currently not enough information to give clear direction on which microorganisms are important to treat. Also, therapeutic tools for managing polymicrobial infections in conjunction with or separate from antibiotics are generally not available. If a clinician has no specific tools to address all the diversity of a polymicrobial infection then is it valuable to get the test in the first place?

4 The Clinical Use of Molecular Methods: Two Eyes

Identifying and quantitating the microorganisms present in an infection are only part of the diagnosis of an infection. Clinical findings play the major role in determining if the microorganisms present are harming the host. It is only through stereoscopic vision of laboratory results and clinical observation that we can clearly see the power of the detailed information provided by molecular methods. Just as when sophisticated imaging technologies emerged such as MRI, the full meaning and nuances of the images provided could not be appreciated until there was clinical application and experience.

Clinicians seem to be divided by the information provided by DNA-based testing. The unfamiliar microbes can both elucidate and complicate the diagnosis of chronic infections produced by biofilm. Through years of use of molecular methods in real-world chronic infections (mainly chronic wounds) several important principles have emerged. Uncommon bugs occur commonly in chronic wounds and many chronic infections. The clinical challenge of treating rare microbes is more difficult but doable. Literature searches usually will yield usable treatment options for the genera that are identified. Even though we like to know the species identification, most antibiotics, biocides, quorum-sensing inhibitors, and ancillary treatments work at the genus or even the family level for many microbes. That is, a treatment that would kill a rat would in general kill a mouse. Therefore, unfamiliar microbes for treatment purposes can be grouped with closely related microbes which are more familiar (e.g., *Raoultella planticola* and *Klebsiella* spp.) or categorized by common groupings such as gram negative, gram positive, anaerobic, etc. But all of the grouping and comparing of microbes to form a treatment plan highlight the main inadequacy of molecular methods, which is the lack of antibiotic sensitivity data similar to that provided by culture methods.

There are several strategies for managing chronic biofilm infections with the lack of antibiotic sensitivity information. First, if the infection is accessible to topical treatment, high concentrations of antibiotics far in excess of resistance factors can overwhelm most mobile genetic element-induced antibiotic resistance. Second, if systemic antibiotics will be necessary then certain mobile genetic elements with limited diversity, such as mecA cassettes, van genes, and others can be identified by real-time PCR. Third, if sensitivity data is still critical, then molecular diagnosis is still very often the quickest and most cost-effective way to proceed because many microbes are not initially grown in routine clinical culture. By first identifying the microbes of interest by molecular methods, custom nutrients and methods can be used to cultivate microbes for sensitivity work or genomic study (Sibley et al. 2011). With the emerging massive increase in capacity per run, advances in bioinformatics and computing, along with steady decreases in costs, it is becoming feasible to evaluate all the genes in a sample which may allow molecular methods to eventually assess resistance directly in the near future.

Dealing with diversity is made easier by the data provided by DNA-based diagnostics, but caveats remain. Sequencing provides a relative abundance for each species identified in the sample; however, it yields no "absolute" quantification for how much microbial material is present. Real-time PCR has the ability to give reproducible estimates of the number of microbes per gram of tissue (such as 10^5/g) which is termed the "microbial load" or "bacterial load." Several factors can fictitiously lower the value for "microbial load," such as inefficient extraction, decreased primer efficiency, and small variations throughout the analysis. As a result, a low "microbial load" should never be discounted as "not a significant infection." The diagnosis of infection is a clinical decision; therefore, chronic infection itself should always dictate treatment. To evaluate the progression or improvement of an infection it may be necessary to have the lab run the initial sample with subsequent samples in the same run to mitigate these variations, which allows for better comparison.

Quantification of microbes in the polymicrobial infections often encountered in biofilm infection is indispensable. For example, if a sample contains just 1 % MRSA but the bacterial load is 10^8/g then there are still 10^6 MRSA even though it is a minor component of the biofilm. So MRSA coverage would be reasonable. But 1 % MRSA with a bacterial load of 10^5/g (10^3 MRSA) requires only observation which can greatly reduce the use of first-line MRSA antibiotics.

The diversity can be daunting at first, but it is amazing how the many disparate microbes resolve down to treatment groups that require only one or two treating agents. For example, a group of microbes in chronic wounds consisting of MRSA, *Streptococcus*, *Peptoniphilus*, *Anaerococcus*, *Bacteroides*, *Pseudomonas*, and *Serratia* can effectively be treated with the use of clindamycin and amikacin. By collapsing the gram positives and anaerobes into one treatment group covered by clindamycin and then covering the gram negatives with amikacin, only two antibiotics are needed. In fact, high-dose (250 times MIC) amikacin can also provide double coverage for MRSA.

One study showed that by just adding the ability to assess chronic wounds with molecular methods (PCR and sequencing), the use of expensive first-line methicillin-resistant *S. aureus* (MRSA) treatments was greatly reduced (Wolcott et al. 2010). Molecular methods identified *S. aureus* along with the mecA cassette in a majority of the wounds evaluated, yet the quantification showed that MRSA was a minor population (less than 1 % of the bacteria present) and therefore was observed and not actively targeted by antibiotic therapy. Wound care outcomes were improved over standard of care with molecular diagnostics used in this manner. The study demonstrates that using currently available treatments directed by a better understanding of the microbial diversity in question improves outcomes.

Now that molecular tools are available to fully define an infection, it will be up to clinicians to develop appropriate solutions. For example, in the companion study to the one noted above, personalized gels to address what were considered the important species identified within the wound biofilm (usually greater than 1 %) were developed to treat each patient. Molecular diagnostics along with multivalent personalized treatment yielded much better healing outcomes (Dowd et al. 2011).

5 Conclusions

Dealing with the complexity of the results is just the beginning—DNA diagnostics face other barriers in routine clinical use. Clinicians must deal with accessibility, choosing the appropriate laboratory for the analysis, and, as always, cost. Yet the cost of DNA extraction, sequencing, bioinformatics, etc., currently rivals cultivation methods and will continue to drop rapidly. Accessibility is still a barrier.

Technologies now exist which very easily could move molecular diagnosis to the bedside in the next several years. Until then, reference laboratories currently offer the best choice of different DNA diagnostic tests utilizing multiple platforms. Nevertheless, the main barrier for general acceptance is the level of enthusiasm of the clinician for translating this technology into managing infections in individual patients. Not until clinicians embrace molecular methods for identifying and quantitating microbes will molecular methods revolutionize the management of chronic infections.

References

Aagaard K, Petrosino J, Keitel W, Watson M, Katancik J, Garcia N, Patel S, Cutting M, Madden T, Hamilton H, Harris E, Gevers D, Simone G, McInnes P, Versalovic J (2012) The human microbiome project strategy for comprehensive sampling of the human microbiome and why it matters. FASEB J 27(3):1012–1022. doi:10.1096/fj.12-220806, fj.12-220806 [pii]

Chain PS, Grafham DV, Fulton RS, Fitzgerald MG, Hostetler J, Muzny D, Ali J, Birren B, Bruce DC, Buhay C, Cole JR, Ding Y, Dugan S, Field D, Garrity GM, Gibbs R, Graves T, Han CS, Harrison SH, Highlander S, Hugenholtz P, Khouri HM, Kodira CD, Kolker E, Kyrpides NC, Lang D, Lapidus A, Malfatti SA, Markowitz V, Metha T, Nelson KE, Parkhill J, Pitluck S, Qin X, Read TD, Schmutz J, Sozhamannan S, Sterk P, Strausberg RL, Sutton G, Thomson NR, Tiedje JM, Weinstock G, Wollam A, Detter JC (2009) Genomics. Genome project standards in a new era of sequencing. Science 326:236–237. doi:10.1126/science.1180614, 326/5950/236

Dalton T, Dowd SE, Wolcott RD, Sun Y, Watters C, Griswold JA, Rumbaugh KP (2011) An in vivo polymicrobial biofilm wound infection model to study interspecies interactions. PLoS One 6:e27317. doi:10.1371/journal.pone.0027317, PONE-D-11-09086 [pii]

Davies CE, Hill KE, Wilson MJ, Stephens P, Hill CM, Harding KG, Thomas DW (2004) Use of 16S ribosomal DNA PCR and denaturing gradient gel electrophoresis for analysis of the microfloras of healing and nonhealing chronic venous leg ulcers. J Clin Microbiol 42:3549–3557

Del Pozo JL, Patel R (2007) The challenge of treating biofilm-associated bacterial infections. Clin Pharmacol Ther 82:204–209. doi:10.1038/sj.clpt.6100247, 6100247 [pii]

Diaz PI, Zilm PS, Rogers AH (2002) Fusobacterium nucleatum supports the growth of Porphyromonas gingivalis in oxygen- and carbon-dioxide-depleted environments. Microbiology 148:467–472

Dowd SE, Wolcott RD, Kennedy J, Jones C, Cox SB (2011) Molecular diagnostics and personalised medicine in wound care: assessment of outcomes. J Wound Care 20:232, 234–232, 239

Elias S, Banin E (2012) Multi-species biofilms: living with friendly neighbors. FEMS Microbiol Rev. doi:10.1111/j.1574-6976.2012.00325.x

Fischbach MA, Sonnenburg JL (2011) Eating for two: how metabolism establishes interspecies interactions in the gut. Cell Host Microbe 10:336–347. doi:10.1016/j.chom.2011.10.002, S1931-3128(11)00295-2 [pii]

Fitzpatrick KA, Kersh GJ, Massung RF (2010) Practical method for extraction of PCR-quality DNA from environmental soil samples. Appl Environ Microbiol 76:4571–4573. doi:10.1128/AEM.02825-09, AEM.02825-09 [pii]

Fodor AA, DeSantis TZ, Wylie KM, Badger JH, Ye Y, Hepburn T, Hu P, Sodergren E, Liolios K, Huot-Creasy H, Birren BW, Earl AM (2012) The "most wanted" taxa from the human microbiome for whole genome sequencing. PLoS One 7:e41294. doi:10.1371/journal.pone.0041294, PONE-D-11-08927 [pii]

Fux CA, Costerton JW, Stewart PS, Stoodley P (2005) Survival strategies of infectious biofilms. Trends Microbiol 13:34–40

Goddard AF, Staudinger BJ, Dowd SE, Joshi-Datar A, Wolcott RD, Aitken ML, Fligner CL, Singh PK (2012) Direct sampling of cystic fibrosis lungs indicates that DNA-based analyses of upper-airway specimens can misrepresent lung microbiota. Proc Natl Acad Sci USA 109:13769–13774. doi:10.1073/pnas.1107435109, 1107435109 [pii]

Haas BJ, Gevers D, Earl AM, Feldgarden M, Ward DV, Giannoukos G, Ciulla D, Tabbaa D, Highlander SK, Sodergren E, Methe B, DeSantis TZ, Petrosino JF, Knight R, Birren BW (2011) Chimeric 16S rRNA sequence formation and detection in Sanger and 454-pyrosequenced PCR amplicons. Genome Res 21:494–504. doi:10.1101/gr.112730.110, gr.112730.110 [pii]

Hager T (2006) The demon under the microscope. Harmony Books, New York, NY, 46 p

Hansen WL, Beuving J, Bruggeman CA, Wolffs PF (2010) Molecular probes for diagnosis of clinically relevant bacterial infections in blood cultures. J Clin Microbiol 48:4432–4438. doi:10.1128/JCM.00562-10, JCM.00562-10 [pii]

Hendricks KJ, Burd TA, Anglen JO, Simpson AW, Christensen GD, Gainor BJ (2001) Synergy between Staphylococcus aureus and Pseudomonas aeruginosa in a rat model of complex orthopaedic wounds. J Bone Joint Surg Am 83-A:855–861

Howell-Jones RS, Wilson MJ, Hill KE, Howard AJ, Price PE, Thomas DW (2005) A review of the microbiology, antibiotic usage and resistance in chronic skin wounds. J Antimicrob Chemother 55:143–149. doi:10.1093/jac/dkh513, dkh513 [pii]

Insa R, Marin M, Martin A, Martin-Rabadan P, Alcala L, Cercenado E, Calatayud L, Linares J, Bouza E (2012) Systematic use of universal 16S rRNA gene polymerase chain reaction (PCR) and sequencing for processing pleural effusions improves conventional culture techniques. Medicine (Baltimore) 91:103–110. doi:10.1097/MD.0b013e31824dfdb0, 00005792-201203000-00005 [pii]

James GA, Swogger E, Wolcott R, Pulcini ED, Secor P, Sestrich J, Costerton JW, Stewart PS (2008) Biofilms in chronic wounds. Wound Repair Regen 16:37–44

Jenkins C, Ling CL, Ciesielczuk HL, Lockwood J, Hopkins S, McHugh TD, Gillespie SH, Kibbler CC (2012) Detection and identification of bacteria in clinical samples by 16S rRNA gene sequencing: comparison of two different approaches in clinical practice. J Med Microbiol 61:483–488. doi:10.1099/jmm.0.030387-0, jmm.0.030387-0 [pii]

Joseph W (2013) http://www.podiatrytoday.com

Krishna NK, Cunnion KM (2012) Role of molecular diagnostics in the management of infectious disease emergencies. Med Clin North Am 96:1067–1078. doi:10.1016/j.mcna.2012.08.005, S0025-7125(12)00153-8 [pii]

Kuboniwa M, Tribble GD, James CE, Kilic AO, Tao L, Herzberg MC, Shizukuishi S, Lamont RJ (2006) Streptococcus gordonii utilizes several distinct gene functions to recruit Porphyromonas gingivalis into a mixed community. Mol Microbiol 60:121–139. doi:10.1111/j.1365-2958.2006.05099.x, MMI5099 [pii]

Laufer AS, Metlay JP, Gent JF, Fennie KP, Kong Y, Pettigrew MM (2011) Microbial communities of the upper respiratory tract and otitis media in children. MBio 2:e00245-10. doi:10.1128/mBio.00245-10, mBio.00245-10 [pii]

Lipsky BA, Itani K, Norden C (2004) Treating foot infections in diabetic patients: a randomized, multicenter, open-label trial of linezolid versus ampicillin-sulbactam/amoxicillin-clavulanate. Clin Infect Dis 38:17–24. doi:10.1086/380449, CID31366 [pii]

Lipsky BA, Itani KM, Weigelt JA, Joseph W, Paap CM, Reisman A, Myers DE, Huang DB (2011) The role of diabetes mellitus in the treatment of skin and skin structure infections caused by methicillin-resistant Staphylococcus aureus: results from three randomized controlled trials. Int J Infect Dis 15:e140–e146. doi:10.1016/j.ijid.2010.10.003, S1201-9712(10)02526-9 [pii]

Madsen JS, Burmolle M, Hansen LH, Sorensen SJ (2012) The interconnection between biofilm formation and horizontal gene transfer. FEMS Immunol Med Microbiol 65(2):183–195. doi:10.1111/j.1574-695X.2012.00960.x

Martens-Uzunova ES, Olvedy M, Jenster G (2013) Beyond microRNA—novel RNAs derived from small non-coding RNA and their implication in cancer. Cancer Lett 340(2):201–211. doi:10.1016/j.canlet.2012.11.058, S0304-3835(13)00081-5 [pii]

Mastropaolo MD, Evans NP, Byrnes MK, Stevens AM, Robertson JL, Melville SB (2005) Synergy in polymicrobial infections in a mouse model of type 2 diabetes. Infect Immun 73:6055–6063. doi:10.1128/IAI.73.9.6055-6063.2005, 73/9/6055 [pii]

Mikamo H, Kawazoe K, Izumi K, Watanabe K, Ueno K, Tamaya T (1998) Studies on the pathogenicity of anaerobes, especially Prevotella bivia, in a rat pyometra model. Infect Dis Obstet Gynecol 6:61–65. doi:10.1155/S1064744998000155

Mikx FH, van der Hoeven JS (1975) Symbiosis of Streptococcus mutans and Veillonella alcalescens in mixed continuous cultures. Arch Oral Biol 20:407–410

Morgan XC, Segata N, Huttenhower C (2013) Biodiversity and functional genomics in the human microbiome. Trends Genet 29:51–58. doi:10.1016/j.tig.2012.09.005, S0168-9525(12)00145-X [pii]

Mothershed EA, Whitney AM (2006) Nucleic acid-based methods for the detection of bacterial pathogens: present and future considerations for the clinical laboratory. Clin Chim Acta 363:206–220. doi:10.1016/j.cccn.2005.05.050, S0009-8981(05)00439-0 [pii]

Mutz KO, Heilkenbrinker A, Lonne M, Walter JG, Stahl F (2013) Transcriptome analysis using next-generation sequencing. Curr Opin Biotechnol 24:22–30. doi:10.1016/j.copbio.2012.09. 004, S0958-1669(12)00131-0 [pii]

O'Meara S, Al-Kurdi D, Ologun Y, Ovington LG (2010) Antibiotics and antiseptics for venous leg ulcers. Cochrane Database Syst Rev: CD003557. doi:10.1002/14651858.CD003557.pub3

Petrosino JF, Highlander S, Luna RA, Gibbs RA, Versalovic J (2009) Metagenomic pyrosequencing and microbial identification. Clin Chem 55:856–866. doi:10.1373/clinchem. 2008.107565, clinchem.2008.107565 [pii]

Post JC, Aul JJ, White GJ, Wadowsky RM, Zavoral T, Tabari R, Kerber B, Doyle WJ, Ehrlich GD (1996) PCR-based detection of bacterial DNA after antimicrobial treatment is indicative of persistent, viable bacteria in the chinchilla model of otitis media. Am J Otolaryngol 17:106–111

Quince C, Lanzen A, Curtis TP, Davenport RJ, Hall N, Head IM, Read LF, Sloan WT (2009) Accurate determination of microbial diversity from 454 pyrosequencing data. Nat Methods 6:639–641. doi:10.1038/nmeth.1361, nmeth.1361 [pii]

Quince C, Lanzen A, Davenport RJ, Turnbaugh PJ (2011) Removing noise from pyrosequenced amplicons. BMC Bioinformatics 12:38. doi:10.1186/1471-2105-12-38, 1471-2105-12-38 [pii]

Ramsey MM, Rumbaugh KP, Whiteley M (2011) Metabolite cross-feeding enhances virulence in a model polymicrobial infection. PLoS Pathog 7:e1002012. doi:10.1371/journal.ppat.1002012

Reddington K, Tuite N, Barry T, O'Grady J, Zumla A (2013) Advances in multiparametric molecular diagnostics technologies for respiratory tract infections. Curr Opin Pulm Med 19 (3):298–304. doi:10.1097/MCP.0b013e32835f1b32

Rhoads DD, Wolcott RD, Sun Y, Dowd SE (2012a) Comparison of culture and molecular identification of bacteria in chronic wounds. Int J Mol Sci 13:2535–2550. doi:10.3390/ijms13032535, ijms-13-02535 [pii]

Rhoads DD, Cox SB, Rees EJ, Sun Y, Wolcott RD (2012b) Clinical identification of bacteria in human chronic wound infections: culturing vs. 16S ribosomal DNA sequencing. BMC Infect Dis 12:321. doi:10.1186/1471-2334-12-321, 1471-2334-12-321 [pii]

Siami G, Christou N, Eiseman I, Tack KJ (2001) Clinafloxacin versus piperacillin-tazobactam in treatment of patients with severe skin and soft tissue infections. Antimicrob Agents Chemother 45:525–531. doi:10.1128/AAC.45.2.525-531.2001

Sibley CD, Grinwis ME, Field TR, Eshaghurshan CS, Faria MM, Dowd SE, Parkins MD, Rabin HR, Surette MG (2011) Culture enriched molecular profiling of the cystic fibrosis airway microbiome. PLoS One 6:e22702. doi:10.1371/journal.pone.0022702, PONE-D-11-00492 [pii]

Stauffer SH, Birkenheuer AJ, Levy MG, Marr H, Gookin JL (2008) Evaluation of four DNA extraction methods for the detection of Tritrichomonas foetus in feline stool specimens by polymerase chain reaction. J Vet Diagn Invest 20:639–641, 20/5/639 [pii]

Stephenson MF, Mfuna L, Dowd SE, Wolcott RD, Barbeau J, Poisson M, James G, Desrosiers M (2010) Molecular characterization of the polymicrobial flora in chronic rhinosinusitis. J Otolaryngol Head Neck Surg 39:182–187

Takahashi N (2003) Acid-neutralizing activity during amino acid fermentation by Porphyromonas gingivalis, Prevotella intermedia and Fusobacterium nucleatum. Oral Microbiol Immunol 18:109–113, 054 [pii]

Tuttle MS, Mostow E, Mukherjee P, Hu FZ, Melton-Kreft R, Ehrlich GD, Dowd SE, Ghannoum MA (2011) Characterization of bacterial communities in venous insufficiency wounds by use of conventional culture and molecular diagnostic methods. J Clin Microbiol 49:3812–3819. doi:10.1128/JCM.00847-11, JCM.00847-11 [pii]

Verhoeven PO, Grattard F, Carricajo A, Lucht F, Cazorla C, Garraud O, Pozzetto B, Berthelot P (2012) Quantification by real-time PCR assay of Staphylococcus aureus load: a useful tool for rapidly identifying persistent nasal carriers. J Clin Microbiol 50:2063–2065. doi:10.1128/JCM. 00157-12, JCM.00157-12 [pii]

Weile J, Knabbe C (2009) Current applications and future trends of molecular diagnostics in clinical bacteriology. Anal Bioanal Chem 394:731–742. doi:10.1007/s00216-009-2779-8

Weimer KE, Juneau RA, Murrah KA, Pang B, Armbruster CE, Richardson SH, Swords WE (2011) Divergent mechanisms for passive pneumococcal resistance to beta-lactam antibiotics in the presence of Haemophilus influenzae. J Infect Dis 203:549–555. doi:10.1093/infdis/jiq087, jiq087 [pii]

Woese CR, Fox GE (1977) Phylogenetic structure of the prokaryotic domain: the primary kingdoms. Proc Natl Acad Sci USA 74:5088–5090

Wolcott RD, Cox SB, Dowd SE (2010) Healing and healing rates of chronic wounds in the age of molecular pathogen diagnostics. J Wound Care 19:272–278, 280–1

Woo PC, Lau SK, Teng JL, Tse H, Yuen KY (2008) Then and now: use of 16S rDNA gene sequencing for bacterial identification and discovery of novel bacteria in clinical microbiology laboratories. Clin Microbiol Infect 14:908–934. doi:10.1111/j.1469-0691.2008.02070.x, CLM2070 [pii]

Imaging Biofilms in Tissue Specimens

Garth James and Alessandra Marçal Agostinho Hunt

Abstract Microscopic imaging can be used to provide direct evidence of the presence of biofilms in tissue samples. However, successful imaging requires a series of well-planned and executed steps as well as adequate controls. The steps can include sample collection, fixation, embedding and sectioning, staining, and imaging. Each of these steps is discussed in this chapter. Selection of an appropriate staining technique is one of the key considerations for sample analysis. A variety of staining techniques ranging from general stains to highly specific molecular probes are available. For fluorescence microscopy, staining complications can include tissue autofluorescence, fixation-induced autofluorescence, and nonspecific binding. To assess the impact of these potential complications, it is best to use multiple complementary staining techniques as well as adequate controls. A variety of microscopic techniques have been used to image biofilms in tissue samples including light microscopy, transmission and scanning electron microscopy, epifluorescent microscopy, and confocal scanning laser microscopy. The latter technique has advantages of minimal sample manipulation, three dimensional imaging, and enables the use of specific fluorescent molecular probes.

1 Introduction

There has been an increasing consensus that a number of diseases are related to biofilms, so the ability to detect biofilms through imaging of tissue represents an important tool not only for biofilm research but also for diagnosis and evaluation of

---

G. James (✉)
Center for Biofilm Engineering, Montana State University, Bozeman, MT, 59717, USA
e-mail: gjames@biofilm.montana.edu

A.M. Agostinho Hunt
Department of Microbiology and Molecular Genetics, Michigan State University, 5177 BPS Building, 567 Wilson Road, East Lansing, MI, 48824, USA
e-mail: alessandraagostinho@gmail.com

treatments. Biofilms consist of tridimensional microbial communities where spatial relationships between bacterial populations and host tissue may be important for biofilm function and interactions with host cells. Thus, it is desirable to image tissue specimens with minimal manipulation to preserve biofilm architecture. Biofilm bacteria are embedded in extracellular polymeric substances (EPS). An ideal technique for imaging biofilms in tissues would allow the visualization of encasing EPS as well as microbial cells.

Imaging of bacteria and fungi in tissue has been performed for many years using methods and dyes developed for conventional pathology. In fact, histopathology of infectious diseases has played an important role on the identification of etiologic agents (Gupta et al. 2009). Classic techniques like Brown and Brenn Gram staining for tissue have been applied for more than 80 years (Engbaek et al. 1979), but despite their proven value they require that the tissue be exposed to a series of chemical rinses. In addition, while bacterial cells can be differentiated from host cells and structures by means of morphology, imaging of EPS poses a challenge.

In this chapter, certain aspects that influence biofilm imaging in tissues will be discussed including sample collection, fixation, embedding and sectioning, staining, and microscopy. Although electron microscopy can be used for high-resolution analysis of tissue samples, this chapter focuses primarily on fluorescence microscopy. Specific examples from published studies of microorganisms and biofilms in tissue are used when possible.

2 Specimen Collection

There are a variety of methods for collecting tissue samples for biofilm analysis and method choice is at the discretion of the clinician. If surgical debridement of tissue is necessary, the debrided material is a convenient source of specimens for analysis. Punch biopsies of various diameters can also be collected. One advantage with both of these methods is that they provide underlying tissue from the wound. This can be helpful for showing biofilm attachment to the tissue and the extent of tissue invasion. However, both of these methods are invasive and the necessity of the procedures and potential risks to the subject must be considered. Wound specimens can also be collected by curettage, which is a less invasive procedure and routinely conducted during standard wound care. Topical anesthetics, such as a lidocaine gel, can be used to prevent or reduce pain during specimen collection. Typically, specimens are collected at the advancing edge of the wound, although the microbial populations can vary at different locations within the wound. Specimens should be placed in an appropriate fixative as soon as possible after collection.

3 Specimen Fixation

Fixation is a chemical process intended to impede autolysis or putrefaction of the specimen by stopping biochemical reactions, thereby preserving tissues. It also inactivates bacteria, fungi, and viruses, providing biosafety for lab personnel (Rubbo et al. 1967). Thus, fixation is an important and necessary step in tissue preparation for microscopy. Certain fixatives, such as glutaraldehyde, are known to induce specimen fluorescence (Collins and Goldsmith 1981). This fixation-induced fluorescence is caused by a reaction between the amines and proteins from the tissue and the aldehyde groups in the fixative generating fluorescent products (Wright Cell Imaging Facility 2013). Thus, for fluorescence microscopy, it is important to select a fixative that will provide adequate preservation of tissue while inducing minimal fluorescence. The authors have found freshly prepared paraformaldehyde (4 %) to be a suitable fixative for fluorescence microscopy.

4 Intrinsic Autofluorescence, Fixation-Induced Fluorescence, and Nonspecific Binding

A certain level of intrinsic fluorescence is expected from tissues due to the presence of flavins and porphyrins. This phenomenon can be useful for orientation purposes during microscopic analysis particularly when non-visible dyes, such as Cy5, are used. However, high levels of background fluorescence can interfere with the visualization of microorganisms by fluorescence microscopy. As discussed above, fixatives can also induce autofluorescence in tissues. The emission spectra of intrinsic or induced autofluorescence is very broad compared to the spectra of typical fluorophores used for fluorescence microscopy, which makes it difficult to separate wanted from unwanted fluorescence by traditional filtering methods (Wright Cell Imaging Facility 2013).

Several techniques using quenching chemical solutions have also been developed with the goal of reducing unwanted tissue fluorescence. These methods include the use of pontamine skye blue, Sudan black B, sodium borohydrate, trypan blue, and ammonia–ethanol (Cowen et al. 1985; Sun et al. 2011; Oliveira et al. 2010; Srivastava et al. 2011; Raghavachari et al. 2003); however, their efficacy is variable. Baschong et al. (2001) studied the effect of three separate reagents on reducing autofluorescence of three different types of tissue. These authors found that no single reagent was able to decrease autofluorescence in all samples and that the selection of an appropriate autofluorescence quenching agent required a trial-and-error search process.

The nonspecific binding of molecular probes in tissue samples is another problem that can occur during imaging of biofilms. While designed to be specific for targeted microorganisms, probes for immunofluorescence and fluorescence in situ hybridization (FISH) can be subject to significant nonspecific binding.

For example, the authors used a general eubacterial probe, EUB 338 tagged with a Cy3 fluorophore, to image mouse tissue infected with *Staphylococcus aureus*. For certain samples, probing revealed red fluorescent aggregates that resembled microcolonies of cocci. However, the same aggregates were observed in control samples (not infected) and cocci were not observed in sections stained using a tissue Gram stain. This led to the suspicion that the probe was nonspecifically binding to mast cells present in the tissue. This example serves to underscore the importance of appropriate controls and complementary alternative staining methods when using species-specific probes. A study published by Wallner and collaborators (1993) investigated the effect of several parameters on fluorescent rRNA-targeted oligonucleotide probe binding and specificity. These authors found that excessively high probe concentrations lead to an increase in nonspecific binding that was likely due to attachment of probes to cell components rather than mispairing of the probe sequence to nontarget sequences.

5 Embedding and Sectioning of Tissues

Embedding is the process of casting a tissue section in a selected medium. In this process, the tissue is infiltrated with a liquid medium, which solidifies, allowing sectioning into thin slices while preserving tissue structure and morphology. The selection of an embedding medium depends on the type of tissue, sectioning technique, and the staining techniques to be used. The medium needs to adhere to the tissue and be elastic enough to withstand the sectioning procedure, while impeding the permanent deformation of the tissue to preserve morphology and structure. A number of embedding media have been developed. Low-temperature agarose gels and gelatin have been used for embedding tissue to be sectioned with vibratomes. These media are useful for tissues that can't stand high temperatures. Plastic resin media, either epoxy or acrylic, have been extensively used for embedding undecalcified tissue, such as bone and teeth and for imaging of tissue-implant combinations such as stented vessels (Rippstein et al. 2006). They also perform well for sections thinner than the normal 4–6 μm usually collected from renal and bone marrow biopsies (Bruce-Gregorios 2006). Epoxy resins allow fast embedding of the tissue and give good contrast for electron microscopy (Luft 1961); however, there are some disadvantages. The epoxide groups in the resin may reduce the antigenicity of the tissue, thus impacting immunohistochemistry. In addition, epoxy resins are toxic, can cause allergic reactions, and the vinylcyclohexane dioxide component is known to be carcinogenic (Bruce-Gregorios 2006). Acrylic resins are made of esters of acrylic or methacrylic acid and have been preferred for embedding hard tissues due to ease of use and good quality of the subsequent staining (Bruce-Gregorios 2006). The most common medium used for tissue embedding is paraffin. This can be conducted using automated systems, resulting in a fast and uniform way of preparing specimens. All water content has to be removed from tissue prior to paraffin embedding because paraffin is not miscible with water.

Dehydration is typically accomplished with a graded ethanol series, followed by infiltration with xylene or a xylene substitute such as SafeClear™.

Cryoembedding and cryosectioning of tissue has been used routinely in the medical field, especially when a fast diagnosis of pathology is required, such as during surgery. These techniques do not provide the best preservation of tissue morphology but are often suitable for specific applications. Cryoembedding of tissues has the advantages of being a fast and simple technique for preparing samples for sectioning and requires virtually no reagents other than the embedding medium. Like other embedding media, the water content of the samples needs to be replaced prior to embedding. In cryoembedding, this prevents ice crystal formation within the tissue during freezing.

For biofilm analysis in tissue specimens, cryosectioning and cryoembedding have proven to be suitable methods (Han et al. 2011). After fixation, the tissue specimen was dehydrated in a 30 % sucrose solution until it sank to the bottom of the flask, which was an indication that the solution had penetrated the entire specimen. The sample was then placed in a small metal or plastic disposable tray with a freezing medium (Tissue-Tek® O.C.T™ Compound) and snap-frozen on a block of dry ice. Sections were then cut with a cryostat, a refrigerated microtome with temperature of about -20 to -30 °C. After cutting, the tissue sections were picked up on plus microscope slides and then stained or frozen for later analysis.

6 Staining

Both conventional and fluorescent stains have been used for staining bacteria in tissue. Hematoxylin and eosin (H&E) is a traditional histological stain that has been used to demonstrate the presence of microorganisms in tissue. Hansen et al. (2006) and Marx and Tursun (2012) applied H&E to biopsies of patients with osteoradionecrosis and detected the presence of large aggregates of bacteria next to areas of necrotic bone. Gram staining of tissue has also been used for many years, and several techniques, such as the Brown and Brenn Gram stain and Brown and Hopps Gram stain, have been tested and compared (Engbaek et al. 1979). Sizemore et al. (1990) used fluorescently labeled wheat germ agglutinin to label Gram-positive bacteria. This lectin binds specifically to N-acetylglucosamine in the peptidoglycan layer of Gram-positive microbes and does not stain Gram-negative bacteria due to their outer membrane. Although conventional Gram staining can provide good differentiation of Gram-positive bacteria, the identification of Gram-negative bacteria can be impaired by coloring of the host tissue. As an alternative, several attempts were made to develop fluorescent dyes to differentially Gram stain microorganisms. Mason et al. (1998) developed a combination of hexidium iodide (HI) and SYTO 13 to Gram label bacteria for both flow cytometry and fluorescence microscopy. When used in combination, Gram negatives were stained green by SYTO 13, while Gram positives were stained by both SYTO 13 and HI, resulting in red–orange fluorescence due to the quenching properties of HI. Han et al. (2011)

used a commercially available fluorescent Gram staining kit for tissue to examine chronic wound biopsies for the presence of biofilms. Fixed tissue biopsies from 15 patients were cryoembeded, cryosectioned, stained with ViaGram™ Red + Bacterial Gram-Stain and Viability Kit (Life Technologies, Carlsbad, CA), and examined using epifluorescence microscopy. The images revealed distributions of microorganisms varying from scattered bacterial cells to dense biofilms.

The use of fluorescent dyes for detecting biofilms in tissues has increased as new fluorochromes for general and species-specific staining have been engineered. However, staining bacteria enmeshed in EPS and attached to tissue remains challenging. As discussed previously, autofluorescence and nonspecific stain binding to tissue components can hinder the detection of bacteria. A further challenge is distinguishing bacterial EPS from the host extracellular matrix (HECM). Often, the detection of EPS around bacterial cells is desired as evidence of biofilm. HECM is present in all animal tissues and organs and consists of a diverse group of proteins, glycoproteins, and proteoglycans that form basement membranes and all interstitial structures of the body (Byron et al. 2013; Hynes 2009). EPS is composed of carbohydrates, proteins, DNA, and may also contain host components. Thus, the overall composition of EPS is similar to HECM and EPS stains also bind HECM. Unfortunately, no reliable stains that can differentiate EPS from HECM have yet been described.

7 Fluorescent Stain Selection

7.1 General Stains

Pilot experiments are highly recommended for determining appropriate staining protocols for a particular set of specimens. The type of tissue, amount of biofilm, and fixation method may influence the outcome of staining, so testing a diverse array of stains, concentrations, and staining times may result in identifying the best approach. The combination of SYTO® 9, a green nucleic acid stain, and propidium iodide, a red nucleic acid stain has been used for staining of biofilms in tissue by several groups. Both stains are part of commercially available LIVE/DEAD® kits from Life Technologies. SYTO® 9 stains the nucleic acids of bacteria with either damaged or intact membranes, while propidium iodide rapidly penetrates bacteria with damaged membranes. Thus, when these two components are used together to stain bacteria for an appropriate contact time, bacteria with intact cell membranes fluoresce green, while bacteria with damaged membranes fluoresce red. Kathju et al. (2012) successfully used SYTO® 9 and propidium iodide to verify the presence of biofilms in tissue from a patient suffering from hidradenitis suppurativa. The specimen was examined by CSLM and the images agreed with the culture results, revealing the presence of cocci and rods which were stained green (live) and red (dead) and arranged in clusters. Anderson et al. (2012) also used SYTO® 9 and propidium iodide to study the effect of different strains of *S. aureus* on vaginal mucosa. Although the dyes also stained the mucosal cells, there was morphological evidence of coccal micro-colonies between the epithelial cells. More recently, Tsai et al. (2013) examined biopsies from patients suffering from nasopharyngeal cancer who developed osteoradionecrosis for the presence of biofilm. Clusters of bacteria were detected using SYTO® 9 and propidium iodide and the authors observed that far more samples were positive for biofilm according to imaging than through culturing techniques.

The authors have observed that the combination of Sytox® green and wheat germ agglutinin conjugated with Texas Red® (WGR-TR) allowed the detection of microorganisms in tissue, regardless of type, fixative, embedding medium, and sectioning technique (Fig. 1). These dyes are components of a commercially available kit, ViaGram™ Red + Bacterial Gram-Stain and Viability Kit (Life Technologies, Carlsbad, CA). Sytox® green is a high affinity nucleic acid stain, and in fixed tissue samples, stains all Gram-negative and Gram-positive microorganisms and the nuclei of mammalian cells, but not HECM. The bacteria are readily differentiated from the mammalian cells and other structures by size and morphology. The fluorescently labeled lectin, WGA-TR, stained HECM and probably also EPS. When both green and red channel images taken separately on a epi-fluorescence or CSLM are combined using imaging software, the resulting image allows the clear visualization of bacterial cells as well as their spatial localization on the tissue.

7.2 Immunofluorescence

Immunofluorescence (IF) utilizes species-specific antibodies labeled with fluorophores to tag microorganisms. This can involve directly labeled primary antibodies or detection of unlabeled primary antibodies with fluorescently labeled secondary antibodies. In order to successfully bind, antibodies need to have specificity and be applied in the correct concentration to antigens with accessible epitopes (Harlow and Lane 1999). Furthermore, to be able to bind to targets within a biofilm, the antibody needs to penetrate the EPS that surrounds the microbial cells. Nonetheless, antibodies against group A Streptococci have been successfully used to image these bacteria in mouse wounds (Connolly et al. 2011) and within the crypts of the palatine tonsils from children (Roberts et al. 2012). The authors have used immunofluorescence and CSLM to image *S. aureus* biofilm cryosections of mouse tissue samples (Figs. 2 and 3). Transmitted light microscopy was used as an alternate imaging method. As with other imaging techniques, it is important to include alternate imaging methods as well as both positive and negative controls to confirm results. For IF, nonspecific binding of both primary and secondary antibodies is possible. Nonspecific binding can be reduced using a variety of blocking agents including serum, albumin, and skim milk powder. The binding of antibodies to nontarget bacteria with similar epitopes to target bacteria is also a possible

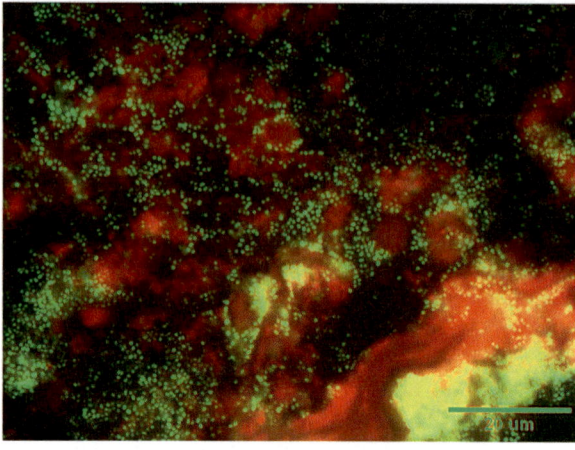

Fig. 1 Epifluorescence micrograph of a thin section from a human wound specimen biopsy stained with Sytox® green and the fluorescent lectin, wheat germ agglutinin conjugated with Texas Red®. The host extracellular matrix (and likely also bacterial extracellular polymeric substances) appears *red* due to lectin binding and is colonized by biofilms of cocci (*green spheres*)

complication in IF. Again, adequate controls and complementary imaging methods can be used to avoid erroneous results. For the detection of particular species within biofilms, IF and FISH are attractive complementary techniques for microscopic analysis.

7.3 Fluorescent In Situ Hybridization

The use of molecular techniques to fluorescently tag specific microorganisms is an attractive approach. Particularly because imaging the spatial distribution of targeted microorganisms in relation to tissue structures and other microorganisms is possible. FISH is a molecular technique that uses fluorescent-labeled rRNA oligonucleotide probes which combined with microscopy or flow cytometry allows the detection of microorganisms. This technique facilitates the rapid and specific identification of microbial cells in their natural environments and can be used in phylogenetic, ecological, diagnostic, and environmental studies (Amann et al. 2001; Bottari et al. 2006; Moter and Göbel 2000). As discussed previously, controlling specimen and fixation-induced autofluorescence as well as nonspecific binding are important aspects of FISH. Moter and Göbel (2000) wrote a comprehensive review of FISH. Although emphasizing the many applications of FISH for detection of microorganisms in systems ranging from water to environment to medicine, the authors described a series of pitfalls of the method. The drawbacks discussed included the occurrence of false positive results due to autofluorescence of bacteria or tissue and the nonspecific binding of the probes. False negative results were also highlighted, particularly the problems with penetration of the probe into bacterial cells (particularly Gram negative bacteria) and the detection of cells with low rRNA contents. Bacteria of low RNA content is of particular concern for analysis of biofilms in tissue samples where the bacteria may have a low metabolic

Imaging Biofilms in Tissue Specimens   39

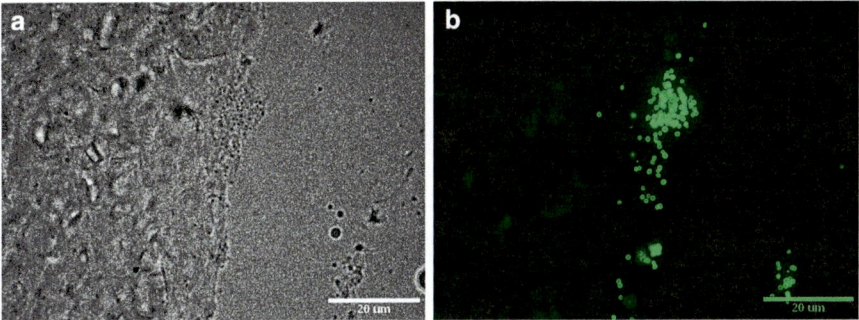

Fig. 2 Transmitted light (**a**) and epifluorescence (**b**) micrographs of a thin section from *Staphylococcus aureus*-infected mouse wound specimen stained with fluorescent anti-*S. aureus* antibodies. The *S. aureus* cells appear as *dark spheres* in the transmitted light image and fluorescent rings in the epifluorescence image. The *ring* appearance is typical of immunofluorescent staining, due to binding of the antibodies to cell surface antigens. The use of complementary imaging techniques help in the interpretation of staining results

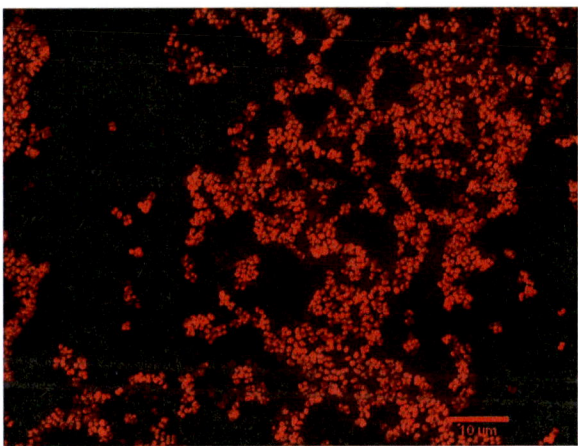

Fig. 3 Epifluorescence micrograph of a thin section from *Staphylococcus xylosis*-infected mouse tissue specimen stained with a Cy3-labeled *S. xylosis* FISH probe (5'-CATGCGGTTCTAAATGTTATCCGGT-3'). This FISH probe was designed by Dr. Elinor deLancey Pulcini, Center for Biofilm Engineering. An *S. xylosis* biofilm is apparent in this micrograph

rate. FISH has been applied to a variety of tissues types. Sunde et al. (2003) used FISH to demonstrate the presence of bacteria in periapical lesions of infected root canals, while Nistico et al. (2009) applied FISH on middle ear and upper respiratory mucosa and Rudkjøbing et al. (2012) utilized FISH to examine lung explants from end-stage cystic fibrosis patients.

A variant of FISH, peptide nucleic acid-based fluorescence in situ hybridization (PNA-FISH) uses DNA analogs with an uncharged polyamide backbone instead of sugar phosphates. Because of the neutral back bone, diffusion of PNA probes through hydrophobic cell walls may be facilitated and help overcome the problem of probe penetration into Gram-negative bacteria. Fazli et al. (2011) applied specific and universal bacterial PNA-FISH probes to chronic venous leg ulcer biopsies to identify wounds containing either *P. aeruginosa* or *S. aureus*. The two groups were then evaluated for the presence of neutrophils. Overall, microorganisms were observed in large aggregates (biofilms) and ulcers harboring *P. aeruginosa* correlated with the presence of higher neutrophil counts. This association between *P. aeruginosa* biofilms and neutrophils was hypothesized to cause the persistent inflammatory response and delayed wound healing in *P. aeruginosa*-infected wounds. The association of *P. aeruginosa* and neutrophils was also observed in experimentally infected mouse wounds using PNA-FISH to detect *P. aeruginosa* and DAPI (4′,6-diamidino-2-phenylindole) to detect neutrophils (Trøstrup et al. 2013). Regardless of the PNA-FISH approach, these examples underscore the potential for the analysis of spatial relationships between microorganisms and host cells to aid in the understanding and diagnosis of disease.

8 Microscopy

The use of microscopy to demonstrate the presence of biofilms in tissue samples has been increasing. Akiyama et al. (1996) used light and transmission electron microscopy of skin biopsy thin sections to image the time course of *S. aureus* growth within inoculated mouse wounds. Microcolonies of *S. aureus* were detected as soon as 3 h after inoculation. James et al. (2008) evaluated samples from chronic and acute human wounds using Gram staining and light microscopy as well as scanning electron microscopy (SEM) and found that chronic wounds were more likely to harbor biofilms. Freeman et al. (2009) also used Gram staining and light microscopy to demonstrate the presence of biofilms in horse wound biopsies.

While light microscopy has been successfully used to detect bacteria in tissue, fluorescence microscopy has been fundamental for the study of biofilms. Not only does fluorescence microscopy provide excellent spatial resolution and the detection of broad emission profiles, it also enables the labeling of specific tissue structures and bacterial cells within a sample (Coling and Kachar 2001). In particular, the application of confocal scanning laser microscopy (CSLM) has revolutionized biofilm imaging. Conventional microscopy techniques require that samples are observed as thin sections on a slide but CSLM allows the examination of fully hydrated relatively thick biofilms. This enables sample preparation with minimal manipulation, helping to ensure that biofilms keep their original morphology and architecture. This attribute is particularly valuable for examining spatial relationships between bacteria within biofilms and host cells and tissue. Confocal microscopy was developed and patented by Marvin Minsky in 1955. It was primarily used in physical and medical sciences until the 1990s when it was first applied for the study of biofilms (Lawrence et al. 1991). CSLM uses optical imaging to create a virtual slice or plane, many micrometers deep, within a sample, without the disturbing light emissions from slices out of the focal plane (Nwaneshiudu et al. 2012). The specimen is moved in the z direction and the stack of images collected is then combined using computer software. The resulting three-dimensional reconstructed image provides a high degree of resolution and detail. Computer image analysis can be combined with CSLM for the determination of various parameters including number of cells, area of coverage, individual dimension measurements, and spatial orientation (Lawrence and Thomas 1999).

For several years CSLM has been used for analysis of suspected bacteria-harboring tissue in addition to culturing and molecular techniques to confirm the presence of biofilms. Neut and collaborators (2003) evaluated joint explants from revision surgeries in which infection was suspected. The analysis involved tissue and biomaterial culturing as well as CSLM of the prosthesis. The samples were stained with LIVE/DEAD® BacLight™ Bacterial Viability Kits and showed the presence of green (live) cell clusters. A CSLM-based study was also conducted by Hall-Stoodley et al. (2006) to test the hypothesis that chronic otitis media is a biofilm-related infection. Fifty-two biopsies from 26 children suffering from otitis media with either effusion or recurrent disease were analyzed in addition to 16 samples from control subjects. The mucosa samples were stained using generic dyes, species-specific probes, and antibodies and underwent thorough examination. All control samples were negative for biofilm presence, while 46 samples from the test group were positive for biofilms according to either a general or specific staining method. Bacteria in the tissue were clearly identifiable in the samples and ranged from micro-colonies to large clusters. A case report published in 2008 also reported the use of CLSM to investigate the cause of a persistent infection following a total joint arthroplasty (Stoodley et al. 2008). The patient had recurring pain and infection episodes for years despite several interventions. CSLM analysis of fluid, tissue, and cement collected at his final surgical revision revealed the presence of biofilm. These examples show that CSLM can be a powerful tool for the diagnosis of biofilm-related diseases. However, as stated by Stoodley et al. (2008), it involves expensive equipment that is not generally available in hospitals for examinations. It also requires an experienced operator and the analysis of samples is a time-consuming process. These drawbacks limit the routine use of CSLM for medical diagnostics.

9 Defining and Quantifying Biofilm

Biofilms are generally defined as communities of microorganisms attached to a surface and enmeshed in an extracellular polymer matrix. For practical purposes related to the microscopic examination of tissue specimens, the presence of microbial aggregates closely associated with tissue is usually accepted as adequate evidence of biofilm. Although demonstrating the presence of EPS would provide further evidence of biofilm, as discussed previously, distinguishing EPS from the extracellular matrix of the host tissue is difficult with currently available staining methods. For quantifying biofilm, most studies have used a simple presence or absence determination. Han et al. (2011) used a 6-point scale to classify the amount of biofilm in wound biopsies, with zero representing the absence of bacterial cells and five indicating the presence of a thick, continuous film. Adoption of a standard classification system for the amount of biofilm present would be an important tool to allow comparisons between multiple examiners in the same or different research groups.

10 Future Prospects

Considering that biofilms are being increasingly implicated in a variety of diseases, particularly those involving chronic infections, the use of imaging to detect biofilms in tissues will be an increasingly important tool. New advances in microscopy equipment, techniques, stains, and probes will likely improve our ability to detect and quantify biofilms. This, in turn, will increase our understanding of biofilm infections and lead to improvements in the diagnosis and treatment of disease.

11 Conclusions

Imaging biofilms in tissue samples is a task that demands a series of well-planned and executed steps that starts with sample collection and culminates in image collection and interpretation. Improper handling of the sample or negligence in any of the steps may directly impact the imaging outcome. Ultimately the detection of biofilms in tissues can be successfully achieved when appropriate staining methods and microscopy techniques are applied and proper controls are utilized. A trained and knowledgeable microscopist is also fundamental for the successful analysis of samples.

References

Akiyama H, Kanzaki H, Tada J, Arata J (1996) *Staphylococcus aureus* infection on cut wounds in the mouse skin: experimental staphylococcal botryomycosis. J Dermatol Sci 11:234–238

Amann R, Fuchs BM, Behrens S (2001) The identification of microorganisms by fluorescence in situ hybridisation. Curr Opin Biotechnol 12:231–236

Anderson MJ, Lin YC, Gillman AN, Parks PJ, Schlievert PM, Peterson ML (2012) Alpha-toxin promotes Staphylococcus aureus mucosal biofilm formation. Front Cell Infect Microbiol 2:64

Baschong W, Suetterlin R, Laeng RH (2001) Control of autofluorescence of archival formaldehyde-fixed, paraffin-embedded tissue in confocal laser scanning microscopy (CLSM). J Histochem Cytochem 49:1565–1572

Bottari B, Ercolini D, Gatti M, Neviani E (2006) Application of FISH technology for microbiological analysis: current state and prospects. Appl Microbiol Biotechnol 73:485–494

Bruce-Gregorios J (2006) Histopathologic techniques, 2nd edn. Goodwill Trading Co, Quezon City

Byron A, Humphries JD, Humphries MJ (2013) Defining the extracellular matrix using proteomics. Int J Exp Pathol 94:75–92

Coling D, Kachar B (2001) Theory and application of fluorescence microscopy. Curr Protoc Neurosci Chapter 2:Unit 2.1

Collins JS, Goldsmith TH (1981) Spectral properties of fluorescence induced by glutaraldehyde fixation. J Histochem Cytochem 29:411–414

Connolly KL, Roberts AL, Holder RC, Reid SD (2011) Dispersal of Group A streptococcal biofilms by the cysteine protease SpeB leads to increased disease severity in a murine model. PLoS One 6:e18984

Cowen T, Haven AJ, Burnstock G (1985) Pontamine sky blue: a counterstain for background autofluorescence in fluorescence and immunofluorescence histochemistry. Histochemistry 82:205–208

Engbaek K, Johansen KS, Jensen ME (1979) A new technique for Gram staining paraffin-embedded tissue. J Clin Pathol 32:187–190

Fazli M, Bjarnsholt T, Kirketerp-Møller K, Jørgensen A, Andersen CB, Givskov M, Tolker-Nielsen T (2011) Quantitative analysis of the cellular inflammatory response against biofilm bacteria in chronic wounds. Wound Repair Regen 19:387–391

Freeman K, Woods E, Welsby S, Percival SL, Cochrane CA (2009) Biofilm evidence and the microbial diversity of horse wounds. Can J Microbiol 55:197–202

Gupta E, Bhalla P, Khurana N, Singh T (2009) Histopathology for the diagnosis of infectious diseases. Indian J Med Microbiol 27:100–106

Hall-Stoodley L, Hu FZ, Gieseke A, Nistico L, Nguyen D, Hayes J, Forbes M, Greenberg DP, Dice B, Burrows A, Wackym PA, Stoodley P, Post JC, Ehrlich GD, Kerschner JE (2006) Direct detection of bacterial biofilms on the middle-ear mucosa of children with chronic otitis media. JAMA 296:202–211

Han A, Zenilman JM, Melendez JH, Shirtliff ME, Agostinho A, James G, Stewart PS, Mongodin EF, Rao D, Rickard AH, Lazarus GS (2011) The importance of a multi-faceted approach to characterizing the microbial flora of chronic wounds. Wound Repair Regen 19:532–541

Hansen T, Kunkel M, Weber A, James Kirkpatrick C (2006) Osteonecrosis of the jaws in patients treated with bisphosphonates – histomorphologic analysis in comparison with infected osteoradionecrosis. J Oral Pathol Med 35:155–160

Harlow E, Lane D (1999) Using antibodies: a laboratory manual. Cold Spring Harbor Laboratory Press, New York, NY

Hynes RO (2009) The extracellular matrix: not just pretty fibrils. Science 326:1216–1219

James GA, Swogger E, Wolcott R, Pulcini Ed, Secor P, Sestrich J, Costerton JW, Stewart PS (2008) Biofilms in chronic wounds. Wound Repair Regen 16:37–44

Kathju S, Lasko LA, Stoodley P (2012) Considering hidradenitis suppurativa as a bacterial biofilm disease. FEMS Immunol Med Microbiol 65:385–389

Lawrence JR, Thomas RN (1999) Confocal laser scanning microscopy for analysis of microbial biofilms. In: Doyle RJ (ed) Methods in enzymology, vol 310. Academic, San Diego, pp p131–p144

Lawrence JR, Korber DR, Hoyle BD, Costerton JW, Caldwell DE (1991) Optical sectioning of microbial biofilms. J Bacteriol 173:6558–6567

Luft JH (1961) Improvements in epoxy resin embedding methods. J Biophys Biochem Cytol 9:409–414

Marx RE, Tursun R (2012) Suppurative osteomyelitis, bisphosphonate induced osteonecrosis, osteoradionecrosis: a blinded histopathologic comparison and its implications for the mechanism of each disease. Int J Oral Maxillofac Surg 41:283–289

Mason DJ, Shanmuganathan S, Mortimer FC, Gant VA (1998) A fluorescent Gram stain for flow cytometry and epifluorescence microscopy. Appl Environ Microbiol 64(7):2681–2685

Moter A, Göbel UB (2000) Fluorescence in situ hybridization (FISH) for direct visualization of microorganisms. J Microbiol Methods 41(2):85–112

Neut D, van Horn JR, van Kooten TG, van der Mei HC, Busscher HJ (2003) Detection of biomaterial-associated infections in orthopaedic joint implants. Clin Orthop Relat Res 413:261–268

Nistico L, Gieseke A, Stoodley P, Hall-Stoodley L, Kerschner JE, Ehrlich GD (2009) Fluorescence "in situ" hybridization for the detection of biofilm in the middle ear and upper respiratory tract mucosa. Methods Mol Biol 493:191–213

Nwaneshiudu A, Kuschal C, Sakamoto FH, Anderson RR, Schwarzenberger K, Young RC (2012) Introduction to confocal microscopy. J Invest Dermatol 132:e3

Oliveira VC, Carrara RC, Simoes DL, Saggioro FP, Carlotti CG Jr, Covas DT, Neder L (2010) Sudan Black B treatment reduces autofluorescence and improves resolution of in situ hybridization specific fluorescent signals of brain sections. Histol Histopathol 25:1017–1024

Raghavachari N, Bao YP, Li G, Xie X, Müller UR (2003) Reduction of autofluorescence on DNA microarrays and slide surfaces by treatment with sodium borohydride. Anal Biochem 312:101–105

Rippstein P, Black MK, Boivin M, Veinot JP, Ma X, Chen YX, Human P, Zilla P, O'Brien ER (2006) Comparison of processing and sectioning methodologies for arteries containing metallic stents. J Histochem Cytochem 54:673–681

Roberts AL, Connolly KL, Kirse DJ, Evans AK, Poehling KA, Peters TR, Reid SD (2012) Detection of group A *Streptococcus* in tonsils from pediatric patients reveals high rate of asymptomatic streptococcal carriage. BMC Pediatr 12:3

Rubbo SD, Gardner JF, Webb RL (1967) Biocidal activities of glutaraldehyde and related compounds. J Appl Bacteriol 30:78–87

Rudkjøbing VB, Thomsen TR, Alhede M, Kragh KN, Nielsen PH, Johansen UR, Givskov M, Høiby N, Bjarnsholt T (2012) The microorganisms in chronically infected end-stage and non-end-stage cystic fibrosis patients. FEMS Immunol Med Microbiol 65:236–244

Sizemore RK, Caldwell JJ, Kendrick AS (1990) Alternate gram staining technique using a fluorescent lectin. Appl Environ Microbiol 56:2245–2247

Srivastava GK, Reinoso R, Singh AK, Fernandez-Bueno I, Hileeto D, Martino M, Garcia-Gutierrez MT, Merino JM, Alonso NF, Corell A, Pastor JC (2011) Trypan Blue staining method for quenching the autofluorescence of RPE cells for improving protein expression analysis. Exp Eye Res 93:956–962

Stoodley P, Nistico L, Johnson S, Lasko LA, Baratz M, Gahlot V, Ehrlich GD, Kathju S (2008) Direct demonstration of viable *Staphylococcus aureus* biofilms in an infected total joint arthroplasty. A case report. J Bone Joint Surg Am 90:1751–1758

Sun Y, Yu H, Zheng D, Cao Q, Wang Y, Harris D, Wang Y (2011) Sudan black B reduces autofluorescence in murine renal tissue. Arch Pathol Lab Med 135:1335–1342

Sunde PT, Olsen I, Göbel UB, Theegarten D, Winter S, Debelian GJ, Tronstad L, Moter A (2003) Fluorescence in situ hybridization (FISH) for direct visualization of bacteria in periapical lesions of asymptomatic root-filled teeth. Microbiology 149:1095–1102

Trøstrup H, Thomsen K, Christophersen LJ, Hougen HP, Bjarnsholt T, Jensen PØ, Kirkby N, Calum H, Høiby N, Moser C (2013) Pseudomonas aeruginosa biofilm aggravates skin inflammatory response in BALB/c mice in a novel chronic wound model. Wound Repair Regen 21:292–299

Tsai YJ, Lin YC, Wu WB, Chiu PH, Lin BJ, Hao SP (2013) Biofilm formations in nasopharyngeal tissues of patients with nasopharyngeal osteoradionecrosis. Otolaryngol Head Neck Surg 148 (4):633–636

Wallner G, Amann R, Beisker W (1993) Optimizing fluorescent in situ hybridization with rRNA-targeted oligonucleotide probes for flow cytometric identification of microorganisms. Cytometry 14:136–143

Wright Cell Imaging Facility, Toronto Western Research Institute (2013) Autofluorescence causes and cures. http://www.uhnresearch.ca/facilities/wcif/PDF/Autofluorescence.pdf. Accessed 3 Nov 2013

Mechanisms of Drug Resistance in Fungi and Their Significance in Biofilms

Rajendra Prasad, Abdul Haseeb Shah, and Sanjiveeni Dhamgaye

Abstract Infections caused by opportunistic human fungal pathogens are very common and have shown steady increase in recent years. The typical hosts, which are prone to fungal infections, are those who possess suppressed immune systems due to conditions such as HIV and transplantation surgery. Due to prolonged chemotherapy, fungal cells also develop tolerance to the most commonly used azole antifungals by employing several strategies. Interestingly, biofilms which are routinely formed by fungal cells on medically implanted devices employ different strategies to become highly resistant to antifungals. Apart from the known tactics, newer approaches have revealed novel mechanisms and regulatory circuits that are responsible for the development of multidrug resistance. Overcoming the major clinical hurdle of fungal resistance demands a great deal of knowledge about the function of fungal machinery that is used under drug stress.

1 Introduction

Fungi are very diverse eukaryotic organisms, which can exist as unicellular or in various multicellular forms. Phylogenetically, they are clustered with higher eukaryotes and hence evolutionary are closer to mammalian systems. Notably, the closeness of fungi to metazoans makes them ideal eukaryotic models; however, their similar cellular machinery poses a challenge in combating their infections.

R. Prasad (✉) • A.H. Shah
Membrane Biology Laboratory, School of Life Sciences, Jawaharlal Nehru University, New Delhi, 110067, India
e-mail: rp47jnu@gmail.com; rp47@mail.jnu.ac.in

S. Dhamgaye
Membrane Biology Laboratory, School of Life Sciences, Jawaharlal Nehru University, New Delhi, 110067, India

Special Centre for Molecular Medicine, Jawaharlal Nehru University, New Delhi, India

While most fungi are nonpathogenic, there are a large number of them, belonging to almost every phylum, which are pathogenic to humans, animals, and plants (Heitman 2011). Fungi as pathogens are involved in many afflictions, which range from superficial to life-threatening disseminated infectious diseases (Odds 1988). Human fungal pathogens are mostly opportunistic, implying that a successful infection depends upon the status of the immune defense system of the host. Advancement in preventive measures like immune suppression during organ transplantations and life threatening diseases like AIDS help opportunistic fungi find suitable hosts in which to thrive. These common opportunistic fungal species are either *Candida* species like *Candida alibicans, C. glabrata, C. tropicals, C. paropsilosis, C. dubliniensis, C. guilliermondii, C. krusei, C. lusitaniae*, or non-*Candida* fungi like *Aspergillus fumigatus, Cryptococcus neoformans, Histoplasma caspulatum, Microsporum canis, Paracoccidioides brasiliensis, Penicillium marneffei, Blastomycoides dermatitidis*, and *Pneumocystis* sp. (Sanglard et al. 2009; Heitman 2011; Gow et al. 2011).

Candida species are one of the most prevalent causes of systemic fungal infections in humans. Among *Candida* species, *C. albicans* is well adapted to thrive in most of the organs and niches of humans, making it the most successful human fungal pathogen. Not surprisingly, *C. albicans* alone contributes to 50–60 % of Candida infections followed by non-albicans species, which mostly include *C. glabrata, C. parapsilosis, C. tropicalis, and C. krusei* (Silva et al. 2011; Prasad et al. 2012). Infections due to uncommon fungi and pathogenic molds like *Trichosporon* species, *Fusarium*, and *Scedosporium* species have also been reported. *Trichosporon* species causes diseases similar to hepatic candidiasis while *Fusarium* and *Scedosporium* species are commonly found in hospital-acquired fungal infections and show high levels of resistance to amphotericin B (AMB) and azoles (Fridkin 2005; Bonatti et al. 2007).

Many opportunistic pathogens pose an additional threat because of their ability to acquire tolerance to antifungal treatments leading to the development of multidrug resistance (MDR). For this, fungi have developed several strategies to tolerate most of the mainstream antifungal drugs like azoles, polyenes, allyamine, and echinocandins. Notably, many of the mechanisms of MDR are also common to multidrug-resistant cancer or bacterial cells and have been discussed and reviewed in recent years (Sanglard and Odds 2002; Prasad and Kapoor 2005; Sanglard et al. 2009; Cannon et al. 2009; Morschhäuser 2010; Prasad and Goffeau 2012). In this chapter we provide a snapshot of the current MDR environment and discuss some of the strategies of azole resistance adopted by fungal cells.

2 MDR Strategies

The limited availability of antifungals is a major impediment for the effective treatment of fungal infections (Ghannoum and Rice 1999). This is further compounded by the fact that the generation of newer antifungals has lagged behind, when compared to the pace of emergence of fungal infections. The components of the fungal cell wall (CW) such as mannans, glucans, and chitins, and a few of the enzymes of the ergosterol biosynthetic pathway, are unique to fungal cells (St Georgiev 2000; Munro et al. 2001) and have been targeted for the development of antifungal agents. Among the enzymes of the ergosterol biosynthetic pathway, squalene epoxidase, P45014DM or CYP51 (*ERG11*), Δ^{14}-reductase (*ERG24*), and Δ^8-Δ^7-isomerase (*ERG2*) have been the targets of most antifungal agents (Sanglard et al. 2003).

Amongst the known antifungals, azole derivatives like fluconazole (FLC), ketoconazole (KTC), and itraconazole (ITC) have been the most widely used triazoles for combating fungal infections. Azoles specifically inhibit the P45014DM enzyme, which results in the accumulation of 14-methylated sterols and results in the disruption of membrane structure and function (Vanden Bossche et al. 1989). Fungal cells have adopted several strategies to cope with incoming drugs (particularly azoles), which are discussed briefly.

2.1 Target Alteration

One of the most common events associated with the development of drug tolerance in fungi relates to the azole target protein. The target of azoles, the P45014DM protein, is modified in resistant cells by the replacement of native amino acids leading to poor binding of the drug without affecting its function. Investigators from different groups compared the sequences of the *ERG11* (encodes P45014DM or Erg11 protein) gene of resistant *C. albicans* strains with the published *ERG11* sequence to that of FLC-susceptible strains and identified several point mutations. Biochemical analysis of these mutations showed that several point mutations in the protein reduce its affinity for azoles. These mutations, single or in combination, do not permit normal binding of FLC to target proteins without affecting ergosterol biosynthesis (Wang et al. 2009; Morio et al 2010). A comparison of sequence disparity of the Erg11 protein between susceptible and resistant clinical isolates of *C. albicans* revealed several mutations in the resistant isolates predominantly localized to certain hot spot regions. Such bunching of critical residues, which probably affects azole binding to target proteins, was subsequently confirmed by different studies (Wang et al. 2009). A few of the Erg11 protein mutations, when expressed in *Saccharomyces cerevisiae*, were shown to confer even higher resistance against azole antifungals (Favre et al. 1999). Notably, the I-helix stretch of the Erg11 protein, which is highly conserved in the cytochrome P450 family, does not show any spontaneous mutations. The exact placement of all the identified mutations in a 3D model of the protein confirmed that these mutations are not randomly distributed but rather are clustered in select hot spot regions (Marichal et al. 1999; Wang et al. 2009).

Mutations in drug target enzymes clearly are an important mechanism resulting in the emergence of FLC resistant *C. albicans* strains. However, sequence comparison between resistant and susceptible isolates also identified point mutations in susceptible isolates as well, which were absent in their resistant counterparts and thus implied that these may not contribute to enhanced drug susceptibility (Morio et al. 2010). While mutations in Erg11 proteins are routinely reported, many also remain to be confirmed, which could be established by expressing each variant in a heterologous system. Some studies point out that a change from heterozygosity to homozygosity for a mutated *ERG11* gene could also contribute to increased resistance to drugs (Ge et al. 2010).

2.2 Overexpression of P45014DM

Resistance to FLC in many clinical isolates is commonly associated with the overexpression of the *ERG11* gene (Hoot et al. 2011; Flowers et al. 2012; Sasse et al. 2012). The zinc cluster transcription factor Upc2p regulates the expression of *ERG11* and other genes involved in ergosterol biosynthesis (White and Silver 2005). It has been observed that an overexpression of *UPC2* increases azole resistance, whereas its disruption results in hypersusceptibility to azoles. A comparison of sequence of *UPC2* between matched pair azole susceptible and resistant isolates resulted in the identification of a base substitution causing a point mutation in the encoded protein. This gain of function (GOF) mutation led to an overexpression of *ERG11* and hyper-resistance to azoles (Heilmann et al. 2010; Hoot et al. 2011; Flowers et al. 2012). In addition, promoter deletion analysis indicated that azole drugs induce the expression of *ERG11* through its azole responsive cis-acting elements (ARE) in the promoter. The presence of ARE alone was not able to stimulate reporter gene expression in the $upc2\Delta/upc2\Delta$ nulls, thus confirming that azoles manifest their effect on *ERG11* expression through transcription factor (TF) Upc2p (Oliver et al. 2007).

2.3 Alterations in Other Enzymes of the Ergosterol Biosynthetic Pathway

Recently, *ERG3* mutations have been found to occur frequently either alone or in combination with *ERG11* mutations, leading to a change in the ratios of various cell sterols and increased resistance to azoles and polyenes. Mutations have been found in *ERG3*, which are imperative in drug resistance even when drug efflux pumps were not the causal factor (Martel et al. 2010a, b; Morio et al. 2012). The cytochrome P450 spectral studies performed in a system reconstituted with purified *ERG5* (Δ^{22}-desaturase or CYP61) of *C. glabrata* revealed interactions between azoles and the heme-protein, implying that *ERG5* could also be a target for azoles and may contribute to antifungal resistance (Lamb et al. 1999). Recently, a

C. albicans drug-resistant clinical isolate was detected with a combination of a single mutation in the *ERG5* gene along with a ten-amino acid duplication in the *ERG11* gene. The mutant, in addition of being resistant to azoles, showed resistance to AMB due to the depletion of membrane ergosterol levels (Martel et al. 2010a, b). *ERG6* in *C. glabrata* is involved in azole resistance due to various base pair alterations leading to missense mutations (Vandeputte et al. 2007). Similarly, an erg6Δ disruptant of the *C. lusitaniae* strain was susceptible to AMB due to decreased membrane ergosterol levels. Coinciding with this, several clinical isolates of *C. lusitaniae* show increased expression of *ERG6* along with a decrease in *ERG3* gene expression and enhanced resistance to AMB (Young et al. 2003).

Together, the azole-induced upregulation of *ERG11*, along with other genes of the ergosterol biosynthetic pathway, suggests the existence of a common mechanism of upregulation in *C. albicans* (Henry et al. 2000). Transcript profiling of ITC treated Candida showed differential regulation of almost 300 genes that included genes involved in transcription, RNA processing, metabolism, CW maintenance, cell cycle control, cell stress, etc. (De Backer et al. 2001). Another study demonstrated that almost 15 % of genes differentially expressed on KTC treatment fall under the category of sterol metabolism including: *NCP1, MCR1, CYB5, ERG2, ERG3, POT14 (ERG10), ERG25, ERG251,* and *ERG11* in a wild-type strain of *C. albicans* (Liu et al. 2005). The other major categories of genes that were affected in that study have roles in small molecule transport (16 %), *CDR1, CDR2, HGT11, HGT12, PRT9*; cell stress (9 %), *DDR49, MCR1, SSA4*; CW maintenance, *ALS4, CRH11*, etc. While a global regulation of *ERG* genes was evident from the transcript profiling, several genes of diverse functions as well as of unknown functions were also either up- or down-regulated by the drug treatment. This reinforces the idea that azoles could contribute to multiple, yet unknown phenotypes that still remain to be identified. The dissection of the mechanisms mediating these phenotypes could provide newer insights into the phenomenon of MDR.

2.4 Drug Import

It is presumed that the hydrophobic nature of drugs permits their easy import by passive diffusion. However, the contribution of drug import to the overall scenario of MDR is not well established. Nonetheless, there are a few studies that showed passive diffusion of drugs was an important determinant of MDR. For example, fluctuations in membrane fluidity affect passive diffusion and susceptibility to drugs. The *erg* mutants of *S. cerevisiae* or of *C. albicans* were shown to possess high membrane fluidity, which led to enhanced diffusion and susceptibility to azoles (Kohli et al. 2002; Prasad et al. 2010). Recently, permeability constrains imposed by Candida cells have been reemphasized towards the development of MDR. In one instance, it was shown that azoles can enter in *C. albicans, C. kruesi,* and *C. neoformans* cells by diffusion (Mansfield et al. 2010). The kinetics of import in de-energized cells established that FLC import proceeds via facilitated diffusion

(FD) through a transporter rather than by passive diffusion. Other azoles compete for FLC import, suggesting that all azoles utilize the same FD mechanism. FLC import was also shown to vary among *C. albicans*-resistant clinical isolates, suggesting that altered FD may be a previously uncharacterized mechanism of resistance to azole drugs (Mansfield et al. 2010). However, the identification of a membrane transporter protein involved in FD of azoles remains elusive (Mansfield et al. 2010). Interestingly, drug inactivation that is a common mechanism in bacteria has not been observed in Candida cells.

2.5 Drug Efflux

Increased efflux, which leads to reduced intracellular accumulation of drugs, is another prominent mechanism of MDR in fungi (Prasad and Kapoor 2005). In *C. albicans*, for example, this is achieved by increasing the efflux of drugs from cells by overproducing the plasma membrane (PM) efflux pump proteins. An induction in the expression levels of genes encoding efflux pump proteins, particularly *A*TP *B*inding *C*assette (ABC) multidrug transporter proteins Cdr1 and Cdr2 or *M*ajor *F*acilitator *S*uperfamily (MFS) efflux pump protein Mdr1, have been commonly observed in azole-resistant clinical isolates of *C. albicans* (White et al. 2002; Prasad and Kapoor 2005; Karababa et al. 2004; Kusch et al. 2004). Invariably, MDR Candida cells, which show enhanced expression of efflux pump encoding genes, also show simultaneous increase in the efflux of drugs, thus implying a causal relationship between efflux pump encoding gene expression levels and intracellular concentration of the drug (Cannon et al. 2009).

2.5.1 ABC Transporters

An inventory of *C. albicans* ABC transporters revealed that there are 28 putative ABC superfamily members, including 12 half transporters, that largely remain uncharacterized (Gaur et al. 2005). These putative ABC proteins could be grouped into five "known" subfamilies, including *C. albicans* Pdr protein (CaPdrp), and a sixth "others" category that includes soluble ABC non-transporter proteins unrelated to the existing fungal subfamilies. The Pdr protein subfamily of *C. albicans* comprises seven full-size members: Cdr1p, Cdr2p, Cdr3p, Cdr4p, Cdr11p, CaSnq2p, and Ca4531. The *C. albicans* Cdr1p and Cdr2p proteins are active multidrug transporters, while Cdr3p and Cdr4p do not efflux drugs and play no apparent role in the development of antifungal resistance (Prasad and Goffeau 2012). Other transporters in related fungi, including *CgCDR1* (Sanglard et al. 1999), *CgCDR2 (PDH1)* (Miyazaki et al. 1998), and *SNQ2* (Torelli et al. 2008) in *C. glabrata*, *ABC1* in *C. krusei* (Katiyar and Edlind 2001), and *AFR1* (Sanguinetti et al. 2006) in *C. neoformans*, are multidrug transporters and play a role in the development of MDR in these pathogenic species.

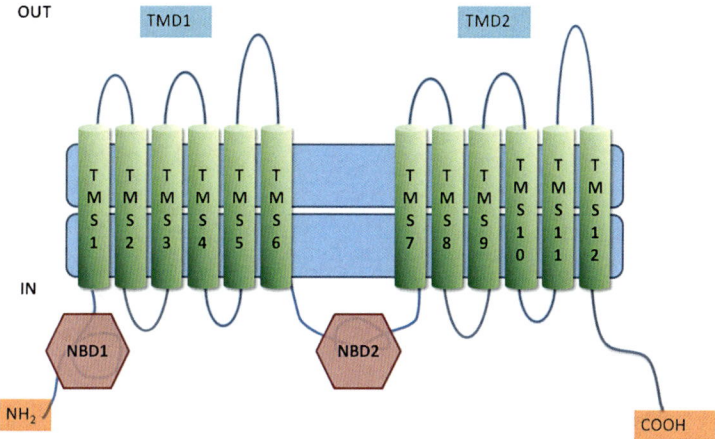

Fig. 1 Schematic representation of ABC transporters involved in multidrug resistance in yeast, depicting two nucleotide binding domains (NBDs) in the cytosolic region and two membrane spanning domains (TMDs) arranged in reverse topology (NBD-TMD)$_2$

Full ABC proteins are made up of two (or three) transmembrane domains (TMDs) and two cytoplasmic nucleotide-binding domains (NBDs) (Fig. 1). Based on biochemical and crystallographic evidence, it appears that the "half proteins," which have only one NBD and one TMD, function as homo- or heterodimers. In the forward topology, the TMDs precede the NBDs (TMD-NBD), whereas the NBDs come first in the reverse topology (NBD-TMD) (Rutledge et al. 2011). NBD's are the nucleotide binding sites, which bind ATP required to power the efflux of substrates bound within TMD's drug binding sites. Each TMD is usually comprised of six transmembrane segments (TMS), which generally are continuous alpha helices arranged to form drug binding sites (Prasad and Goffeau 2012).

2.5.2 MFS Transporters

MFS transporters, which are also called uniporter–symporter–antiporter family, are the second major superfamily of transporters divided into 29 families (Saier et al. 1999). A phylogentic analysis identified 95 potential MFS transporters in *C. albicans* (Gaur et al. 2008). Most MFS transporters consist of two domains of six TMSs within a single polypeptide chain with few exceptions as shown in Fig. 2 (Stephanie et al. 1998). On the basis of hydropathy and phylogenetic analysis, the drug efflux MFS proteins can be divided into two distinct types: Drug: H$^+$ Antiporter-1 (DHA1), consisting of 12 TMSs, and Drug: H$^+$ Antiporter-2 (DHA2) that contains 14 TMSs. Homologues of *MDR1* have been identified from *C. dubliniensis* and *C. glabrata,* which are designated as *CdMDR1* and *CgMDR1*, respectively (Moran et al. 1998; Sanglard et al. 1999). It appears that increased expression of *CdMDR1* is one of the main mechanisms of FLC resistance in

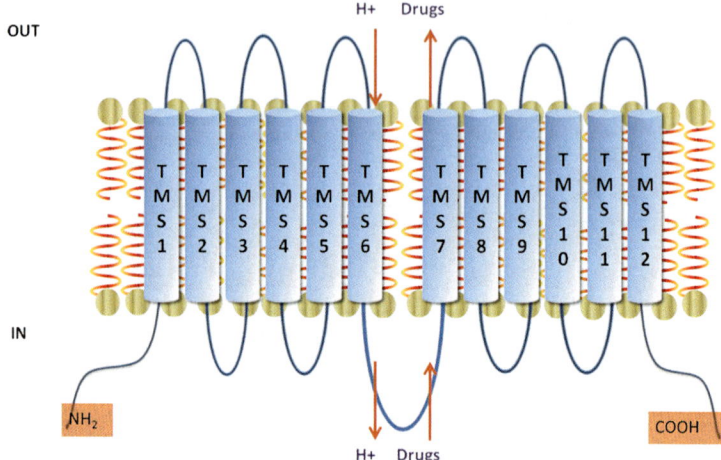

Fig. 2 Predicted topology of MFS drug: H^+ antiporter involved in drug transport. It has 12 (DHA1) or 14 (DHA2) TMSs and does not contain NBDs, but TMS5 harbors an H^+-antiporter motif, which couples electrochemical gradient of protons to drug transport

C. dubliniensis clinical isolates (Moran et al. 1998). Since *CgMDR1* confers specific resistance to FLC, its constitutive expression in *C. glabrata* may be responsible for the intrinsically low susceptibility of this yeast species to FLC (Sanglard et al. 1999).

Among all the MFS proteins, only one member, *MDR1*, has been implicated clinically to be involved in azole resistance. *FLU1*, a close homologue of *MDR1*, has also been isolated by its ability to confer fluconazole resistance in hypersusceptible *S. cerevisiae* transformants. However, overexpression of *FLU1* has not been detected in FLC-resistant clinical isolates of *C. albicans*. None of the other 95 members of this superfamily are implicated in MDR (Gaur et al. 2008). As an important MDR gene of the MFS family, *MDR1* has been extensively studied for its role in drug resistance. The functional evaluation of critical amino acid residues of the Mdr1 protein revealed that the residues of TMS5 which harbor antiporter motifs (G(X6)G(X3)GP(X2)GP(X2)G) are potentially significant for their functionality and contribute to drug: H^+ transport. Independent of the substrate specificity of the antiporter, the antiporter motif in the predicted TMS5 is conserved in all of the functionally related subgroups in bacteria and plants. Multiple-sequence analysis of the MFS transporters revealed that proteins within this family share greater similarity between their N-terminal halves than their C-terminal halves, and it is assumed that the later half is responsible for substrate recognition (Paulsen et al. 1996).

3 Regulation of MDR Genes

As discussed above, the MDR strains overexpress two ABC transporter encoding genes, *CDR1* and *CDR2*, which are homologous to the pleiotropic drug-resistance transporter Pdr5 from *S. cerevisiae* (Coste et al. 2004). In *C. albicans*, *FCR1* and *TAC1* were identified as Sc*PDR1–PDR3* homologs. While in *S. cerevisiae* the *FCR1* gene product behaved as a transcriptional activator, in *C. albicans* it acted as a negative regulator (Talibi and Raymond 1999). Only Tac1p has been experimentally proven to function as a transcriptional activator of *C. albicans* drug efflux transporter genes, such as *CDR1* and *CDR2* (Coste et al. 2004). Most of the MDR strains present a loss of heterozygocity and/or aneuploidy at the *TAC1* locus, combined to GOF mutations of *TAC1* (Coste et al. 2004).

Another TF, *CaNDT80*, has diverse roles as it controls the expression of not only *CDR* genes but of *ERG* and *MDR1* genes as well (Sasse et al. 2011). *CaNDT80* is required for constitutive overexpression of the *CDR1* and *ERG* genes but is dispensable for *MDR1* and *CDR2* drug-mediated induction, indicating its contribution in nearly all the predominant mechanisms of drug resistance in *C. albicans*. Notably, there was a slight increase in *CDR2* expression in *CaNDT80* mutants indicating the gene's repressible nature. The role of *CaNDT80* in FLC resistance was not prominent as *CaNDT80* mutant cells were more resistant to FLC than the wild-type strain (Sasse et al. 2011).

Another family of TFs controlling drug transporter genes belongs to bZip family (Alarco and Raymond 1999). *C. albicans* harbors a homologue of the YAP protein family designated as Cap1 protein which is involved in oxidative stress and also regulates expression of genes encoding members of both the MFS and ABC transporter superfamilies (Alarco and Raymond 1999). There is an intricate interplay between *MRR1*, *CAP1*, and *UPC2* TFs that governs the induction and constitutive overexpression of *MDR1* (Schubert et al. 2011). Interestingly, a mutation in *MRR1* resulted in the constitutive overexpression of *MDR1* even in the absence of *UPC2* or *CAP1*, whereas a GOF mutation in *UPC2* slightly activated *MDR1* expression, which was dependent on the presence of *MRR1*. The activated form of *CAP1* was also partially dependent on *MRR1* for expression of *MDR1*. On the other hand, induction of *MDR1* by drugs was also shown to be independent of *UPC2* but required *MRR1* and was partially dependent on *CAP1* (Schubert et al. 2011). In addition to *MDR1*, the global regulator *UPC2* also regulates other mediators of MDR such as *CDR1/CDR2* and ergosterol biosynthesis genes including the FLC target *ERG11*. Yet another TF that has been implicated in the regulation of *MDR1* expression is Mcm1p, a member of the MADS box TF family. The MDRE/BRE in the *MDR1* promoter contains a putative Mcm1p binding site (Riggle and Kumamoto 2006). Table 1 lists some of the known regulators of MDR genes.

Three additional zinc cluster TFs were identified by Sanglard and coworkers (*CTA4, ASG1,* and *CTF1*) and all complemented the FLC hypersusceptibility of *S. cerevisiae* nulls of pdr1 pdr3 and restored PDR5 expression via the cis-acting

Table 1 MDR regulators in *C. albicans* and *S. cerevisiae* and their target genes

TFs	Target genes	References
	Candida albicans	
TAC1	CDR1, CDR2, IFU5, HSP12, RTA3, GPX1, CHK1, LCB4, NDH2, SOU1, etc.	Liu et al. (2007)
MRR1	MDR1	Morschhauser et al. (2007)
NDT80	CDR1, ERG genes, MDR1, CDR2	Sasse et al. (2011)
MCM1	MDR1, EFG1, WOR1, WOR2, CZF1	Tuch et al. (2008)
CAP1	CAP1, GLR1, TRX1, SOD1, CAT1, PDR16, MDR1, FLU1, YCF1, FCR1	Znaidi et al. (2009)
UPC2	CDR1, MDR1, YOR1, MET6, ERG genes	Znaidi et al. (2008)
	Saccharomyces cerevisiae	
PDR1	PDR5, PDR15, PDR10, SNQ2, YOR1, HXT9, HXT11	Bauer et al. (1999)
PDR3	SNQ2, HXT9, HXT11, PDR5, PDR15, YOR1	Bauer et al. (1999)
YAP1	SNQ2, YCF1	Bauer et al. (1999)

sequence, the pleiotropic drug resistance responsive element (PDRE) used by Pdr1 and Pdr3. Notably, the deletion of their counterparts in Candida neither affected the drug-induced expression of *CDR1*, *CDR2*, and *MDR1* nor their level of resistance. Therefore, the role of these TFs in MDR in Candida is yet to be uncovered (Coste et al. 2008).

4 Novel Mechanisms of MDR

In addition to the well-known mechanisms and circuitry implicated in drug resistance, there are various novel pathways or unconventional mechanisms, which are emerging as new MDR determinants in Candida cells. Some of these mechanisms are discussed briefly in the succeeding text (Table 2).

4.1 Mitochondria and Cell Wall Integrity Affects Drug Susceptibility

Protein kinase C (PKC) regulates CW integrity during growth, morphogenesis, and response to stress. The genetic impairment of Pkc1 confers hypersensitivity to multiple drugs that target synthesis of the key cell membrane sterol ergosterol, including azoles, allylamines, and morpholines. Deletion of *C. albicans PKC1* in turn makes fungistatic ergosterol biosynthesis inhibitors fungicidal and attenuates virulence (LaFayette et al. 2010). Notably, Pkc1 enables survival of cell membrane stress at least in part via the mitogen-activated protein kinase (MAPK) cascade in *S. cerevisiae* and *C. albicans* through distinct downstream effectors. Strikingly,

Table 2 List of TFs with novel roles in MDR

TFs	Original function	Implication in drug resistance	References
CZF1	Regulates white opaque switching and hyphal growth	Upregulated in drug-resistant strains	Brown et al. (1999), Dhamgaye et al. (2012a)
RCA1	Regulates carbonic anhydrases	Homozygous nulls resistant to FLC	Cottier et al. (2012), Vandeputte et al. (2012)
STP2	Amino acid regulated TF, regulates amino acid permease genes	Homozygous nulls are susceptible to antifungal malachite green and other compounds	Martinez and Ljungdahl (2005), Dhamgaye et al. (2012b)
STP4	C2H2 TF (not clear)	Induced under core caspofungin response	Cheng et al. (2006), Blankenship et al. (2010)
ZCF3	Required for filamentous growth	Resistance to rapamycin and flucytosine	Nobile et al. (2012)
AHR1	Regulates adhesion genes, involved in white opaque switching	Upregulated in drug-resistant strain	Askew et al. (2011), Dhamgaye et al. (2012a)
ZFU2	Required for yeast form adherence	Upregulated in drug-resistant strains	Finkel et al. (2012)

inhibition of Pkc1 phenocopies inhibition of the molecular chaperone Hsp90 or its client protein calcineurin. PKC signaling is also required for calcineurin activation in response to drug exposure in *S. cerevisiae*. In contrast, Pkc1 and calcineurin independently regulate drug resistance via a common target in *C. albicans* (LaFayette et al. 2010).

Mitochondrial dysfunction in pathogenic fungi or model yeast causes altered susceptibilities to antifungal drugs. In *C. galbrata* and *S. cerevisiae* loss of mitochondrial DNA (mtDNA) is linked to increased resistance to azoles by upregulation of *ScPDR1* and *CgPDR1*, respectively, which upregulates their target genes *ScPDR5* or *CgCDR1* and *CgCDR2* (Shingu-Vazquez and Traven 2011). The mutants defective in mitochondrial electron transport display increased susceptibility to FLC in *C. albicans* cells. The null mutants *goa1*Δ and *ndh51*Δ, which are defective in electron transport, show decreased resistance to FLC along with the downregulation of *CDR1* and *CDR2* (Sun et al. 2013). Thus, implying that cell energy is required for azole susceptibility and that downregulation of efflux genes may be an outcome of this dysfunction. The mitochondrial dysfunction is also been shown to interact with CW stress pathways. A recent study showed that the Ccr4-Pop2 mitochondrial mRNA deadenylase was required for CW integrity, tolerance to echinocandin caspofungin and virulence in a mouse model. The study provided evidence that the CW defect of the deadenylase mutants (Ccr4-Pop2) is not only linked to mitochondrial dysfunction but also impact phosholipid homeostasis. Recently, lipidomics of MDR clinical isolates of *C. albicans* also suggest a link between mitochondrial dysfunction and CW integrity (Dagley et al. 2011). Data presented suggested that a decrease in the mitochondrial lipid phospahtidyl glycerol

(PG) in azole-resistant isolates could be linked to compromised CW integrity and drug resistance in C. *albicans* cells (Singh et al. 2013).

Evidence of cross talk between CW and MDR comes from another unrelated study. RNAseq of an MDR strain and its isogenic drug susceptible counterpart led to the identification of an upregulated TF encoding gene *CZF1*, which was involved in hyphal transition and white/opaque switching (Vinces et al. 2006). Notably, *CZF1* was also co-induced with *CDR1* and *CDR2* in MDR isolates. Interestingly, the inactivation of *CZF1* increased the resistance of the cells to CW perturbing agents, through the overexpression of beta glucan synthesis genes. The study proposed a positive role of *CZF1* on MDR and a negative role on CW integrity (Dhamgaye et al. 2012a). The mechanism of cross talk between CW stresses, mitochondrial dysfunction, and MDR is not completely lucid, but these examples reiterate their interdependence to mitigate cellular stresses.

4.2 Lipids in MDR

On the basis of several studies, a close interaction between membrane lipids and drug-extrusion pump proteins has been recognized (Marie and White 2009). For example, the drug-extrusion pump proteins, particularly belonging to ABC superfamily, are predominantly localized within microdomains of PM (rafts) and thus are sensitive to the nature and the physical state of the membrane lipids (Pasrija et al. 2008). For example, any imbalance in the main constituents of membrane rafts, such as sphingolipids or ergosterol levels, result in abrogated functionality of the drug extrusion pumps (Prasad et al. 2006). It has also been observed that the ABC drug-efflux proteins in yeast (Pdr5p and Yor1p in *S. cerevisiae*, and Cdr1p and Cdr2p in *C. albicans*) can translocate phospholipids between the two monolayers of the PM (Smriti et al. 2002).

The adaptation of *C. albicans* to tolerate antifungals is accompanied by many specific and global changes in lipids (Singh et al. 2013). Recently, the detailed lipidomics of several genetically matched (isogenic) as well as select sequential azole sensitive and resistant clinical isolates of *C. albicans* provided a comprehensive evaluation of lipids as the determinants of drug resistance and showed that each resistant isolate possessed a characteristic lipid composition. Development of azole tolerance also impelled the remodeling of molecular species of lipids. The fact that lipidomic response of match pair isolates was associated with simultaneous overproduction of efflux pump membrane proteins suggested a possible common regulatory mechanism between the two phenomena (Singh et al. 2013). Such a common link has already been observed in *S. cerevisiae* and *C. glabrata,* where genes encoding efflux pumps, such as ScPdr5 or CgCdr1 and CgCdr2, play an important role in regulating lipid levels (Shahi and Moye-Rowley 2009).

4.3 Iron Levels Affect MDR

For its survival in host cells, Candida, like many other pathogens, has also adapted many complex strategies to scavenge depleted iron from the host environment (Nyilassi et al. 2005). In fact, the availability of iron can serve as a common adaptive signal for pathogens to induce the expression of virulence traits (Mekalanos 1992). Recent studies have already established a role for iron in systemic infections, whereby the requirement of a high-affinity iron transporter (CaFtr1p) for infection in a mouse model was shown (Ramanan and Wang 2000). Similarly, the requirement of a siderophore transporter (Arn1p) for epithelial invasion (Heymann et al. 2002) and iron dependent endothelial cell injury (Fratti et al. 1998) suggests that iron plays a vital role in the virulence of *C. albicans*. The role of iron in MDR is well established in mammalian cells, where *HIF1* is activated under low iron concentrations that in turn induce the expression of its target gene *MDR1* (Epsztejn et al. 1999).

Iron depletion in *C. albicans* with bathophenanthrolene disulfonic acid and ferrozine as chelators enhanced its susceptibility to FLC and several drugs (Prasad et al. 2006). Several other species of Candida also display increased sensitivity to FLC because of iron restriction. Iron uptake mutants, namely *ftr1* and *ftr2*, as well as the copper transporter mutant *ccc2*, which affects high-affinity iron uptake in Candida, showed increased susceptibility to FLC. The effect of iron depletion on drug sensitivity appeared to be independent of the efflux pump proteins Cdr1p and Cdr2p. This study showed that iron deprivation led to the lowering of membrane ergosterol and an increase in membrane fluidity, resulting in enhanced passive diffusion of drugs. Transcriptome analysis of iron deprived cells showed a connection between calcineurin signaling and iron homeostatsis. Notably, iron-deprived cells phenocopy deletion of calcineurin pathway genes by showing susceptibility to alkaline pH, membrane perturbing agents, and salinity stress (Hameed et al. 2011).

5 Biofilms and MDR

Biofilms are complex microbial conglomerates which are predominantly surface attached and enclosed by a thick layer of polysaccharides, making them extremely drug resistant and difficult to eradicate (Kumamoto 2002; Fanning and Mitchell 2012). Using an in vitro model, Mukherjee et al. demonstrated that *C. albicans* biofilm formation proceeds in three developmental phases: (1) early phase involving adhesion of fungal cells, (2) intermediate phase during which the blastospores coaggregate, proliferate, and produce a carbohydrate-rich extracellular matrix (ECM), and (3) maturation phase, in which the fungal cells are completely encased in a thick ECM (Mukherjee et al. 2003). Fungal biofilms show an increase in drug resistance with MIC values ranging from 30- to 2,000-fold higher than their corresponding planktonic cells, which is also developmental phase dependent

(Hawser and Douglas 1994). Because of the very high level of resistance to azoles, biofilms are a serious clinical threat as they can develop robustly on most surgically implanted synthetic devices (Douglas 2003). Currently, echinocandins and liposomal AMB are the only drug formulations that show some efficacy against fungal biofilms (Ramage et al. 2002; Kuhn et al. 2002; Uppuluri et al. 2011).

Since *C. albicans* is the fungi most widely associated with human infections, and because biofilm formation is so important to its pathogenesis, it has gained enormous focus (Kumamoto 2002; Seneviratne et al. 2008). In addition to *C. albicans* many of non-albican *Candida* species (NACS), including *C. parapsilosis, C. tropicalis, C. dubliensis,* and *C. glabrata,* also form drug-resistant biofilms to different extents and in different conditions. Although there are many reports of biofilm formation by NACS, those formed by *C. albicans* are most extensive and complex (Kumamoto 2002). Biofilms formed by different *Candida* species differ largely in their chemical composition, extent of extracellular matrix (ECM), structure and thickness. For example, *C. parapsilosis* and *C. tropicalis* form robust biofilms while *C. glabrata* biofilms are generally underdeveloped (Silva et al. 2011). *A. fumigatus* biofilms, like Candida, produce large amounts of ECM and are highly drug resistant with galactosaminogalactan and galactomannan being their major polysaccharide components (Seidler et al. 2008; Loussert et al. 2010). *C. neoformans* develops biofilms on medically implanted devices. In addition to showing enhanced resistance to various drugs, cryptococcal biofilms also impact resistance against many different environmental stress conditions (Martinez and Casadevall 2006, 2007). The MDR biofilms on medically implanted devices are also seen with *Trichosporon asahii* which is the major cause of disseminated trichosporonosis (Di Bonaventura et al. 2006). It has been demonstrated that biofilms formed by *Pneumocystis* species are sensitive to the quorum sensing molecule farnesol similar to *Candida* species, thus suggesting the possible involvement of a more closely related pathway in these two different fungi (Cushion et al. 2009).

6 Major Contributors of MDR in Biofilms

There are many mechanisms which are prevalent in the formation of biofilms that contribute to enhanced antifungal resistance. As compared to planktonic cells where an increased expression of drug efflux pump genes like *CDR1, CDR2, CaMDR1* (Prasad et al. 1995; Sanglard et al. 1997; Pasrija et al. 2007) have an active role in azole tolerance, in sessile cells the efflux pump dependent decreased susceptibility to drugs occurs only in early stages of biofilm development. At later stages of biofilm formation Candida strains lacking these transporters are found to be equally resistant to azole antifungals (Mukherjee et al. 2003). Although drug efflux pumps do not seem to be an important contributor of drug resistance in Candida biofilms, they seem to play a role in the development of azole drug resistance in *A. fumigates* biofilms. Azole drug treatment also led to an increase in *AfMDR4* drug efflux pump gene expression in these biofilms and a subsequent increase in drug resistance, which could be reversed by the presence of an efflux pump inhibitor (Rajendran et al. 2011).

Several studies show that ECM comprised predominantly of β-1, 3 glucans is most predominantly linked to antifungal resistance in biofilm. ECM acts as a barrier which imposes restriction on free diffusion of drugs within the biofilm. Therefore, the nature and extent of ECM determines the drug diffusion and threshold of resistance for the particular biofilm community (Seneviratne et al. 2008). There are several studies where glucan levels were linked to MDR of biofilms. For example, by exploiting *FKS1* mutant strains (defective in β-1, 3 glucans synthase), it was shown that glucan production in the mutant strain was reduced to 60 % to that of wild-type cells rendering biofilms susceptible to azoles (Nett et al. 2010). Similarly, HSP90 in addition to having a role in biofilm dispersion has also been linked to biofilm hyper resistance to antifungals since it also regulates the production of ECM components (Robbins et al. 2011). The presence of ECM has also been shown to be responsible for the development of persister cells within fungal biofilms. Persister cells evolve from similar susceptible cells but escape the drug assault by being located within the biofilm matrix, grow slowly and develop tolerance to drugs (Lewis 2005).

Although, the presence of ECM seems to be the major contributor of MDR in biofilms, in some cases where ECM is underdeveloped (like early stages of biofilms), sessile cells grown planktonically still show enhanced drug resistance. It seems there are several diverse cellular pathways leading to increased drug resistance in biofilms (Baillie and Douglas 2000; Blankenship and Mitchell 2006). In most of the tested conditions, no one mechanism of biofilm resistance was the sole determinant of increased resistance, as each controls specific developmental stages of biofilm.

In conclusion, all the evidence related to MDR and its mechanism in fungal cells, including those in biofilms, suggests that it is a multifactorial phenomenon, where not only MDR efflux pumps but also drug target alteration and transcriptional activation have an active role. Several other factors also contribute to the development of drug tolerance in fungi, in this context emerging roles of mitochondria, CW and lipids in MDR deserve more attention. The blocking of efflux pumps to expel incoming drugs is a common strategy pursued which has resulted in the identification of a few select inhibitors and modulators. However, such approaches, while promising, may not be sufficient to sensitize MDR cells, particularly when one considers the existence of several efflux proteins which may still be drug transporters and the contribution of yet unknown factors. Multi-target therapeutic strategies would be an ideal comprehensive approach in combating not only fungal infections but also in preventing the development of MDR.

References

Alarco AM, Raymond M (1999) The bZip transcription factor Cap1p is involved in multidrug resistance and oxidative stress response in *Candida albicans*. J Bacteriol 181:700–708

Askew C, Sellam A, Epp E, Mallick J, Hogues H, Mullick A et al (2011) The zinc cluster transcription factor Ahr1p directs Mcm1p regulation of *Candida albicans* adhesion. Mol Microbiol 79:940–953

Baillie GS, Douglas LJ (2000) Matrix polymers of Candida biofilms and their possible role in biofilm resistance to antifungal agents. J Antimicrob Chemother 46:397–403

Bauer BE, Wolfger H, Kuchler K (1999) Inventory and function of yeats ABC proteins: about sex, stress, and pleiotropic drug and heavy metal resistance. Biochim Biophys Acta 1461:217–236

Blankenship JR, Mitchell AP (2006) How to build a biofilm: a fungal perspective. Curr Opin Microbiol 9:588–594

Blankenship JR, Fanning S, Hamaker JJ, Mitchell AP (2010) An extensive circuitry for cell wall regulation in Candida albicans. PLoS Pathog 6(2):e1000752

Bonatti H, Goegele H, Tabarelli D, Muehlmann G, Sawyer R, Margreiter R et al (2007) *Pseudallescheria boydii* infection after liver retransplantation. Liver Transpl 13:1068–1069

Brown DH Jr, Giusani AD, Chen X, Kumamoto CA (1999) Filamentous growth of *Candida albicans* in response to physical environmental cues and its regulation by the unique *CZF1* gene. Mol Microbiol 34:651–662

Cannon RD, Lamping E, Holmes AR, Niimi K, Baret PV, Keniya MV, Tanabe K, Niimi M, Goffeau A, Monk BC (2009) Efflux-mediated antifungal drug resistance. Clin Microbiol Rev 22:291–321

Cheng G, Yeater KM, Hoyer LL (2006) Cellular and molecular biology of *Candida albicans* estrogen response. Eukaryot Cell 5:180–191

Coste AT, Karababa M, Ischer F, Bille J, Sanglard D (2004) *TAC1*, transcriptional activator of CDR genes, is a new transcription factor involved in the regulation of *Candida albicans* ABC transporters *CDR1* and *CDR2*. Eukaryot Cell 3:1639–1652

Coste AT, Ramsdale M, Ischer F, Sanglard D (2008) Divergent functions of three *Candida albicans* zinc-cluster transcription factors (*CTA4*, *ASG1* and *CTF1*) complementing pleiotropic drug resistance in *Saccharomyces cerevisiae*. Microbiology 154:1491–1501

Cottier F, Raymond M, Kurzai O, Bolstad M, Leewattanapasuk W, Jiménez-López C et al (2012) The bZIP transcription factor Rca1p is a central regulator of a novel CO_2 sensing pathway in yeast. PLoS Pathog 8:e1002485

Cushion MT, Collins MS, Linke MJ (2009) Biofilm formation by Pneumocystis spp. Eukaryot Cell 8:197–206

Dagley MJ, Gentle IE, Beilharz TH, Pettolino FA, Djordjevic JT, Lo TL, Uwamahoro N, Rupasinghe T, Tull DL et al (2011) Cell wall integrity is linked to mitochondria and phospholipid homeostasis in *Candida albicans* through the activity of the post-transcriptional regulator Ccr4-Pop2. Mol Microbiol 79:968–989

De Backer MD, Ilyina T, Ma XJ, Vandoninck S, Luyten WH, Vanden Bossche H (2001) Genomic profiling of the response of *Candida albicans* to itraconazole treatment using a DNA microarray. Antimicrob Agents Chemother 45(6):1660–1670

Dhamgaye S, Bernard M, Lelandais G, Sismeiro O, Lemoine S, Coppée JY et al (2012a) RNA sequencing revealed novel actors of the acquisition of drug resistance in *Candida albicans*. BMC Genomics 13:396

Dhamgaye S, Devaux F, Manoharlal R, Vandeputte P, Shah AH, Singh A et al (2012b) In vitro effect of malachite green on *Candida albicans* involves multiple pathways and transcriptional regulators *UPC2* and *STP2*. Antimicrob Agents Chemother 56:495–506

Di Bonaventura G, Pompilio A, Picciani C, Iezzi M, D'Antonio D, Piccolomini R (2006) Biofilm formation by the emerging fungal pathogen *Trichosporon asahii*: development, architecture, and antifungal resistance. Antimicrob Agents Chemother 50:3269–3276

Douglas LJ (2003) Candida biofilms and their role in infection. Trends Microbiol 11:30–36

Epsztejn S, Glickstein H, Picard V, Slotki IN, Breuer W, Beaumont C, Cabantchik ZI (1999) H-ferritin subunit overexpression in erythroid cells reduces the oxidative stress response and induces multidrug resistance properties. Blood 94:3593–3603

Fanning S, Mitchell AP (2012) Fungal biofilms. PLoS Pathog 8:e1002585

Favre B, Didmon M, Ryder NS (1999) Multiple amino acid substitutions in lanosterol 14alpha-demethylase contribute to azole resistance in *Candida albicans*. Microbiology 145:2715–2725

Finkel JS, Xu W, Huang D, Hill EM, Desai JV, Woolford CA, Nett JE, Taff H, Norice CT, Andes DR, Lanni F, Mitchell AP (2012) Portrait of *Candida albicans* adherence regulators. PLoS Pathog 8:e1002525

Flowers SA, Barker KS, Berkow EL, Toner G, Chadwick SG, Gygax SE et al (2012) Gain-of-function mutations in *UPC2* are a frequent cause of *ERG11* upregulation in azole-resistant clinical isolates of *Candida albicans*. Eukaryot Cell 11:1289–1299

Fratti RA, Belanger PH, Ghannoum MA, Edwards JE Jr, Filler SG (1998) Endothelial cell injury caused by *Candida albicans* is dependent on iron. Infect Immun 66:191–196

Fridkin SK (2005) The changing face of fungal infections in health care settings. Clin Infect Dis 41:1455–1460

Gaur M, Choudhury D, Prasad R (2005) Complete inventory of ABC proteins in human pathogenic yeast, *Candida albicans*. J Mol Microbiol Biotechnol 9:3–15

Gaur M, Puri N, Manoharlal R, Rai V, Mukhopadhayay G, Choudhury D et al (2008) MFS transportome of the human pathogenic yeast *Candida albicans*. BMC Genomics 9:579

Ge SH, Wan Z, Li J, Xu J, Li RY, Bai FY (2010) Correlation between azole susceptibilities, genotypes, and ERG11 mutations in *Candida albicans* isolates associated with vulvovaginal candidiasis in China. Antimicrob Agents Chemother 54:3126–3131

Ghannoum MA, Rice LB (1999) Antifungal agents: mode of action, mechanisms of resistance, and correlation of these mechanisms with bacterial resistance. Clin Microbiol Rev 12:501–517

Gow NA, van de Veerdonk FL, Brown AJ, Netea MG (2011) *Candida albicans* morphogenesis and host defence: discriminating invasion from colonization. Nat Rev Microbiol 10:112–122

Hameed S, Dhamgaye S, Singh A, Goswami SK, Prasad R (2011) Calcineurin signaling and membrane lipid homeostasis regulates iron mediated multidrug resistance mechanisms in *Candida albicans*. PLoS ONE 6(4):e18684

Hawser SP, Douglas LJ (1994) Biofilm formation by Candida species on the surface of catheter materials in vitro. Infect Immun 62(3):915–921

Heilmann CJ, Schneider S, Barker KS, Rogers PD, Morschhäuser J (2010) An A643T mutation in the transcription factor Upc2p causes constitutive *ERG11* upregulation and increased fluconazole resistance in *Candida albicans*. Antimicrob Agents Chemother 54:353–359

Heitman J (2011) Microbial pathogens in the fungal kingdom. Fungal Biol Rev 25:48–60

Henry KW, Nickels JT, Edlind TD (2000) Upregulation of ERG genes in Candida species by azoles and other sterol biosynthesis inhibitors. Antimicrob Agents Chemother 44:2693–2700

Heymann P, Gerads M, Schaller M, Dromer F, Winkelmann G, Ernst JF (2002) The siderophore iron transporter of *Candida albicans* (Sit1p/Arn1p) mediates uptake of ferrichrome-type siderophores and is required for epithelial invasion. Infect Immun 70:5246–5255

Hoot SJ, Smith AR, Brown RP, White TC (2011) An A643V amino acid substitution in Upc2p contributes to azole resistance in well-characterized clinical isolates of *Candida albicans*. Antimicrob Agents Chemother 55:940–942

Karababa M, Coste AT, Rognon B, Bille J, Sanglard D (2004) Comparison of gene expression profiles of *Candida albicans* azole-resistant clinical isolates and laboratory strains exposed to drugs inducing multidrug transporters. Antimicrob Agents Chemother 48:3064–3079

Katiyar SK, Edlind TD (2001) Identification and expression of multidrug resistance related ABC transporter genes in *Candida krusei*. Med Mycol 39:109–116

Kohli A, Smirti, Mukhopadhyay K, Rattan A, Prasad R (2002) In vitro low-level resistance to azole in *Candida albicans* is associated with changes in membrane fluidity and asymmetry. Antimicrob Agents Chemother 46:1046–1052

Kuhn DM, George T, Chandra J, Mukherjee PK, Ghannoum MA (2002) Antifungal susceptibility of Candida biofilms: unique efficacy of amphotericin B lipid formulations and echinocandins. Antimicrob Agents Chemother 46:1773–1780

Kumamoto CA (2002) Candida biofilms. Curr Opin Microbiol 5:608–611

Kusch H, Biswas K, Schwanfelder S, Engelmann S, Rogers PD, Hecker M et al (2004) A proteomic approach to understanding the development of multidrug-resistant *Candida albicans* strains. Mol Genet Genomics 271:554–565

LaFayette SL, Collins C, Zaas AK, Schell WA, Betancourt-Quiroz M et al (2010) PKC signaling regulates drug resistance of the fungal pathogen *Candida albicans* via circuitry comprised of Mkc1, calcineurin, and Hsp90. PLoS Pathog 6:e1001069

Lamb DC, Maspahy S, Kelly DE, Manning NJ, Geber A, Bennett JE et al (1999) Purification, reconstitution, and inhibition of cytochrome P-450 sterol delta22-desaturase from the pathogenic fungus *Candida glabrata*. Antimicrob Agents Chemother 43:1725–1728

Lewis K (2005) Persister cells and the riddle of biofilm survival. Biochemistry (Mosc) 70:267–274

Liu TT, Lee REB, Barker KS, Lee RE, Wei L, Homayouni R et al (2005) Genome-wide expression profiling of the response to azole, polyene, echinocandin, and pyrimidine antifungal agents in *Candida albicans*. Antimicrob Agents Chemother 49:2226–2236

Liu TT, Znaidi S, Barker KS, Xu L, Homayouni R et al (2007) Genome-wide expression and location analyses of the Candida albicans Tac1p regulon. Eukaryot Cell 6:2122–2138

Loussert C, Schmitt C, Prevost MC, Balloy V, Fadel E, Philippe B et al (2010) In vivo biofilm composition of Aspergillus fumigatus. Cell Microbiol 12:405–410

Mansfield BE, Oltean HN, Oliver BG, Hoot SJ, Leyde SE et al (2010) Azole drugs are imported by facilitated diffusion in *Candida albicans* and other pathogenic fungi. PLoS Pathog 6(9): e1001126

Marichal P, Koymans L, Willemsens S, Bellens D, Verhasselt P, Luyten W et al (1999) Contribution of mutations in the cytochrome P450 14alpha-demethylase (Erg11p, Cyp51p) to azole resistance in *Candida albicans*. Microbiology 145:2701–2713

Marie C, White TC (2009) Genetic basis of antifungal drug resistance. Curr Fungal Infect Rep 3:123–131

Martel CM, Parker JE, Bader O, Weig M, Gross U, Warrilow AG et al (2010a) Identification and characterization of four azole-resistant erg3 mutants of *Candida albicans*. Antimicrob Agents Chemother 54:4527–4533

Martel CM, Parker JE, Bader O, Weig M, Gross U, Warrilow AG et al (2010b) A clinical isolate of *Candida albicans* with mutations in *ERG11* (encoding sterol 14alpha-demethylase) and *ERG5* (encoding C22 desaturase) is cross resistant to azoles and amphotericin B. Antimicrob Agents Chemother 54:3578–3583

Martinez LR, Casadevall A (2006) Susceptibility of *Cryptococcus neoformans* biofilms to antifungal agents in vitro. Antimicrob Agents Chemother 50:1021–1033

Martinez LR, Casadevall A (2007) *Cryptococcus neoformans* biofilm formation depends on surface support and carbon source and reduces fungal cell susceptibility to heat, cold, and UV light. Appl Environ Microbiol 73:4592–4601

Martínez P, Ljungdahl PO (2005) Divergence of *Stp1* and *Stp2* transcription factors in *Candida albicans* places virulence factors required for proper nutrient acquisition under amino acid control. Mol Cell Biol 25:9435–9446

Mekalanos JJ (1992) Environmental signals controlling expression of virulence determinants in bacteria. J Bacteriol 174:1–7

Miyazaki H, Miyazaki Y, Geber A, Parkinson T, Hitchcock C et al (1998) Fluconazole resistance associated with drug efflux and increased transcription of a drug transporter gene, *PDH1*, in *Candida glabrata*. Antimicrob Agents Chemother 42:1695–1701

Moran GP, Sanglard D, Donnelly SM, Shanley DB, Sullivan DJ, Coleman DC (1998) Identification and expression of multidrug transporters responsible for fluconazole resistance in *Candida dubliniensis*. Antimicrob Agents Chemother 42:1819–1830

Morio F, Loge C, Besse B, Hennequin C, Le Pape P (2010) Screening for amino acid substitutions in the *Candida albicans* Erg11 protein of azole-susceptible and azole-resistant clinical isolates: new substitutions and a review of the literature. Diagn Microbiol Infect Dis 66:373–384

Morio F, Pagniez F, Lacroix C, Miegeville M, Le Pape P (2012) Amino acid substitutions in the *Candida albicans* sterol Δ5,6-desaturase (Erg3p) confer azole resistance: characterization of two novel mutants with impaired virulence. J Antimicrob Chemother 67:2131–2138

Morschhäuser J (2010) Regulation of multidrug resistance in pathogenic fungi. Fungal Genet Biol 47:94–106

Morschhauser J, Barker KS, Liu TT, Blaß-Warmuth J, Homayouni R, Rogers PD (2007) The transcription factor Mrr1p controls expression of the *MDR1* efflux pump and mediates multidrug resistance in *Candida albicans*. PLoS Pathog 3:e164

Mukherjee PK, Chandra J, Kuhn DK, Ghannoum MA (2003) Mechanism of fluconazole resistance in *Candida albicans* biofilms: phase-specific role of efflux pumps and membrane sterols. Infect Immun 71:4333–4340

Munro CA, Winter K, Buchan A, Henry K, Becker JM, Brown AJ et al (2001) Chs1 of *Candida albicans* is an essential chitin synthase required for synthesis of the septum and for cell integrity. Mol Microbiol 39:1414–1426

Nett JE, Sanchez H, Cain MT, Andes DR (2010) Genetic basis of Candida biofilm resistance due to drug-sequestering matrix glucan. J Infect Dis 202:171–175

Nobile CJ, Fox EP, Nett JE, Sorrells TR, Mitrovich QM, Hernday AD et al (2012) A recently evolved transcriptional network controls biofilm development in *Candida albicans*. Cell 148:126–138

Nyilassi I, Papp T, Tako M, Nagy E, Vagvolgyi C (2005) Iron gathering of opportunistic pathogenic fungi: a mini review. Acta Microbiol Immunol Hung 52:185–197

Odds FC (1988) Candida and candidosis: a review and bibliography, 2nd edn. Bailliere Tindall, London

Oliver BG, Song JL, Choiniere JH, White TC (2007) cis-Acting elements within the *Candida albicans ERG11* promoter mediate the azole response through transcription factor Upc2p. Eukaryot Cell 6:2231–2239

Pasrija R, Banerjee D, Prasad R (2007) Structure and function analysis of CaMdr1p, a major facilitator superfamily antifungal efflux transporter protein of *Candida albicans*: identification of amino acid residues critical for drug/H+ transport. Eukaryot Cell 6:443–453

Pasrija R, Panwar SL, Prasad R (2008) Multidrug transporter CaCdr1p and CaMdr1p of *Candida albicans* display different lipid specificites: both ergosterol and sphingolipids are essential for targeting of CaCdr1p to membrane rafts. Antimicrob Agents Chemother 52:694–704

Paulsen IT, Brown MH, Skurray RA (1996) Proton-dependent multidrug efflux systems. Micobiol Rev 60:575–608

Prasad R, Goffeau A (2012) Yeast ATP-binding cassette transporters conferring multidrug resistance. Annu Rev Microbiol 66:39–63

Prasad R, Kapoor K (2005) Multidrug resistance in yeast Candida. Int Rev Cytol 242:215–248

Prasad R, De Wergifosse P, Goffeau A, Balzi E (1995) Molecular cloning and characterization of a novel gene of *Candida albicans*, *CDR1*, conferring multiple resistance to drugs and antifungals. Curr Genet 27:320–329

Prasad T, Chandra A, Mukhopadhyay CK, Prasad R (2006) Unexpected link between iron and drug resistance of Candida spp.: iron depletion enhances membrane fluidity and drug diffusion, leading to drug-susceptible cells. Antimicrob Agents Chemother 2006(50):3597–3606

Prasad T, Hameed S, Manoharlal R, Biswas S, Mukhopadhyay CK, Goswami SK, Prasad R (2010) Morphogenic regulator *EFG1* affects the drug susceptibilities of pathogenic *Candida albicans*. FEMS Yeast Res 10:587–596

Prasad R, Sharma M, Rawal MK (2012) Functionally relevant residues of Cdr1p: a multidrug ABC transporter of human pathogenic *Candida albicans*. J Amino Acids 2011:531412

Rajendran R, Mowat E, McCulloch E, Lappin DF, Jones B, Lang S et al (2011) Azole resistance of *Aspergillus fumigatus* biofilms is partly associated with efflux pump activity. Antimicrob Agents Chemother 55:2092–2097

Ramage G, Vande Walle K, Bachmann SP, Wickes BL, López-Ribot JL (2002) In vitro pharmacodynamic properties of three antifungal agents against preformed *Candida albicans* biofilms determined by time-kill studies. Antimicrob Agents Chemother 46:3634–3636

Ramanan N, Wang Y (2000) A high-affinity iron permease essential for *Candida albicans* virulence. Science 288:1062–1064

Riggle PJ, Kumamoto CA (2006) Transcriptional regulation of *MDR1*, encoding a drug efflux determinant, in fluconazole-resistant *Candida albicans* strains through an Mcm1p binding site. Eukaryot Cell 5:1957–1968

Robbins N, Uppuluri P, Nett J, Rajendran R, Ramage G, Lopez-Ribot JL et al (2011) Hsp90 governs dispersion and drug resistance of fungal biofilms. PLoS Pathog 7:e1002257

Rutledge RM, Esser L, Ma J, Xia D (2011) Toward understanding the mechanism of action of the yeast multidrug resistance transporter Pdr5p: a molecular modeling study. J Struct Biol 173:333–344

Saier MH Jr, Beatty JT, Goffeau A, Harley KT, Heijne WHM, Huang SC et al (1999) The major facilitator superfamily. J Mol Microbiol Biotechnol 1:257–279

Sanglard D, Odds FC (2002) Resistance of Candida species to antifungal agents: molecular mechanisms and clinical consequences. Lancet Infect Dis 2:73–85

Sanglard D, Ischer F, Monod M, Bille J (1997) Cloning of *Candida albicans* genes conferring resistance to azole antifungal agents: characterization of *CDR2*, a new multidrug ABC transporter gene. Microbiology 143:405–416

Sanglard D, Ischer F, Calabrese D, Majcherczyk PA, Bille J (1999) The ATP binding cassette transporter gene *CgCDR1* from *Candida glabrata* is involved in the resistance of clinical isolates to azole antifungal agents. Antimicrob Agents Chemother 43:2753–2765

Sanglard D, Ischer F, Parkinson T, Falconer D, Bille J (2003) *Candida albicans* mutations in the ergosterol biosynthetic pathway and resistance to several antifungal agents. Antimicrob Agents Chemother 47:2404–2412

Sanglard D, Alix Coste A, Ferrari S (2009) Antifungal drug resistance mechanisms in fungal pathogens from the perspective of transcriptional gene regulation. FEMS Yeast Res 9:1029–1050

Sanguinetti M, Posteraro B, La Sorda M, Torelli R, Fiori B, Santangelo R et al (2006) Role of *AFR1*, an ABC transporter-encoding gene, in the in vivo response to fluconazole and virulence of *Cryptococcus neoformans*. Infect Immun 74:1352–1359

Sasse C, Schillig R, Dierolf F, Weyler M, Schneider S, Mogavero S et al (2011) The transcription factor Ndt80 does not contribute to Mrr1-, Tac1-, and Upc2-mediated fluconazole resistance in *Candida albicans*. PLoS ONE 6(9):e25623

Sasse C, Schillig R, Reimund A, Merk J, Morschhäuser J (2012) Inducible and constitutive activation of two polymorphic promoter alleles of the Candida albicans multidrug efflux pump MDR1. Antimicrob Agents Chemother 56:4490–4494

Schubert S, Barker KS, Znaidi S, Schneider S, Dierolf F, Dunkel N, Aïd M et al (2011) Regulation of efflux pump expression and drug resistance by the transcription factors Mrr1, Upc2, and Cap1 in *Candida albicans*. Antimicrob Agents Chemother 55:2212–2223

Seidler MJ, Salvenmoser S, Müller FM (2008) *Aspergillus fumigatus* forms biofilms with reduced antifungal drug susceptibility on bronchial epithelial cells. Antimicrob Agents Chemother 52:4130–4136

Seneviratne CJ, Jin L, Samaranayake LP (2008) Biofilm lifestyle of Candida: a mini review. Oral Dis 14:582–590

Shahi P, Moye-Rowley WS (2009) Coordinate control of lipid composition and drug transport activities is required for normal multidrug resistance in fungi. Biochim Biophys Acta 1794:852–859

Shingu-Vazquez M, Traven A (2011) Mitochondria and fungal pathogenesis: drug tolerance, virulence, and potential for antifungal therapy. Eukaryot Cell 10:1376–1383

Silva S, Negri M, Henriques M, Oliveira R, Williams DW, Azeredo J (2011) Adherence and biofilm formation of non-*Candida albicans* Candida species. Trends Microbiol 19:241–247

Singh A, Mahto KK, Prasad R (2013) Lipidomics and in vitro azole resistance in *Candida albicans*. OMICS 17(2):84–93

Smriti, Krishnamurthy S, Dixit, BL, Gupta CM, Milewski S, Prasad R (2002) ABC transporters Cdr1p, Cdr2p and Cdr3p of a human pathogen *Candida albicans* are general phospholipid translocators. Yeast 19:303–318

St Georgiev V (2000) Membrane transporters and antifungal drug resistance. Curr Drug Targets 1:261–284

Stephanie SP, Paulsen IT, Saier MH Jr (1998) Major facilitator superfamily. Microbiol Mol Biol Rev 62:1–34

Sun N, Fonzi W, Chen H, She X, Zhang L, Zhang L et al (2013) Azole susceptibility and transcriptome profiling in *Candida albicans* mitochondrial electron transport chain complex I mutants. Antimicrob Agents Chemother 57:532–542

Talibi D, Raymond M (1999) Isolation of a putative *Candida albicans* transcriptional regulator involved in pleiotropic drug resistance by functional complementation of a pdr1 pdr3 mutation in *Saccharomyces cerevisiae*. J Bacteriol 181:231–240

Torelli R, Posteraro B, Ferrari S, La Sorda M, Fadda G, Sanglard D et al (2008) The ATP-binding cassette transporter-encoding gene *CgSNQ2* is contributing to the *CgPDR1*-dependent azole resistance of *Candida glabrata*. Mol Microbiol 68:186–201

Tuch BB, Galgoczy DJ, Hernday AD, Li H, Johnson AD (2008) The evolution of combinatorial gene regulation in fungi. PLoS Biol 6:e38

Uppuluri P, Srinivasan A, Ramasubramanian A, Lopez-Ribot JL (2011) Effects of fluconazole, amphotericin B, and caspofungin on *Candida albicans* biofilms under conditions of flow and on biofilm dispersion. Antimicrob Agents Chemother 55:3591–3593

Vanden Bossche H, Marichal P, Gorrens J, Coene MC, Willemsens G, Bellens D et al (1989) Biochemical approaches to selective antifungal activity: focus on azole antifungals. Mycoses 32:35–52

Vandeputte P, Tronchin G, Bergès T, Hennequin C, Chabasse D, Bouchara JP (2007) Reduced susceptibility to polyenes associated with a missense mutation in the ERG6 gene in a clinical isolate of *Candida glabrata* with pseudohyphal growth. Antimicrob Agents Chemother 51:982–990

Vandeputte P, Pradervand S, Ischer F, Coste AT, Ferrari S, Harshman K et al (2012) Identification and functional characterization of Rca1, a transcription factor involved in both antifungal susceptibility and host response in *Candida albicans*. Eukaryot Cell 11:916–931

Vinces MD, Haas C, Kumamoto CA (2006) Expression of the *Candida albicans* morphogenesis regulator gene *CZF1* and its regulation by Efg1p and Czf1p. Eukaryot Cell 5:825–835

Wang H, Kong F, Sorrell TC, Wang B, McNicholas P, Pantarat N et al (2009) Rapid detection of *ERG11* gene mutations in clinical *Candida albicans* isolates with reduced susceptibility to fluconazole by rolling circle amplification and DNA sequencing. BMC Microbiol 14:167

White TC, Silver PM (2005) Regulation of sterol metabolism in *Candida albicans* by the *UPC2* gene. Biochem Soc Trans 33:1215–1218

White TC, Holleman S, Dy F, Mirels LF, Stevens DA (2002) Resistance mechanisms in clinical isolates of *Candida albicans*. Antimicrob Agents Chemother 46:1704–1713

Young LY, Hull CM, Heitman J (2003) Disruption of ergosterol biosynthesis confers resistance to amphotericin B in *Candida lusitaniae*. Antimicrob Agents Chemother 47:2717–2724

Znaidi S, Weber S, Al-Abdin OZ, Bomme P, Saidane S, Drouin S et al (2008) Genomewide location analysis of *Candida albicans* Upc2p, a regulator of sterol metabolism and azole drug resistance. Eukaryot Cell 7:836–847

Znaidi S, Barker KS, Weber S, Alarco AM, Liu TT, Boucher G et al (2009) Identification of the *Candida albicans* Cap1p regulon. Eukaryot Cell 8:806–820

Horizontal Gene Transfer in Planktonic and Biofilm Modes

Melanie Broszat and Elisabeth Grohmann

Abstract Horizontal gene transfer (HGT) is an important means to obtain and maintain plasticity of microbial genomes. Basically, bacteria apply three different modes to horizontally exchange genetic material: (1) conjugative transfer mediated by mobile genetic elements (MGE), (2) DNA uptake via transformation, and (3) transduction. The three modes rely on different prerequisites of the participating cells: conjugative transfer depends on close cell to cell contact between a donor and a recipient cell and is mediated through multi-protein complexes, denominated type IV secretion systems (T4SS), and DNA transformation does not rely on cell–cell contact but is the uptake of free DNA from the environment by a competent bacterial cell. In some bacteria it is also mediated by a T4SS. The third mechanism depends on the presence of a bacteriophage, which can transfer genomic DNA from one host cell to another. Experimental evidence exists that all three modes occur in planktonic cultures and recent data have also been provided for the occurrence of all three ways in biofilms. Regulation of these HGT events and their consequences for the acting microbes and the biofilms they live in are discussed in this chapter. Additionally, we focus on modern techniques to visualize and to quantify HGT in planktonic and biofilm modes.

1 Introduction

HGT is the most important means for the spread of antimicrobial resistance and virulence genes between related but also among unrelated bacteria. If antimicrobial resistance traits are taken up by pathogenic bacteria, multiple-resistant pathogens can evolve which then can further disseminate their resistance factors among the microbial population. Conjugative transfer mediated by MGEs is the most

---

M. Broszat • E. Grohmann (✉)
Division of Infectious Diseases, University Medical Centre Freiburg, University Hospital Freiburg, Hugstetter Strasse 55, Freiburg, 79106, Germany
e-mail: melanie.broszat@uniklinik-freiburg.de; elisabeth.grohmann@uniklinik-freiburg.de

important mechanism for this efficient horizontal gene spread. Various excellent recent reviews exist on the mechanistic details of conjugative plasmid transfer among bacteria mediated by T4SS, with emphasis on mechanistic details for Gram-negative bacteria (Alvarez-Martinez and Christie 2009; Fronzes et al. 2009; Waksman and Fronzes 2010; Wallden et al. 2010; Thanassi et al. 2012; Zechner et al. 2012). Therefore, there is no particular need to focus on it in this chapter. Instead, we will put emphasis on the occurrence of all three modes of HGT, as well as on specialized modes of HGT, for which only recently some mechanistic details have been elucidated, such as for the intra- and intergenomic transfer of pathogenicity islands, their regulation and consequences for survival, and fitness of the participating microbes, in different environments.

Biofilms are the predominant mode of life for bacteria in nature. Bacteria living in biofilms have been shown to be better protected from harmful impacts from their environment than their planktonic counterparts. Indeed, biofilm-associated bacteria exhibit increased resistance to antimicrobials, water stress, osmotic pressure, or grazing by protozoans (Costerton et al. 1999; Hogan and Kolter 2002) and adapt more readily to environmental changes via specialized communication systems, denominated quorum-sensing (Parsek and Greenberg 2005; Schuster et al. 2013).

The evolution, adaptation, and ecology of bacteria are intertwined mechanisms. Genes that are transferred horizontally between bacteria contribute essentially to bacterial evolution; interspecies HGT may lead to entirely new genetic combinations, which occasionally impose serious threats to human health (Madsen et al. 2012). Biofilm formation is essentially a product of interbacterial relations. Biofilms can consist of only one species, but in most cases natural biofilms contain different species, characterized by primary colonizers, which start "conquering" the habitat of choice, followed by secondary colonizers which establish in the matrix of the biofilm. In any case, the formation of a stable mature biofilm is the product of social interactions that have evolved through adaptations (Madsen et al. 2012). For several decades, both HGT and biofilms have been central areas of microbiological research in environmental as well as medical microbiology, resulting in the recognition of their high significance for bacterial adaptation and evolution. A growing number of studies showed that plasmid biology (in particular of conjugative plasmids) and biofilm community structure and functions are intertwined through many complex interactions, ranging from the genetic level to the community level. This fact points towards a principal role of the concerted action of these activities in socio-microbiology and bacterial evolution (Fig. 1; Madsen et al. 2012).

There is growing evidence that conjugative plasmids can promote the formation of biofilms or at least increase or accelerate their formation through genetic traits encoded on their genomes (May and Okabe 2008; Yang et al. 2008; D'Alvise et al. 2010; Madsen et al. 2012). This chapter will review the state of the art of interconnections between MGEs, MGE-mediated HGT, and biofilm formation.

However, not only conjugative transfer of plasmids and integrative conjugative elements (ICE) has been shown to take place in biofilms, but experimental evidence also exists for the occurrence of bacterial transformation via DNA uptake from the

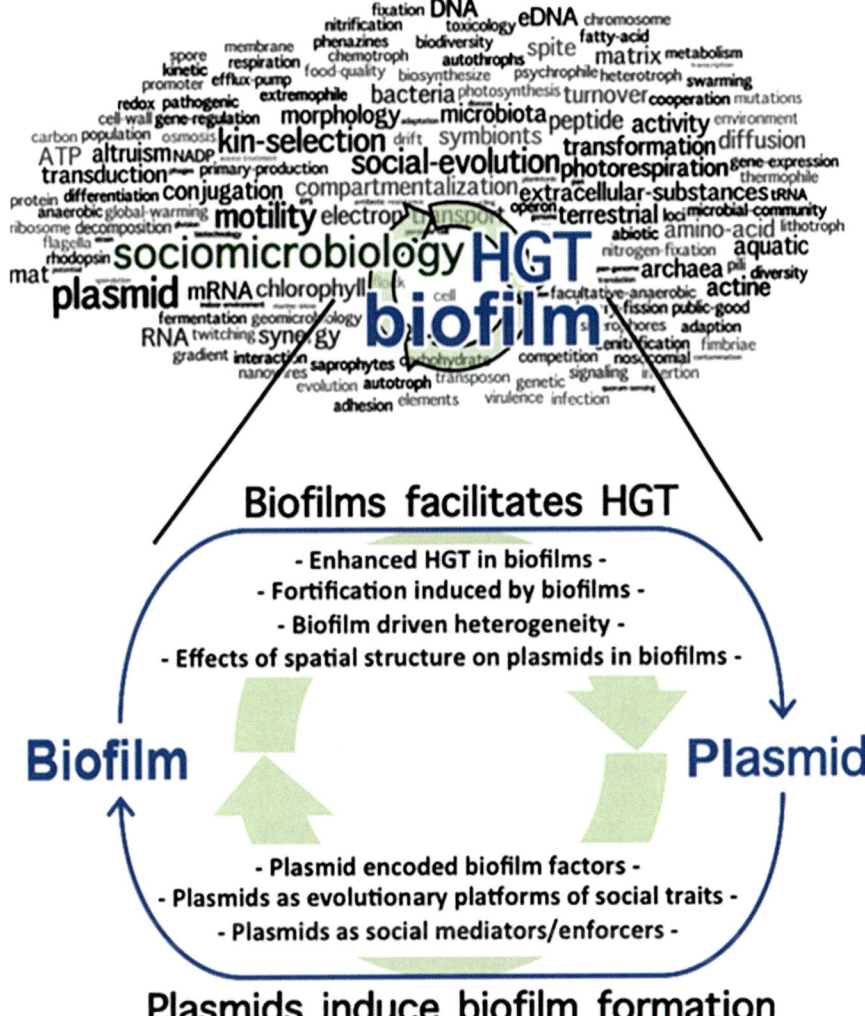

Fig. 1 The interconnection between biofilm formation and horizontal gene transfer (Madsen et al. 2012)

environment and for phage-mediated transduction. Recent knowledge on these processes and their regulatory networks will be summarized.

A variety of classical and molecular tools are available to monitor and quantify HGT in biofilms; most elaborated tools exist for the assessment of conjugative plasmid transfer in biofilms and in planktonic modes, ranging from classical selection for transconjugants (bacterial recipients which have acquired the plasmid via conjugation) on selective plates to (confocal) fluorescence microscopy to detect fluorescently labeled transconjugants or fluorescence-activated cell sorting (FACS)

(Nancharaiah et al. 2003; Arends et al. 2012; Reisner et al. 2012). Further more sophisticated tools are currently being developed to distinguish via distinct fluorescence labels of the plasmids involved between plasmid donors and transconjugants (P. Modrie and J. Mahillon, unpublished data; Broszat and Grohmann, unpublished data).

2 Modes of Horizontal Gene Transfer

As outlined above, there exist three major mechanisms of HGT between and among bacterial populations; all three of them contribute significantly to the horizontal dissemination of bacterial traits and the extraordinary adaptability of microbes to changing environmental conditions. Some special types of HGT, such as the excision, transfer, and reinsertion of genomic islands, particularly of pathogenicity islands, into a distinct genome, will be also discussed.

2.1 Conjugative Transfer

The conjugative plasmid systems are the largest and most widely distributed subfamily of T4SSs, with systems described for most species of the *Bacteria* and some members of the *Archaea* (Alvarez-Martinez and Christie 2009). The overall process of conjugative DNA transfer can be dissected into three biochemical reactions: DNA substrate processing, substrate recruitment, and DNA translocation (Pansegrau and Lanka 1996; Ding et al. 2003; Christie et al. 2005; Schröder and Lanka 2005; Alvarez-Martinez and Christie 2009). In the DNA processing reaction, DNA transfer and replication (Dtr) proteins initiate processing by binding a cognate origin of transfer (*oriT*) sequence on the conjugative element. The Dtr proteins include a relaxase and one or more accessory factors (for some plasmid systems, such as the broad-host-range plasmid pIP501, no accessory factors have been found so far (Kopec et al. 2005; Kurenbach et al. 2006), and when bound to *oriT*, the resulting DNA–protein complex is termed the relaxosome (Alvarez-Martinez and Christie 2009). Accompanying the nicking reaction, relaxase remains bound to the 5′-end of the transferred plasmid strand (T strand). The bound relaxase, probably together with other relaxosome components, mediates recognition of the DNA substrate by a cognate T4SS. The relaxase guides the T strand through the translocation channel. In the recipient cell, the relaxase catalyzes the re-circularization of the T strand and may also be involved in second-strand synthesis or recombination into the chromosome (Draper et al. 2005; César et al. 2006; Alvarez-Martinez and Christie 2009). The self-transmissible plasmids are only one of two major subgroups of conjugative elements. The second group of conjugative elements, originally denominated "conjugative transposons" and more recently termed ICEs (Integrative and Conjugative Elements), is also present in many bacterial and archaeal species (Burrus et al. 2002; Burrus and Waldor 2004; Juhas et al. 2007, 2008; Alvarez-Martinez and Christie 2009). These elements are excised from the chromosome through the action of a recombinase/excisionase complex and followed by the formation of a circular intermediate. Then the circularized intermediate is processed at *oriT* in the same way as described for conjugative plasmids. In the recipient cell, ICEs reintegrate into the chromosome (or plasmid) by homologous recombination or through the action of an integrase encoded by the ICE itself (Alvarez-Martinez and Christie 2009). Conjugative plasmids and ICEs are recruited to the transfer machine through interactions between the relaxosome or processed DNA transfer intermediate and a highly conserved ATPase termed the type IV coupling protein. This protein interacts with the translocation channel, which consists of the mating-pair formation proteins (Christie 2004; Schröder and Lanka 2005; Alvarez-Martinez and Christie 2009). Two types of mating-pair formation proteins, an ATPase and a polytopic membrane subunit, are associated with all T4SSs, whereas other mating-pair formation proteins are less phylogenetically conserved. In Gram-negative bacteria, the mating-pair formation proteins build the secretion channel, as well as a pilus or other surface filaments, to achieve attachment to target cells (Lawley et al. 2003; Christie and Cascales 2005; Alvarez-Martinez and Christie 2009). In Gram-positive bacteria, surface adhesins rather than conjugative pili apparently mediate attachment (Grohmann et al. 2003; Alvarez-Martinez and Christie 2009). For the majority of Gram-positive bacteria the origin and nature of the surface adhesins or other surface located factors involved in attachment and/or recognition of the recipient cell have not been elucidated so far.

2.2 DNA Uptake via Transformation

DNA transformation is based on the uptake of free DNA from the environment and therefore does not rely on MGEs; it is only encoded by the acceptor bacterium. Natural competence is the developmental state of the bacterium in which it is capable of taking up external DNA and to recombine this DNA into the chromosome, thereby undergoing natural transformation (Seitz and Blokesch 2013). A wide variety of bacterial species can develop natural competence and consequently take up external DNA (for recent reviews, see Lorenz and Wackernagel 1994; Chen and Dubnau 2004). The main steps to uptake external DNA include (1) binding of double-stranded DNA outside the cell to a (pseudo-) pilus structure elaborated by the acceptor cell, (2) extension and retraction of the pilus, driven by ATP-dependent motor proteins, that mediate the uptake of the double-stranded DNA through the secretin pore, which spans the outer membrane of the acceptor cell, (3) binding of the double-stranded DNA by the DNA-binding protein ComEA which occurs in the periplasmic space, (4) transport across the inner membrane which is carried out by ComEC concomitantly with the degradation of one strand by an unidentified nuclease, (5) single-stranded DNA reaches the cytoplasm and is decorated by

DNA processing protein A (DprA) and a single-strand binding protein to protect it against degradation, and (6) DprA recruits RecA, which catalyzes homologous recombination with the genomic DNA of the acceptor cell (Seitz and Blokesch 2013). Details of the DNA-uptake complexes have been reviewed in Averhoff and Friedrich (2003), Claverys et al. (2009), Burton and Dubnau (2010), and Allemand et al. (2012). However, less is known about the initiation of competence, particularly in Gram-negative bacteria. Current knowledge of environmental signals, which drive natural competence and transformation in Gram-negative bacteria, has been summarized by Seitz and Blokesch (2013). For Gram-positive bacteria, the signals triggering competence have been recently reviewed by Claverys et al. (2006).

2.3 DNA Transduction

Transduction is the process in which bacterial DNA gets erroneously packaged into the heads of phages; when the phage infects another bacterial cell the packaged DNA is incorporated into the new host genome (Roberts and Mullany 2010).

Phages are often highly specific to their bacterial hosts, able to infect even after significant periods of hiatus, and reproduce rapidly when their ecosystem permits. The viral genome is stored in safety, usually DNA encapsulated in the protein "head," until the virion attaches itself to a bacterial host cell for genome insertion (Brabban et al. 2005). This attachment process is specific involving the precise recognition of cell surface receptors, such as proteins and lipopolysaccharide elements, by specialized phage recognition structures (anti-receptors). When the viral genome has been introduced into the host, the life cycles of the lytic or temperate phages diverge. This divergence is determined by both the phage biology (lytic phages can only reproduce via a lytic life cycle while temperate phages can either reproduce lytically or enter lysogeny) and the cellular environment. Phages are grossly classified based on their life cycle (lytic vs. temperate), although finer subdivisions are based on their morphological characteristics (tailless vs. tailed), nature of the genome (e.g., DNA vs. RNA or single-stranded vs. double-stranded), and other factors (Brabban et al. 2005). In the last two decades, it has become more common to classify phages at a molecular level through the comparison of specific genes with the well-characterized T-4-like phages (Tétart et al. 2001).

2.4 Transfer of Genomic Islands

Genomic islands (GEI) are in essence discrete DNA segments differing widely between closely related bacterial strains to which usually some past or present mobility is attributed (Juhas et al. 2008). GEIs represent a broad and diverse group of DNA elements with a large variety of sizes and abundance in bacterial genomes

(Dobrindt et al. 2004). The coding capacity of GEIs is not limited to pathogenicity functions, but can be very diverse, including traits such as symbiosis (Sullivan et al. 2002), sucrose and aromatic compound metabolism (Gaillard et al. 2006), mercury resistance, and siderophore synthesis (Larbig et al. 2002). Bioinformatics studies have shown that GEIs tend to carry more novel genes (i.e., genes that do not have orthologs in other species) than the rest of the genome (Hsiao et al. 2005). This suggests that GEIs have become strongly selected for adaptive and auxiliary functions. Juhas et al. (2009) proposed that the term GEI should be used for the overarching family of discrete "DNA elements" which are part of a cell chromosome and can drive or have driven strain differentiation.

As not all GEIs contain the same components, it is difficult to define a unifying mode of GEI functioning or lifestyle (referring to the functions required for maintenance, excision, transfer, or integration). Interestingly, many GEIs for which self-mobility has been shown can excise from the chromosome, encode the full capacity for horizontal self-transfer to another cell, and reintegrate into the target site in the new host chromosome (Juhas et al. 2009). GEIs that encode all these features and self-transfer by conjugation are part of a well-defined group of elements that have been named ICEs (Burrus and Waldor 2004). A wide variety of GEIs are intimately connected to phages and conjugative plasmids through their evolutionary origins. As a consequence, besides transformation, their transfer often occurs via conjugation and transduction (Jain et al. 2002; Chen et al. 2005; Juhas et al. 2009). GEIs do not necessarily encode the whole genetic information for self-transfer, and several cases are known in which GEIs are packaged by another co-residing lysogenic phage or mobilized by a plasmid or the conjugative system of an ICE (Shoemaker et al. 2000).

Hall (2010) recently published an excellent review on *Salmonella* genomic islands and their dissemination. *Salmonella* genomic island 1 (SGI1), the first island of this type, was found in *S. enterica* serovar Typhimurium DT104 isolates, which are resistant to seven different antibiotics. Early studies by Schmieger and Schicklmaier (1999) demonstrated that SGI1 was moved into new hosts by transduction via a phage produced by the DT104 isolates. SGI1 cannot transfer itself into a new host because it does not encode a full set of conjugative transfer genes, but it is mobilizable (Doublet et al. 2005). It can be transferred into *Salmonella* spp. or *Escherichia coli* hosts if an IncA/C plasmid is present in the donor to supply the conjugative transfer machinery (Doublet et al. 2005). SGI1 is found in many different *S. enterica* serovars. It carries class 1 integrons containing five antibiotic resistance genes conferring resistance to seven antibiotics, namely, ampicillin, chloramphenicol, florfenicol, streptomycin, spectinomycin, sulfamethoxazole, and tetracycline.

3 Horizontal Gene Transfer in Liquid, on Surfaces, and in Biofilm Mode

It has been known for a long time that conjugative DNA transfer takes place on surfaces as well as in liquid medium in the laboratory. For transfer to take place in liquid medium the mating partners need to have special surface structures evolved such as conjugative pili in Gram-negative bacteria (for recent reviews, refer to Schröder and Lanka 2005; Silverman and Clarke 2010) and adhesins in Gram-positive bacteria, such as enterococci (for a recent review, consult Dunny 2007; Palmer et al. 2010). For conjugative transfer on surfaces, which in the laboratory in essence is performed on filters placed on top of agar plates, close contact between the mating partners is acquired through high cell densities of donor and recipient cell. Conjugative DNA transfer in biofilms has been shown for the first time by Hausner and Wuertz (1999) for conjugative plasmids and by Roberts et al. (1999) for conjugative transposons, now termed ICEs. Ghigo was the first to claim that conjugative plasmids per se can encode traits that induce biofilm formation of planktonic bacteria (Ghigo 2001).

3.1 Interconnection Between HGT and Biofilm Formation

Recent research has revealed that HGT and biofilm formation are connected processes (Madsen et al. 2012). Biofilm formation depends on interbacterial relations and bacterial interactions: A biofilm is a gathering of bacterial cells embedded in a self-produced polymeric matrix consisting of extracellular polymeric substances (EPS), mainly exopolysaccharides, proteins, and nucleic acids. Biofilms may adhere to an inert or biotic surface or exist as free-floating communities. Biofilm cells often show an altered phenotype regarding growth rate and gene transcription, and they exhibit enhanced tolerance towards antibiotics and immune responses. Biofilms offer excellent conditions for bacterial interactions because of (1) the high-density and well-organized diverse microbial community allowing physical contact between the cells and (2) the matrix that concentrates various chemical compounds (e.g., communication signals and extracellular DNA). Furthermore, environmental biofilms are generally multispecies communities. A characteristic feature is their organization of cells into matrix-enclosed structures, varying in size from smaller microcolonies to large and sometimes "mushroom-shaped" structures, which enable nutrient supply and waste product removal for cells resident in the deeper biofilm layers. Nowadays it is generally accepted that the biofilm mode is the predominant mode of growth in natural bacterial habitats (Madsen et al. 2012).

In the following sections we will provide evidence and arguments for biofilms as microbial community structures that can promote plasmid transfer and stability.

Conversely, we will also show that conjugative plasmids can in turn promote biofilm formation.

3.2 Conjugative Transfer in Biofilms

Numerous experimental data provide evidence that conjugative transfer occurs at higher frequencies between members of biofilm communities than when bacteria are in a planktonic state (Hausner and Wuertz 1999; Sørensen et al. 2005; Madsen et al. 2012). This is very well illustrated by the fact that more transconjugants can be found after mating on a filter compared with mating in liquid culture. This is typically explained by the fact that biofilms are dense communities that accelerate the spread of MGEs (Madsen et al. 2012). It has been demonstrated that high HGT frequencies of plasmids can enable them to persist as molecular parasites (Bahl et al. 2007), whilst other MGEs are only transmitted vertically. It is likely that a trade-off exists between horizontal and vertical transmission of MGEs—a trade-off that may be facilitated by the costs that the MGEs impose on the host (Andersson and Levin 1999; Bergstrom et al. 2000; Madsen et al. 2012). Plasmids that are only maintained through high transfer frequencies may thus only be able to persist in biofilms (Lili et al. 2007).

In the following sections, examples of efficient conjugative plasmid transfer in biofilms are given.

3.2.1 Conjugative Plasmid Transfer Among Gram-Negative Bacteria in Biofilms

Although biofilms represent the most common bacterial lifestyle in clinically and environmentally important habitats, there is rare information on the extent of HGT in the spatially structured populations in biofilms (Król et al. 2011). Król and coworkers studied the factors that affect transfer of the promiscuous multidrug IncP-1 resistance plasmid pB10 in *E. coli* biofilms grown under varying conditions: in closed flow cells, plasmid transfer in surface-attached submerged biofilms was very low, whereas a high plasmid transfer frequency was observed in a biofilm floating at the air–liquid interface in an open flow cell with low flow rates. Extensive plasmid transfer was detected only in the narrow zone near the interface; much lower transfer frequencies in the lower zones/deeper biofilm layers coincided with rapidly decreasing oxygen concentrations. Król et al. (2011) concluded that the air–liquid interface could be a hot spot for plasmid-mediated HGT due to high densities of juxtaposed donor and recipient cells. However, conjugative transfer was not limited to the air–liquid interphase when the *E. coli* recipient strain was a good biofilm former, which is the case for many pathogenic *E. coli* strains (Król et al. 2011; Beloin et al. 2008; Kaper et al. 2004). In these cases conjugative transfer also occurred efficiently in submerged biofilms.

Hennequin et al. (2012) studied the plasmid transfer capacity of a CTX-M-15 beta lactamase producing *Klebsiella pneumoniae* isolate in both planktonic and biofilm conditions. Plasmid transfer frequencies in biofilms reached very high frequencies of about 0.5 per donor cell, in comparison to the planktonic mode where they amounted to 10^{-3}/donor. Ma and Bryers (2013) quantified conjugative transfer of the *Pseudomonas* TOL plasmid in biofilms as a function of limiting nutrient concentrations. Frequencies of plasmid transfer within biofilm populations were affected by limiting substrate loading in the following way: low concentrations of the limiting substrate, in this case, glucose, generated thinner biofilms comprised of more porous biofilm clusters that allowed greater penetration of donor cells throughout the clusters with more exposure of recipient population to donor cells, which resulted in an increase of plasmid transfer frequencies (Ma and Bryers 2013). The opposite held true at high substrate concentrations that produced very dense compact biofilm clusters, with corresponding low plasmid transfer efficiencies.

ICE*fe*1 is a 291-kbp ICE that was identified in the genome of the booming bacterium *Acidithiobacillus ferrooxidans*. Bustamante et al. (2012) investigated the excision of the element and expression of relevant genes under normal and DNA-damaging growth conditions. Both basal and mitomycin C-inducible excision as well as expression and induction of the genes for integration/excision were observed, suggesting that ICE*fe*1 is an actively excising SOS-regulated MGE (Bustamante et al. 2012). The presence of a complete set of genes for self-transfer functions that are induced in response to DNA damage additionally suggested that ICE*fe*1 is capable of conjugative transfer to suitable recipients. Transfer of ICE*fe*1 may provide selective advantages to other acidophilic bacteria in the ecological niche through dissemination of gene clusters expressing CRISPRs and exopolysaccharide biosynthesis enzymes, probably by resistance to phage infection and biofilm formation, respectively (Bustamante et al. 2012). The presence of a number of genes predicted to encode proteins involved in the synthesis of exopolysaccharides that contribute to biofilm formation could enhance the persistence of these bacteria in the environment and/or help to improve mineral dissolution (Rohwerder et al. 2003; Bustamante et al. 2012).

3.2.2 Conjugative Plasmid Transfer Among Gram-Positive Bacteria in Biofilms

Roberts et al. (1999) used a constant depth film fermenter to demonstrate that transient bacteria may be able to act as a donor to oral bacteria in an oral biofilm community in the short time that they are present in the oral cavity. Tn*5397* (a conjugative transposon carrying *tetM*) in a *B. subtilis* donor was shown to transfer to a streptococcal recipient growing as part of an artificial oral biofilm in the constant depth film fermenter. The *B. subtilis* donor was not recovered from the biofilm 24 h after inoculation, showing that even though the donor bacteria are no longer present, their genetic information can persist (summarized in Roberts and

Mullany 2010). Tn916-like conjugative transposons have been shown to be common in tetracycline-resistant *Veillonella* spp. and some of them were transmissible within a mixed-species consortium in the oral cavity consisting of 21 tetracycline-sensitive members (Ready et al. 2006). Sedgley and coworkers (2008) also provided evidence that the conjugative plasmid pAM81 was able to transfer between *Streptococcus gordonii* and *E. faecalis* in the root canals of human teeth in vivo.

Cook and colleagues demonstrated that growth in biofilms alters the induction of conjugative transfer by a sex pheromone in *E. faecalis* harboring the conjugative plasmid pCF10. Mathematical modeling suggested that a higher pCF10 copy number in biofilm cells would enhance a switch-like behavior in the pheromone response of donor cells with a delayed but increased response to the mating signal (Cook et al. 2011). Variations in plasmid copy number and a bimodal response to induction of conjugative transfer in populations of plasmid-harboring donor cells were both observed in biofilms, which is consistent with the predictions of the model. The pheromone system may have evolved such that donors in biofilms are only induced to transfer when they are in extremely close proximity to potential recipients in the biofilm (Cook et al. 2011). In contrast to the popular notion of biofilms being the optimal niche for conjugation (Hausner and Wuertz 1999), Cook and coworkers observed reduced efficiency of pCF10 transfer in biofilms. Their mating experiments employed biofilms grown in vitro with inducing pheromone produced by the recipient cells. The authors argued that their results reflect the anatomy of enterococci and differences in the cell attachment mechanisms used by the conjugative transfer machines of Gram-positive vs. Gram-negative bacteria: *E. faecalis* cells are not motile. When *E. faecalis* cells colonize a surface and initiate biofilm growth or attach and become part of a preexisting biofilm, they probably remain in the same location until they die or detach to reenter the planktonic phase (Cook et al. 2011). There is very low probability that donor and recipient cells get into close/intimate contact for the exchange of pCF10. In the pCF10 T4SS, mating pair formation is mediated by the surface adhesin Asc10 (encoded by the *prgB* gene on pCF10) which can stably bind the surfaces of cells that randomly collide; there are no sex pili that could attach cells that are not in direct wall-to-wall contact. In planktonic cultures of sufficient population density, however, random diffusion increases the probability of collision between donors and recipients, and induced donors can form stable mating pairs extremely efficiently (Cook et al. 2011).

Ghosh et al. (2011) investigated the enterococcal populations of dogs leaving the veterinary intensive care unit (ICU) for multidrug resistance, the capacity for biofilm formation, and HGT. The enterococcal diversity based on 210 isolates was low as represented by *E. faecium* (54.6 %) and *E. faecalis* (45.4 %). Most isolates were resistant to various antibiotics. All *E. faecalis* strains were biofilm formers in vitro. In vitro intra-species conjugation assays demonstrated that *E. faecium* were capable of transferring tetracycline, doxycycline, streptomycin, gentamicin, and erythromycin resistance traits to human clinical strains. High transfer rates (10^{-5}–10^{-4}) for both tetracycline and doxycycline resistance traits were observed, indicating potential involvement of Tn916 (Ghosh et al. 2011). The study demonstrated that companion animals after release from the ICU and on antibiotic treatment harbor a large multidrug-resistant enterococcal community. Genotyping of *E. faecium* strains revealed very low clonal diversity, their possible nosocomial origin, and close relatedness to human clinical isolates. Ghosh and coworkers recommended restricted contact after release from ICU between treated dogs and their owners to avoid health risks.

Due to the particular structure of their cell wall mycobacteria are neither considered as Gram-negative nor as Gram-positive bacteria. Nguyen and coworkers (2010) investigated conjugative transfer of chromosomal DNA between different strains of *Mycobacterium smegmatis*. They showed that efficient DNA transfer between different strains occurred in a mixed biofilm and that the process required expression of the *lsr2* gene, a gene product involved in biofilm formation, in the donor but not in the recipient strain. Transfer occurred predominantly at the biofilm liquid–air interface.

As demonstrated above, higher HGT frequencies in biofilms is the general observation. Nevertheless, there are also examples of spatial constraints within biofilms that may hinder the spread of plasmids in an already-established biofilm (Merkey et al. 2011). Król and coworkers (2011) showed how the transfer of an IncP-1 plasmid has spatial and nutritional constraints and occurred predominantly in the aerobic zone in an *E. coli* biofilm. Madsen et al. (2012) speculated that a prerequisite for successful transfer of certain plasmids in a biofilm community is that the plasmid is present in the initial phases of biofilm formation. This can be fulfilled if the biofilm priming probabilities are encoded by the plasmid itself (Madsen et al. 2012).

3.3 Transformation in Biofilms

Successful transformation of a bacterial cell depends on physicochemical factors of the DNA molecules, such as their size, conformation, and concentration of the transforming DNA and other environmental factors, such as UV light, salt, pH, temperature, and the presence of extracellular nucleases (Roberts and Mullany 2010). Furthermore, there are genetic obstacles to overcome for successful transformation, such as the presence of restriction systems and the ability of the incoming DNA to either replicate autonomously or integrate into the recipient genome (Ogunseitan 1995; Roberts and Mullany 2010).

Transformation has no requirement for live donor cells as the DNA released upon cell death is the principal source of transforming DNA. In addition, some bacteria, such as *Neisseria gonorrhoeae*, can actively release DNA into their environment. Therefore, one of the rate-limiting steps for transformation of bacteria growing in a biofilm is the longevity of DNA molecules in both the biofilm and the cytoplasm of the transformed cell (Roberts and Mullany 2010). Roberts and Mullany (2010) summarized recent experimental data on transformation in oral biofilms: Mercer et al. (1999) studied the persistence of *Lactococcus lactis* chromosomal and plasmid DNA in human saliva. They found, although the DNA was partially degraded, that it was still visible on an agarose gel after 3.5 min incubation. Furthermore, the presence of plasmid DNA of about 500 bp length was demonstrated by PCR after 24 h of incubation in human saliva. Duggan et al. (2000) demonstrated that pUC18 could transform *E. coli* to ampicillin resistance following 24 h incubation in clarified ovine saliva. A recent study by Hannan and coworkers showed that the genomic DNA of a *Veillonella dispar* strain carrying the conjugative transposon Tn*916* could transform *Streptococcus mitis* to tetracycline resistance within an oral biofilm grown in a fermenter (Hannan et al. 2010).

Tribble et al. (2012) studied chromosomal DNA transfer between *Porphyromonas gingivalis*, a Gram-negative anaerobe residing exclusively in the human oral cavity. Their results revealed that natural competence mechanisms are present in multiple strains of *P. gingivalis*, and DNA uptake is not sensitive to DNA source or modification status. Tribble and coworkers (2012) were the first to observe extracellular (e) DNA in *P. gingivalis* biofilms and predicted it to be the major DNA source for HGT and allelic exchange between strains. They proposed that exchange of DNA in plaque biofilms by a transformation-like process is of major ecological importance in the survival and persistence of *P. gingivalis* in the challenging oral environment.

Certain oral streptococci produce hydrogen peroxide under aerobic growth conditions to inhibit competing species like *Streptococcus mutans*. By using *Streptococcus gordonii* as a model organism Itzek and coworkers demonstrated hydrogen peroxide-dependent eDNA release (Itzek et al. 2011). Under defined growth conditions, the eDNA release was shown to be entirely dependent on hydrogen peroxide. Chromosomal DNA damage seemed to act as the intrinsic signal for the release. Interestingly, the generation of eDNA was found to be coupled with the induction of the *S. gordonii* natural competence system. Consequently, the production of hydrogen peroxide triggered the transfer of antimicrobial resistance genes (Itzek et al. 2011). Thus, the eDNA found in the oral cavity can serve as a pool for novel genetic information, since hydrogen peroxide production by one species can induce the release of eDNA in other species. Itzek et al. (2011) argued that hydrogen peroxide is potentially much more than a toxic metabolic by-product; rather, it could serve as an important environmental signal that facilitates species evolution by HGT of genetic information and an increase in the mutation rate.

The action of a competence-specific murein hydrolase, CbpD, strongly increases the rate of HGT between pneumococci. CbpD is the key component of a bacteriolytic mechanism termed fratricide. It is secreted by competent pneumococci and mediates the release of donor DNA from sensitive streptococci present in the same environment (Wei and Håvarstein 2012). Recent data from Wei and Håvarstein demonstrated that the fratricide mechanism has a strong positive effect on intrabiofilm HGT, indicating that it is important for active acquisition of homologous donor DNA under natural conditions. Additionally, they found that competent biofilm cells of *S. pneumoniae* acquire a resistance marker much more efficiently from neighboring cells than from the growth medium. This could be explained by the fact that externally added DNA is not able to penetrate into the biofilm and is therefore available only to competent cells that are exposed to the growth medium.

Vibrio cholerae, the causative agent of cholera and a natural inhabitant of aquatic environments, regulates various behaviors by a quorum-sensing system conserved among many members of the genus *Vibrio*. The quorum-sensing system is mediated by two extracellular autoinducers, CAI-1, which is secreted only by vibrios, and AI-2, which is produced by many bacteria (Antonova and Hammer 2011). In marine biofilms on chitinous surfaces, quorum-sensing-proficient *V. cholerae* cells become naturally competent to take up eDNA. It could be hypothesized that *V. cholerae* can switch to the competent state in a chitinous environmental biofilm by responding to autoinducers derived from members of the multispecies bacterial consortium (Antonova and Hammer 2011). Antonova and Hammer (2011) also showed that *comEA* transcription and the horizontal uptake of DNA by *V. cholerae* are induced in response to purified CAI-1 and AI-2, and also by autoinducers originating from other vibrios cocultured with *V. cholerae* within a mixed-species biofilm.

3.4 Transduction in Biofilms

Roberts and Mullany (2010) summarized recent data on transduction in the human oral cavity. One of the main barriers to the activity of phage in oral biofilms is the access to the cells within the EPS secreted by the biofilm residents themselves (Sutherland 2001). Little is known about the effect of phage and the extent to which transduction contributes to genetic exchange within oral biofilms. There are some studies that indicate transduction may be occurring within the oral cavity. Although these works demonstrated the isolation of distinct phages from human saliva (Bachrach et al. 2003; Hitch et al. 2004), DNA transduction mediated by these phages to bacteria resident in oral biofilms could not be demonstrated so far. Evidence for the involvement of phage in the HGT of DNA among residents of oral biofilms only comes from studies carried out in vitro: for instance, tetracycline resistance encoded on Tn*916* and chloramphenicol resistance present on plasmid pKT210 have been transferred between *Actinobacillus actinomycetemcomitans* by the generalized transducing phageAaØ23 (Willi et al. 1997). Additional evidence that this phage may be transducing DNA between bacteria in the oral cavity was obtained by isolation of phage particles from subgingival plaque from periodontitis patients (Sandmeier et al. 1995; Willi et al. 1997).

Dissemination of Shiga toxin (Stx)-encoding phages is the most likely mechanism for the spread of Stx-encoding genes and the emergence of new Stx-producing *E. coli* (STEC) (Solheim et al. 2013). Solheim and coworkers observed transfer of Stx-encoding phages to potentially pathogenic *E. coli* in biofilm at both 20 °C and 37 °C, with the infection rates being higher at 37 °C than at 20 °C. The study of Solheim and coworkers is the first to show HGT in a laboratory grown biofilm mediated by a temperate phage.

3.4.1 Evidence for Transduction as an Important Means to Disseminate Antibiotic Resistance

Brabban et al. (2005) summarized the current knowledge of the role of temperate phages in the dissemination of antibiotic resistance. As transformation and conjugation are not common modes of HGT in *Salmonella*, phage-mediated transduction has been suggested as the most important mode. The phages ES18 and PDT17 have been shown to transduce antibiotic resistance genes in *S. Typhimurium* DT104, with PDT17 having been found integrated into the genome of all strains of DT104 so far studied (Schmieger and Schicklmaier 1999). This observation is consistent with the fact that the core resistance genes in *S. Typhimurium* DT104 are chromosomally encoded in a tight cluster as part of *Salmonella* genomic island I (43 kb), well within the size that a phage could package and transduce (Schmieger and Schicklmaier 1999; Cloeckaert and Schwarz 2001; Brabban et al. 2005). Most of the other studied strains of *Salmonella* carry complete prophages within their genomes, many of which are capable of generalized transduction upon induction, spontaneous or otherwise (Schicklmaier et al. 1999; Mmolawa et al. 2002; Bossi et al. 2003; Brabban et al. 2005).

Brabban and coworkers also summarized recent observations on the induction of phages from the lysogenic state to the lytic pathway. In particular, antibiotics affecting DNA metabolism (such as the quinolones trimethoprim, norfloxacin, and ciprofloxacin) can induce phages to leave their prophage state and reproduce lytically, even at sub-inhibitory concentrations. Matsushiro and coworkers observed a 1,000-fold increase in phage titers and a 60-fold increase of Shiga toxin production within 6 h of in vitro ciprofloxacin exposure (Matsushiro et al. 1999). Experiments using growth-promoting antibiotics typically used in animal husbandry found that olaquindox and carbadox (both DNA targeting agents) increased both phage and Stx production (Köhler et al. 2000). This phenomenon is not limited to *E. coli* O157:H7, but it has been also reported that antibiotic resistance transfer occurred in *V. cholerae* at much higher efficiency when the SOS response was induced by antibiotics (Beaber et al. 2004; Hastings et al. 2004). Brabban et al. concluded that the use of antibiotics, whether therapeutically or as a growth promoter, not only provides a selective pressure on bacterial populations that favors resistant strains, but it also potentially increases the number of transduction and lysogenic conversion events within the population (Brabban et al. 2005).

The poultry industry faces a significant challenge in dealing with antibiotic resistant strains of *S. typhimurium*, *P. aeruginosa*, and *E. coli* that infect and can alter productivity of poultry flocks (Vandemaele et al. 2002; Walker et al. 2002; Brabban et al. 2005). Thus, the continued presence of antibiotic resistant strains of these common food-borne pathogens among poultry and other livestock coupled with the prevalence of transducing phages in the gastrointestinal tract of these animals is a matter of concern for public health officials worldwide (Brabban et al. 2005). For example, multiple antibiotic resistance cassettes can be transduced into antibiotic susceptible strains of *Pseudomonas aeruginosa* by phages released from multiple antibiotic resistant lysogenic strains of *P. aeruginosa* or by generalized transducing phages that already carry multiple antibiotic resistance markers (Blahova et al. 1999, 2000; Brabban et al. 2005). Resistance to imipenem, ceftazidime, and cefotaxime was transduced into antibiotic susceptible *P. aeruginosa* by a phage released from a lysogenic strain (Blahova et al. 1999; Brabban et al. 2005).

3.5 Transfer of Genomic Islands

Recently, a novel T4SS has been identified in *N. gonorrhoeae* that secretes chromosomal DNA in the surrounding environment in a non-contact-dependent manner (Hamilton et al. 2005; Juhas et al. 2009). This T4SS is localized in the large, horizontally acquired gonococcal genetic island present in the chromosome of *N. gonorrhoeae*, thus, by facilitating chromosomal DNA secretion this GEI also encodes the mechanism of its own dissemination by transformation (Juhas et al. 2009).

A new conjugation type GEI-encoded T4SS has been also described. It is evolutionarily distant from all previously described T4SSs and plays a key role in the horizontal transfer of a wide variety of GEIs originating from a broad spectrum of bacteria, including *Haemophilus* spp., *Pseudomonas* spp., *Erwinia carotovora* (*Pectobacterium carotovorum*), *Salmonella enterica* serovar Typhi, *Legionella pneumophila*, and others (Juhas et al. 2007, 2008, 2009) by conjugation.

The 153-kb *E. faecalis* pathogenicity island (PAI) was first described by Shankar et al. (2002). It encodes several pathogenicity factors, among them the enterococcal surface protein (esp) conferring increased biofilm formation and colonization, a cytolysin with hemolytic, cytolytic, and antibacterial activity, the aggregation substance, surface proteins, and general stress proteins (Shankar et al. 2002). The *E. faecalis* PAI is widely distributed among isolates of different origins, clonal types, and complexes, and it probably evolved by modular gain and loss of internal gene clusters (McBride et al. 2009). Laverde Gomez et al. (2011) demonstrated for the first time precise excision, circularization, and horizontal transfer of the entire PAI from the chromosome of *E. faecalis* strain UW3114. This PAI (ca. 200 kb) contained some deletions and insertions as compared to the PAI of the reference strain MMH594, transferred precisely and integrated site specifically into the chromosome of *E. faecalis* and *E. faecium*. The internal PAI structure was maintained after transfer. Biofilm formation and cytolytic activity were enhanced in *E. faecalis* transconjugants after acquisition of the PAI. A 66-kb conjugative pheromone-responsive erythromycin resistance plasmid (pLG2) that was transferred in parallel with the PAI was sequenced. It contains complete replication and conjugation modules of enterococcal origin in a mosaic-like composition; it is likely that it promotes horizontal transfer of the PAI (Laverde Gomez et al. 2011).

Many phages are able to transfer GEIs, as passengers in their genomes. Members of the *Staphylococcus* SaPI island family were shown to be induced to excise and replicate by certain resident temperate phages that are also involved in their packaging into small phage-like particles (Maiques et al. 2007; Ubeda et al. 2007; Juhas et al. 2009) that are transferred from donor to recipient cells at frequencies commensurate with the plaque-forming titer of the phage (Ruzin et al. 2001; Juhas et al. 2009). The high-pathogenicity island of *Yersinia pseudotuberculosis* (Lesic et al. 2004) and GEIs of the marine cyanobacterium *Prochlorococcus* (Coleman et al. 2006) have also been reported to be transferred by phages (Juhas et al. 2009).

Self-transfer via conjugation has also been described for some GEIs. For the ICE*Hin*1056, transfer frequencies of 10^{-1}–10^{-2} (transconjugants/recipient) were reported between two *H. influenzae* strains (Juhas et al. 2007). ICE*clc* of *Pseudomonas* sp. strain B13, a distant member of the same ICE*Hin*1056 subfamily, was shown to be self-transferable at similar frequencies to *P. putida*, *Cupriavidus necator*, and *P. aeruginosa* (Gaillard et al. 2006; Juhas et al. 2009).

Although GEIs basically can do the same job as self-transmissible plasmids, Juhas and coworkers argued that GEIs conceptually may have a number of advantages over a plasmid, one of the most notable being that GEIs are integrated in the host chromosome. Thus, unlike replicating plasmid molecules, GEIs do not need to continuously ensure coordinated replication, partitioning, or maintenance, and as there is often only a single copy of the GEI present per genome, its replication cost may not be as heavy a burden to the host cell (Gaillard et al. 2008).

4 Factors Promoting Biofilm Formation

4.1 Environmental Factors Promoting Biofilm Formation

Many studies provide evidence that sub-inhibitory concentrations of antibiotics promote biofilm formation. Hennequin et al. (2012) showed that the presence of the antibiotic cefotaxime at subminimal inhibitory concentrations enhanced biofilm formation of a *K. pneumoniae* isolate that is highly resistant to this antibiotic.

Two further studies showed that sublethal doses of antibiotic could exacerbate biofilm formation. Bagge et al. (2004) reported that sublethal concentrations of the beta-lactam antibiotic, imipenem, not only affected fivefold gene regulation of 34 genes in *P. aeruginosa* but also resulted in a 20-fold increase in alginate matrix synthesis that translated to an increase in biofilm volume and two orders of magnitude higher cell numbers (Ma and Bryers 2013). Hoffman et al. (2005) showed similar results for both *P. aeruginosa* and *E. coli* biofilms exposed to the aminoglycoside, tobramycin.

Lebreton et al. (2012) identified an oxidative stress sensor and response regulator in the important multidrug-resistant nosocomial pathogen *E. faecium* belonging to the MarR family; it was denominated AsrR, *a*ntibiotic and *s*tress *r*esponse *r*egulator. Deletion of *asrR* led to overexpression of two major adhesins, *acm* and *ecbA*, which resulted in enhanced in vitro adhesion to human intestinal cells, increased biofilm formation, and enhanced Tn*916* DNA transfer frequencies (Lebreton et al. 2012).

Nguyen and coworkers (2010) investigated the influence of the presence of heavy metals on biofilm formation and DNA uptake capacity of *Mycobacterium*. Cu^{2+} was shown to stimulate biofilm formation. In addition, a small but reproducible increase in DNA transfer efficiency (up to tenfold) was detected in the presence of 100 µM Cu^{2+}.

4.2 Mobile Genetic Element-Encoded Traits Promoting Biofilm Formation

As already indicated, Ghigo was the first to demonstrate that natural conjugative plasmids can express factors that induce planktonic bacteria to form or enter biofilm communities, which favor the infectious transfer of the plasmid (Ghigo 2001).

The *Enterococcus* PAI whose intra- and interspecies genomic transfer has been demonstrated by Laverde Gomez et al. (2011) encodes the enterococcal surface protein (esp), which is involved in enterococcal biofilm formation (Shankar et al. 2002).

Burmølle et al. (2012) determined the nucleotide sequence of three newly isolated conjugative IncX plasmids from *Enterobacteriaceae*. Their sequences revealed a remarkable occurrence of gene cassettes that promote biofilm formation in *K. pneumoniae* or *E. coli*. Two of the plasmids were shown to induce biofilm formation in a crystal violet retention assay in *E. coli*. Sequence comparisons revealed that all these plasmids contain the *mrkABCDF* gene cassette coding for type 3 fimbriae, which have been shown to promote cell attachment and biofilm formation on abiotic surfaces (Norman et al. 2008; Ong et al. 2008). The type 3 fimbriae gene cassette was demonstrated to originate from pathogenic *K. pneumoniae* (Burmølle et al. 2012). Thus, these data suggested an apparent ubiquity of a mobile form of an important virulence factor and are an illuminating example of the recruitment, evolution, and dissemination of genetic traits through plasmid-mediated HGT (Burmølle et al. 2012).

5 Monitoring of Horizontal Gene Transfer

A variety of monitoring techniques, either PCR-based or fluorescence microscopy based, have been designed in the last two decades to quantitatively follow the horizontal spread of metabolic traits or virulence factors under different conditions and in distinct environments.

Jussila et al. (2007) designed a molecular profiling method for HGT of aromatics-degrading plasmids. The method was successfully applied during rhizomediation and conjugation in vivo. It is based on the PCR detection of the TOL plasmid-specific *xylE* gene.

With increasing areas of transgenic crops during the last decades and the necessity of biological control agents as alternative to chemical pesticides, the establishment of techniques for environmental risk assessment has become necessary for the evaluation of biological control microorganisms released into the environment (Kim et al. 2012). Kim and coworkers investigated the possible HGT between released recombinant agricultural microorganisms and indigenous soil microorganisms. A recombinant *B. subtilis* strain and a recombinant plant growth-promoting *P. fluorescens* strain were used as model microorganisms (Kim et al. 2012). Soils of cucumber or tomato plants cultivated in the greenhouse were inoculated with the recombinant bacteria. For a 6-month period the soils were investigated for the presence of the recombinant bacteria by PCR, real-time PCR, Southern hybridization, and terminal restriction fragment length polymorphism fingerprinting. No positive signals for the recombinant *B. subtilis* and *P. fluorescens* strains were detected in the soils suggesting that horizontal gene flow from *B. subtilis* or *P. fluorescens* to soil bacteria in the greenhouse did not occur during the 6-month period (Kim et al. 2012).

In complex microbial communities with a high background of antibiotic resistance genes detection of HGT of resistance genes is challenging. One option to overcome the problem is labeling the antibiotic resistance gene. This approach was carried out by Haug et al. (2010). The conjugative multiresistance plasmid pRE25, originating from *E. faecalis*, was tagged with a 34-bp random sequence marker spliced by *tet*(M). The plasmid construct, denominated pRE25*, was introduced into *E. faecalis* CG110/*gfp*, a strain containing a *gfp* gene as chromosomal marker. The plasmid pRE25* was shown to be fully functional compared with its parental pRE25 and could be transferred to *Listeria monocytogenes* and *Listeria innocua* at frequencies of 6×10^{-6} to 8×10^{-8} transconjugants per donor. Different markers on the chromosome and the plasmid enabled independent quantification of donor and plasmid via specific quantitative PCR, even if antibiotic resistance genes occurred at high numbers in the background ecosystem. Haug and coworkers concluded that *E. faecalis* CG110/*gfp*/pRE25* is a potent tool for the study of horizontal antibiotic resistance transfer in complex environments such as biofilms, food matrices, or colonic models (Haug et al. 2010).

Over the last decade, advances in reporter gene technology have provided new insights into the extent and spatial frequency of HGT in vitro and in natural environments (Sørensen et al. 2005; Reisner et al. 2012). This methodology involves integration of genes encoding reporter proteins such as GFP in the conjugative plasmid of interest. In this way, the fate of plasmids in a bacterial community can be monitored in situ nondestructively. By this approach, spread of different IncP-1 and IncP-9 plasmids was monitored in a variety of environments including agar surface-grown colonies (Christensen et al. 1998; Krone et al. 2007; Fox et al. 2008), biofilm model systems (Christensen et al. 1998; Hausner and Wuertz 1999; Król et al. 2011; Seoane et al. 2011), freshwater microcosms (Dahlberg et al. 1998), or plant leaves (Normander et al. 1998).

Nancharaiah and coworkers were the first to use a dual-labeling technique involving GFP and the red fluorescent protein (DsRed) for in situ monitoring of HGT via conjugation (Nancharaiah et al. 2003). A GFPmut3b-tagged derivative of narrow-host-range TOL plasmid (pWWO) was delivered to *P. putida* KT2442, which was chromosomally labeled with *dsRed* by transposon insertion via biparental mating. GFP and DsRed were coexpressed in donor *P. putida* cells (Nancharaiah et al. 2003). Donors and transconjugants in mixed culture or sludge samples were discriminated on the basis of their fluorescence through confocal laser scanning microscopy. Conjugative transfer frequencies on agar surfaces and in sludge microcosms were determined microscopically without cultivation. The new method worked well for in situ monitoring of HGT in addition to tracking the fate of microorganisms released into a laboratory sequencing batch biofilm reactor treating synthetic wastewater (Nancharaiah et al. 2003).

Interestingly, spatial analysis of green fluorescence in various studies conducted by different laboratories revealed that invasive spread of IncP plasmids was neither detectable in recipient colonies on agar surfaces nor in recipient microcolonies in flow-chamber biofilms suggesting that local factors limit plasmid transfer (Christensen et al. 1998; Fox et al. 2008). Reisner et al. (2012) aimed to reveal the local distribution of IncF plasmid transfer in agar surface-grown colonies. To this goal, they developed a dual-color labeling strategy: *E. coli* donor cells expressing a chromosomally encoded LacI repressor and carrying $P_{A1/04/03}$-*cfp**-tagged conjugative plasmids were combined with recipient cells lacking a functional LacI protein. Since *cfp**expression from $P_{A1/04/03}$ is under tight control of the LacI repressor in the donor cells, cyan fluorescence can only emerge after transfer of the tagged conjugative plasmid to a recipient cell. To differentiate donor from recipient cells a $P_{A1/04/03}$-*yfp**-tagged *E. coli* CSH26 strain was utilized as recipient strain. Transconjugant cells are therefore distinguishable by expressing both cyan and yellow fluorescence (Reisner et al. 2012). Reisner and coworkers investigated two different plasmids, R1 and R1*drd*19. High-resolution in situ analysis through epifluorescence and confocal microscopy revealed that plasmid invasion did not reach beyond the first five recipient cell layers at the donor/recipient interface for both plasmids (Reisner et al. 2012). Extension of in situ analysis to other prototypical plasmids of the IncF, IncI, and IncW families revealed similarly limited levels of recipient colony invasion. The results were in agreement with previous studies monitoring IncP plasmid invasion in agar colonies and biofilm setups (Christensen et al. 1998; Fox et al. 2008). Fox and coworkers found that replenishment of

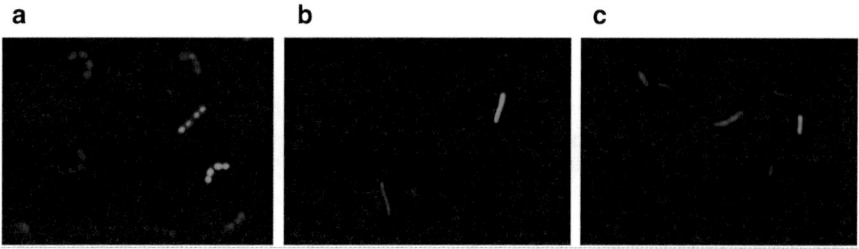

Fig. 2 Fluorescent microscopy images of *E. faecalis* T9 (**a**), *B. subtilis* subsp. *natto* (**b**), and *E. coli* XL10 (**c**) transconjugants expressing GFP

nutrients increased IncP plasmid invasion and regular disturbance of the spatial organization in the biofilm strongly improved plasmid invasiveness (Fox et al. 2008).

Arends and coworkers also used a dual fluorescence approach differing from that described above in the way that two differently labeled plasmids, a mobilizable GFP-labeled plasmid based on the transfer region of broad-host-range Inc18 plasmid pIP501 and a RFP-labeled non-mobilizable plasmid, were used (Arends et al. 2012; Arends, Schiwon, and Grohmann, unpublished data). Donors and transconjugants were distinguished by using fluorescence microscopy: donors exhibited green and red fluorescence (GFP and RFP), whereas recipients, which had acquired the mobilizable plasmid, showed only green fluorescence. Using this approach conjugative transfer among distinct Gram-positive bacteria and from the Gram-positive *E. faecalis* to the Gram-negative *E. coli* could be visualized (Fig. 2). A similar approach was applied for monitoring conjugative plasmid transfer among Gram-positive bacteria using GFP-labeled mobilizable plasmid, CFPopt-labeled non-mobilizable plasmid, and YFP-labeled conjugative plasmid (P. Modrie and J. Mahillon, unpublished results).

6 Conclusions and Perspectives

HGT mediated via all three major modes, conjugation, transduction, and transformation occurs efficiently in planktonic cultures and in microbial biofilms. A general feature is that HGT occurs with higher frequencies in biofilm mode. HGT enables bacteria to obtain and maintain the extraordinary plasticity of their genomes and is an important mechanism for the enormous adaptability of bacteria to changing environmental conditions.

Many questions concerning the trigger of HGT in different environments still remain unanswered. In their recent review on GEIs, Juhas and coworkers suggested that we are unwittingly, by changing the conditions for bacteria in hospitals—via antibiotic stress—and in the environment—via pollution—generating selective conditions which promote the success of self-transferable and stress-responsive

GEIs. Undoubtedly, as shown throughout the chapter, GEIs and other MGEs play a crucial role in the evolution of a broad spectrum of pathogenic or environmental bacteria (Juhas et al. 2008). Hence, deciphering the environmental signals promoting HGT of the diverse sets of MGEs could help establishing conditions in hospitals and healthcare centers which are less favorable to HGT of antibiotic resistance and virulence traits and/or enable the design of novel potential transfer inhibitors ("so-called eco-evo drugs," as summarized in Baquero et al. 2011).

Furthermore, the development of new monitoring tools for the quantification of HGT events in complex environments has enabled to estimate the extent of horizontal gene transmission in natural habitats through mimicking natural conditions in laboratory microcosms or biofilm models. However, there is still urgent need of the design of in situ tools, which can be applied to measure HGT events nondestructively in diverse real-life aquatic and terrestrial environments. One major methodological obstacle to overcome will be the high background signals, e.g., in the case of fluorescence labels, of real-life environments.

Acknowledgments We sincerely thank Jacques Mahillon for critical reading of the manuscript. We regret that not all valuable contributions of colleagues in the field could be included in this chapter due to space limitation.

References

Allemand J, Maier B, Smith DE (2012) Molecular motors for DNA translocation in prokaryotes. Curr Opin Biotechnol 23(4):503–509. doi:10.1016/j.copbio.2011.12.023

Alvarez-Martinez CE, Christie PJ (2009) Biological diversity of prokaryotic type IV secretion systems. Microbiol Mol Biol Rev 73(4):775–808. doi:10.1128/MMBR.00023-09

Andersson DI, Levin BR (1999) The biological cost of antibiotic resistance. Curr Opin Microbiol 2 (5):489–493

Antonova ES, Hammer BK (2011) Quorum-sensing autoinducer molecules produced by members of a multispecies biofilm promote horizontal gene transfer to Vibrio cholerae. FEMS Microbiol Lett 322(1):68–76. doi:10.1111/j.1574-6968.2011.02328.x

Arends K, Schiwon K, Sakinc T, Hübner J, Grohmann E (2012) Green fluorescent protein-labeled monitoring tool to quantify conjugative plasmid transfer between Gram-positive and Gram-negative bacteria. Appl Environ Microbiol 78(3):895–899. doi:10.1128/AEM.05578-11

Averhoff B, Friedrich A (2003) Type IV pili-related natural transformation systems: DNA transport in mesophilic and thermophilic bacteria. Arch Microbiol 180(6):385–393. doi:10.1007/s00203-003-0616-6

Bachrach G, Leizerovici-Zigmond M, Zlotkin A, Naor R, Steinberg D (2003) Bacteriophage isolation from human saliva. Lett Appl Microbiol 36(1):50–53

Bagge N, Schuster M, Hentzer M, Ciofu O, Givskov M, Greenberg EP, Høiby N (2004) Pseudomonas aeruginosa biofilms exposed to imipenem exhibit changes in global gene expression and beta-lactamase and alginate production. Antimicrob Agents Chemother 48(4):1175–1187

Bahl MI, Hansen LH, Sørensen SJ (2007) Impact of conjugal transfer on the stability of IncP-1 plasmid pKJK5 in bacterial populations. FEMS Microbiol Lett 266(2):250–256. doi:10.1111/j.1574-6968.2006.00536.x

Baquero F, Coque TM, de la Cruz F (2011) Ecology and evolution as targets: the need for novel eco-evo drugs and strategies to fight antibiotic resistance. Antimicrob Agents Chemother 55 (8):3649–3660. doi:10.1128/AAC.00013-11

Beaber JW, Hochhut B, Waldor MK (2004) SOS response promotes horizontal dissemination of antibiotic resistance genes. Nature 427(6969):72–74. doi:10.1038/nature02241

Beloin C, Roux A, Ghigo JM (2008) Escherichia coli biofilms. Curr Top Microbiol Immunol 322:249–289

Bergstrom CT, Lipsitch M, Levin BR (2000) Natural selection, infectious transfer and the existence conditions for bacterial plasmids. Genetics 155(4):1505–1519

Blahová J, Králiková K, Krcméry V, Jezek P (1999) Transduction of antibiotic resistance in Pseudomomas aeruginosa: relationship between lytic and transducing activity of phage isolate AP-423. Acta Virol 43(6):395–398

Blahová J, Králiková K, Krcméry V, Jezek P (2000) Low-frequency transduction of imipenem resistance and high-frequency transduction of ceftazidime and aztreonam resistance by the bacteriophage AP-151 isolated from a Pseudomonas aeruginosa strain. J Chemother 12 (6):482–486. doi:10.1179/joc.2000.12.6.482

Bossi L, Fuentes JA, Mora G, Figueroa-Bossi N (2003) Prophage contribution to bacterial population dynamics. J Bacteriol 185(21):6467–6471

Brabban AD, Hite E, Callaway TR (2005) Evolution of foodborne pathogens via temperate bacteriophage-mediated gene transfer. Foodborne Pathog Dis 2(4):287–303. doi:10.1089/fpd. 2005.2.287

Burmølle M, Norman A, Sørensen SJ, Hansen LH, Bruggemann H (2012) Sequencing of IncX-plasmids suggests ubiquity of mobile forms of a biofilm-promoting gene cassette recruited from Klebsiella pneumoniae. PLoS One 7(7):e41259. doi:10.1371/journal.pone.0041259

Burrus V, Waldor MK (2004) Shaping bacterial genomes with integrative and conjugative elements. Res Microbiol 155(5):376–386. doi:10.1016/j.resmic.2004.01.012

Burrus V, Pavlovic G, Decaris B, Guédon G (2002) Conjugative transposons: the tip of the iceberg. Mol Microbiol 46(3):601–610

Burton B, Dubnau D (2010) Membrane-associated DNA transport machines. Cold Spring Harb Perspect Biol 2(7):a000406. doi:10.1101/cshperspect.a000406

Bustamante P, Covarrubias PC, Levicán G, Katz A, Tapia P, Holmes D, Quatrini R, Orellana O (2012) ICE Afe 1, an actively excising genetic element from the biomining bacterium Acidithiobacillus ferrooxidans. J Mol Microbiol Biotechnol 22(6):399–407. doi:10.1159/ 000346669

César CE, Machón C, de la Cruz F, Llosa M (2006) A new domain of conjugative relaxase TrwC responsible for efficient oriT-specific recombination on minimal target sequences. Mol Microbiol 62(4):984–996. doi:10.1111/j.1365-2958.2006.05437.x

Chen I, Dubnau D (2004) DNA uptake during bacterial transformation. Nat Rev Microbiol 2 (3):241–249. doi:10.1038/nrmicro844

Chen I, Christie PJ, Dubnau D (2005) The ins and outs of DNA transfer in bacteria. Science 310 (5753):1456–1460. doi:10.1126/science.1114021

Christensen BB, Sternberg C, Andersen JB, Eberl L, Moller S, Givskov M, Molin S (1998) Establishment of new genetic traits in a microbial biofilm community. Appl Environ Microbiol 64(6):2247–2255

Christie PJ (2004) Type IV secretion: the Agrobacterium VirB/D4 and related conjugation systems. Biochim Biophys Acta 1694(1–3):219–234. doi:10.1016/j.bbamcr.2004.02.013

Christie PJ, Cascales E (2005) Structural and dynamic properties of bacterial type IV secretion systems (review). Mol Membr Biol 22(1–2):51–61

Christie PJ, Atmakuri K, Krishnamoorthy V, Jakubowski S, Cascales E (2005) Biogenesis, architecture, and function of bacterial type IV secretion systems. Annu Rev Microbiol 59 (1):451–485

Claverys J, Prudhomme M, Martin B (2006) Induction of competence regulons as a general response to stress in gram-positive bacteria. Annu Rev Microbiol 60:451–475. doi:10.1146/ annurev.micro.60.080805.142139

Claverys J, Martin B, Polard P (2009) The genetic transformation machinery: composition, localization, and mechanism. FEMS Microbiol Rev 33(3):643–656. doi:10.1111/j.1574-6976.2009.00164.x

Cloeckaert A, Schwarz S (2001) Molecular characterization, spread and evolution of multidrug resistance in Salmonella enterica typhimurium DT104. Vet Res 32(3–4):301–310. doi:10.1051/vetres:2001126

Coleman ML, Sullivan MB, Martiny AC, Steglich C, Barry K, Delong EF, Chisholm SW (2006) Genomic islands and the ecology and evolution of Prochlorococcus. Science 311(5768):1768–1770. doi:10.1126/science.1122050

Cook L, Chatterjee A, Barnes A, Yarwood J, Hu W, Dunny G (2011) Biofilm growth alters regulation of conjugation by a bacterial pheromone. Mol Microbiol 81(6):1499–1510. doi:10.1111/j.1365-2958.2011.07786.x

Costerton JW, Stewart PS, Greenberg EP (1999) Bacterial biofilms: a common cause of persistent infections. Science 284(5418):1318–1322

D'Alvise PW, Sjøholm OR, Yankelevich T, Jin Y, Wuertz S, Smets BF (2010) TOL plasmid carriage enhances biofilm formation and increases extracellular DNA content in Pseudomonas putida KT2440. FEMS Microbiol Lett 312(1):84–92. doi:10.1111/j.1574-6968.2010.02105.x

Dahlberg C, Bergström M, Hermansson M (1998) In situ detection of high levels of horizontal plasmid transfer in marine bacterial communities. Appl Environ Microbiol 64(7):2670–2675

Ding Z, Atmakuri K, Christie PJ (2003) The outs and ins of bacterial type IV secretion substrates. Trends Microbiol 11(11):527–535

Dobrindt U, Hochhut B, Hentschel U, Hacker J (2004) Genomic islands in pathogenic and environmental microorganisms. Nat Rev Microbiol 2(5):414–424. doi:10.1038/nrmicro884

Doublet B, Boyd D, Mulvey MR, Cloeckaert A (2005) The Salmonella genomic island 1 is an integrative mobilizable element. Mol Microbiol 55(6):1911–1924. doi:10.1111/j.1365-2958.2005.04520.x

Draper O, César CE, Machón C, de la Cruz F, Llosa M (2005) Site-specific recombinase and integrase activities of a conjugative relaxase in recipient cells. Proc Natl Acad Sci USA 102(45):16385–16390. doi:10.1073/pnas.0506081102

Duggan PS, Chambers PA, Heritage J, Forbes JM (2000) Survival of free DNA encoding antibiotic resistance from transgenic maize and the transformation activity of DNA in ovine saliva, ovine rumen fluid and silage effluent. FEMS Microbiol Lett 191(1):71–77

Dunny GM (2007) The peptide pheromone-inducible conjugation system of Enterococcus faecalis plasmid pCF10: cell-cell signalling, gene transfer, complexity and evolution. Philos Trans R Soc Lond B Biol Sci 362(1483):1185–1193. doi:10.1098/rstb.2007.2043

Fronzes R, Christie PJ, Waksman G (2009) The structural biology of type IV secretion systems. Nat Rev Microbiol 7(10):703–714. doi:10.1038/nrmicro2218

Fox RE, Zhong X, Krone SM, Top EM (2008) Spatial structure and nutrients promote invasion of IncP-1 plasmids in bacterial populations. ISME J 2(10):1024–1039. doi:10.1038/ismej.2008.53

Gaillard M, Vallaeys T, Vorhölter FJ, Minoia M, Werlen C, Sentchilo V, Pühler A, van der Meer, Jan Roelof (2006) The clc element of Pseudomonas sp. strain B13, a genomic island with various catabolic properties. J Bacteriol 188(5):1999–2013. doi:10.1128/JB.188.5.1999-2013.2006

Gaillard M, Pernet N, Vogne C, Hagenbüchle O, van der Meer JR (2008) Host and invader impact of transfer of the clc genomic island into Pseudomonas aeruginosa PAO1. Proc Natl Acad Sci USA 105(19):7058–7063. doi:10.1073/pnas.0801269105

Ghigo JM (2001) Natural conjugative plasmids induce bacterial biofilm development. Nature 412(6845):442–445. doi:10.1038/35086581

Ghosh A, Dowd SE, Zurek L, Webber MA (2011) Dogs leaving the ICU carry a very large multidrug resistant enterococcal population with capacity for biofilm formation and horizontal gene transfer. PLoS One 6(7):e22451. doi:10.1371/journal.pone.0022451

Grohmann E, Muth G, Espinosa M (2003) Conjugative plasmid transfer in gram-positive bacteria. Microbiol Mol Biol Rev 67(2):277–301, table of contents

Hall RM (2010) Salmonella genomic islands and antibiotic resistance in Salmonella enterica. Future Microbiol 5(10):1525–1538. doi:10.2217/fmb.10.122

Hamilton HL, Domínguez NM, Schwartz KJ, Hackett KT, Dillard JP (2005) Neisseria gonorrhoeae secretes chromosomal DNA via a novel type IV secretion system. Mol Microbiol 55(6):1704–1721. doi:10.1111/j.1365-2958.2005.04521.x

Hannan S, Ready D, Jasni AS, Rogers M, Pratten J, Roberts AP (2010) Transfer of antibiotic resistance by transformation with eDNA within oral biofilms. FEMS Immunol Med Microbiol 59(3):345–349. doi:10.1111/j.1574-695X.2010.00661.x

Hastings PJ, Rosenberg SM, Slack A (2004) Antibiotic-induced lateral transfer of antibiotic resistance. Trends Microbiol 12(9):401–404. doi:10.1016/j.tim.2004.07.003

Haug MC, Tanner SA, Lacroix C, Meile L, Stevens MJ (2010) Construction and characterization of Enterococcus faecalis CG110/gfp/pRE25*, a tool for monitoring horizontal gene transfer in complex microbial ecosystems. FEMS Microbiol Lett 313(2):111–119. doi:10.1111/j.1574-6968.2010.02131.x

Hausner M, Wuertz S (1999) High rates of conjugation in bacterial biofilms as determined by quantitative in situ analysis. Appl Environ Microbiol 65(8):3710–3713

Hennequin C, Aumeran C, Robin F, Traore O, Forestier C (2012) Antibiotic resistance and plasmid transfer capacity in biofilm formed with a CTX-M-15-producing Klebsiella pneumoniae isolate. J Antimicrob Chemother 67(9):2123–2130. doi:10.1093/jac/dks169

Hitch G, Pratten J, Taylor PW (2004) Isolation of bacteriophages from the oral cavity. Lett Appl Microbiol 39(2):215–219. doi:10.1111/j.1472-765X.2004.01565.x

Hoffman LR, D'Argenio DA, MacCoss MJ, Zhang Z, Jones RA, Miller SI (2005) Aminoglycoside antibiotics induce bacterial biofilm formation. Nature 436(7054):1171–1175. doi:10.1038/nature03912

Hogan D, Kolter R (2002) Why are bacteria refractory to antimicrobials? Curr Opin Microbiol 5(5):472–477

Hsiao WW, Ung K, Aeschliman D, Bryan J, Finlay BB, Brinkman FS (2005) Evidence of a large novel gene pool associated with prokaryotic genomic islands. PLoS Genet 1(5):e62. doi:10.1371/journal.pgen.0010062

Itzek A, Zheng L, Chen Z, Merritt J, Kreth J (2011) Hydrogen peroxide-dependent DNA release and transfer of antibiotic resistance genes in Streptococcus gordonii. J Bacteriol 193(24):6912–6922. doi:10.1128/JB.05791-11

Jain R, Rivera MC, Moore JE, Lake JA (2002) Horizontal gene transfer in microbial genome evolution. Theor Popul Biol 61(4):489–495

Juhas M, Power PM, Harding RM, Ferguson DJP, Dimopoulou ID, Elamin AR, Mohd-Zain Z, Hood DW, Adegbola R, Erwin A, Smith A, Munson RS, Harrison A, Mansfield L, Bentley S, Crook DW (2007) Sequence and functional analyses of Haemophilus spp. genomic islands. Genome Biol 8(11):R237. doi:10.1186/gb-2007-8-11-r237

Juhas M, Crook DW, Hood DW (2008) Type IV secretion systems: tools of bacterial horizontal gene transfer and virulence. Cell Microbiol 10(12):2377–2386. doi:10.1111/j.1462-5822.2008.01187.x

Juhas M, Der Meer V, Roelof J, Gaillard M, Harding RM, Hood DW, Crook DW (2009) Genomic islands: tools of bacterial horizontal gene transfer and evolution. FEMS Microbiol Rev 33(2):376–393. doi:10.1111/j.1574-6976.2008.00136.x

Jussila MM, Zhao J, Suominen L, Lindström K (2007) TOL plasmid transfer during bacterial conjugation in vitro and rhizoremediation of oil compounds in vivo. Environ Pollut 146(2):510–524. doi:10.1016/j.envpol.2006.07.012

Kaper JB, Nataro JP, Mobley HL (2004) Pathogenic Escherichia coli. Nat Rev Microbiol 2(2):123–140. doi:10.1038/nrmicro818

Kim SE, Moon JS, Choi WS, Lee SH, Kim SU (2012) Monitoring of horizontal gene transfer from agricultural microorganisms to soil bacteria and analysis of microbial community in soils. J Microbiol Biotechnol 22(4):563–566

Kopec J, Bergmann A, Fritz G, Grohmann E, Keller W (2005) TraA and its N-terminal relaxase domain of the Gram-positive plasmid pIP501 show specific oriT binding and behave as dimers in solution. Biochem J 387(Pt 2):401–409. doi:10.1042/BJ20041178

Köhler B, Karch H, Schmidt H (2000) Antibacterials that are used as growth promoters in animal husbandry can affect the release of Shiga-toxin-2-converting bacteriophages and Shiga toxin 2 from Escherichia coli strains. Microbiology 146(Pt 5):1085–1090

Krone SM, Lu R, Fox R, Suzuki H, Top EM (2007) Modelling the spatial dynamics of plasmid transfer and persistence. Microbiology 153(Pt 8):2803–2816. doi:10.1099/mic.0.2006/004531-0

Król JE, Nguyen HD, Rogers LM, Beyenal H, Krone SM, Top EM (2011) Increased transfer of a multidrug resistance plasmid in Escherichia coli biofilms at the air-liquid interface. Appl Environ Microbiol 77(15):5079–5088. doi:10.1128/AEM.00090-11

Kurenbach B, Kopeć J, Mägdefrau M, Andreas K, Keller W, Bohn C, Abajy MY, Grohmann E (2006) The TraA relaxase autoregulates the putative type IV secretion-like system encoded by the broad-host-range Streptococcus agalactiae plasmid pIP501. Microbiology 152(Pt 3):637–645. doi:10.1099/mic.0.28468-0

Larbig KD, Christmann A, Johann A, Klockgether J, Hartsch T, Merkl R, Wiehlmann L, Fritz H, Tummler B (2002) Gene islands integrated into tRNAGly genes confer genome diversity on a Pseudomonas aeruginosa clone. J Bacteriol 184(23):6665–6680. doi:10.1128/JB.184.23.6665-6680.2002

Laverde Gomez JA, Hendrickx AP, Willems RJ, Top J, Sava I, Huebner J, Witte W, Werner G (2011) Intra- and interspecies genomic transfer of the Enterococcus faecalis pathogenicity island. PLoS One 6(4):e16720. doi:10.1371/journal.pone.0016720

Lawley TD, Klimke WA, Gubbins MJ, Frost LS (2003) F factor conjugation is a true type IV secretion system. FEMS Microbiol Lett 224(1):1–15

Lebreton F, van Schaik W, Sanguinetti M, Posteraro B, Torelli R, Le Bras F, Verneuil N, Zhang X, Giard J, Dhalluin A, Willems RJ, Leclercq R, Cattoir V (2012) AsrR is an oxidative stress sensing regulator modulating Enterococcus faecium opportunistic traits, antimicrobial resistance, and pathogenicity. PLoS Pathog 8(8):e1002834. doi:10.1371/journal.ppat.1002834

Lesic B, Bach S, Ghigo J, Dobrindt U, Hacker J, Carniel E (2004) Excision of the high-pathogenicity island of Yersinia pseudotuberculosis requires the combined actions of its cognate integrase and Hef, a new recombination directionality factor. Mol Microbiol 52(5):1337–1348. doi:10.1111/j.1365-2958.2004.04073.x

Lili LN, Britton NF, Feil EJ (2007) The persistence of parasitic plasmids. Genetics 177(1):399–405. doi:10.1534/genetics.107.077420

Lorenz MG, Wackernagel W (1994) Bacterial gene transfer by natural genetic transformation in the environment. Microbiol Rev 58(3):563–602

Ma H, Bryers JD (2013) Non-invasive determination of conjugative transfer of plasmids bearing antibiotic-resistance genes in biofilm-bound bacteria: effects of substrate loading and antibiotic selection. Appl Microbiol Biotechnol 97(1):317–328. doi:10.1007/s00253-012-4179-9

Madsen JS, Burmølle M, Hansen LH, Sørensen SJ (2012) The interconnection between biofilm formation and horizontal gene transfer. FEMS Immunol Med Microbiol 65(2):183–195. doi:10.1111/j.1574-695X.2012.00960.x

Maiques E, Ubeda C, Tormo MA, Ferrer MD, Lasa I, Novick RP, Penadés JR (2007) Role of staphylococcal phage and SaPI integrase in intra- and interspecies SaPI transfer. J Bacteriol 189(15):5608–5616. doi:10.1128/JB.00619-07

Matsushiro A, Sato K, Miyamoto H, Yamamura T, Honda T (1999) Induction of prophages of enterohemorrhagic Escherichia coli O157:H7 with norfloxacin. J Bacteriol 181(7):2257–2260

May T, Okabe S (2008) Escherichia coli harboring a natural IncF conjugative F plasmid develops complex mature biofilms by stimulating synthesis of colanic acid and Curli. J Bacteriol 190(22):7479–7490. doi:10.1128/JB.00823-08

McBride SM, Coburn PS, Baghdayan AS, Willems RJ, Grande MJ, Shankar N, Gilmore MS (2009) Genetic variation and evolution of the pathogenicity island of Enterococcus faecalis. J Bacteriol 191(10):3392–3402. doi:10.1128/JB.00031-09

Mercer DK, Scott KP, Bruce-Johnson WA, Glover LA, Flint HJ (1999) Fate of free DNA and transformation of the oral bacterium Streptococcus gordonii DL1 by plasmid DNA in human saliva. Appl Environ Microbiol 65(1):6–10

Merkey BV, Lardon LA, Seoane JM, Kreft J, Smets BF (2011) Growth dependence of conjugation explains limited plasmid invasion in biofilms: an individual-based modelling study. Environ Microbiol 13(9):2435–2452. doi:10.1111/j.1462-2920.2011.02535.x

Mmolawa PT, Willmore R, Thomas CJ, Heuzenroeder MW (2002) Temperate phages in Salmonella enterica serovar Typhimurium: implications for epidemiology. Int J Med Microbiol 291 (8):633–644. doi:10.1078/1438-4221-00178

Nancharaiah YV, Wattiau P, Wuertz S, Bathe S, Mohan SV, Wilderer PA, Hausner M (2003) Dual labeling of Pseudomonas putida with fluorescent proteins for in situ monitoring of conjugal transfer of the TOL plasmid. Appl Environ Microbiol 69(8):4846–4852

Nguyen KT, Piastro K, Gray TA, Derbyshire KM (2010) Mycobacterial biofilms facilitate horizontal DNA transfer between strains of Mycobacterium smegmatis. J Bacteriol 192 (19):5134–5142. doi:10.1128/JB.00650-10

Norman A, Hansen LH, She Q, Sørensen SJ (2008) Nucleotide sequence of pOLA52: a conjugative IncX1 plasmid from Escherichia coli which enables biofilm formation and multidrug efflux. Plasmid 60(1):59–74. doi:10.1016/j.plasmid.2008.03.003

Normander B, Christensen BB, Molin S, Kroer N (1998) Effect of bacterial distribution and activity on conjugal gene transfer on the phylloplane of the bush bean (Phaseolus vulgaris). Appl Environ Microbiol 64(5):1902–1909

Ogunseitan OA (1995) Bacterial genetic exchange in nature. Sci Prog 78(Pt 3):183–204

Ong CY, Ulett GC, Mabbett AN, Beatson SA, Webb RI, Monaghan W, Nimmo GR, Looke DF, McEwan AG, Schembri MA (2008) Identification of type 3 fimbriae in uropathogenic Escherichia coli reveals a role in biofilm formation. J Bacteriol 190(3):1054–1063. doi:10.1128/JB.01523-07

Palmer KL, Kos VN, Gilmore MS (2010) Horizontal gene transfer and the genomics of enterococcal antibiotic resistance. Curr Opin Microbiol 13(5):632–639. doi:10.1016/j.mib.2010.08.004

Pansegrau W, Lanka E (1996) Enzymology of DNA transfer by conjugative mechanisms. Prog Nucleic Acid Res Mol Biol 54:197–251

Parsek MR, Greenberg EP (2005) Sociomicrobiology: the connections between quorum sensing and biofilms. Trends Microbiol 13(1):27–33. doi:10.1016/j.tim.2004.11.007

Ready D, Pratten J, Roberts AP, Bedi R, Mullany P, Wilson M (2006) Potential role of Veillonella spp. as a reservoir of transferable tetracycline resistance in the oral cavity. Antimicrob Agents Chemother 50(8):2866–2868. doi:10.1128/AAC.00217-06

Reisner A, Wolinski H, Zechner EL (2012) In situ monitoring of IncF plasmid transfer on semi-solid agar surfaces reveals a limited invasion of plasmids in recipient colonies. Plasmid 67 (2):155–161. doi:10.1016/j.plasmid.2012.01.001

Roberts AP, Mullany P (2010) Oral biofilms: a reservoir of transferable, bacterial, antimicrobial resistance. Expert Rev Anti Infect Ther 8(12):1441–1450. doi:10.1586/eri.10.106

Roberts AP, Pratten J, Wilson M, Mullany P (1999) Transfer of a conjugative transposon, Tn5397 in a model oral biofilm. FEMS Microbiol Lett 177(1):63–66

Rohwerder T, Gehrke T, Kinzler K, Sand W (2003) Bioleaching review part A: progress in bioleaching: fundamentals and mechanisms of bacterial metal sulfide oxidation. Appl Microbiol Biotechnol 63(3):239–248. doi:10.1007/s00253-003-1448-7

Ruzin A, Lindsay J, Novick RP (2001) Molecular genetics of SaPI1–a mobile pathogenicity island in Staphylococcus aureus. Mol Microbiol 41(2):365–377

Sandmeier H, van Winkelhoff AJ, Bär K, Ankli E, Maeder M, Meyer J (1995) Temperate bacteriophages are common among Actinobacillus actinomycetemcomitans isolates from periodontal pockets. J Periodontal Res 30(6):418–425

Schicklmaier P, Wieland T, Schmieger H (1999) Molecular characterization and module composition of P22-related Salmonella phage genomes. J Biotechnol 73(2–3):185–194

Schmieger H, Schicklmaier P (1999) Transduction of multiple drug resistance of Salmonella enterica serovar typhimurium DT104. FEMS Microbiol Lett 170(1):251–256. doi:10.1016/S0378-1097(98)00553-9

Schröder G, Lanka E (2005) The mating pair formation system of conjugative plasmids-A versatile secretion machinery for transfer of proteins and DNA. Plasmid 54(1):1–25. doi:10.1016/j.plasmid.2005.02.001

Schuster M, Sexton DJ, Diggle SP, Greenberg EP (2013) Acyl-homoserine lactone quorum sensing: from evolution to application. Annu Rev Microbiol 67:43–63. doi:10.1146/annurev-micro-092412-155635

Sedgley CM, Lee EH, Martin MJ, Flannagan SE (2008) Antibiotic resistance gene transfer between Streptococcus gordonii and Enterococcus faecalis in root canals of teeth ex vivo. J Endod 34(5):570–574. doi:10.1016/j.joen.2008.02.014

Seitz P, Blokesch M (2013) Cues and regulatory pathways involved in natural competence and transformation in pathogenic and environmental Gram-negative bacteria. FEMS Microbiol Rev 37(3):336–363. doi:10.1111/j.1574-6976.2012.00353.x

Seoane J, Yankelevich T, Dechesne A, Merkey B, Sternberg C, Smets BF (2011) An individual-based approach to explain plasmid invasion in bacterial populations. FEMS Microbiol Ecol 75(1):17–27. doi:10.1111/j.1574-6941.2010.00994.x

Shankar N, Baghdayan AS, Gilmore MS (2002) Modulation of virulence within a pathogenicity island in vancomycin-resistant Enterococcus faecalis. Nature 417(6890):746–750. doi:10.1038/nature00802

Shoemaker NB, Wang G, Salyers AA (2000) Multiple gene products and sequences required for excision of the mobilizable integrated bacteroides element NBU1. J Bacteriol 182(4):928–936. doi:10.1128/JB.182.4.928-936.2000

Silverman PM, Clarke MB (2010) New insights into F-pilus structure, dynamics, and function. Integr Biol 2(1):25–31. doi:10.1039/b917761b

Solheim HT, Sekse C, Urdahl AM, Wasteson Y, Nesse LL (2013) Biofilm as an environment for dissemination of stx genes by transduction. Appl Environ Microbiol 79(3):896–900. doi:10.1128/AEM.03512-12

Sørensen SJ, Bailey M, Hansen LH, Kroer N, Wuertz S (2005) Studying plasmid horizontal transfer in situ: a critical review. Nat Rev Microbiol 3(9):700–710. doi:10.1038/nrmicro1232

Sullivan JT, Trzebiatowski JR, Cruickshank RW, Gouzy J, Brown SD, Elliot RM, Fleetwood DJ, McCallum NG, Rossbach U, Stuart GS, Weaver JE, Webby RJ, Bruijn D, Frans J, Ronson CW (2002) Comparative sequence analysis of the symbiosis island of Mesorhizobium loti strain R7A. J Bacteriol 184(11):3086–3095

Sutherland IW (2001) The biofilm matrix–an immobilized but dynamic microbial environment. Trends Microbiol 9(5):222–227

Thanassi DG, Bliska JB, Christie PJ (2012) Surface organelles assembled by secretion systems of Gram-negative bacteria: diversity in structure and function. FEMS Microbiol Rev 36(6):1046–1082. doi:10.1111/j.1574-6976.2012.00342.x

Tétart F, Desplats C, Kutateladze M, Monod C, Ackermann HW, Krisch HM (2001) Phylogeny of the major head and tail genes of the wide-ranging T4-type bacteriophages. J Bacteriol 183(1):358–366. doi:10.1128/JB.183.1.358-366.2001

Tribble GD, Rigney TW, Dao DV, Wong CT, Kerr JE, Taylor BE, Pacha S, Kaplan HB (2012) Natural competence is a major mechanism for horizontal DNA transfer in the oral pathogen Porphyromonas gingivalis. MBio 3(1). doi:10.1128/mBio.00231-11

Ubeda C, Barry P, Penadés JR, Novick RP (2007) A pathogenicity island replicon in Staphylococcus aureus replicates as an unstable plasmid. Proc Natl Acad Sci USA 104(36):14182–14188. doi:10.1073/pnas.0705994104

Vandemaele F, Assadzadeh A, Derijcke J, Vereecken M, Goddeeris BM (2002) Aviaire pathogene Escherichia coli (APEC) (Avian pathogenic Escherichia coli (APEC)). Tijdschr Diergeneeskd 127(19):582–588

Waksman G, Fronzes R (2010) Molecular architecture of bacterial type IV secretion systems. Trends Biochem Sci 35(12):691–698. doi:10.1016/j.tibs.2010.06.002

Walker SE, Sander JE, Cline JL, Helton JS (2002) Characterization of Pseudomonas aeruginosa isolates associated with mortality in broiler chicks. Avian Dis 46(4):1045–1050

Wallden K, Rivera-Calzada A, Waksman G (2010) Type IV secretion systems: versatility and diversity in function. Cell Microbiol 12(9):1203–1212. doi:10.1111/j.1462-5822.2010.01499.x

Wei H, Håvarstein LS (2012) Fratricide is essential for efficient gene transfer between pneumococci in biofilms. Appl Environ Microbiol 78(16):5897–5905. doi:10.1128/AEM.01343-12

Willi K, Sandmeier H, Kulik EM, Meyer J (1997) Transduction of antibiotic resistance markers among Actinobacillus actinomycetemcomitans strains by temperate bacteriophages Aa phi 23. Cell Mol Life Sci 53(11–12):904–910

Yang X, Ma Q, Wood TK (2008) The R1 conjugative plasmid increases Escherichia coli biofilm formation through an envelope stress response. Appl Environ Microbiol 74(9):2690–2699. doi:10.1128/AEM.02809-07

Zechner EL, Lang S, Schildbach JF (2012) Assembly and mechanisms of bacterial type IV secretion machines. Philos Trans R Soc B Biol Sci 367(1592):1073–1087. doi:10.1098/rstb.2011.0207

The Role of Quorum Sensing in Biofilm Development

Kendra P. Rumbaugh and Andrew Armstrong

Abstract Quorum-sensing (QS) systems have been discovered in over 100 microbial species, many of which use these cell-to-cell signaling mechanisms for the coordinated production of biofilms. While our understanding of QS dynamics in laboratory culture conditions has dramatically expanded over the last few decades, we still understand very little about how these systems govern bacterial behavior in complex, natural settings. What we do know is that QS can influence every stage of biofilm formation; however, this influence is dependent on the microorganism and the type of QS system it employs. Furthermore, QS can both positively and negatively regulate biofilm formation in different environmental conditions. Investigations of QS in situ have been hampered by a lack of experimental tools; however, innovative new strategies are being developed that should help shed light on the involvement of QS in complex, polymicrobial biofilms. While developing agents that modulate QS to control microbial biofilm formation has faced significant hurdles, there are some promising agents in development and a more complete understanding of the role QS plays in biofilm formation should help drive future advances.

1 Overview of QS

Biological fitness and success in an environment are often dependent on adaptations for cooperation within a species as well as communication between multiple species competing for resources. Amongst multicellular organisms these behaviors have been studied in depth; however, much less understanding has been obtained regarding similar strategies in unicellular organisms. Traditional microbiological investigations, focused on monoculture studies, have elucidated vast amounts of

K.P. Rumbaugh (✉) • A. Armstrong
Department of Surgery, Texas Tech University Health Sciences Center, Lubbock, TX, 79430, USA
e-mail: Kendra.rumbaugh@ttuhsc.edu

information about growth and virulence characteristics, yet recent discoveries on the capacity of bacteria to communicate with one another have revealed a complexity and fitness that rival that of multicellular species (Haruta et al. 2009). This communication is accomplished primarily through chemical cell-to-cell signaling, or QS, which is population dependent and influences gene expression within a community of cells, often creating cooperation and allowing for successful colonization in a variety of environments, as well as initiation of pathogenicity and infectious disease processes (Miller and Bassler 2001).

QS systems have been discovered in over 100 microbial species and are associated with dozens of different signals, receptors, and effector molecules. Here we give a very general overview of the most common types of QS systems. The basic QS system consists of a constitutively expressed signal molecule, or autoinducer (AI), and a corresponding receptor that, when cell density and signal concentration have reached a threshold, regulates gene expression (Fig. 1). In general, AIs either diffuse in and out of cells and activate intracellular receptors, which directly influence gene expression, or are pumped out of cells and bind to transmembrane receptors (usually two-component), resulting in downstream changes in gene expression (Fig. 1). Many Gram-negative bacterial QS systems feature *a*cylated *h*omoserine *l*actones (AHLs, an abbreviation often used synonymously with HSL for *h*omoserine *l*actone) as signal molecules produced by an AI synthase LuxI homolog. Typically, these signal molecules bind an intracellular LuxR homolog receptor that is a ligand-activated transcriptional regulator (Bassler 1999). For example, two systems have been identified in *Pseudomonas aeruginosa* that follow this pattern. The LasI/R and RhlI/R systems are activated by *N*-3-oxododecanoyl homoserine lactone ($3OC_{12}$-HSL) and *N*-butanoyl-L-homoserine lactone (C_4-HSL), respectively, and influence a range of behaviors including biofilm formation and production of virulence factors (Pearson et al. 1997; de Kievit 2009).

Characteristically, Gram-positive QS systems feature small peptide AIs that are actively exported from the cell and bind to transmembrane receptors, which are often two-component signal transduction systems, as in the AgrC/A system of *Staphylococcus aureus* (Otto 2013). What is unique about the Agr system is the utilization of a regulatory RNA molecule as the global gene effector (Novick and Geisinger 2008). Alternatively, some peptide AIs are actively transported into the bacterial cell by specific oligopeptide permeases and bind to cognate intracellular regulatory proteins, which in turn regulate the transcription of target genes (Williams 2007). A third archetypical QS system utilizes derivatives of 4,5-dihydroxy-2,3-pentanedione, collectively known as autoinducer-2 (AI-2) and produced by the synthase LuxS, as its quorum signal. AI-2 is produced by a large variety of both Gram-negative and Gram-positive bacterial species and influences many behaviors such as bioluminescence, virulence factor production, and secretion (Taga et al. 2001).

Observations of QS systems have not only allowed for a greater understanding of communication within a single bacterial species, but also of competition and cooperation between multiple species. Dozens of unique signaling molecules have been identified within many different classes of chemicals including AHLs,

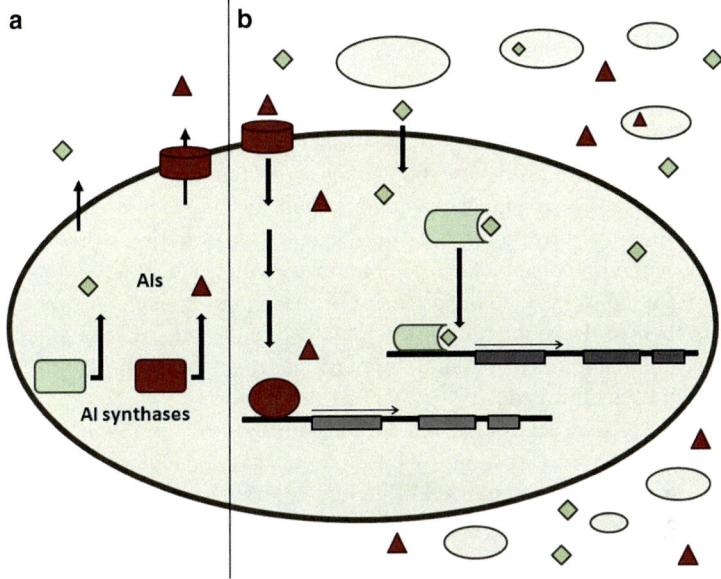

Fig. 1 Simplified diagram of AHL-mediated QS in Gram-negative bacteria and oligopeptide-mediated signaling in Gram-positive bacteria. (**a**) AIs are constitutively synthesized by AI synthase proteins. AHLs (*green*) typically diffuse out of the cell, while oligopeptides (*red*) are actively exported from the cell. (**b**) As the bacterial population increases, the quorum signals accumulate in the outside environment. Once a threshold concentration is reached, AHLs will freely enter into neighboring cells and bind to their cognate intracellular receptors, which then become activated transcriptional regulators and directly affect gene transcription. Alternatively, oligopeptide AIs either bind to transmembrane receptors, which are often two-component signal transduction systems, or are actively transported into the bacterial cell by specific oligopeptide permeases and bind to cognate intracellular regulatory proteins. These regulatory proteins may act directly to initiate the transcription of target genes or may influence the expression of secondary gene effectors, such as regulatory RNAs

quinolones, and small peptides. Many of these molecules were previously disregarded as metabolic by-products; however, it has since been demonstrated that they play roles in chemical communication and transcriptional regulation. While it is known that some QS systems upregulate the production of molecules that can be used as antimicrobials in the presence of other bacteria, many autoinducers, such as AI-2, are not unique to a single bacterial species and are a potential mechanism of interspecies communication and subsequent cooperation (Taga et al. 2001; Williams et al. 2007).

2 The Connection Between QS and Biofilms

Many factors influence the ability of microbes to construct biofilms. Environmental conditions such as the availability of nutrients, composition of the microbial population, surface properties, flow, and other physical forces will all affect the temporal sequence of biofilm developmental stages. For many microbes, the ability to "communicate" is advantageous for building biofilms. Intuitively this makes sense. It would be almost impossible for a team of construction workers to erect a skyscraper without the ability to communicate with each other. In the biblical Tower of Babylon story, God thwarted the people's plan to build a tower to heaven by cursing them with the inability to communicate. Similarly, it is assumed that the ability of many microbes to build biofilms can also be thwarted if microbial communication is inhibited.

The first published study that linked QS with biofilm formation demonstrated that *P. aeruginosa lasI* mutants, who were unable to synthesize $3OC_{12}$-HSL, formed biofilms that were flat, undifferentiated, and less resistant to dispersion with sodium-dodecyl sulfate than biofilms made by their wild-type parent strain (Davies et al. 1998). In the interim 15 years, many reports have been published that link QS and biofilm formation by several different organisms (see Table 1 for examples). Parsek and Greenberg (2005) proposed that QS and biofilms have been inextricably linked because both areas of study consider social phenomena exhibited by bacteria and thus suggested the term "sociomicrobiology" be used to encompass both. However, while QS and biofilms tend to be lumped together, our understanding of *how,* and under what circumstances, QS influences biofilm formation is still very limited. And while it may be intuitive that biofilm formation relies on bacterial communication, this is not always the case. The QS systems of several different microorganisms either negatively influence different stages of biofilm formation (Table 1) or have no effect on it. *Y. pestis* is an example of a bacterial species whose QS system is apparently not involved in biofilm formation (Jarrett et al. 2004).

3 QS-Dependent Biofilm Processes

It is generally accepted that for many microbes QS plays an important, if not essential, role in their ability to construct optimal biofilms. This notion is based on numerous studies with several different microbial species that have demonstrated altered biofilm formation by QS mutants. However, observing that a mutant forms a biofilm with less or more biomass, or an altered three-dimensional shape, than a wild-type strain after a defined amount of time does not tell us what processes were deficient or what fitness consequences this has for the mutant. The formation of a biofilm is a dynamic process. In general, biofilm formation involves five different stages (Stoodley et al. 2002) (Fig. 2): (1) Reversible attachment of the

Table 1 Examples of QS-dependent biofilm-related processes in different microbes

Microorganism	Implicated QS genes and/or signals	QS-regulated, biofilm-related process	References
Aeromonas hydrophila	*ahyI* Endogenous signal: C_4-HSL	Maturation	Lynch et al. (2002)
Bacillus cereus	*plcR, papR* Endogenous signal: AI-2	Promotes dispersal	Hsueh et al. (2006), Auger et al. (2006)
Burkholderia cepacia	*cciI/R, cepI/R* Endogenous signals: C_6-HSL, C_8-HSL	Attachment maturation	Tomlin et al. (2005), Huber et al. (2001, 2002)
Campylobacter jejuni	*luxS* Endogenous signal: AI-2	Not defined	Reeser et al. (2007)
Candida albicans	Endogenous signal: farnesol; exogenous signals: $3OC_{12}$-HSL, BDSF, cis-2-decenoic acid, CSP, SDSF, AI-2	Promotes dispersal; promote hyphal formation	Ramage et al. (2002), Hogan et al. (2004), Davies and Marques (2009), Boon et al. (2008), Jarosz et al. (2009), Vilchez et al. (2010), Bamford et al. (2009)
Helicobacter pylori	*luxS* Endogenous signal: AI-2	Negatively regulates attachment	Cole et al. (2004)
Klebsiella pneumonia	*luxS* Endogenous signal: AI-2; exogenous signal: cis-2-decenoic acid	Development of microcolonies; promotes dispersal	Balestrino et al. (2005), Davies and Marques (2009)
Listeria monocytogenes	*luxS; agrA, C* and *D*	Not defined; attachment	Challan Belval et al. (2006), Rieu et al. (2007)
Lactobacillus plantarum	*lam* operon	Attachment	Sturme et al. (2005)
Pantoea stewartii	*esaI/R* Endogenous signal: $3OC_6$-HSL	Negatively regulates attachment	von Bodman et al. (1998), Koutsoudis et al. (2006)
Pseudomonas aeruginosa	*lasI/R* Endogenous signals: $3OC_{12}$-HSL; cis-2-decenoic acid; exogenous signal: DSF	Maturation; promotes dispersal; maturation	Davies et al. (1998), Davies and Marques (2009), Ryan et al. (2008)
Rhodobacter sphaeroides	*cerI/R* Endogenous signal: 7,8-cis-*N*-(tetradecenoyl) homoserine lactone	Negatively regulates aggregation	Puskas et al. (1997)
Serratia liquefaciens	*swrI* Endogenous signals: C_4-HSL, C_6-HSL	Maturation	Labbate et al. (2004)
Serratia marcescens	*swrI* Endogenous signals: C_4-HSL	Attachment	Labbate et al. (2007)

(continued)

Table 1 (continued)

Microorganism	Implicated QS genes and/or signals	QS-regulated, biofilm-related process	References
Sinorhizobium meliloti	*sinI, expR*	Exopolysaccharide synthesis	Gao et al. (2012), Sorroche et al. (2010)
Staphylococcus aureus	*agr* Endogenous signal: AIP	Negatively regulates attachment maturation, dispersal	Reviewed in Otto (2013)
Staphylococcus epidermidis	*agr; luxS*	Negatively regulate attachment maturation; dispersal	Yao et al. (2006), reviewed in Otto (2013), Xu et al. (2006)
Streptococcus gordonii	*luxS* Endogenous signal:AI-2	Maturation; polymicrobial interactions	Blehert et al. (2003), McNab et al. (2003)
Streptococcus mutans	*luxS*	Maturation	Merritt et al. (2003), Wen and Burne (2004)
Streptococcus pneumonia	*luxS* Endogenous signal:AI-2	Attachment or microcolony formation	Vidal et al. (2011)
Vibrio cholera	*hapR, cqsA*	Maturation; promotes dispersal	Hammer and Bassler (2003), Zhu and Mekalanos (2003)
Vibrio scophthalmi	*luxS/R*	Not defined	Garcia-Aljaro et al. (2012)
Vibrio vulnificus	*smcR*	Shift to biofilm phenotype, maturation	McDougald et al. (2001, 2006)
Xanthomonas campestris	*rpfF, rpfC,* and *rpfG* Endogenous signal: DSF	Dispersal	Dow et al. (2003), Slater et al. (2000)
Yersinia pseudotuberculosis	*ypsI/R* and *ytbI/R* Endogenous signals: multiple AHLs	Not defined	Atkinson et al. (2011)

microbe to a surface mediated by pili, flagella, or other surface appendages or specific receptors; (2) the secretion of exopolymeric material, which results in irreversible attachment; (3) cell proliferation, resulting in the formation of a microcolony; (4) growth of the microcolony and differentiation of the biofilm, culminating in a "mature" biofilm community with characteristic structural features such as water channels and towering clusters of cells; and (5) active dispersion or passive detachment of biofilm cells. So at what stages of biofilm formation is QS important? Collectively, QS has been shown to be important during all five stages of biofilm development, but the specific QS-controlled stages differ between microbes, which employ different mechanisms of QS.

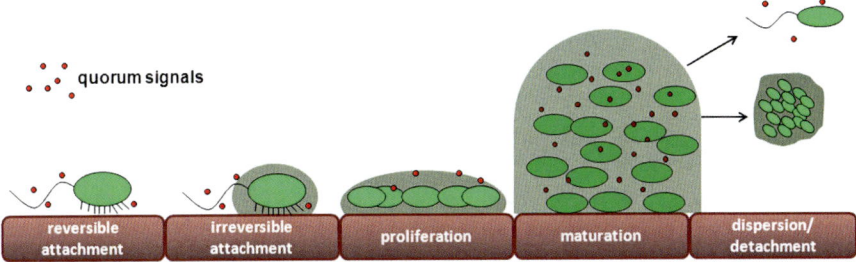

Fig. 2 Stages of biofilm development. Reversible attachment of the microbe to a surface is mediated by pili, flagella, or other surface appendages or receptors. This is followed by the secretion of exopolymeric material, which results in irreversible attachment. The attached cells proliferate, resulting in the formation of a microcolony. These microcolonies grow and differentiate into a mature biofilm community with characteristic structural features such as water channels and towering clusters of cells. Active dispersion or passive detachment of biofilm cells that can adhere at other locations results in cyclic periods of biofilm growth. Collectively QS has been shown to be important at every stage of biofilm development as indicated by the *red dots*, depicting quorum signals

3.1 Agr QS System-Dependent Biofilm Processes

QS can influence the attachment of a microbe to a surface by influencing the expression or functions of appendages that help anchor the microbe. For *S. aureus*, *S. epidermidis*, *Listeria plantarum*, and *L. monocytogenes* cell attachment in the earliest stages of biofilm development is regulated, at least in part, by their *agr* QS systems, which control the production of several adhesion proteins in addition to others factors (Cramton et al. 1999; Mack et al. 2007; Sturme et al. 2005; Rieu et al. 2007). However, the *agr* systems of *S. aureus* and *S. epidermidis* negatively control attachment, while the *L. monocytogenes agr* system and the homologous *lam* system in *L. plantarum* positively control attachment. The *agr* system also regulates biofilm maturation and dispersal in *S. aureus* and *epidermidis* through its control of phenol-soluble modulins (PSMs), which are amphipathic, α-helical peptides that act as surfactants thought to disrupt noncovalent interactions between biofilm cells and matrix components (reviewed in Otto 2013).

3.2 HSL-Mediated Biofilm Processes

For many Gram-negative bacterial species, QS is mediated by the production of HSL-based signals. These AIs are synthesized by members of the LuxI family of AI synthases and typically bind to intracellular receptors, which are LuxR homologs. Several examples of the HSL-based QS systems that are involved in biofilm formation are listed in Table 1 and include AhyI/R in *A. hydrophila*, LasI/R, and

RhlI/R in *P. aeruginosa*, SwrI/R in *S. liquefaciens*, and YpsI/R and YtbI/R in *Y. pseudotuberculosis*. While precise roles of the endogenous HSL signals have not been elucidated for all of these microbes, they are typically involved in some aspect of biofilm maturation.

R. sphaeroides, *Serratia marcescens*, *Burkholderia cepacia*, and *Pantoea stewartii* are examples of bacterial species that use HSL-based QS systems to control attachment. Mutations in the Cep and Cci QS systems of *B. cepacia* (Tomlin et al. 2005) and the Swr system in *S. marcescens* (Labbate et al. 2007) inhibited the microbes' ability to adhere to surfaces, resulting in thinner, less dense biofilms. However, mutations in the Esa QS system of *Pantoea stewartii* resulted in significantly better adhesion due to higher expression of EPS (von Bodman et al. 1998; Koutsoudis et al. 2006). Inactivation of the Cer system in *R. sphaeroides* resulted in more pronounced self-aggregation in liquid cultures, which could be reversed by adding the native HSL signal.

3.3 *LuxS or AI-2 QS System-Dependent Biofilm Processes*

The LuxS-based QS systems of many bacteria are also involved in varied stages of biofilm formation (Table 1). For *H. pylori luxS*-mediated QS appears to promote the planktonic lifestyle, as mutations in *luxS* promoted *H. pylori* attachment (Cole et al. 2004). The same is true for *S. epidermidis* where *luxS* appears to negatively regulate biofilm formation presumably through inhibiting exopolysaccharide expression, thus reducing adhesion (Xu et al. 2006). Interestingly, *S. epidermidis* also possesses an *agr*-type QS system, which regulates biofilm formation in a similar way, but through different pathways. Conversely, *luxS* appears to promote attachment of *S. pneumoniae*, as strains with mutations in *luxS* formed biofilms with 80 % less biomass (Vidal et al. 2011). In *S. mutans* and *S. gordonii luxS* is involved in biofilm maturation. *S. mutans luxS* mutants formed biofilms with lower biomass and altered structures that proved more sensitive to acid killing (Merritt et al. 2003; Wen and Burne 2004). Mutations in *luxS* also altered microcolony formation by *S. gordonii* and disrupted its ability to form polymicrobial biofilms with *luxS*-deficient *P. gingivalis* (Blehert et al. 2003; McNab et al. 2003).

4 Methods for Studying QS in Biofilms

QS by planktonic cells has been intensely studied, and, for many microbial species, much is known about the timing and conditions required for optimal QS. The structures of many signals and the details of several QS signaling pathways have been elucidated. However, in most cases, we have very little understanding of how these processes occur in biofilm settings. Furthermore, QS studies have only been performed in a limited number of laboratory culturing conditions. We know that environment dictates bacterial behavior, so applying the QS rules of engagement seen in planktonic cell cultures to all populations is intrinsically flawed. Therefore, it is important that investigators study biofilms in as close to their native state as possible. In addition, most of the studies discussed above were based on the behavior displayed by strains harboring mutations in different QS regulators. As these QS regulators can control a significant portion of the transcriptome, teasing out exactly what role cell-to-cell communication plays can be challenging. As Parsek and Greenberg noted, "Perhaps the best way to evaluate the role of quorum sensing is to monitor the signaling process in situ in a developing biofilm of a wild-type strain and determine if the onset of quorum sensing corresponds to any discernible transition in development, such as changes in structure or an increase in antimicrobial tolerance" (Parsek and Greenberg 2005). The methods described below have been developed to meet this goal.

4.1 QS Monitor Strains

Probably the most straightforward way to determine when and where QS is involved in biofilm formation is to use QS reporter strains to monitor the activity of QS in situ. De Kievit et al. used this approach to study the roles of the *las* and *rhl* QS systems in biofilm formation by *P. aeruginosa* (De Kievit et al. 2001). Biofilms grown in vitro with *P. aeruginosa* strains harboring *lasI* and *rhlI* transcriptional fusions to green fluorescent protein (GFP) were imaged after 4, 6, and 8 days of growth in polycarbonate flowcells. Prior to imaging, biofilm cells were stained with propidium iodide-Syto85 so that all cells could be visualized in the red spectra, while only those producing GFP could be visualized in the green. The investigators found *lasI* expression was maximal at 4 days and then decreased over time, while *rhlI* expression was more stable but occurred in fewer cells. They also found that bacterial cells closest to the polycarbonate surface maximally expressed both genes and expression of the QS genes decreased with increasing biofilm height. This study clearly showed that QS occurred in biofilms, but the percentage of cells expressing *lasI* or *rhlI* was very limited, both in number and in the area of the biofilm in which they resided.

Similar strategies have been used to monitor QS in *P. aeruginosa* lung infections in animal models and measure the effects of QS inhibiting agents in situ (Hentzer et al. 2003; Wu et al. 2004). Yarwood et al. also used a GFP-fusion strain to examine the contribution of the *agr* QS system to *S. aureus* biofilm formation in vitro (Yarwood et al. 2004). The investigators found that *agr*'s influence on biofilms was highly dependent on growth conditions, with *agr* expression enhancing biofilm development under some conditions and having no effect or inhibiting biofilm development under other conditions. They also noted that Agr-dependent expression occurred in patches within cell clusters in the biofilm and sometimes inversely correlated with detachment of cells from the biofilm.

4.2 Biofabricated Microenvironments to Study Signaling Dynamics

While being able to visualize when and where the expression of QS-related genes occurs within a given biofilm community, and test what environmental factors influence this expression, is of extreme value, these data do not give us much information about the fate of the signals themselves. Many analytical methods have been used to detect and quantify AHLs from cultures and extracts including liquid chromatography-mass spectrometry (Lepine and Deziel 2011), thin-layer chromatography (Ravn et al. 2001), proton nuclear magnetic resonance (Cao and Meighen 1989; Schaefer et al. 2000), and electrospray ionization-ion trap mass spectrometry (Frommberger et al. 2003). While surface-enhanced Raman spectroscopy holds promise for measuring AIs in situ (Pearman et al. 2007), at present only the use of live monitor strains, as described above, effectively allows for the detection of AHLs in situ.

The fate of quorum signals will be influenced by physical, chemical, and biological factors in the biofilm environment and subject to diffusion limitation, nonspecific binding, and signal interference. While experiments addressing how these factors affect microbial communities are much harder to conduct than simply determining when genes are turned on or off, some novel bioengineering approaches have been employed to try to better understand the dynamics of signaling in biofilms. Luo et al. used a microfluidics approach to preassemble or biofabricate biofilm-like environments in which pH and chemical gradients could be locally generated for the assembly of cell-polysaccharide composites in spatially localized and physically separated hydrogel layers (Luo et al. 2012). Populations of *E. coli* strains that were constructed to either "transmit" an AI-2 signal or "receive" the signal were placed strategically into these biofabricated biofilms so temporal and spatial aspects of AI-2 production, diffusion, and cell uptake could be studied. Visualization of active "transmitting" and "receiving" by the two populations was accomplished by epifluorescence microscopy as the transmitting population was engineered to express AI-2 and GFP simultaneously, whereas RFP production by the receiving population was induced at a threshold AI-2 concentration. The investigators found that signaling could be postponed or completely inhibited by adjusting flow conditions, independent of cell density.

To study the effects of bacterial density and environmental forces on QS in community structures, Connell et al. used mask-based multiphoton lithography to construct three-dimensional picoliter-scale microcavities, which they called lobster traps (Connell et al. 2010). The extremely small volume and versatility of design allowed the investigators to capture a single bacterium in traps of different volumes and monitor their proliferation and QS behavior. As the traps were constructed with cross-linked bovine serum albumin they were permeable to small molecules, allowing for controlled exposure to antibiotics and the study of antibiotic tolerance within the populations. Recent modifications to this approach were described by the same group, which allowed for strategic "3D printing" of different species of

Fig. 3 Gelatin-based micro-3D printed polymicrobial community. Cut-away 3D mask reconstruction (*upper*) and bright-field image (*lower*) of a nested polymicrobial community with an *S. aureus* microcluster confined in a 1 pL hemispherical cavity surrounded by a high-density *P. aeruginosa* population confined in a 30 pL square cavity. The 5-μm-thick roofs used to seal inner cavities and the outer chambers are not visible in the bright-field images (scale bars, 10 μm). Images courtesy of J. Connell and M. Whiteley

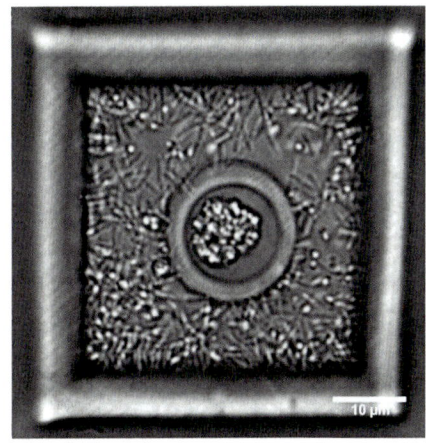

bacteria within highly porous, cross-linked gelatin structures (Connell et al. 2013) (Fig. 3). These innovative strategies should prove invaluable for studying cell-to-cell signaling interactions between polymicrobial populations and provide new insights into the dynamics of QS in biofilm formation.

5 Efficacy of QS Modulators as Antibiofilm Agents

Academically, it is of interest to understand how QS impacts biofilm formation by different microbes, but in regard to developing antibiofilm agents that target QS, it is essential. For example, if QS in a specific species is only required for attachment or at early stages of biofilm formation, then treating already mature biofilms with a QS inhibitor (QSI) would be ineffective. Similarly, treating bacteria that use QS to negatively regulate biofilm formation (e.g., *S. aureus*) with a QSI could be detrimental. Unfortunately, as the information presented above illustrates, different classes of signals can positively or negatively regulate different stages of biofilm formation in different species. Therefore, developing a universal LuxS inhibitor, for example, would be unrealistic. Furthermore, potential effects of any QS modulating agent on off-target microbial species or on host cell signaling pathways would have to be carefully studied. On a positive note, as most QS-modulators are not bacteriocidal, they should pose less of a risk to the development of resistance in target microbes, although there is recent evidence that challenges this theory (Garcia-Contreras et al. 2013).

While universal QS modulators (QSMs) are unlikely to be developed due to the high level of variability in the ways QS influences biofilm formation by different microbes, one agent has shown promise as a broad spectrum QSM. A fatty acid compound produced by *P. aeruginosa*, called cis-2-decenoic acid, was shown to induce the dispersion of biofilms formed by *E. coli*, *Klebsiella pneumoniae*, *Proteus mirabilis*, *Streptococcus pyogenes*, *Bacillus subtilis*, *Staphylococcus aureus*, *Candida albicans*, and *P. aeruginosa* itself, at nanomolar concentrations in vitro (Davies and Marques 2009). This compound was found to be structurally similar to the DSF, BDSF, and SDSF biofilm-related signals made by *X. campestris*, *B. cenocepacia*, and *S. mutans*, respectively (Dow et al. 2003; Boon et al. 2008; Vilchez et al. 2010). While not all of these related compounds induce dispersal, they may represent a class of QSMs that could be used to fight biofilms across a wide range of microbes.

Of course, fighting microbes with QSMs is not a new idea. Mother Nature evolved ways to do this long before we conceived of it. Although biological sources are being heavily screened for natural compounds that modulate QS, the most cited examples are the halogenated furanones produced by the Australian macroalga *Delisea pulchra*. It is thought that the macroalga evolved to produce these compounds as an innate defense against biofilm-related infections. The furanones, which are structurally similar to AHLs, inhibited QS-controlled processes in several Gram-negative bacterial species, including *Serratia liquefaciens*, *Vibrio fischeri*, and *P. aeruginosa* (Givskov et al. 1996). These early studies led to an abundance of work examining the efficacy of synthetic furanones, which showed efficacy against *P. aeruginosa*, when used in combination with antibiotics (Hentzer et al. 2003; Wu et al. 2004). In the intervening years dozens of QSMs have been synthesized or purified and have demonstrated efficacy against dozens of different microbes. Many of these QSMs are discussed in other chapters of this book and there have been several excellent recent reviews written on the topic (see e.g., LaSarre and Federle 2013; Zhu and Kaufmann 2013).

Many investigators and companies have pursued development of QSMs, and while several have been shown to work very well in vitro, for the most part that efficacy has not translated to in vivo studies (with the exception of the synthetic furanones discussed above). There are several explanations for this; for example, it is possible that QS is not as important in vivo as it appears to be in vitro. As mentioned above, most studies have only been performed in laboratory cultures using strains that have mutations in global regulators. It is possible that observations seen in these simplified model systems do not represent what happens in complex natural environments. Even more likely is that the biodelivery of these QSMs has not been optimized for in vivo applications. However, recent work by O'Loughlin et al. demonstrated the efficacy of meta-bromo-thiolactone, an analog of the native *P. aeruginosa* AHL autoinducers, to inhibit biofilm formation and virulence in a nematode model and in human cell culture (O'Loughlin et al. 2013). Although there is still a long path to demonstrating clinical efficacy and commercialization, reports such as this indicate that progress is being made in the development of these alternative antimicrobials.

6 Conclusions

Important advances have been made in the field of microbiology over the last three decades. Our understanding of bacteria as social organisms that live in communities and communicate with each other has challenged the ways in which we view microbes. Along with the basic science advances in sociomicrobiology have come translational studies aimed at developing new agents to fight biofilms. Many of these endeavors have focused on modulating QS in hopes of disrupting environmental or medically related biofilms. Unfortunately the urgent need to develop antibiotic alternatives has somewhat put the cart before the horse in regard to our incomplete understanding of QS in biofilm formation. Although our perceptions of microbial communities have evolved, in many ways our methods for studying them have lagged behind. For example, while we know that the environment clearly affects QS, biofilm formation, and the link between the two, most biofilm studies are still performed under a small number of in vitro laboratory culture conditions with laboratory-adapted strains. It is unrealistic to infer that observations seen in vitro occur in natural environments.

When it comes to understanding the role of QS in biofilm formation, many important questions still need to be addressed. Many QS mutants display reduced or altered biofilm formation, but what consequences does this have? Does it make them easier to eradicate with conventional agents? What is the role of communication in population homeostasis, and who is capable of communicating with whom? Does the ability of different microbes to "communicate" with each other contribute to cooperative or synergistic community interactions versus antagonistic interactions between non-communicators? Furthermore, we still do not fully understand how signals are perceived between microbial species or by the host. While there are still many unknowns, the good news is that the sociomicrobiology field continues to grow and diversify, incorporating other disciplines such as evolutionary biology, bioengineering, computer science, and physics. The resulting interdisciplinary approaches will undoubtedly result in new and exciting ways in which to study microbial social interactions within biofilms.

References

Atkinson S, Goldstone RJ, Joshua GW, Chang CY, Patrick HL, Camara M, Wren BW, Williams P (2011) Biofilm development on Caenorhabditis elegans by Yersinia is facilitated by quorum sensing-dependent repression of type III secretion. PLoS Pathog 7:e1001250

Auger S, Krin E, Aymerich S, Gohar M (2006) Autoinducer 2 affects biofilm formation by Bacillus cereus. Appl Environ Microbiol 72:937–941

Balestrino D, Haagensen JA, Rich C, Forestier C (2005) Characterization of type 2 quorum sensing in Klebsiella pneumoniae and relationship with biofilm formation. J Bacteriol 187:2870–2880

Bamford CV, D'mello A, Nobbs AH, Dutton LC, Vickerman MM, Jenkinson HF (2009) Streptococcus gordonii modulates Candida albicans biofilm formation through intergeneric communication. Infect Immun 77:3696–3704

Bassler BL (1999) How bacteria talk to each other: regulation of gene expression by quorum sensing. Curr Opin Microbiol 2:582–587

Blehert DS, Palmer RJ Jr, Xavier JB, Almeida JS, Kolenbrander PE (2003) Autoinducer 2 production by Streptococcus gordonii DL1 and the biofilm phenotype of a luxS mutant are influenced by nutritional conditions. J Bacteriol 185:4851–4860

Boon C, Deng Y, Wang LH, He Y, Xu JL, Fan Y, Pan SQ, Zhang LH (2008) A novel DSF-like signal from Burkholderia cenocepacia interferes with Candida albicans morphological transition. ISME J 2:27–36

Cao JG, Meighen EA (1989) Purification and structural identification of an autoinducer for the luminescence system of Vibrio harveyi. J Biol Chem 264:21670–21676

Challan Belval S, Gal L, Margiewes S, Garmyn D, Piveteau P, Guzzo J (2006) Assessment of the roles of LuxS, S-ribosyl homocysteine, and autoinducer 2 in cell attachment during biofilm formation by Listeria monocytogenes EGD-e. Appl Environ Microbiol 72:2644–2650

Cole SP, Harwood J, Lee R, She R, Guiney DG (2004) Characterization of monospecies biofilm formation by Helicobacter pylori. J Bacteriol 186:3124–3132

Connell JL, Wessel AK, Parsek MR, Ellington AD, Whiteley M, Shear JB (2010) Probing prokaryotic social behaviors with bacterial "lobster traps". MBio 1

Connell JL, Ritschdorff ET, Whiteley M, Shear JB (2013) 3D printing of microscopic bacterial communities. Proc Natl Acad Sci USA 110:18380–18385

Cramton SE, Gerke C, Schnell NF, Nichols WW, Gotz F (1999) The intercellular adhesion (ica) locus is present in Staphylococcus aureus and is required for biofilm formation. Infect Immun 67:5427–5433

Davies DG, Marques CN (2009) A fatty acid messenger is responsible for inducing dispersion in microbial biofilms. J Bacteriol 191:1393–1403

Davies DG, Parsek MR, Pearson JP, Iglewski BH, Costerton JW, Greenberg EP (1998) The involvement of cell-to-cell signals in the development of a bacterial biofilm [see comments]. Science 280:295–298

de Kievit TR (2009) Quorum sensing in Pseudomonas aeruginosa biofilms. Environ Microbiol 11:279–288

de Kievit TR, Gillis R, Marx S, Brown C, Iglewski BH (2001) Quorum-sensing genes in Pseudomonas aeruginosa biofilms: their role and expression patterns. Appl Environ Microbiol 67:1865–1873

Dow JM, Crossman L, Findlay K, He YQ, Feng JX, Tang JL (2003) Biofilm dispersal in Xanthomonas campestris is controlled by cell-cell signaling and is required for full virulence to plants. Proc Natl Acad Sci USA 100:10995–11000

Frommberger M, Schmitt-Kopplin P, Menzinger F, Albrecht V, Schmid M, Eberl L, Hartmann A, Kettrup A (2003) Analysis of N-acyl-L-homoserine lactones produced by Burkholderia cepacia with partial filling micellar electrokinetic chromatography–electrospray ionization-ion trap mass spectrometry. Electrophoresis 24:3067–3074

Gao M, Coggin A, Yagnik K, Teplitski M (2012) Role of specific quorum-sensing signals in the regulation of exopolysaccharide II production within Sinorhizobium meliloti spreading colonies. PLoS One 7:e42611

Garcia-Aljaro C, Melado-Rovira S, Milton DL, Blanch AR (2012) Quorum-sensing regulates biofilm formation in Vibrio scophthalmi. BMC Microbiol 12:287

Garcia-Contreras R, Maeda T, Wood TK (2013) Resistance to quorum-quenching compounds. Appl Environ Microbiol 79:6840–6846

Givskov M, DE Nys R, Manefield M, Gram L, Maximilien R, Eberl L, Molin S, Steinberg PD, Kjelleberg S (1996) Eukaryotic interference with homoserine lactone-mediated prokaryotic signalling. J Bacteriol 178:6618–6622

Hammer BK, Bassler BL (2003) Quorum sensing controls biofilm formation in Vibrio cholerae. Mol Microbiol 50:101–104

Haruta S, Kato S, Yamamoto K, Igarashi Y (2009) Intertwined interspecies relationships: approaches to untangle the microbial network. Environ Microbiol 11:2963–2969

Hentzer M, Wu H, Andersen JB, Riedel K, Rasmussen TB, Bagge N, Kumar N, Schembri MA, Song Z, Kristoffersen P, Manefield M, Costerton JW, Molin S, Eberl L, Steinberg P, Kjelleberg S, Hoiby N, Givskov M (2003) Attenuation of Pseudomonas aeruginosa virulence by quorum sensing inhibitors. EMBO J 22:3803–3815

Hogan DA, Vik A, Kolter R (2004) A Pseudomonas aeruginosa quorum-sensing molecule influences Candida albicans morphology. Mol Microbiol 54:1212–1223

Hsueh YH, Somers EB, Lereclus D, Wong AC (2006) Biofilm formation by Bacillus cereus is influenced by PlcR, a pleiotropic regulator. Appl Environ Microbiol 72:5089–5092

Huber B, Riedel K, Hentzer M, Heydorn A, Gotschlich A, Givskov M, Molin S, Eberl L (2001) The cep quorum-sensing system of Burkholderia cepacia H111 controls biofilm formation and swarming motility. Microbiology 147:2517–2528

Huber B, Riedel K, Kothe M, Givskov M, Molin S, Eberl L (2002) Genetic analysis of functions involved in the late stages of biofilm development in Burkholderia cepacia H111. Mol Microbiol 46:411–426

Jarosz LM, Deng DM, van der Mei HC, Crielaard W, Krom BP (2009) Streptococcus mutans competence-stimulating peptide inhibits Candida albicans hypha formation. Eukaryot Cell 8:1658–1664

Jarrett CO, Deak E, Isherwood KE, Oyston PC, Fischer ER, Whitney AR, Kobayashi SD, Deleo FR, Hinnebusch BJ (2004) Transmission of Yersinia pestis from an infectious biofilm in the flea vector. J Infect Dis 190:783–792

Koutsoudis MD, Tsaltas D, Minogue TD, Von Bodman SB (2006) Quorum-sensing regulation governs bacterial adhesion, biofilm development, and host colonization in Pantoea stewartii subspecies stewartii. Proc Natl Acad Sci USA 103:5983–5988

Labbate M, Queck SY, Koh KS, Rice SA, Givskov M, Kjelleberg S (2004) Quorum sensing-controlled biofilm development in Serratia liquefaciens MG1. J Bacteriol 186:692–698

Labbate M, Zhu H, Thung L, Bandara R, Larsen MR, Willcox MD, Givskov M, Rice SA, Kjelleberg S (2007) Quorum-sensing regulation of adhesion in Serratia marcescens MG1 is surface dependent. J Bacteriol 189:2702–2711

Lasarre B, Federle MJ (2013) Exploiting quorum sensing to confuse bacterial pathogens. Microbiol Mol Biol Rev 77:73–111

Lepine F, Deziel E (2011) Liquid chromatography/mass spectrometry for the detection and quantification of N-acyl-L-homoserine lactones and 4-hydroxy-2-alkylquinolines. Methods Mol Biol 692:61–69

Luo X, Wu HC, Tsao CY, Cheng Y, Betz J, Payne GF, Rubloff GW, Bentley WE (2012) Biofabrication of stratified biofilm mimics for observation and control of bacterial signaling. Biomaterials 33:5136–5143

Lynch MJ, Swift S, Kirke DF, Keevil CW, Dodd CE, Williams P (2002) The regulation of biofilm development by quorum sensing in Aeromonas hydrophila. Environ Microbiol 4:18–28

Mack D, Davies AP, Harris LG, Rohde H, Horstkotte MA, Knobloch JK (2007) Microbial interactions in Staphylococcus epidermidis biofilms. Anal Bioanal Chem 387:399–408

Mcdougald D, Rice SA, Kjelleberg S (2001) SmcR-dependent regulation of adaptive phenotypes in Vibrio vulnificus. J Bacteriol 183:758–762

Mcdougald D, LIN WH, Rice SA, Kjelleberg S (2006) The role of quorum sensing and the effect of environmental conditions on biofilm formation by strains of Vibrio vulnificus. Biofouling 22:133–144

Mcnab R, Ford SK, El-Sabaeny A, Barbieri B, Cook GS, Lamont RJ (2003) LuxS-based signaling in Streptococcus gordonii: autoinducer 2 controls carbohydrate metabolism and biofilm formation with Porphyromonas gingivalis. J Bacteriol 185:274–284

Merritt J, Qi F, Goodman SD, Anderson MH, Shi W (2003) Mutation of luxS affects biofilm formation in Streptococcus mutans. Infect Immun 71:1972–1979

Miller MB, Bassler BL (2001) Quorum sensing in bacteria. Annu Rev Microbiol 55:165–199

Novick RP, Geisinger E (2008) Quorum sensing in staphylococci. Annu Rev Genet 42:541–564

O'Loughlin CT, Miller LC, Siryaporn A, Drescher K, Semmelhack MF, Bassler BL (2013) A quorum-sensing inhibitor blocks Pseudomonas aeruginosa virulence and biofilm formation. Proc Natl Acad Sci USA 110:17981–17986

Otto M (2013) Staphylococcal infections: mechanisms of biofilm maturation and detachment as critical determinants of pathogenicity. Annu Rev Med 64:175–188

Parsek MR, Greenberg EP (2005) Sociomicrobiology: the connections between quorum sensing and biofilms. Trends Microbiol 13:27–33

Pearman WF, Lawrence-Snyder M, Angel SM, Decho AW (2007) Surface-enhanced Raman spectroscopy for in situ measurements of signaling molecules (autoinducers) relevant to bacteria quorum sensing. Appl Spectrosc 61:1295–1300

Pearson JP, Pesci EC, Iglewski BH (1997) Roles of Pseudomonas aeruginosa las and rhl quorum-sensing systems in control of elastase and rhamnolipid biosynthesis genes. J Bacteriol 179:5756–5767

Puskas A, Greenberg EP, Kaplan S, Schaefer AL (1997) A quorum-sensing system in the free-living photosynthetic bacterium Rhodobacter sphaeroides. J Bacteriol 179:7530–7537

Ramage G, Saville SP, Wickes BL, Lopez-Ribot JL (2002) Inhibition of Candida albicans biofilm formation by farnesol, a quorum-sensing molecule. Appl Environ Microbiol 68:5459–5463

Ravn L, Christensen AB, Molin S, Givskov M, Gram L (2001) Methods for detecting acylated homoserine lactones produced by Gram-negative bacteria and their application in studies of AHL-production kinetics. J Microbiol Methods 44:239–251

Reeser RJ, Medler RT, Billington SJ, Jost BH, Joens LA (2007) Characterization of Campylobacter jejuni biofilms under defined growth conditions. Appl Environ Microbiol 73:1908–1913

Rieu A, Weidmann S, Garmyn D, Piveteau P, Guzzo J (2007) Agr system of Listeria monocytogenes EGD-e: role in adherence and differential expression pattern. Appl Environ Microbiol 73:6125–6133

Ryan RP, Fouhy Y, Garcia BF, Watt SA, Niehaus K, Yang L, Tolker-Nielsen T, Dow JM (2008) Interspecies signalling via the Stenotrophomonas maltophilia diffusible signal factor influences biofilm formation and polymyxin tolerance in Pseudomonas aeruginosa. Mol Microbiol 68:75–86

Schaefer AL, Hanzelka BL, Parsek MR, Greenberg EP (2000) Detection, purification, and structural elucidation of the acylhomoserine lactone inducer of Vibrio fischeri luminescence and other related molecules. Methods Enzymol 305:288–301

Slater H, Alvarez-Morales A, Barber CE, Daniels MJ, Dow JM (2000) A two-component system involving an HD-GYP domain protein links cell-cell signalling to pathogenicity gene expression in Xanthomonas campestris. Mol Microbiol 38:986–1003

Sorroche FG, Rinaudi LV, Zorreguieta A, Giordano W (2010) EPS II-dependent autoaggregation of Sinorhizobium meliloti planktonic cells. Curr Microbiol 61:465–470

Stoodley P, Sauer K, Davies DG, Costerton JW (2002) Biofilms as complex differentiated communities. Annu Rev Microbiol 56:187–209

Sturme MH, Nakayama J, Molenaar D, Murakami Y, Kunugi R, Fujii T, Vaughan EE, Kleerebezem M, DE Vos WM (2005) An agr-like two-component regulatory system in Lactobacillus plantarum is involved in production of a novel cyclic peptide and regulation of adherence. J Bacteriol 187:5224–5235

Taga ME, Semmelhack JL, Bassler BL (2001) The LuxS-dependent autoinducer AI-2 controls the expression of an ABC transporter that functions in AI-2 uptake in Salmonella typhimurium. Mol Microbiol 42:777–793

Tomlin KL, Malott RJ, Ramage G, Storey DG, Sokol PA, CERI H (2005) Quorum-sensing mutations affect attachment and stability of Burkholderia cenocepacia biofilms. Appl Environ Microbiol 71:5208–5218

Vidal JE, Ludewick HP, Kunkel RM, Zahner D, Klugman KP (2011) The LuxS-dependent quorum-sensing system regulates early biofilm formation by Streptococcus pneumoniae strain D39. Infect Immun 79:4050–4060

Vilchez R, Lemme A, Ballhausen B, Thiel V, Schulz S, Jansen R, Sztajer H, Wagner-Dobler I (2010) Streptococcus mutans inhibits Candida albicans hyphal formation by the fatty acid signaling molecule trans-2-decenoic acid (SDSF). Chembiochem 11:1552–1562

von Bodman SB, Majerczak DR, Coplin DL (1998) A negative regulator mediates quorum-sensing control of exopolysaccharide production in Pantoea stewartii subsp. stewartii. Proc Natl Acad Sci USA 95:7687–7692

Wen ZT, Burne RA (2004) LuxS-mediated signaling in Streptococcus mutans is involved in regulation of acid and oxidative stress tolerance and biofilm formation. J Bacteriol 186:2682–2691

Williams P (2007) Quorum sensing, communication and cross-kingdom signalling in the bacterial world. Microbiology 153:3923–3938

Williams P, Winzer K, Chan WC, camara M (2007) Look who's talking: communication and quorum sensing in the bacterial world. Philos Trans R Soc Lond B Biol Sci 362:1119–1134

Wu H, Song Z, Hentzer M, Andersen JB, Molin S, Givskov M, Hoiby N (2004) Synthetic furanones inhibit quorum-sensing and enhance bacterial clearance in Pseudomonas aeruginosa lung infection in mice. J Antimicrob Chemother 53:1054–1061

Xu L, Li H, Vuong C, Vadyvaloo V, Wang J, Yao Y, Otto M, Gao Q (2006) Role of the luxS quorum-sensing system in biofilm formation and virulence of Staphylococcus epidermidis. Infect Immun 74:488–496

Yao Y, Vuong C, Kocianova S, Villaruz AE, Lai Y, Sturdevant DE, Otto M (2006) Characterization of the Staphylococcus epidermidis accessory-gene regulator response: quorum-sensing regulation of resistance to human innate host defense. J Infect Dis 193:841–848

Yarwood JM, Bartels DJ, Volper EM, Greenberg EP (2004) Quorum sensing in Staphylococcus aureus biofilms. J Bacteriol 186:1838–1850

Zhu J, Kaufmann GF (2013) Quo vadis quorum quenching? Curr Opin Pharmacol 13:688–698

Zhu J, Mekalanos JJ (2003) Quorum sensing-dependent biofilms enhance colonization in Vibrio cholerae. Dev Cell 5:647–656

Part II
Strategies for Biofilm Control

Current and Emergent Control Strategies for Medical Biofilms

Mohd Sajjad Ahmad Khan, Iqbal Ahmad, Mohammad Sajid, and Swaranjit Singh Cameotra

Abstract In nature, microorganisms prefer to live in structured microbial communities rather than as free-floating planktonic cells. These dynamic microbial communities are termed biofilms, in which transitions between planktonic and sessile modes of growth occur interchangeably in response to different environmental cues. Such phenomenas are advantageous for microbial pathogens but disadvantageous for human health. Due to the increased resistance/tolerance of biofilm cells to antimicrobial treatment, it becomes difficult to eradicate pathogens, which results in relapses of infections even after appropriate therapy. In clinically relevant biofilms, *Pseudomonas* spp., *Staphylococcus* spp., and *Candida* spp. are the most frequently isolated microorganisms. These microorganisms are able to adhere to and colonize surfaces of medical devices such as central venous catheters, intrauterine devices, voice prostheses, and prosthetic joints, resulting in the development of a biofilm. Many antimicrobial agents are now being used against microbial biofilms. However, inappropriate use of conventional antibiotic therapy may also contribute to inefficient biofilm control and to the dissemination of resistance. Consequently, new control strategies are constantly emerging to control biofilm-associated infections, such as the antifungal lock therapy, improved drug delivery, penetration of matrix-attacking extracellular polymetric substances, and regulation

---

M.S.A. Khan (✉)
Environmental Biotechnology and Microbial Biochemistry Laboratory, Institute of Microbial Technology, Chandigarh 160036, India

Department of Agricultural Microbiology, Aligarh Muslim University, Aligarh 202002, India
e-mail: khanmsa@hotmail.com; msajjadakhan@rediffmail.com

I. Ahmad
Department of Agricultural Microbiology, Aligarh Muslim University, Aligarh 202002, India
e-mail: ahmadiqbal8@yahoo.co.in

M. Sajid • S.S. Cameotra
Environmental Biotechnology and Microbial Biochemistry Laboratory, Institute of Microbial Technology, Chandigarh 160036, India
e-mail: sajid.zilli387@gmail.com; swaranjitsingh@yahoo.com

of biofilm inhibition/disruption by manipulating small molecules. The present chapter is focused on describing the clinical aspects of biofilm formation and deleterious effects associated with their presence. This chapter will highlight current and emergent control strategies for biofilms.

1 Introduction

While microbes are often thought to be multiplying and growing as free floating cells, most microbes live in aggregations and form complex structures termed biofilms. These organized structures are communities of microorganisms that form on solid or liquid interfaces and provide protection to individual cells by producing extracellular polymeric substances (EPS). The cells in the biofilms exhibit an altered phenotype compared with corresponding planktonic cells, especially in regard to gene transcription, and in interacting with each other (Donlan 2002; Hall-Stoodley et al. 2004). Biofilms result from a natural tendency of microbes to attach to biotic or abiotic surfaces. The formation of biofilms starts by irreversible attachment of microorganisms to a surface, which can vary from mineral surfaces and mammalian tissues to synthetic polymers and indwelling medical devices, followed by the production of extracellular substances by one or more of the attached microorganisms (Nikolaev and Plankunov 2007; Dongari-Bagtzoglou 2008).

Typically, most of the research on infectious microorganisms is conducted on single-celled (planktonic forms) of bacteria and fungi because of ease of study and manipulation. Consequently, most of the drugs developed have efficacy against planktonic forms of microbes, and unfortunately these drugs do not work or work poorly against the same organisms in their biofilm form. Moreover, the failure of antibacterial and antifungal drugs to combat such infections is due to the increased resistance and/or tolerance of the organisms in their biofilm state. The National Institute of Health estimates that biofilms cause more than 80 % of infections, which have imposed an enormous cost on human health (Sachachter 2003). Most infections on biomedical devices and mucosal surfaces, including oral and uro-genital tracts, are reportedly caused by the biofilm growth of *Escherichia coli*, *Pseudomonas aeruginosa*, *Staphylococcus aureus*, *Streptococcus pyrogens*, and *Candida albicans* (Donlan 2001; Wilson 2001; Douglas 2002).

Development of effective strategies to control or prevent biofilm-associated infections requires a thorough understanding of the biofilm development process (Jain et al. 2007). The adhesion of bacteria to a surface depends on a number of microbiological, physical, chemical, and material-related parameters. Biofilms may consist of mono or mixed species, are highly interactive, and employ a range of cell-to-cell communication or "quorum sensing" (QS) systems (Hogan 2006; Jayaraman and Wood 2008). This phenomenon for promoting collective behavior within a population is important for ensuring survival and propagation by enhancing access to nutrients and niches, as well as for providing protection (Nikolaev and

Plankunov 2007). The dense population structure in biofilms also increases the opportunity of gene transfer between the species which can convert a previously avirulent commensal organism into a highly virulent pathogen (Molin and Tolker-Nielson 2003). The enhanced efficiency of gene transfer in biofilms induces enhanced stabilization of the biofilm structure but, more importantly, also facilitates the spread of antibiotic resistance (Molin and Tolker-Nielson 2003; Wuertz and Hausner 2004). The increasing emergence of drug resistance to commonly used antibiotics and antifungals has increased the need for the identification of novel therapeutics and approaches. Therefore, understanding how antibiotic resistance develops is a prerequisite to the design of intervention strategies intended to minimize the threat of biofilm-associated infections. This chapter outlines our understanding and current state of knowledge of the nature of microbial biofilms in clinical context with emphasis on novel prophylactic and therapeutic strategies targeting prevention and management of biofilms.

2 Clinical Significance of Biofilms

It is estimated that the majority of clinical infections exist as biofilms rather than as planktonic cells. In medical settings, biofilms can occur in several places, such as the intestinal brush border (e.g., *Vibrio cholerae*), urethral lining (e.g., *Neisseria gonorrhoeae*), lymphoid patches in the intestine (e.g., *Salmonella typhimurium*) (Costerton et al. 1999), antibiotic-recalcitrant acne (Coates et al. 2003), chronically infected tonsils (Chole and Faddis 2003), cystic fibrosis (lungs) (Prince 2002), urinary and central venous catheters, and mechanical heart valves (Donlan 2002). A list of microorganisms causing infection due to biofilm growth on tissues or medical devices is given in Table 1. Intravascular administration of antibiotics is used to prevent surgical site and other infections, but the formation of a biofilm makes antibiotic therapy ineffective at eradicating the bacteria or fungi. The formation of biofilms with low sensitivity to antibiotics in the course of chronic infections, such as cystic fibrosis, is a matter of great concern (Il'ina et al. 2004).

A range of mucosal to systemic fungal infections have been reported to be caused by opportunistic pathogen *Candida* spp. such as oral candidiasis, vaginitis, and candidemia. Vulvovaginal infections are among the most common infections caused by *C. albicans*. Most women experience a vaginal *Candida* infection at some point in their lifetimes (Mardh et al. 2002). Oropharangeal candidiasis occurs most commonly in immunocompromised individuals, especially people infected with HIV and cancer patients (De Repentigny et al. 2004; Davies et al. 2006). Recent evidence suggests that the majority of such diseases produced by this pathogen are associated with biofilm growth (Ramage et al. 2005; Hasan et al. 2009; Dongari-Bagtzoglou et al. 2009).

The polymicrobial nature of oral biofilms associated with dental plaque and periodontitis has made them a pioneering model of interspecies interactions and highlights the level of complexity in biofilm research (Kuramitsu et al. 2007;

Table 1 Microorganisms that commonly cause biofilm-associated infection on tissues and indwelling medical devices (Donlan 2001; Wilson 2001; Donlan and Costerton 2002; Chuang et al. 2006; Kokare et al. 2009; Muller et al. 2011)

Microorganism	Sites of biofilm formation	
	Indwelling devices	Organs
Enterococcus spp.	Artificial hip prosthesis, central venous catheter, intrauterine device, prosthetic heart valve, urinary catheter	Intestinal tract
Coagulase-negative staphylococci	Artificial hip prosthesis, artificial voice prosthesis, central venous catheter, intrauterine device, prosthetic heart valve, urinary catheter, contact lenses	Skin, respiratory, gastrointestinal mucosa, middle ear
Klebsiella pneumoniae	Central venous catheter, urinary catheter	Pyogenic liver abscess, endopthalmitis
Pseudomonas aeruginosa	Artificial hip prosthesis, central venous catheter, urinary catheter, contact lenses	Lungs of cystic fibrosis patients, wounds, burns
Staphylococcus aureus	Artificial hip prosthesis, central venous catheter, intrauterine device, prosthetic heart valve	Skin wounds, burns
Streptococcus spp.	Endocarditis valve	Teeth
Lactobacillus sp.	Intrauterine devices	Vagina and teeth
Actinomyces sp.	Lenses	Teeth
Candida albicans	Artificial voice prosthesis, central venous catheter, intrauterine device, contact lenses	Vaginal mucosa, oral mucosa, nail bed
Aspergillus fumigatus	Endotracheal tubes, pace makers	Bronchial tract, lungs of cystic fibrosis patients

Shirtliff et al. 2009). Such a complex interaction can be seen in the biofilms formed between *C. albicans* and *S. epidermidis* (Fig. 1). A study conducted by Harriott and Noverr (2009, 2010) on polymicrobial versus monomicrobial biofilms suggested that *S. aureus* may become coated in the matrix secreted by *C. albicans*. The enhancement in *S. aureus* resistance to vancomycin within the polymicrobial biofilm required viable *C. albicans*, and this was in part facilitated by *C. albicans* matrix. However, the growth or sensitivity to amphotericin B (AMB) of *C. albicans* was not altered in the polymicrobial biofilm. Peters et al. (2010) reported that the pathogenicity of *S. aureus* was increased due to its interaction with *C. albicans* in a mixed biofilm.

Overall, biofilms are increasingly being recognized by the public health community as an important source of bacterial and fungal pathogens for all classes of patients, especially immunocompromised individuals and those with indwelling medical devices. Biofilm device infections can lead to significant morbidity and mortality, and may impair device function. Removal and/or replacement of devices are often the only treatment options, which can be very costly and also risky to the patients.

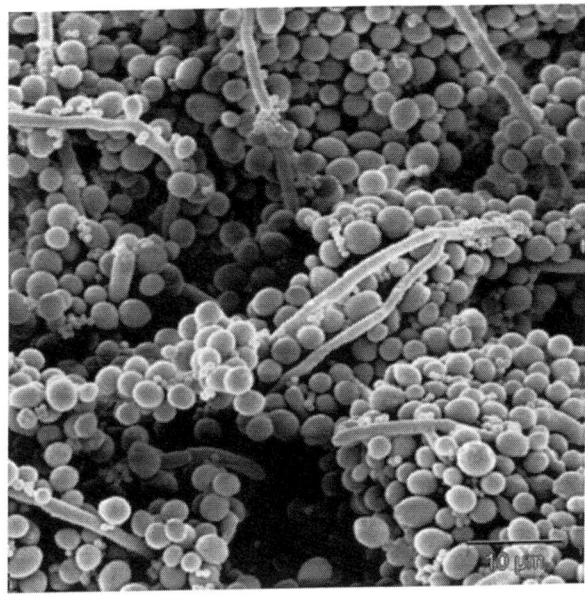

Fig. 1 Scanning electron micrograph of a mixed-species biofilm of *Candida albicans* and *Staphylococcus epidermidis*. Smaller bacterial cells can be seen adherent to both yeasts and hyphae (Shirtliff et al. 2009)

3 Biofilms: A Challenge for Antibiotic Therapy

The pathogenicity of biofilms is amplified by two of their major characteristics: (1) their increased tolerance to antimicrobials; (2) their protection of cells against the host's defense mechanisms. Overall, the combined action of different mechanisms is believed to contribute to increased resistance and tolerance in biofilms: slow growth; phenotypic variation and differential regulation of the cell metabolic activity caused by nutrient limitation, stress, and cell density; over-expression of resistance genes and amplified expression of efflux pumps; a changing sterol composition in the membrane; limited diffusion of antibiotics and immunological molecules through the extracellular matrix; and presence of persisters in the biofilm, which are able to tolerate high concentrations of antibiotics.

Microbes within biofilms are significantly more resistant to standard antibiotic therapy and may require up to 1,000 times the antibiotic dose to achieve efficacy (Davies 2003; Lewis 2005). Therefore, the doses of antibiotics used effectively against planktonic cells are usually not enough to tackle biofilms, leading to resistant subpopulations remaining in the biofilm and causing recurring infections. This has led to a more judicious approach for antibiotic use in order to limit further development of resistant strains. The emergence of biofilm-associated infections and rise in resistant strains has threatened the efficacy of current antimicrobial agents, and therefore newer antimicrobial tools and strategies are needed to combat such infections. Here, we have reviewed some of the novel and successful strategies and approaches being used to prevent and control the biofilm infections.

4 Novel Strategies to Combat Biofilm-Associated Infections

Efforts to develop successful treatments for biofilm-associated infections are urgently needed in clinical practice. These new strategies must take into account the differences in physiology and antibiotic/host defense susceptibility of biofilm embedded microorganisms. The genetic and phenotypic versatility of the cells within biofilms represent a challenge for discovering new methods of treatment and prevention of biofilm-associated infections. Biofilm penetration by biocides or antibiotics is typically strongly hindered. To increase the efficiency of new treatment strategies against bacterial and fungal infections, factors that lead to inhibition of biofilm growth, disruption, or eradication of biofilms are being sought (Francolini and Donelli 2010). These factors include microbial products, enzymes, sodium salts, metal nanoparticles, antibiotics, acids, chitosan and its derivatives, or plant products. All of these factors influence biofilm structure via various mechanisms and with different efficiencies.

4.1 Use of Combination Therapy

Conventional therapies target individual microbial species without consideration that most biofilms are polymicrobial. However, a careful attempt should be made to identify the causative microorganisms in a biofilm community. Appropriate management of mixed infections requires the administration of antimicrobials that are effective against all the components of the biofilms. Many nosocomial infections involve microbial biofilms and persistence of chronic infections is attributed to the persistence of polymicrobial biofilms (Brogden and Guthmiller 2002; Hall-Stoodley and Stoodley 2009). The standard treatment for such infections involves two or more antibiotics, referred to as combination therapy (Brook 2002). The use of novel antibiotic combinations may increase the effectiveness of antibiotic therapies.

Another potential strategy could be to sensitize the bacteria or fungal biofilms by synthetic or natural compounds (other than antibiotics). For example, Jabra-Rizk et al. (2006) reported the sensitization of *S. aureus* biofilms by farnesol, a fungal QS molecule. The combined effect of gentamicin at 2.5 times the MIC and farnesol at 100 µM (22 µg/mL) was able to reduce bacterial populations by more than 2 log units and demonstrated a synergy between the two agents. This observed sensitization of resistant strains to antimicrobials and the observed synergistic effect with gentamicin indicates a potential application for farnesol as an adjuvant therapeutic agent for the prevention of biofilm-related infections. Using a combination approach we have demonstrated that the phenolic compounds eugenol and phenyl aldehyde cinnamdehyde potentiate the activity of flucoanzole against biofilm forming drug-resistant strains of *C. albicans* (Khan and Ahmad 2012a).

4.2 Prevention Against Catheter-Related Blood Stream Infections

Biofilms play a pivotal role in healthcare-associated infections, especially those related to the implantation of medical devices, such as intravascular catheters, urinary catheters, and orthopaedic implants. Implants act as passive surfaces prone to bacterial adhesion and biofilm formation. This tendency can result in implant-associated infection of the surgical site. In spinal surgery, implant-associated deep body infections are still a major problem (Trampuz and Widmer 2006). Some bacteria produce slime, which is responsible for bacterial adhesion and formation of biofilms on artificial surfaces. This slime is composed of proteins, hexosamines, neutral sugars, and phosphorus-containing compounds. If slime-forming bacteria colonize an artificial surface and develop a biofilm, this layer protects the bacteria from antibiotic agents. Thus, treatment against implant-associated infection must target the development of a biofilm (Secinti et al. 2011).

The most successful approaches for the control and prevention of infections due to adhesion, colonization, and biofilm formation on medical devices have been described in a review article by Francolini and Donelli (2010). Readers are suggested to go through this article for more detailed strategies currently in use for preventing biofilm formation on medical implants. In this chapter we will be reviewing developments in novel strategies to prevent biofilm infection of implants and tissues.

4.2.1 Lock Therapy Approach

Nosocomial infections associated with medical devices represent a large proportion of all cases of hospital-acquired infections (Bell 2001). In particular, insertion of any vascular catheter can result in a catheter-related infection, as microorganisms can colonize external and internal catheter surfaces. Adherence to the catheter surface is facilitated by host proteins such as fibronectin and fibrinogen, which can then lead to biofilm formation (Christner et al. 2010). Such problems can be overcome by one of the approaches termed lock therapy. This approach is currently recommended and employed in treating catheter-related bloodstream infections (CRBSI), in particular for long-term catheters, according to the Infectious Diseases Society of America's guidelines (Mermel et al. 2009). The choice of antibiotics used in the lock technique is dependent on the pathogen suspected of infecting the catheter lumen, characteristics of the organism (i.e., ability to produce slime, adherence to host proteins), and the pharmacodynamic properties of the antimicrobial agent. Lock therapy involves the coating of high doses of an antimicrobial agent [from 100- to 1,000-fold the minimal inhibitory concentration, (MIC)] directly into the catheter in order to "lock" it for a certain period of time (from hours to days) (Carratala 2002). If host proteins such as fibronectin, fibrinogen, and fibrin are present in the catheter lumen, heparin may increase the efficacy of the antibiotics. Liposomal AMB and echinocandisn have been used successfully in a rabbit model of *C. albicans* biofilm infection (Schinabeck et al. 2004; Donlan 2008). While these results are promising for potential use of the lock technique to treat infected catheters, 100 % biofilm inhibition could not be achieved (Tournu and Dijck 2012). Synergistic antibiofilm combinations, between classical antimicrobial agents and other compounds such as the mucolytic agent *N*-acetylcysteine, ethanol, or the chelating agent EDTA, are being used as lock solutions and appeared to be very effective against *S. epidermidis* and *C. albicans* individual and mixed biofilms (Venkatesh et al. 2009). In a similar approach, recent findings suggest that the combination of antibacterial agents with Gram-positive activity, including doxycycline and tigecycline, with known antifungals, such as AMB, caspofungin, and fluconazole, can be useful for the treatment of *C. albicans* biofilms (Miceli et al. 2009; Ku et al. 2010).

The prevention of CRBSI has also been the focus of research and randomized controlled trials. The clinical effectiveness of central venous catheters (CVCs) treated with anti-infective agents (AI-CVC) in preventing CRBSI has been shown by Hockenhull et al. (2009). Antifungal impregnated CVCs have also been tested in animal models. Caspofungin was employed to prevent *C. albicans* biofilm formation in a murine biofilm model. *C. albicans* biofilm formation was reported to be greatly reduced in CVCs that had been pretreated for 24 h with high doses of caspofungin (Lazzell et al. 2009). The antibiofilm potential of liposomal AMB as a lock solution to inhibit *C. albicans*, *Candida glabrata*, and *Candida parapsilosis* biofilms in vitro has been reported by Toulet et al. (2012). Thus, the use of the lock technique or preventive impregnation of antifungals in combating catheter-associated infection seems promising, but not yet convincing from a cost-effective point of view, as huge doses are still needed to eradicate microbial growth.

4.2.2 Material Coatings and Novel Antibiofilm Surfaces

Among the promising approaches to combat biofilm infections is the generation of surface modification of devices to reduce microbial attachment and biofilm development. Typically, this strategy uses the incorporation of antimicrobial agents to prevent colonization (Smith 2005). Implanted materials are prone to biofilm formation affecting health in general and duration of the implant in particular. Surface characteristics, such as roughness, free energy, and chemistry, can influence the type and the feature of the biofilms (Teughels et al. 2006). For example, *C. albicans* adhesion is enhanced if the roughness of denture materials is increased (Radford et al. 1998). Currently, coatings may be engineered to promote selective adhesion to cells or tissue in bone implants but not to microbes. They may also address the second phase of biofilm development involving QS, by inhibiting cell–cell communication signals (Bruellhoff et al. 2010; Xiong and Liu 2010). Biomaterial modifications are a way to prevent biofilm development and have been the focus of intense research. While most research has focused on bacterial biofilms, the efficacy of biomaterial modifications also appears to inhibit *Candida* biofilms (Tournu and Dijck 2012).

Surface Modifications

The surface properties of medical devices constitute a major factor contributing not only to their stability in the body but also to their performance and lifetime in vivo and their colonization by microorganisms. Accordingly, albumin adhesion to surfaces is potentially beneficial since it has been shown to prevent binding of microorganisms, while fibrinogen has the opposite effect (Anderson et al. 2008). Chemical grafting of polyethylene and polypropylene surfaces with functionalized cyclodextrins changes the protein adsorption profile of these polymers by promoting adsorption of albumin and reducing the adhesion of fibrinogen to the material surface (Nava-Ortiz et al. 2010). These modified substrates were able to incorporate the antifungal agent miconazole very well and retarded biofilm formation by *C. albicans*. Modified polyethylene and silicone rubbers proved to be very efficient in inhibiting *C. albicans* biofilm formation (Contreras-Garcia et al. 2011). These materials are cytocompatible and also capable of releasing considerable amounts of nalidixic acid for several hours. This may further potentiate efficacy of treated surfaces to prevent formation of biofilms.

Biofilms on voice prostheses consist of mixed populations. Modification of the silicone surface of the prostheses has been employed to limit *C. albicans* colonization, as opposed to incorporation of antimicrobial agents in order to avoid the occurrence of resistance (De Prijck et al. 2010a). Silicone disks grafted with C1 and C8 alkyl side chains demonstrated reduced microbial adherence and inhibited biofilm formation by *C. albicans* by up to 92 %. Similarly, grafting of silicone rubber with cationic peptides, such as the salivary peptide Hst5 and synthetic variants, inhibited biofilm formation by up to 93 %, in a peptide-dependent manner (De Prijck et al. 2010b).

Preconditioning surfaces with surfactants also has potential to prevent bacterial adhesion and inhibit formation of biofilms. Splendiani et al. (2006) screened 22 surfactants for their potential to increase the cell wall charge of a *Burkholderia* sp. strain and reduce the ability to attach and form biofilms. The authors demonstrated that some surfactants affected the development of flagella, demonstrating significant changes in the ability of bacteria to attach in the presence of the surfactant. In addition to surfactants, biosurfactants synthesized by microbes have also been used as coating agents for medical implants leading to a reduction in hospital infections caused by biofilm growth (Rodrigues et al. 2006).

Surface Coatings

Microbicidal or static materials have been employed to fabricate or coat the surfaces of medical devices and have a great potential in reducing or eliminating the incidence of biofilm-related infections. Studies have reported the use of several compounds and synthetic analogues to prevent biofilm formation such as farnesol, quaternary ammonium salts, and silver ions, which were shown to effectively inhibit both bacterial and fungal biofilm formation (Gottenbos et al. 2001; Hashimoto 2001; Jabra-Rizk et al. 2006; Shirtliff et al. 2009). One particular study highlighted the role of two quaternary ammonium silanes (QAS) to coat silicone rubber tracheoesophageal shunt prostheses, yielding a positively charged surface. One QAS coating [(trimethoxysilyl)-propyldimethyloctadecylammonium chloride] was applied through chemical bonding, while the other coating, Biocidal ZF, was sprayed onto the silicone rubber surface. This was the first report on the inhibitory effects of positively charged coatings of tracheoesophageal shunt prostheses on the viability of yeasts and bacteria in mixed biofilms (Oosterhof et al. 2006). Although the study initially aimed at reducing voice prosthetic biofilms, its relevance extends to all biomedical surfaces where mixed biofilms develop and become problematic. Similarly, dental resin material coated with thin-film polymer formulations containing the polyene antifungal nystatin, AMB, or the antiseptic agent chlorhexidine were used in *C. albicans* and mixed biofilm prevention (Redding et al. 2009).

The polysaccharide dextran is widely used in medicine and is also one of the main components of dental plaque. Cross-linked dextran disks soaked with AMB solutions, described as amphogel, killed fungi within 2 h of contact and could be reused for almost 2 months without losing their efficacy against *C. albicans* (Zumbuehl et al. 2007). This antifungal material is biocompatible and could be used to coat medical devices to prevent microbial attachment.

Another option is to coat biomaterial surfaces with organic molecules to prevent protein adsorption which may also inhibit biofilm formation (Njoroge and Sperandio 2009). Coating of medical material surfaces has been employed and tested with several types of coating molecules, including the naturally occurring polymer chitosan and antimicrobial peptides such as Histatin 5 (Hst5). Histatins, a family of histidine-rich cationic peptides, are secreted by the major salivary glands in humans, especially histatin 5, which possess significant antifungal properties. A recent study demonstrated that histatin 5 exhibited antifungal activity against *C. albicans* biofilms and to a lesser extent against *C. glabrata* biofilms developed on denture acrylic (Konopka et al. 2010).

Naturally occurring antimicrobial peptides are promising therapeutic agents against pathogens such as *C. albicans*. But they are difficult and expensive to produce in large quantities and are also often sensitive to protease digestion. Therefore, their development as coating agents has been hampered. The search for new and improved antimicrobial peptides has led to the study of peptide mimetics. Synthetic analogs that mimic the properties of these peptides have many advantages and exhibit potent and selective antimicrobial activity (Tew et al. 2002). New classes of antimicrobial peptides were designed to mimic transmembrane segments of integral membrane proteins and were tagged with lysine residues to facilitate solubilization in aqueous media. These peptides, designated kaxins, have a non-amphipathic hydrophobic core segment, which distinguishes them from many natural linear cationic antimicrobial peptides. With this peptide

Stark et al. (2002) showed that placing all of the K residues on the N-terminus and generating all-D enantiomeric versions, in combination with decreasing the length of the hydrophobic segment, resulted in shorter peptides that generally displayed increased antimicrobial activity. Generation of these shorter peptides is cost-effective and has shown potential applications in surface coatings. Karlsson and coworkers showed in 2009 that β-peptides (β-amino acid oligomers), at a concentration near the MIC, completely inhibited *C. albicans* planktonic cells from forming a biofilm by a toxicity mechanism involving membrane disruption. The same group reported in 2010 that fabrication of multilayered polyelectrolyte thin films promoted the surface-mediated release of an antifungal β-peptide. These films inhibited the growth of *C. albicans* on film-coated surfaces. In addition, β-peptide-containing films inhibited hyphal elongation by 55 %. This approach could ultimately be used to coat the surfaces of catheters, surgical instruments, and other devices to inhibit drug-resistant *C. albicans* biofilm formation in clinical settings (Karlsson et al. 2010). The utility and potential of selected peptides as therapeutic molecules, including the β-glucan synthesis inhibitors, the histidine-rich peptides, and the LL-37 cathelicidin family, are being determined and could be used as coating compounds against adherence and biofilm formation (Matejuk et al. 2010; Tsai et al. 2011).

Chitosan, a polymer isolated from crustacean exoskeletons, recently proved to be active against *Candida* biofilms in vitro. Surfaces coated with chitosan reduced the viable cell number in biofilms by more than 95 % in the case of *C. albicans* and also for many bacteria such as *S. aureus* (Carlson et al. 2008). Chitosan is a hydrophilic biopolymer that is industrially obtained by means of *N*-deacetylation of crustacean chitin. It is active against a wide range of pathogenic microbes including fungi, bacteria, and viruses (Rabea et al. 2003) by disrupting cell membranes as cells settle on to its surface. The use of such polymers offers a biocompatible tool for coating medical devices. Chitosan has been used to pretreat catheters and prevent *C. albicans* biofilm formation as validated in an in vivo CVC biofilm model by Martinez et al. (2010). The investigators demonstrated that mature *C. albicans* and *C. parapsilosis* biofilms were susceptible to chitosan in vitro. Chitosan decreased the metabolic activity and survival of *Candida* species biofilms, with more than 95 % killing of the sessile cells after 0.5 h treatment with 2.5 mg/mL chitosan.

Use of Nanoparticles

Nanotechnology is providing new ways to manipulate the structure and chemistry of surfaces to inhibit bacterial colonization. It is a new discipline with many applications in biological sciences and medicine, and is discussed in detail in other chapters included in this book. Nanomaterials are applied as coating materials, as well as in treatment and diagnosis (Colvin 2003). The advantages of nanoparticles are their high surface-to-volume ratios, quantum confinement, and nanoscale sizes. These properties allow more active sites of nanoparticles to interact with biological systems, including bacteria and fungi. This is the most important difference between nanoparticles and typical antimicrobial agents and could minimize the risk of developing antimicrobial resistance (Hernandez-Delgadillo et al. 2012). The mechanism of antimicrobial activity for nanoparticles is not completely understood. However, the positive charge of metal ions is known to be critical for antimicrobial activity, because it allows for their electrostatic attraction with the negative charge of the bacterial cell membrane. It has been reported that silver nanoparticles can damage DNA, alter gene expression, and affect membrane-bound respiratory enzymes (Kim et al. 2007). Nanoparticles of titanium, silver, copper oxide, selenium diamond, iron oxide, carbon nanotubes, and biodegradable polymers have also been studied for their use in diagnosis and treatment and their reported antimicrobial activities are summarized below.

As shown in Fig. 2, incorporation of cross-linked quaternary ammonium polyethylenimine (QPEI) nanoparticles in dental resin composite at a low concentration exerted a significant in vivo antibiofilm activity and potent broad spectrum antibacterial activity against salivary bacteria (Beyth et al. 2010). The antibacterial and antifungal effects of silver ions have long been known, and silver seems to inhibit biofilm formation by *S. epidermidis*, *P. aeruginosa*, and *Candida* spp. as evident from various studies (Kalishwaralal et al. 2010; Secinti et al. 2011; Monteiro et al. 2011a, b). These findings highlighted that nanoparticle silver ion-coated titanium implants are safe and provide a means to treat *Candida*-associated denture stomatitis. Recently, inhibition of biofilm formation by a *S. aureus* clinical isolate, with silver nanoparticle-coated catheters, was reported by Namasivayam et al. (2012). They also found that these naonoparticles exhibited synergistic effects to eradicate biofilms with the antibiotics ofloxacin, cephalexin, and neoflaxin.

In another study, glass slides coated with zinc oxide (ZnO) nanoparticles restricted the biofilm formation of common bacterial pathogens. The generation of hydroxyl radicals, originating from the coated surface, was found to play a key role in antibiofilm activity (Applerot et al. 2012). Functionalized magnetite (Fe_3O_4/C18) nanoparticles have the potential to improve the antibiofilm properties of textile dressings against *C. albicans* biofilms (Anghel et al. 2012). In addition, these functionalized surface-based approaches are very useful in the prevention of wound microbial contamination and subsequent biofilm development on viable tissues or implanted devices. Recently, zerovalent stable colloidal bismuth nanoparticles were shown to possess antimicrobial activity against *S. mutans* and *C. albicans* growth and completely inhibited their biofilm formation. The results are similar to those obtained with chlorhexidine, the most commonly used oral antiseptic agent, and suggest that zerovalent bismuth nanoparticles could be an interesting antimicrobial agent to incorporate into an oral antiseptic preparation (Hernandez-Delgadillo et al. 2012, 2013a, b).

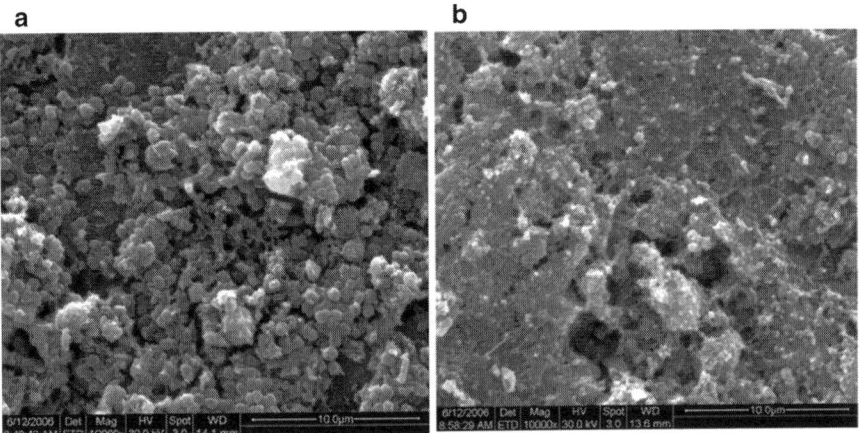

Fig. 2 Biofilms formed on resin composite incorporating QPEI nanoparticles and on nonmodified resin composite. Scanning electron micrographs (×10,000) of biofilms formed on resin composite (**a**) and resin composite with incorporated QPEI nanoparticles (**b**) (Beyth et al. 2010)

4.3 Disruption of Biofilms

Since biofilms must release and disperse cells into the environment in order to colonize new sites (Kaplan 2010), biofilm dispersal is another promising area of research that may lead to the development of novel agents to promote biofilm cell detachment. Furthermore, since the biofilm matrix also contains polysaccharides and DNA, a promising strategy could be the use of enzymes (e.g., DNase and alginate lyase) that can disrupt and dissolve biofilms by attacking surface polysaccharides and the extracellular DNA which is critical for the early development of biofilms (Arciola 2009; Taraszkiewicz et al. 2013).

Xavier et al. (2005) proposed a kinetic model to assess the feasibility of strategies for the removal of biofilms by using substances that induce detachment by affecting the cohesiveness of the EPS. Detachment-promoting agents are enzymes, chelating agents, or any other agents that reduce EPS cohesiveness through a variety of mechanisms. Promoting detachment is the least investigated of the possible strategies to remove unwanted biofilms. However, the use of substances to induce biofilm removal directly by destroying the physical integrity of the biofilm matrix would be an attractive alternative for both medical and industrial applications where complete biofilm removal is essential. This approach could also overcome the problem of recalcitrant infections of biofilms due to persister cells.

4.3.1 Biofilm-Disrupting Enzymes

The two most well-studied biofilm-dispersing enzymes are deoxyribonuclease I (DNase I) and dispersin B (DspB) (Kaplan 2009), but other extracellular enzymes have also been explored as antibiofilm agents.

Deoxyribonuclease I. Deoxyribonuclease I (DNase I) degrades extracellular DNA (eDNA), a newly highlighted structural component of biofilms that confers firmness and stability. Tetz et al. (2009) reported a strong negative impact of DNase I on the structures of biofilms formed by *Acinetobacter baumannii*, *Haemophilus influenzae*, *K. pneumoniae*, *E. coli*, *P. aeruginosa*, *S. aureus*, and *S. pyogenes*. Using DNase I at a concentration of 10 µg/mL, they observed degradation of mature 24 h-old biofilms by 53.85, 52.83, 50.24, 53.61, 51.64, 47.65, and 49.52 %, respectively. Moreover, bacterial susceptibility to selected antibiotics (azithromycin, rifampin, levofloxacin, ampicillin, and cefotaxime) increased in the presence of 5 µg/mL DNase I.

The antibiofilm activity of DNase I (130 µg/mL) in combination with selected antibiotics toward *C. albicans* biofilms has been estimated by Martins et al. (2012). In their study, reduction of viable counts by 0.5 log10 units was observed for *C. albicans* biofilms incubated with DNase I. Treatment of *C. albicans* with AMB alone (1 µg/mL) resulted in a 1 log10 unit reduction in cell viability, which increased to 3.5 log10 units in combination with DNase I. At higher concentrations of AMB (>2 µg/mL) and DNase I, cell viability was reduced by 5 log10 units.

DispersinB. DispersinB is a naturally occurring *N*-acetylglucosaminidase enzyme produced by a periodontal disease-associated oral bacterium, *Aggregatibacter actinomycetemcomitans*. This 41 kDa enzyme consists of a single chain containing 361 amino acid residues and is a highly active and stable glycoside hydrolase that functions in a narrow pH range. DispersinB specifically hydrolyses the glycosidic linkages of poly-β-1, 6-*N*-acetylglucosamine in the polysaccharide adhesins of bacteria, which are needed for biofilm formation, and are present in the polysaccharide matrix of mature biofilms, without affecting bacterial growth (Itoh et al. 2005). Thus, it inhibits as well as disperses bacterial biofilms and has been reported to be active against biofilms produced by various organisms such as *E. coli*, *S. aureus*, *S. epidermidis*, and *P. fluorescence* (Itoh et al. 2005; Rohde et al. 2007).

Lysostaphin. Lysostaphin is a natural staphylococcal endopeptidase that can penetrate bacterial biofilms (Belyansky et al. 2011). Promising antibiofilm results have been obtained for lysostaphin. The antimicrobial properties of lysostaphin were analyzed by Walencka et al. (2005), who reported the biofilm inhibitory concentration (BIC) of the enzyme for various *S. aureus* and *S. epidermidis* clinical strains. In addition, the combined use of lysostaphin with oxacillin resulted in increased susceptibility of the biofilm-growing bacteria to the antibiotic. Likewise, Aguinaga et al. (2011) reported a synergistic effect of lysostaphin in combination with doxycycline leading to significantly increased antibiotic susceptibility against methicillin-resistant *S. aureus* (MRSA) and methicillin-sensitive *S. aureus* (MSSA) strains.

Other Enzymes. Alkawash et al. (2006) showed application for lyase in the destruction of biofilms made by two mucoid *P. aeruginosa* strains. Treatment of the biofilms with gentamycin (64 µg/mL) in combination with alginate lyase (20 U/mL) resulted in biofilm matrix liquefaction. Incubation of the biofilm with lyase and gentamycin for 96 h resulted in the complete eradication of the biofilm structure and living bacteria. The antibiofilm activity of α-amylases against strains of *S. aureus* was analyzed by Craigen et al. (2011). This enzyme effectively reduced biofilm formation in the case of *S. aureus*. Time-course experiments for *S. aureus* showed that biofilms were degraded by 79 % within 5 min and by 89 % within 30 min of incubation with α-amylases. Amylase at doses of 10, 20, and 100 mg/mL reduced biofilms by 72, 89, and 90 %, respectively, and inhibited matrix formation by 82 %. In addition, they also investigated antibiofilm activities of amylases from different biological sources. The most effective biofilm reduction was reported for the α-amylase isolated from *Bacillus subtilis*. Although enzymes derived from human saliva and sweet potato had no effect against preformed biofilms, all of the tested enzymes, regardless of origin, were highly effective in inhibiting biofilm formation. Lactonase was also identified as a potential antibiofilm agent. Kiran et al. (2011) showed that biofilms formed by *P. aeruginosa* strains exhibited growth inhibition of 68.8–76.8 % in the presence of the enzyme (1 U/mL). They also found that 0.3 U/mL of the enzyme disrupted the biofilm structure and led to increased ciprofloxacin and gentamycin penetration and antimicrobial activity.

4.3.2 Photodynamic Therapy

Another innovative approach to disrupt biofilms is to expose them to photodynamic substances (Njoroge and Sperandio 2009). Antimicrobial Photodynamic Therapy (APDT) consists of three major components: light, a chemical molecule known as a photosensitizer, and oxygen. Photodynamic therapy (PDT) is based on the concept that a certain nontoxic photoactivatable compound or photosensitizer (PS) can be preferentially localized in certain tissues. These photosensitizers can be excited by absorbing a certain amount of energy from light of the appropriate wavelength. After excitation, photosensitizers usually form a long-lived triplet-excited state that will then generate reactive oxygen species (ROS), such as singlet oxygen and superoxide from which energy can be transferred to biomolecules or directly to molecular oxygen, depending on the reaction type. This results in oxidation of biomolecules, especially proteins involved in transport and membrane structure, in microorganisms leading to cell damage and death (Hamblin and Hasan 2004). Recent studies have shown that antimicrobial effects can be obtained with the use of photosensitizers belonging to different chemical groups such as phenothiazine dyes [methylene blue (MB) and toluidine blue O (TBO)]; porphyrin and its derivatives, TMPyP (5-,10-,15-,20-tetrakis (1-methylpyridinium-4-yl)-porphyrin), tetra *p*-toluenesulfonate; fullerenes; and cyanines and its derivatives (Taraszkiewicz et al. 2013).

The phenothiazinium salts are most commonly used in the clinic, and combinations of MB or TBO together with red light are used to disinfect blood products, sterilize dental cavities and root canals, and treat periodontitis (Wainwright 2003) and also have been actively investigated for the eradication of bacterial biofilm growing on dental plaques and oral implants (Saino et al. 2010). Tri-meso (*N*-methyl-pyridyl), meso (*N*-tetradecyl-pyridyl) porphine (C14) was exploited for inactivation of two structurally distinct *S. epidermidis* biofilms grown on Ti6Al4V alloy by Saino et al. (2010). They also compared its photosensitizing efficiency with that of the parent molecule, tetra-substituted *N*-methyl-pyridyl-porphine (C1). Their data suggested that C14 is a potential photosensitizer for the inactivation of staphylococcal biofilms for many device-related infections which are accessible to visible light.

Kishen et al. (2010) evaluated the ability of a cationic, phenothiazinium photosensitizer, methylene blue (MB), and an anionic, xanthene photosensitizer, rose bengal (RB), to inactivate and disrupt biofilms produced by *E. faecalis* (OGIRF and FA 2-2). The role of a specific microbial efflux pump inhibitor (EPI), verapamil hydrochloride in the MB-mediated antimicrobial photodynamic inactivation (aPDI) of *E. faecalis* biofilms, was also investigated. Their results showed that APDT with cationic MB produced superior inactivation of *E. faecalis* strains in a biofilm along with significant destruction of the biofilm structure when compared to anionic RB ($P < 0.05$). The ability to inactivate biofilm bacteria was further enhanced when the EPI was used with MB ($P < 0.001$). These experiments demonstrated the advantage of a cationic phenothiazinium photosensitizer combined with an EPI to inactivate biofilm bacteria and disrupt biofilm structure as shown in Fig. 3.

Collins et al. (2010) studied the effect of TMP on *P. aeruginosa* biofilms. In their study, a significant decrease in biofilm density was observed, and the majority of the cells within the biofilm were nonviable when 100 μM TMP and 10 min of irradiation (mercury vapor lamp, 220–240 J/cm^2) were used. Moreover, the use of 225 μM TMP and the same light dose resulted in almost complete disruption and clearance of the biofilm. Biel et al. (2011a, b) demonstrated that MB-mediated APDT was highly effective in the photo-eradication of multispecies bacterial biofilms (multidrug-resistant *P. aeruginosa* and MRSA). They observed a significant decrease in CFU/mL (>6 log10 units) when 300 μg/mL MB and a light dose of 60 J/cm^2 (diode laser, 664 nm) were used. The reduction was >7 log10 units when 500 μg/mL MB and two light doses of 55 J/cm^2 separated by a 5-min break were used. Recently, Meire et al. (2012) observed a statistically significant 1.9 log10 reduction in the viable counts of *E. faecalis* biofilms treated with 10 mg/mL MB and exposed to a soft laser at an output power of 75 mW (660 nm) for 2 min.

Since APDT represents an alternative method of killing resistant pathogens, efforts have been made to develop delivery systems for hydrophobic drugs to improve the photokilling. In this regard a study was conducted by Ribeiro et al. (2013) to evaluate the photodynamic effect of chloro-aluminum phthalocyanine (ClAlPc) encapsulated in nanoemulsions (NE) on MRSA and MSSA

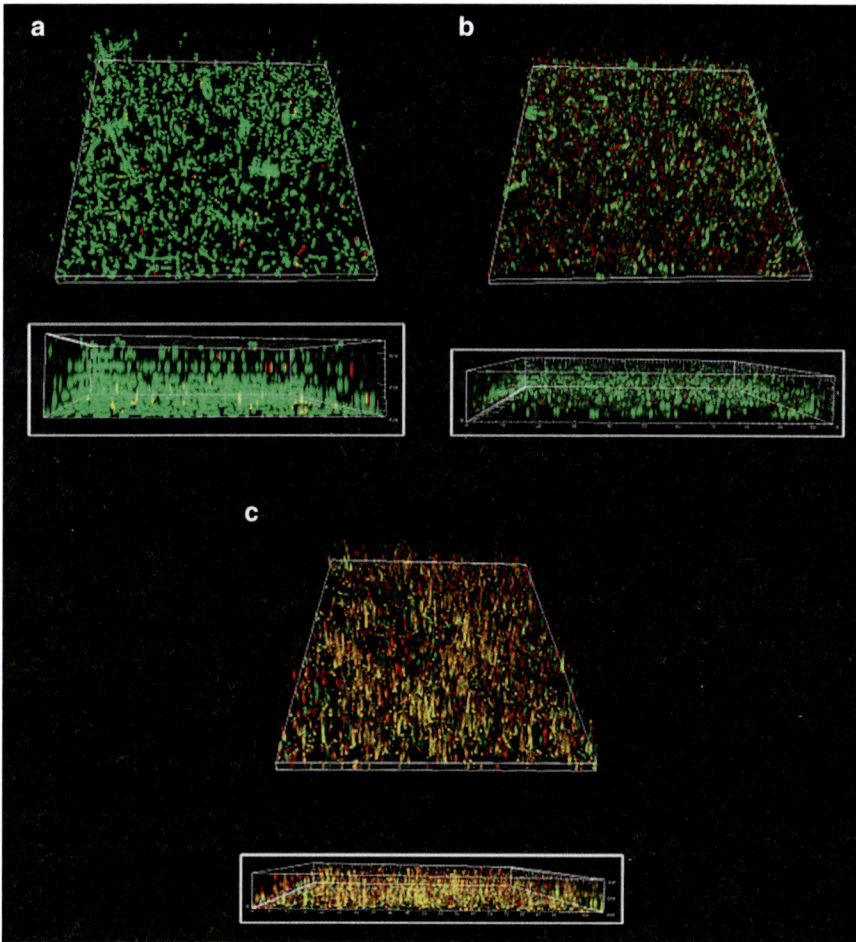

Fig. 3 The three-dimensional CLSM reconstruction of *E. faecalis* biofilms subjected to aPDI (*inset* shows the sagittal section). (**a**) The untreated biofilm. (**b**) The biofilm incubated with 100 μM RB followed by irradiation at 40 J/cm^2. (**c**) The biofilm incubated with 100 μM MB followed by irradiation at 40 J/cm^2 (The *colors* represent: *green* viable, *red* dead, *yellow* intermediate) (Kishen et al. 2010)

suspensions and biofilms. Suspensions and biofilms were treated with different delivery systems containing ClAlPc. For biofilms, cationic NE-ClAlPc reduced cell metabolism by 80 and 73 % of susceptible and resistant strains, respectively. Although anionic NE-ClAlPc caused a significant CFU/mL reduction for MSSA and MRSA, it was not capable of reducing MRSA biofilm metabolism. Moreover, a very recent study has shown improved efficacy of twofold positively charged porphyrin (XF-73) in comparison to fourfold positively charged porphyrin [5,10,15,20-tetrakis(1-methyl-4-pyridyl)-21H,23H-porphine, tetra-*p*-tosylate salt] against *C. albicans* planktonic cells and biofilms (Gonzales et al. 2013). Overall, this therapy is a very effective means of biofilm disruption and may represent an alternative treatment for eradicating resistant strains.

4.4 Biofilm Control Through Microbial Interactions or Interference

The existence of multiple interspecies interactions or the simple production of a metabolite can interfere with biofilm formation and development (Rossland et al. 2005; Valle et al. 2006). Competition for substrates is considered to be one of the major evolutionary driving forces in the bacterial world, and numerous experimental data obtained in the laboratory, under controlled conditions, have shown how different microorganisms may effectively outcompete others because they are better able to utilize a given energy source (Simoes et al. 2007). *P. aeruginosa* and *Candida* in a dual species environment mutually suppress biofilm development, both quantitatively and qualitatively (Bandara et al. 2010). In their study, Bandara et al. found that *P. aeruginosa* attached to *C. albicans* hyphae in a mixed-species biofilm and killed the fungi, whereas the yeast forms could not be killed. Isolation and purification of microbial compounds that mediate these types of interactions could lead to the development of new antibiofilm agents.

Commensal bacteria are known to inhibit pathogen colonization; however, complex host–microbe and microbe–microbe interactions have made it difficult to gain a detailed understanding of the mechanisms involved in the inhibition of colonization (Wertheim et al. 2005). In an attempt to understand these relationships, Iwase et al. (2010) found that the serine protease Esp, secreted by a subset of *S. epidermidis*, which are commensals, inhibited biofilm formation and nasal colonization by *S. aureus*. Furthermore, Esp enhanced the susceptibility of *S. aureus* biofilms to immune system components. In vivo studies have also shown that Esp-secreting *S. epidermidis* eliminates *S. aureus* nasal colonization. These findings indicate that Esp hinders *S. aureus* colonization in vivo through a novel mechanism of bacterial interference, which could lead to the development of novel therapeutics to prevent *S. aureus* colonization and infection. Two of the most commonly used strategies, designed to exploit microbe–microbe interactions, are discussed below.

4.4.1 Use of Probiotics

A novel mechanism for prophylactic or therapeutic management of biofilm-associated diseases is by microbial interference, through the use of probiotics. Probiotics are live microbial supplements which beneficially affect the host by improving its microbial balance, producing metabolites which inhibit the colonization or growth of other microorganisms, or by competing with them for resources such as nutrients or space. The use of antibiotics and immunosuppressive drugs often causes alterations in the composition of host microflora particularly in the oral cavity and intestinal and urogenital tracts. Therefore, the introduction of beneficial microbial species is a very attractive option to reestablish the microbial equilibrium and prevent disease (Gupta 2009).

The most commonly used genera in probiotic preparations are *Lactobacillus*, *Bifidobacterium*, *Escherichia*, *Enterococcus*, *Bacillus*, *Streptococcus*, and *Saccharomyces*. The use of probiotics creates a biofilm niche less conducive to proliferation of pathogens and their virulence factors via immune modulation and pathogen displacement activity. This phenomenon has been shown to be effective in varied clinical conditions such as antibiotic-associated diarrhea and *Helicobacter pylori* infections (Gupta 2009; Dobrogosz et al. 2010). One example is use of lactobacilli to improve urogenital health in women. Four probiotic strains *Lactobacillus rhamnosus* GG, *L. plantarum* 299v, and *L. reuteri* strains PTA 5289 and SD2112 were shown to interfere with the biofilms of salivary *Streptococcus mutans*. This antimicrobial activity against *S. mutans* was found to be pH dependent (Soderling et al. 2011).

There have also been reports of the inhibitory effect of probiotic *Enterococcus faecium* WB2000 on biofilm formation by cariogenic streptococci. Dental caries is a very common chronic disease arising from the interplay among the oral flora, teeth, and dietary factors. The major etiological players are the two α-hemolytic "mutans group" streptococci: *S. mutans* and *Streptococcus sobrinus*. The effect of *Lactobacillus acidophilus DSM 20079* as a probiotic strain on the adhesion of some of the selected streptococcal strains was reported by Tahmourespour and Kermanshahi (2011). This strain was used for its ability to inhibit biofilm formation among mutans and non-mutans oral streptococci. In the presence of the probiotic strain, streptococcal adhesion was reduced and this reduction was not significantly higher if the probiotic strain was inoculated before the oral bacteria. The *Lactobacillus acidophilus* had a significantly higher effect on adherence of mutans streptococci than non-mutans streptococci ($p < 0.05$). It is expected that adhesion reduction is likely due to bacterial interactions and colonization of adhesion sites by the probiotic strain before the streptococci. Adhesion reduction can be an effective way to decrease the cariogenic potential of oral streptococci. Moreover, the ability of *E. faecium* WB2000 and JCM5804 and *Enterococcus faecalis* JCM5803 to inhibit biofilm formation by seven laboratory oral streptococcal strains and 13 clinical mutans streptococcal strains was assayed by Suzuki et al. (2011). *E. faecium* WB2000 inhibited biofilm formation by 90.0 % (9/10) of the clinical *S. mutans* strains and 100 % (3/3) of the clinical *S. sobrinus* strains.

4.4.2 Use of Phages as Antibiofilm Agents

Phages are ubiquitous in nature. Bacteriophages are viruses that infect bacteria and may provide a natural, highly specific, nontoxic, feasible approach for controlling several microorganisms involved in biofilm formation (Kudva et al. 1999). The ability of these phages to inhibit and/or eradicate biofilms has been demonstrated for biofilms of several pathogens including *P. aeruginosa, K. pneumonia, E. coli, Proteus mirabilis,* and *S. epidermidis,* and these studies are summarized here briefly.

Biofilms of *E. coli* strains 3000 XIII developed on the surfaces of polyvinylchloride coupons in a modified Robbins device were infected and lysed using bacteriophage T4D. Similar studies with phage E79 infecting *P. aeruginosa* indicated that phages were infecting the surface organisms but access to the cells deep in the biofilm was restricted (Doolittle et al. 1995). Investigators have demonstrated the use of bacteriophages in killing *S. aureus* and *P. fluorescens* biofilms; however, the infection of biofilm cells by phages is extremely conditional on their chemical composition and environmental factors such as temperature, growth stage, media, and phage concentration (Sillankorva et al. 2004; Chaignon et al. 2007).

The crucial role of titers of specifically selected phages with a proper virion-associated exopolysaccharide (EPS) depolymerase in the development of phage therapy were shown by Cornelissen et al. (2011). They carried out an experiment to investigate the in vitro degradation of single-species *Pseudomonas putida* biofilms, PpG1 and RD5PR2, by the novel phage Q15, a "T7-like virus" EPS depolymerase. Phage Q15 formed plaques surrounded by growing opaque halo zones, on seven out of 53 *P. putida* strains. This has happened because of EPS degradation. Since halos were absent on infection-resistant strains, they suggested that the EPS probably acts as a primary bacterial receptor for phage infection. EPS degrading activity of recombinantly expressed viral tail spike was also confirmed by capsule staining.

Application of bacteriophages in controlling mixed biofilms of *Pseudomonas fluorescens* and *Staphylococcus lentus* has also been reported. Sillankorva et al. (2010) challenged the biofilms with phage phiIBB-PF7A, specific for *P. fluorescens*, and the results obtained showed that phiIBB-PF7A readily reached the target host and caused a significant population decrease. This phage was also capable of causing partial damage to the biofilms leading to the release of the non-susceptible host (*S. lentus*) from the dual species biofilms.

Phage therapy has been successfully employed in the treatment of lung infections of cystic fibrosis caused by colonization of *S. aureus* and further predominant growth of *P. aeruginosa* biofilms. The treatment is very difficult with antibiotics due to several fold increased drug resistance (Brussow 2012). Applications of bacteriophages φMR299-2 and φNH-4 eliminated *P. aeruginosa* in the murine lung and cystic fibrosis lung airway cells (Alemayehu et al. 2012).

Recently, potential of the bacteriophage-derived peptidase, $CHAP_K$, for the rapid disruption of biofilm was reported against staphylococci, associated with the bovine mastitis (Fenton et al. 2013). Purified $CHAP_K$ was able to prevent biofilm formation and also completely eliminated biofilms of *S. aureus* DPC5246 within 4 h. The $CHAP_K$ lysin also reduced *S. aureus* in a skin decolonization model. Furthermore, Shen et al. (2013) found rapid degradation of *S. pyogenes* biofilms by PlyC, a bacteriophage-encoded endolysin. Laser scanning confocal microscopy revealed that lytic action of PlyC destroys the biofilm as it diffuses through the matrix in a time-dependent fashion, and biofilm rapidly become refractory to traditional antibiotics.

Phage therapy is very effective in killing drug-resistant strains because of its specificity toward particular bacterial populations. Formation of a protected biofilm environment is one of the major causes of the increasing antibiotic resistance development. These facts emphasize the need to develop alternative antibacterial strategies, like phage therapy (Cornelissen et al. 2011).

4.5 Nature's Own Biofilm Inhibitors

Interest in studying natural products derived from plant sources for the discovery of new biologically active compounds is not uncommon as many traditional medicines have been rooted. Some of the most active antibiofilm compounds discovered to date have been based upon the molecular scaffolds of natural products isolated from marine natural products (Worthington et al. 2012).

4.5.1 Plant Products

The prevention or control of biofilms by interfering with QS systems is one possible strategy; however, other studies have indicated that phytochemicals can inhibit interspecies coaggregation (Weiss et al. 1998), prevent bacterial adhesion (Kuzma et al. 2007), and inactivate mature single and multispecies biofilms (Niu and Gilbert 2004; Knowles et al. 2005). There is a novel trend in the antibiofilm research area toward the identification of natural products, such as plants and their extracts with antibiofilm activity. Plants offer a virtually inexhaustible and sustainable resource of very interesting classes of biologically active, low-molecular weight compounds. Several microbes in complex ecological niches or in association with biofilms produce compounds that act as antibiofilm agents to gain advantage over others. Certain marine plants are known to produce compounds that inhibit biofilm formation in order to prevent microbes from attaching and blocking the sunlight. The best characterized example is the red algae *Delisea pulchra* that produces halogenated furanones to ward off bacterial biofilms (Ren et al. 2004). Several marine plants and microbes have been shown to inhibit biofilm formation.

Plant extracts and essential oils from several medicinal plants have been exploited as antibiofilm agents for pathogenic biofilm forming bacteria and fungi. In this respect, xanthorrhizol isolated from *Curcuma xanthorrhiza* (Rukayadi et al. 2011) and the oil of *Boesenbergia pandurata* rhizomes (Taweechaisupapong et al. 2010) and *Ocimum americanum* (Thaweboon and Thaweboon 2009) showed potent in vitro activity against *Candida* biofilms. Nostro et al. (2007) studied the effect of oregano essential oil, carvacrol, and thymol on biofilm made by *S. aureus* and *Staphylococcus epidermidis* strains. They found that sub-inhibitory concentrations of the oils attenuated biofilm formation by *S. aureus* and *S. epidermidis* strains on polystyrene microtiter plates. Agarwal et al. (2008) studied 30 plant oils for their activity against *C. albicans* biofilms. Peppermint, eucalyptus, ginger grass, and clove oils resulted in a reduction in *C. albicans* biofilm formation. Dalleau et al. (2008) performed a study on 10 terpenic derivatives, corresponding to major components of essential oils, for their activity against *C. albicans* biofilms. Almost all the studied terpenic derivatives showed antibiofilm activity; however, carvacrol, geraniol, and thymol exhibited the strongest activity. Moreover, these compounds also proved to be efficient against biofilms made by *C. glabrata* and *C. parapsilosis*. In addition, Hendry et al. (2009) have shown potent antibiofilm activity from the main component of eucalyptus oil, 1,8-cineole, against *C. albicans* biofilms.

Harjai et al. (2010) reported anti-QS activity by fresh *Allium sativum* extract [fresh garlic extract (FGE)] and subsequently inhibited *P. aeruginosa* biofilm formation by 6 log10 units. Moreover, in vivo prophylactic treatment in a mouse model of kidney infection with FGE (35 mg/mL) for 14 days resulted in a 3 log10 unit decrease in the bacterial load on the fifth day after infection compared to untreated animals. They found that FGE also protected renal tissue from bacterial adherence and resulted in a milder inflammatory response and histopathological changes in infected tissues. FGE inhibited expression of *P. aeruginosa* virulence factors such as pyoverdin, hemolysin, and phospholipase C. Moreover, killing efficacy and phagocytic uptake of bacteria by peritoneal macrophages was enhanced by administration of garlic extract.

Issac Abraham et al. (2011) reported efficacy of *Capparis spinosa* (caper bush) extract to inhibit biofilm formation by 73 %, at a concentration of 2 mg/mL, in *E. coli*. Also, for the pathogens *Serratia marcescens*, *P. aeruginosa*, and *P. mirabilis*, biofilm biomass was reduced by 79, 75, and 70 %, respectively. Moreover, the mature biofilm structure was disrupted for all of the studied pathogens. Furthermore, the addition of *C. spinosa* extract (100 µg/mL) to a bacterial culture resulted in swimming and swarming inhibition. Similarly, *Melia dubia* (bead tree) bark extracts were examined by Ravichandiran et al. (2012) at a concentration of 30 mg/mL. In their study, these extracts reduced *E. coli* biofilm formation by 84 % and inhibited expression of virulence factors, such as hemolysins, by 20 %. Bacterial swarming regulated by QS was inhibited by 75 %, resulting in decreased biofilm expansion. Recently, our group (Khan and Ahmad 2012a, b) has shown antibiofilm activity by *Cymbopogon citratus* and *Syzygium aromaticum* essential oils and active compounds, namely cinnamaldehyde and eugenol, in drug-resistant strains of *C. albicans* (Fig. 4).

Interference with Quorum Sensing

A new drug target is to interfere with the process of QS, a phenomenon of communication cross talk. This phenomenon is used by many pathogenic microorganisms to establish a biofilm and control much of their virulence arsenal.

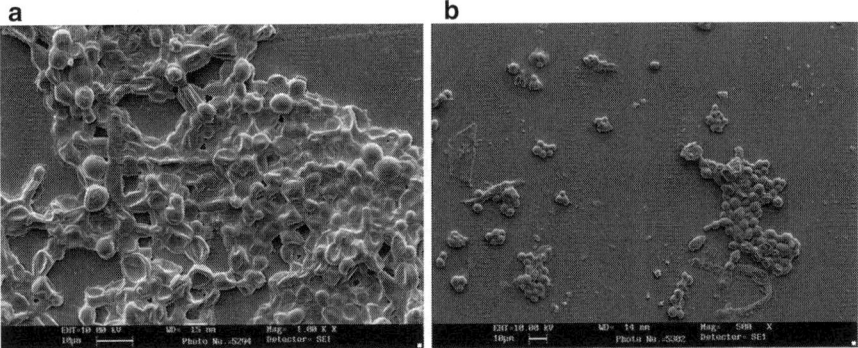

Fig. 4 Scanning electron micrograph of the 48 h-old *C. albicans* 04 biofilm formed on catheter discs in the absence and presence of eugenol or cinnamaldehyde. In (**a**) biofilm formed in the absence of active compounds, dense network of cells and hyphae along with exopolysaccharide material are observed, and in (**b**) biofilm formed in the presence of eugenol at 100 mg/L, no exopolysaccharide material and aggregation of cells are observed (Khan and Ahmad 2012b)

The process is regulated by means of extracellular signal molecules (Rasmussen et al. 2005). Although the exact role of QS in various stages of biofilm formation, maturation, and dispersal and in biofilm resistance is not entirely clear, the use of QS inhibitors (QSI) has been proposed as a potential antibiofilm strategy. It is conceivable that QS inhibition may represent a natural, widespread, antimicrobial strategy with significant impact on biofilm formation (Dong et al. 2002). Acting on biofilms by interfering with their command language, QS can provide an alternative to the ineffective conventional biofilm control strategies (Rasmussen and Givskov 2006).

QS has been shown to be responsible for the development of resistance to various antimicrobial agents and immune modulation in biofilm entities. Several organisms seem to have evolved the ability to interrupt this process. Examples include plants (e.g., tomato, rice, and pea) and soil bacteria that secrete compounds that alter homoserine lactone activity and *Delisea pulchra*, which secretes a halogenated furanone that inhibits QS signaling (Bauer and Robinson 2002). This suggests that synthetic analogs of such substances, or novel compounds from drug discovery efforts, could interrupt QS in one or more (Stewart 2003) ways. QS signaling can be interrupted in several manners like targeting to ligand-receptor pathways, i.e., by inhibiting ligand synthesis, transport, or release; inhibiting receptor synthesis and processing; and perhaps most analogous to current pharmacotherapy, inhibiting enzyme activity or ligand-receptor binding (Raffa et al. 2005). Furthermore, the use of QS inhibitors may control biofilm formation by making biofilms more susceptible to antibiotics as well as to host defenses (Bjarnsholt and Givskov 2007; Jayaraman and Wood 2008; Hoiby et al. 2010). Attenuation of bacterial virulence or biofilms by QSI rather than by antibiotics is a very interesting concept, which could prove to be a new target with less risk of inducing resistance

(Lazar 2011). This strategy could lead to the development of new and efficient natural products for biofilm control.

Adonizio et al. (2008) have shown that certain plant extracts from Southern Florida caused the inhibition of QS genes and QS-controlled factors, with marginal effects on the growth of *P. aeruginosa*. Lonn-Stensrud et al. (2009) have shown that furanones may inhibit biofilm formation through interference with QS and thus represent promising agents for protecting surfaces from being colonized by *S. epidermidis*. Our group has also shown that pea seedling inhibits QS in *P. aeruginosa* PA01 and *Cromobacterium violaceum* CV12472 (Fatima et al. 2010).

Ding et al. (2011) screened 46 active components found in traditional Chinese medicines (TCMs) that inhibit bacterial biofilm formation. Six of 46 active components found in TCMs were identified as putative QSIs based on molecular docking studies. Of these, three compounds inhibited biofilm formation by *P. aeruginosa* and *Stenotrophomonas maltophilia* at a concentration of 200 μM. A fourth compound (emodin) significantly inhibited biofilm formation at 20 μM and induced proteolysis of the QS signal receptor TraR in *Escherichia coli* at a concentration of 3–30 mM. Emodin also increased the activity of ampicillin against *P. aeruginosa*. Therefore, they suggested that emodin might be suitable for development into an antivirulence and antibacterial agent based on disruption of biofilms. Brackman et al. (2011) have shown that QSI (baicalin hydrate, cinnamaldehyde, and hamamelitannin) increased the success of the antibiotics vancomycin, tobramycin, and clindamicin by increasing the susceptibility of bacterial biofilms and/or by increasing host survival following infection. Damte et al. (2013) screened for anti-QS activity in 97 indigenous plant extracts from Korea, through biomonitor bacterial strains, *Chromobacterium violaceum* (CV12472) and *P. aeruginosa* (PAO1), and found 18 plant extracts to exhibit anti-QS activity against both reporter systems. Sarabhai et al. (2013) have published a first report on the anti-QS activity of ellagic acid derivative compounds from *T. chebula* fruit. They found that these compounds downregulated the expression of the *lasI/R* and *rhlI/R* genes with concomitant decreases in *N*-acyl homoserine lactones (AHLs) in *P. aeruginosa* PAO1 causing attenuation of its virulence and enhanced sensitivity of its biofilm to tobramycin.

These data confirm that plant and microbial products have anti-QS, antiseptic, and antivirulence factor properties and can easily inhibit biofilm formation as well as disrupt the mature biofilm structure. These natural products represent a virtually inexhaustible and sustainable source of biocide-free antibiofilm agents with novel targets, unique modes of action, and proprieties with potential for utilization in clinical perspectives. Testing sublethal concentrations of plant-derived compounds for disrupting microbial biofilms could be of great importance to reveal mechanisms other than killing activity to overcome the emergence of drug-resistant strains. This strategy could offer an elegant way to develop novel biocide-free antibiofilm strategies. The studies conducted in this regard and significance and future prospects are well reviewed by Villa and Cappitelli (2013), and readers are directed to this chapter to get more insight on plant-derived products as effective antibiofilm agents.

4.5.2 Microbial Metabolites

The secondary metabolites of several microorganisms, ranging from furanone to exo-polysaccharides, have been suggested to have antibiofilm activity in various recent studies. Among these, *E. coli* group II capsular polysaccharides were shown to inhibit biofilm formation of a wide range of organisms, and marine *Vibrio* sp. were found to secrete complex exopolysaccharides having the potential for broad-spectrum biofilm inhibition and disruption (Abu Sayem et al. 2011). Extracts from coral associated *Bacillus horikoshii* (Thenmozhi et al. 2009) and actinomycetes (Nithyanand et al. 2010) inhibit biofilm formation of *S. pyogenes*. The exoproducts of marine *Pseudoalteromonas* impair biofilm formation by a wide range of pathogenic strains (Dheilly et al. 2010). Most recently, exopolysaccharides from the marine bacterium *Vibrio* sp. QY101 were shown to control biofilm-associated infections (Jiang et al. 2011). Abu Sayem et al. (2011) reported antibiofilm activity from a newly identified ca. 1,800 kDa polysaccharide, which has simple monomeric units of α-D-galactopyranosyl-(1→2)-glycerol-phosphate, against a number of both pathogenic and nonpathogenic strains without bactericidal effects. This polysaccharide was extracted from a *Bacillus licheniformis* strain associated with the marine organism *Spongia officinalis*. Musthafa et al. (2011) have shown the antibiofilm potential of ethyl acetate extract of marine *Bacillus* sp. SS4 using a static biofilm ring assay. Their study showed a concentration-dependent reduction in the biofilm-forming ability of PAO1 by these compounds. Members of the actinomycetes family are a rich source of bioactive compounds including diverse antibiotics.

Role of QS Molecules

Gram-negative bacteria predominantly use AHLs as autoinducers, which show variation in the length and oxidation state of the acyl side chain. In *V. fischeri*, AHL synthesis occurs when the *luxI* gene is activated to produce the AHL synthase enzyme LuxI. When these AHLs reach a threshold intracellular concentration, they bind to the transcriptional activator LuxR and lead to activation of the luxR gene set. AHLs are able to freely diffuse in and out of bacterial cells, allowing the total AHL concentration to correlate to the total bacterial concentration, thus enabling population density-based control of gene expression. This cascade of events ultimately leads to the control of gene expression resulting in the control of virulence factor production and biofilm formation and maintenance (Finch et al. 1998). A huge amount of work has been reported involving the biological consequences of chemically modified AHL derivatives in a variety of QS systems. Work from the Blackwell group has documented the synthesis and identification of a number of natural and synthetic AHLs with the ability to modulate QS in *P. aeruginosa* and *Agrobacterium tumefaciens* (Geske et al. 2005). They also found two of their most active synthetic AHLs retarded biofilm formation in *P. aeruginosa* PA01.

QS in Gram-positive bacteria is predominantly mediated by autoinducing peptides (AIPs) but may not exclusively utilize peptide signaling molecules for communication. Small molecules known as γ-butyrolactones have been identified as signaling molecules in some species of *Streptomyces* (Takano et al. 2001). The *agr* and TRAP (target of RNA-III activating peptide) QS systems in *S. aureus* regulate a number of virulence phenotypes, including biofilm formation. The RNA-III activating protein (RAP) activates TRAP via phosphorylation, leading to increased cell adhesion and biofilm formation, in addition to inducing expression of the *agr* operon (Fux et al. 2003). It has been demonstrated that the RNA-III inhibiting peptide (RIP) inhibits phosphorylation of TRAP, leading to reduced biofilm formation (Giacometti et al. 2003).

A small molecule termed autoinducer-2 (AI-2) is one of the putative universal QS mechanisms shared by both Gram-negative and Gram-positive bacteria. AI-2 molecules are derived from the precursor molecule (S)-4,5-dihydroxy-2,3-pentanedione (DPD), and the synthase enzyme that drives DPD production has been found to be conserved in over 55 bacterial species (Waters and Bassler 2005). An adenosine analogue of DPD was found to block AI-2-based QS without interfering with bacterial growth. This compound was subsequently shown to affect biofilm formation in *Vibrio anguillarum*, *Vibrio vulnificus*, and *V. cholerae* (Brackman et al. 2009).

Indole is a putative universal intercellular signal molecule amongst diverse bacteria that plays a direct role in the control of biofilm formation (Lee et al. 2007). Another attractive target for control of biofilm formation is interfering with c-di-GMP signaling using either c-di-GMP analogues or with small molecules that interfere with the synthesis or degradation of c-di-GMP (Sintim et al. 2010). Bis-(3'5')-cyclic di-guanylic acid (c-di-GMP) is a second messenger signaling molecule that is thought to be ubiquitous in bacteria. Diguanylate cyclases (DGCs) and phosphodiesterases (PDEs) are responsible for the synthesis and breakdown of c-di-GMP, respectively (Yan and Chen 2010). There is increasing evidence that the transition between the planktonic and biofilm lifestyle of *P. aeruginosa* is regulated via proteins with DGC or PDE activities through control of c-di-GMP levels (Tamayo et al. 2007). It has also been observed that exopolysaccharide synthesis (and thus the exopolysaccharide-dependent formation of biofilms) is regulated by c-di-GMP in various proteobacterial species such as *V. cholera*, *P. aeruginosa*, *P. fluorescens*, *A. tumefaciens*, *E. coli*, and *Salmonella enterica* (Ryjenkov et al. 2005).

Role of Biosurfactants

Many bacteria are capable of synthesizing and excreting biosurfactants with anti-adhesive properties (Rodrigues et al. 2004; van Hamme et al. 2006). Biosurfactants are amphihilic biological compounds that are produced extracellularly or intracellularly by a wide variety of microorganisms, which include bacteria, yeasts, and filamentous fungi (Cameotra and Makkar 2004). These biosurfactants have promising applications in biomedical sciences (Singh and Cameotra 2004). Biosurfactants produced by *Lactococcus lactis* impaired biofilm formation on silicone rubber (Rodrigues et al. 2004). Surfactin from *Bacillus subtilis* dispersed biofilms without affecting cell growth and prevented biofilm formation by microorganisms such as *Salmonella enterica*, *E. coli*, and *P. mirabilis* (Mireles et al. 2001). Valle et al. (2006) demonstrated that *E. coli* expressing group II capsules released a soluble polysaccharide into their environment that induced physicochemical surface alterations, which prevented biofilm formation by a wide range of Gram-positive and Gram-negative bacteria. Many other researchers have demonstrated the potential for biofilm control by various other biosurfactants made by bacteria and fungi (Davey et al. 2003; Walencka et al. 2008). Two lipopeptide biosurfactants produced by *B. subtilis* and *B. licheniformis* have been shown by Rivardo et al. 2009 to exhibit anti-adhesive activity by selectively inhibiting biofilm formation of two human pathogenic strains, *E. coli* CFT073 and *S. aureus* ATCC29213. Davies and Marques (2009) found that *P. aeruginosa* produces cis-2-decenoic acid, which is capable of inducing the dispersion of established biofilms and of inhibiting biofilm development by *B. subtilis*, *E. coli*, *S. aureus*, *Klebsiella pneumoniae*, *P. aeruginosa*, *P. mirabilis*, *S. pyogenes*, and the yeast *C. albicans,* when applied exogenously. The authors also suggested that this molecule is functionally and structurally related to a class of short-chain fatty acid signaling molecules.

Fracchia et al. (2010) reported biofilm inhibitory activity by a *Lactobacillus*-derived biosurfactant against human pathogenic *C. albicans*. Rufino et al. (2011) have isolated a biosurfactant rufisan from the yeast *Candida lipolytica* UCP0988 that exhibited antimicrobial and anti-adhesive activities against many *Streptococcus* spp. Monteiro et al. (2011a, b) have evaluated the effects of a glycolipid-type biosurfactant produced by *Trichosporon montevideense* CLOA72 in the formation of biofilms in polystyrene plate surfaces by *C. albicans* CC isolated from the apical tooth canal. Biofilm formation was reduced up to 87.4 % with use of this biosurfactant at a 16 mg/mL concentration. This biomolecule did not present any cytotoxic effects in a HEK 293A cell line at concentrations of 0.25–1 mg/mL. Their studies indicated a possible application of the referred biosurfactant in inhibiting the formation of biofilms on plastic surfaces by *C. albicans*.

Recently, Padmapriya and Suganthi (2013) found antimicrobial and anti-adhesive activity of a biosurfactant produced by *Candida tropicalis* and *C. albicans* against a variety of urinary and clinical pathogens such as *Bacillus*, *C. albicans*, *Citrobacter*, *E. coli*, *K. pneumoniae*, *P. mirabilis*, *P. aeruginosa*, *Salmonella*, and *S. aureus*. A study from our group, Singh et al. (2013) has demonstrated *Candida* biofilm disrupting ability by di-rhamnolipid (RL-2) produced by *P. aeruginosa* DSVP20.

The antibiofilm activity of a glycolipid biosurfactant isolated from the marine actinobacterium *Brevibacterium casei* MSA19 was evaluated by Kiran et al. (2010) against pathogenic biofilms in vitro. The disruption of the biofilm by the MSA19 glycolipid was consistent against mixed pathogenic biofilm bacteria. Therefore, it could be suggested that the glycolipid biosurfactant can be used as a lead compound for the development of novel antibiofilm agents.

Janek et al. (2012) have recently identified a biosurfactant, Pseudofactin II, secreted by *Pseudomonas fluorescens* BD5, the strain obtained from freshwater from the Arctic Archipelago of Svalbard. Pseudofactin II showed anti-adhesive activity against several pathogenic microorganisms (*E. coli, E. faecalis, Enterococcus hirae, S. epidermidis, P. mirabilis*, and two *C. albicans* strains), which are potential biofilm formers on catheters, implants, and internal prostheses. Up to 99 % prevention was achieved by 0.5 mg/mL pseudofactin II. In addition, pseudofactin II dispersed preformed biofilms. Pseudofactin II can be used as a disinfectant or surface coating agent against microbial colonization of different surfaces, e.g., implants or urethral catheters.

An overview of all the above-discussed strategies to control biofilms is given in Fig. 5. The targets of each approaches at different stages of biofilms and their interrelation are depicted.

4.6 Small Molecule Control of Biofilms

Given the prominence of biofilms in infectious diseases, there has been an increased effort toward the development of small molecules that inhibit and/or disperse bacterial biofilms through non-microbicidal mechanisms. It will be meaningful to distinguish molecules that have the ability to affect biofilm development via non-microbicidal mechanisms, as the pressure on bacteria or fungi to evolve resistance to these agents will be significantly reduced or even eliminated (Worthington et al. 2012). Due to the scarcity of known molecular scaffolds that inhibit/disperse bacterial biofilms, high throughput screening (HTS) has been employed in attempts to discover leads for new anti-biofilm modulators. Here, we have briefly summarized the application of chemical databases for the discovery of lead small molecules, using HTS approaches, which mediate biofilm development.

These approaches are grouped into three steps:

1. The identification and development of small molecules that target one of the bacterial signaling pathways involved in biofilm regulation
2. Chemical library screening for compounds with antibiofilm activity
3. The identification of natural products that possess antibiofilm activity, and the chemical manipulation of these natural products to obtain analogues (using structure activity relationship (SAR) method) with increased activity.

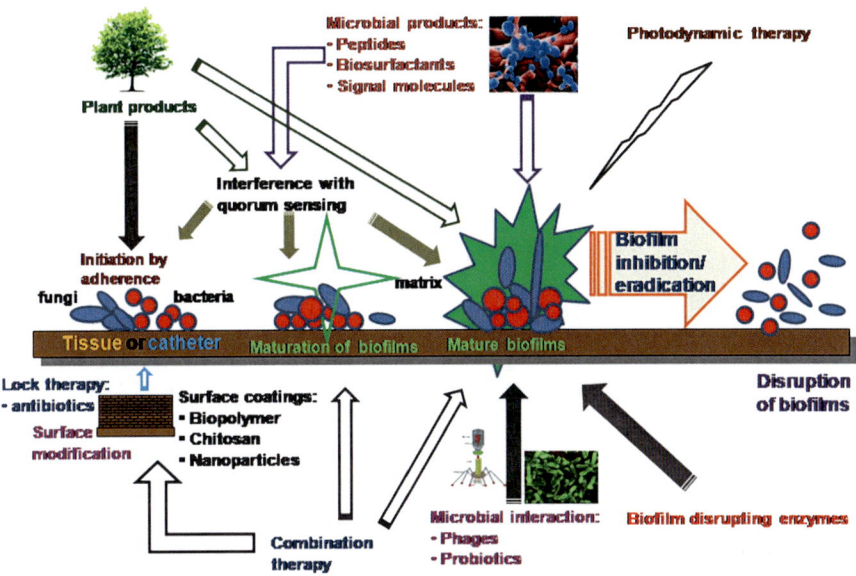

Fig. 5 Depiction of targets of various approaches acting at different stages of biofilm growth and their interrelationship to control biofilm infections

Natural products provide a diverse array of chemical structures and possess a plethora of biological activities. A number of natural products that possess the ability to inhibit or disperse bacterial biofilms have been used as the starting points for medicinal chemistry programs in which synthetic manipulation of the natural product scaffold has allowed for the design of more efficacious compounds. Much of the natural product inspiration for these programs has come from compounds isolated from plants and marine organisms. It is known that QS pathways heavily influence the formation of biofilms, in addition to the production of other virulence factors. A diverse range of biomolecules serve as the facilitators for QS systems in bacteria. Therefore, extensive research in this area has produced a number of analogues with the ability to modulate QS-dependent enzymes. These molecules compose the vast majority of compounds thus far investigated for biofilm control. AHLs have served as one of the primary scaffolds studied over the past 30 years for the design of potential biofilm inhibitors (Geske et al. 2008). A considerable amount of work has been published involving the biological consequences of chemically modified AHL derivatives in a variety of QS systems and reviewed by Worthington et al. (2012). Here, we focus on using small molecules to derive novel compounds capable of controlling biofilm-associated infections (Fig. 6).

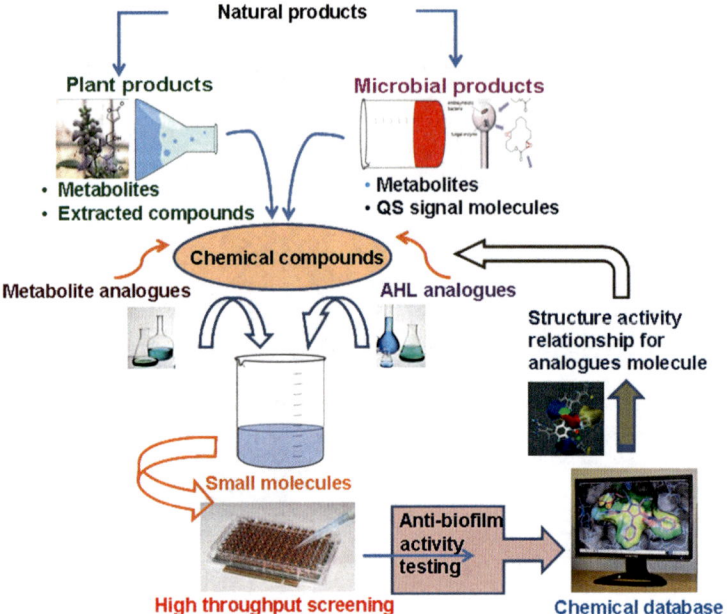

Fig. 6 Principle of small molecule high throughput screening from natural products for development of anti-biofilm agents

4.6.1 Application of Chemical Library Screening

One of the first reports detailing the screening of a large library of compounds with the objective of identifying novel small molecules that possessed antibiofilm activity was reported from Biosignals (Sydney, Australia). It has developed over 200 furanone-like compounds and evaluated them as biofilm inhibitors in preventive therapy. Other natural product compounds from plants make up the material for Sequoia Sciences (San Diego, CA) to design biofilm inhibitors. They have developed a high-throughput strategy for extracting, purifying, and structurally characterizing libraries of natural product compounds from plants. They generated a library of over 150,000 natural compounds for evaluation as antibiofilm compounds (Sachachter 2003). In 2005, workers from the Wood group (Ren et al. 2005) screened 13,000 compounds. The study revealed a hit (0.08 %), identified as ursolic acid 57, as a compound which effectively inhibited *E. coli* biofilm formation at concentrations as low as 22 µM without affecting growth. In the same year, the Hergenrother group reported the identification of iron salts as effective nonantibiotic inhibitors and disruptors of *P. aeruginosa* biofilms from a screen of over 4,500 compounds which belonged to the University of Illinois Marvel Library Compound Collection (MLCC) (Musk et al. 2005).

Work from the Blackwell group has documented the synthesis and identification of a number of natural and unnatural AHLs with the ability to modulate QS in

P. aeruginosa and *Agrobacterium tumefaciens* (Geske et al. 2005). They also demonstrated that two of their most active synthetic AHLs could retard biofilm formation in *P. aeruginosa* PA01. Other research that includes the modification of AHLs to discern their effects on QS and biofilm formation in *P. aeruginosa* comes from the Suga group. The work exploited the synthesis of a 96-member library constructed through solid phase protocols to mimic AHLs by replacing the homoserine lactone moiety with a variety of functionalities. A noteworthy compound identified within this study was AHL derivative 8, which had no effect on biofilm growth, yet elicited a noticeable change in the biofilm morphology of *P. aeruginosa* PA01 (Smith et al. 2003a, b). Analogues of *P. aeruginosa* AHLs in which the lactone functionality was replaced by a ketone had additional difluorination between the β-keto amide positions (Glansdorp et al. 2004).

Junker and Clardy (2007) have developed a HTS method for small molecule inhibitor of *P. aeruginosa* biofilms at the Institute of Chemistry and Cell Biology-Longwood (ICCB-L) at Harvard Medical School, Boston, MA (http://iccb.med.harvard.edu/). They have obtained 66,095 compounds, from natural products of microbial or plant origin and also some commercial chemical compounds, to identify those that prevent biofilm formation without affecting planktonic bacterial growth. The screen is a luminescence-based attachment assay that has been validated with several strains of *P. aeruginosa* and compared to a well-established but low-throughput crystal violet staining biofilm assay. They have determined the potencies of 61 compounds against biofilm attachment and have identified 30 compounds that fall into different structural classes as biofilm attachment inhibitors with 50 % effective concentrations of less than 20 μM. The most active compound discovered was shown to possess an IC50 value of 530 nM for biofilm inhibition. This makes this compound as one of the most active biofilm modulators ever disclosed against either Gram-positive or Gram-negative bacteria. Their study has highlighted these small-molecule inhibitors for identification of their relevant biofilm targets or potential therapeutics for *P. aeruginosa* infections.

A structure-based virtual screen (SB-VS) for the identification of putative QS inhibitors was carried out using a focused database comprising compounds that possess structural similarities to the known QS inhibitors furanone C30, patulin, the *P. aeruginosa* LasR natural ligand (3-oxo-C12-AHL 5), and a known QS receptor agonist TP-1, (Yang et al. 2009). This screen led to the discovery of three compounds, which were all recognized drugs, salicylic acid, nifuroxazide, and chlorzoxazone, and were subsequently shown to significantly inhibit QS-regulated gene expression at concentrations at which they did not affect bacterial growth. In addition to affecting QS regulated virulence factor production, these compounds were shown to affect biofilm formation by PA01. Screening of approximately 66,000 compounds and natural product extracts from the Center for Chemical Genomics at the University of Michigan to identify compounds that affected induction of a *V. cholerae* c-di-GMP-inducible transcriptional fusion led to the discovery of a novel benzimidazole (Sambanthamoorthy et al. 2011). This compound was examined for its ability to inhibit biofilm formation by a number of pathogenic bacterial strains. Compound 61 was shown to be a broad spectrum inhibitor of biofilm formation, significantly inhibiting biofilm formation by *P. aeruginosa* (CF-145), *K. pneumoniae*, *Erwinia amylovora*, and *Shigella boydii* at 100 μM, by MRSA USA300, and by *S. aureus* Newman at 25 μM, using the minimum biofilm eradication concentration (MBEC) static assay, without affecting bacterial growth. These signaling molecule derivatives are particularly important because the biological activity of nearly every compound in this class is not driven by microbicidal properties.

5 Conclusions

Biofilms have great importance for public health because of their role in certain infectious diseases and their importance in a variety of device-related infections. Most of our understanding of infections is based on research that has examined free-living organisms. The results do not necessarily apply to biofilm organisms, since metabolic and synthetic characteristics of free-living organisms can change when they assume the biofilm mode of growth. Microbial adhesion and biofilm formation are major concerns in control strategies. Drug resistance, virulence, and pathogenicity of microorganisms are often enhanced when growing in a biofilm, and new strategies are therefore required to control biofilm formation and development. A greater understanding of biofilm processes should lead to novel, effective control strategies for biofilm control and a resulting improvement in disease management. The similarity of QS processes to ligand-receptor binding could be exploited as a guide to direct novel antibiotic drug design efforts based on standard pharmacologic principles and drug discovery processes. The unique nature of their mechanism should provide these new antibiotics with greater activity against currently resistant bacteria. In addition, plant and microbial products in combination with other antimicrobial strategies such as antibiotics or photodynamic inactivation could provide an effective bactericidal tool for the treatment of various bacterial and yeast infections. Furthermore, development of high throughput methods to identify natural compounds or their analogues will be a promising strategy to overcome the problem of biofilm management.

Acknowledgments We would like to thank University Grants Commission, and Indian Council of Medical Research, New Delhi for their financial support.

References

Abu Sayem AM, Manzo E, Ciavatta L, Tramice A, Cordone A, Zanfardino A, De Felice M, Varcamonti M (2011) Anti-biofilm activity of an exopolysaccharide from a sponge-associated strain of *Bacillus licheniformis*. Microb Cell Fact 10:74

Adonizio A, Kong KF, Mathee K (2008) Inhibition of quorum sensing-controlled virulence factor production in *Pseudomonas aeruginosa* by South Florida plant extracts. Antimicrob Agents Chemother 52:198–203

Agarwal V, Lal P, Pruthi V (2008) Prevention of *Candida albicans* biofilm by plant oils. Mycopathologia 165:13–19

Aguinaga A, Francés ML, Del Pozo JL, Alonso M, Serrera A, Lasa I, Leiva J (2011) Lysostaphin and clarithromycin: a promising combination for the eradication of *Staphylococcus aureus* biofilms. Int J Antimicrob Agents 37:585–587

Alemayehu D, Casey PG, McAuliffe O, Guinane CM, Martin JG, Shanahan F, Coffey A, Ross RP, Hill C (2012) Bacteriophages φMR299-2 and φNH-4 can eliminate *Pseudomonas aeruginosa* in the murine lung and on cystic fibrosis lung airway cells. mBio 3:e00029-12

Alkawash MA, Soothill JS, Schiller NL (2006) Alginate lyase enhances antibiotic killing of mucoid *Pseudomonas aeruginosa* in biofilms. Acta Pathol Microbiol Immunol Scand 114:131–138

Anderson JM, Rodriguez A, Chang DT (2008) Foreign body reaction to biomaterials. Semin Immunol 20:86–100

Anghel I, Grumezescu AM, Andronescu E, Anghel AG, Ficai A, Saviuc C, Grumezescu V, Vasile BS, Chifiriuc MC (2012) Magnetite nanoparticles for functionalized textile dressing to prevent fungal biofilms development. Nanoscale Res Lett 7:501

Applerot G, Lellouche J, Perkas N, Nitzan Y, Gedanken A, Banin E (2012) ZnO nanoparticle-coated surfaces inhibit bacterial biofilm formation and increase antibiotic susceptibility. RSC Adv 2:2314–2321

Arciola C (2009) New concepts and new weapons in implant infections. Int J Artif Organs 32:533–536

Bandara HMHN, Yau JYY, Watt RM, Jin LJ, Samaranayake LP (2010) *Pseudomonas aeruginosa* inhibits *in-vitro Candida* biofilm development. BMC Microbiol 10:125

Bauer WD, Robinson JB (2002) Disruption of bacterial quorum sensing by other organisms. Curr Opin Biotechnol 13:234–237

Bell M (2001) Biofilms: a clinical perspective. Curr Infect Dis Rep 3:483–486

Belyansky I, Tsirline VB, Martin TR, Klima DA, Heath J, Lincourt AE, Satishkumar R, Vertegel A, Heniford BT (2011) The addition of lysostaphin dramatically improves survival, protects porcine biomesh from infection, and improves graft tensile shear strength. J Surg Res 171:409–415

Beyth N, Yudovin-Farber I, Perez-Davidi M, Domb AJ, Weiss EI (2010) Polyethyleneimine nanoparticles incorporated into resin composite cause cell death and trigger biofilm stress in vivo. Proc Natl Acad Sci USA 107:22038–22043

Biel MA, Sievert C, Usacheva M, Teichert M, Balcom J (2011a) Antimicrobial photodynamic therapy treatment of chronic recurrent sinusitis biofilms. Int Forum Allergy Rhinol 1:329–334

Biel MA, Sievert C, Usacheva M, Teichert M, Wedell E, Loebel N, Rose A, Zimmermann R (2011b) Reduction of endotracheal tube biofilms using antimicrobial photodynamic therapy. Laser Surg Med 43:586–590

Bjarnsholt T, Givskov M (2007) Quorum-sensing blockade as a strategy for enhancing host defences against bacterial pathogens. Philos Trans R Soc B 362:1213–1222

Brackman G, Celen S, Baruah K, Bossier P, Van Calenbergh S, Nelis HJ, Coenye T (2009) AI-2 quorum-sensing inhibitors affect the starvation response and reduce virulence in several Vibrio species, most likely by interfering with LuxPQ. Microbiology 155:4114–4122

Brackman G, Cos P, Maes L, Nelis HJ, Coenye T (2011) Quorum sensing inhibitors increase the susceptibility of bacterial biofilms to antibiotics in vitro and in vivo. Antimicrob Agents Chemother 55:2655–2661

Brogden KA, Guthmiller JM (2002) Polymicrobial diseases. ASM Press, Washington, DC

Brook I (2002) Microbiology of polymicrobial abscesses and implications for therapy. J Antimicrob Chemother 50:805–810

Bruellhoff K, Fiedler J, Oller MM, Groll J, Brenner RE (2010) Surface coating strategies to prevent biofilm formation on implant surfaces. Int J Artif Organs 33:646–653

Brussow H (2012) *Pseudomonas* biofilms, cystic fibrosis, and phage: a silver lining? mBio 3: e00061-12. doi:10.1128/mBio.00061-12

Cameotra SS, Makkar RS (2004) Recent applications of biosurfactants as biological and immunological molecules. Curr Opin Microbiol 7:262–266

Carlson RP, Taffs R, Davison WM, Stewart PS (2008) Anti-biofilm properties of chitosan-coated surfaces. J Biomat Sci 19:1035–1046

Carratala J (2002) The antibiotic-lock technique for therapy of "highly needed" infected catheters. Clin Microbiol Infect 8:282–289

Chaignon P, Sadovskaya I, Ragunah C, Ramasubbu N, Kaplan JB, Jabbouri S (2007) Susceptibility of staphylococcal biofilms to enzymatic treatments depends on their chemical composition. Appl Microbiol Biotechnol 75:125–132

Chole R, Faddis B (2003) Anatomical evidence of microbial biofilms in tonsillar tissues, a possible mechanism to explain chronicity. Arch Otolaryngol Head Neck Surg 129:634–636

Christner M, Franke GC, Schommer NN et al (2010) The giant extracellular matrix-binding protein of *Staphylococcus epidermidis* mediates biofilm accumulation and attachment to fibronectin. Mol Microbiol 75:187–207

Chuang YP, Fang CT, Lai SY, Chang SC, Wang JT (2006) Genetic determinants of capsular serotype K1 of *Klebsiella pneumoniae* causing primary pyogenic liver abscess. J Infect Dis 193:645–654

Coates T, Eady A, Cove J (2003) Propionibacterial biofilms cannot explain antibiotic resistance but might contribute to some cases of antibiotic recalcitrant acne. Br J Dermatol 148:366–367

Collins TL, Markus EA, Hassett DJ, Robinson JB (2010) The effect of a cationic porphyrin on *Pseudomonas aeruginosa* biofilms. Curr Microbiol 61:411–416

Colvin VL (2003) The potential environmental impact of engineered nanomaterials. Nat Biotechnol 21:1166–1170

Contreras-Garcia A, Bucio E, Brackmanc G, Coenye T, Concheiro A, Alvarez-Lorenzo C (2011) Biofilm inhibition and drug-eluting properties of novel DMAEMA-modified polyethylene and silicone rubber surfaces. Biofouling 27:123–135

Cornelissen A, Ceyssens PJ, T'Syen J, Van Praet H, Noben JP, Shaburova OV, Krylov VN, Volckaert G, Lavigne R (2011) The T7-related *Pseudomonas putida* phage Q15 displays virion-associated biofilm degradation properties. PLoS ONE 6:e18597. doi:10.1371/journal.pone.001 8597

Costerton JW, Stewart PS, Greenberg EP (1999) Bacterial biofilms: a common cause of persistent infections. Science 284:1318–1322

Craigen B, Dashiff A, Kadouri DE (2011) The use of commercially available alpha-amylase compounds to inhibit and remove *Staphylococcus aureus* biofilms. Open Microbiol J 5:21–31

Dalleau S, Cateau E, Berges T, Berjeaud JM, Imbert C (2008) In vitro activity of terpenes against *Candida* biofilms. Int J Antimicrob Agents 31:572–576

Damte D, Gebru E, Lee SJ, Suh JW, Park SC (2013) Evaluation of anti-quorum sensing activity of 97 indigenous plant extracts from Korea through bioreporter bacterial strains *Chromobacterium violaceum* and *Pseudomonas aeruginosa*. J Microb Biochem Technol 5:42–46

Davey ME, Caiazza NC, O'Toole GA (2003) Rhamnolipid surfactant production affects biofilm architecture in *Pseudomonas aeruginosa* PAO1. J Bacteriol 185:1027–1036

Davies D (2003) Understanding biofilm resistance to antibacterial agents. Nat Rev Drug Discov 2:114–122

Davies DG, Marques CN (2009) A fatty acid is responsible for inducing dispersion in microbial biofilms. J Bacteriol 191:1393–1403

Davies AN, Brailsford SR, Beighton D (2006) Oral candidosis in patients with advanced cancer. Oral Oncol 42:698–702

De Prijck K, Smet D, Coenye T, Schacht E, Nelis HJ (2010a) Prevention of *Candida albicans* biofilm formation by covalently bound dimethylaminoethylmethacrylate and polyethylenimine. Mycopathologia 170:213–221

De Prijck K, Smet D, Rymarczyk-Machal M, Van Driessche G, Devreese B, Coenye T, Schacht E, Nelis HJ (2010b) *Candida albicans* biofilm formation on peptide functionalized polydimethylsiloxane. Biofouling 26:269–275

Dheilly A, Soum-Soutera E, Klein GL, Bazire A, Compère C, Haras D, Dufour A (2010) Antibiofilm activity of the marine bacterium *Pseudoalteromonas* sp. strain 3J6. Appl Environ Microbiol 76:3452–3461

Ding X, Yin B, Qian L, Zeng Z, Yang Z, Li H, Lu Y, Zhou S (2011) Screening for novel quorum-sensing inhibitors to interfere with the formation of *Pseudomonas aeruginosa* biofilm. J Med Microbiol 60:1827–1834

Dobrogosz WJ, Peacock TJ, Hassan HM, Allen I. Laskin SS, Geoffrey MG (2010) Evolution of the probiotic concept: from conception to validation and acceptance in medical science. Adv Appl Microbiol 72:1–41

Dong YH, Gusti AR, Zhang Q, Xu JL, Zhang LH (2002) Identification of quorum-sensing N-acyl homoserine lactonases from *Bacillus* species. Appl Environ Microbiol 68:1754–1759

Dongari-Bagtzoglou A (2008) Pathogenesis of mucosal biofilm infections: challenges and progress. Expert Rev Anti Infect Ther 6:201–208

Dongari-Bagtzoglou A, Kashleva H, Dwivedi P, Diaz P, Vasilakos J (2009) Characterization of mucosal *Candida albicans* biofilms. PLoS ONE. doi:10.1371/ journal.pone.0007967

Donlan RM (2001) Biofilm formation: a clinically relevant microbiological process. Clin Infect Dis 33:1387–1392

Donlan RM (2002) Biofilms: microbial life on surfaces. Emerg Infect Dis 8:881e890

Donlan RM (2008) Biofilms on central venous catheters: is eradication possible? Curr Top Microbiol Immunol 322:133–161

Donlan RM, Costerton JW (2002) Biofilms: survival mechanisms of clinically relevant microorganisms. Clin Microbiol Rev 15:167–193

Doolittle MM, Cooney JJ, Caldwell DE (1995) Lytic infection of *Escherichia coli* biofilms by bacteriophage-T4. Can J Microbiol 41:12–18

Douglas LJ (2002) Medical importance of biofilms in *Candida* infections. Rev Iberoam Micol 19:139–143

Fatima Q, Zahin M, Khan MSA, Ahmad I (2010) Modulation of quorum sensing controlled behaviour of bacteria by growing seedling, seed and seedling extracts of leguminous plants. Indian J Microbiol 50:238–242

Fenton M, Keary R, McAuliffe O, Ross RP, O'Mahony J, Coffey A (2013) Bacteriophage-derived peptidase CHAPK eliminates and prevents staphylococcal biofilms. Int J Microbiol. Article ID 625341. doi:10.1155/2013/625341

Finch RG, Pritchard DI, Bycroft BW, Williams P, Stewart GS (1998) Quorum sensing: a novel target for anti-infective therapy. J Antimicrob Chemother 42:569–571

Fracchia L, Cavallo M, Allegrone G, Martinotti MG (2010) A *Lactobacillus*-derived biosurfactant inhibits biofilm formation of human pathogenic *Candida albicans* biofilm producers. In: Mendez-Vilas A (ed) Current research, technology and education topics in applied microbiology and microbial biochemistry. Formatex, Spain, pp 827–837

Francolini I, Donelli G (2010) Prevention and control of biofilm-based medical-device-related infections. FEMS Immunol Med Microbiol 59:227–238

Fux CA, Stoodley P, Hall-Stoodley L, Costerton JW (2003) Bacterial biofilms: a diagnostic and therapeutic challenge. Expert Rev Anti Infect Ther 1:667–683

Geske GD, Wezeman RJ, Siegel AP, Blackwell HE (2005) Small molecule inhibitors of bacterial quorum sensing and biofilm formation. J Am Chem Soc 127:12762–12763

Geske GD, O'Neill JC, Blackwell HE (2008) Expanding dialogues: from natural autoinducers to non-natural analogues that modulate quorum sensing in Gram-negative bacteria. Chem Soc Rev 37:1432–1447

Giacometti A, Cirioni O, Gov Y, Ghiselli R, Del Prete MS, Mocchegiani F, Saba V, Orlando F, Scalise G, Balaban N, Dell'Acqua G (2003) RNA III inhibiting peptide inhibits in vivo biofilm formation by drug-resistant *Staphylococcus aureus*. Antimicrob Agents Chemother 47:1979–1983

Glansdorp FG, Thomas GL, Lee JJ, Dutton JM, Salmond GP, Welch M, Spring DR (2004) Synthesis and stability of small molecule probes for *Pseudomonas aeruginosa* quorum sensing modulation. Org Biomol Chem 2:3329–3336

Gonzales FP, Felgentrager A, BaumLer W, Maisch T (2013) Fungicidal photodynamic effect of two fold positively charged porphyrin against *Candida albicans* planktonic cells and biofilms. Future Microbiol 8:785–797

Gottenbos B, van der Mei HC, Klatter F, Nieuwenhuis P, Busscher HJ (2001) In vitro and in vivo antimicrobial activity of covalently coupled quaternary ammonium silane coatings on silicone rubber. Biomaterials 23:1417–1423

Gupta VGR (2009) Probiotics. Indian J Med Microbiol 27:202–209

Hall-Stoodley L, Stoodley P (2009) Evolving concepts in biofilm infections. Cell Microbiol 11:1034–1043

Hall-Stoodley L, Costerton JW, Stoodley P (2004) Bacterial biofilms: from the natural environment to infectious diseases. Nat Rev Microbiol 2:95e108

Hamblin MR, Hasan T (2004) Photodynamic therapy: a new antimicrobial approach to infectious disease? Photochem Photobiol Sci 3:436–450

Harjai K, Kumar R, Singh S (2010) Garlic blocks quorum sensing and attenuates the virulence of *Pseudomonas aeruginosa*. FEMS Immunol Med Microbiol 58:161–168

Harriott MM, Noverr MC (2009) *Candida albicans* and *Staphylococcus aureus* form polymicrobial biofilms: effects on antimicrobial resistance. Antimicrob Agents Chemother 53:3914–3922

Harriott MM, Noverr MC (2010) Ability of *Candida albicans* mutants to induce *Staphylococcus aureus* vancomycin resistance during polymicrobial biofilm formation. Antimicrob Agents Chemother 54:3746–3755

Hasan F, Xess I, Wang X, Jain N, Fries BC (2009) Biofilm formation in clinical *Candida* isolates and its association with virulence. Microbes Infect 11:753–761

Hashimoto H (2001) Evaluation of the anti-biofilm effect of a new antibacterial silver citrate/lecithin coating in an in-vitro experimental system using a modified Robbins device. J Jpn Assoc Infect Dis 75:678–685

Hendry ER, Worthington T, Conway BR, Lambert PA (2009) Antimicrobial efficacy of eucalyptus oil and 1,8-cineole alone and in combination with chlorhexidine digluconate against microorganisms grown in planktonic and biofilm cultures. J Antimicrob Chemother 64:1219–1225

Hernandez-Delgadillo R, Velasco-Arias D, Diaz D, Arevalo-Niño K, Garza-Enriquez M, De la Garza-Ramos MA, Cabral-Romero C (2012) Zerovalent bismuth nanoparticles inhibit *Streptococcus mutans* growth and formation of biofilm. Int J Nanomed 7:2109–2113

Hernandez-Delgadillo R, Velasco-Arias D, Diaz D, Arevalo-Nino K, Garza-Enriquez M, Garza-Ramos MAD, Cabral-Romero C (2013a) Zerovalent bismuth nanoparticles inhibit *Streptococcus mutans* growth and formation of biofilm. Int J Nanomed 7:2109–2113

Hernandez-Delgadillo R, Velasco-Arias D, Diaz D, Arevalo-Nino K, Garza-Enriquez M, Garza-Ramos MAD, Cabral-Romero C (2013b) Bismuth oxide aqueous colloidal nanoparticles inhibit *Candida albicans* growth and biofilm formation. Int J Nanomed 8:1645–1652

Hockenhull JC, Dwan KM, Smith GW, Gamble CL, Boland A, Walley TJ, Dickson RC (2009) The clinical effectiveness of central venous catheters treated with anti infective agents in preventing catheter-related bloodstream infections: a systematic review. Crit Care Med 37:702–712

Hogan DA (2006) Talking to themselves: autoregulation and quorum sensing in fungi. Eukaryot Cell 5:613–619

Hoiby N, Bjarnsholt T, Givskov M, Molin S, Ciofu O (2010) Antibiotic resistance of bacterial biofilms. Int J Antimicrob Agents 35:322–332

Il'ina TS, Romanova YM, Gintsburg AL (2004) Biofilms as a mode of existence of bacteria in external environment and host body: the phenomenon, genetic control, and regulation systems of development. Genetika 40:1445–1456

Issac Abraham SV, Palani A, Ramaswamy BR, Shunmugiah KP, Arumugam VR (2011) Antiquorum sensing and antibiofilm potential of *Capparis spinosa*. Arch Med Res 42:658–668

Itoh Y, Wang X, Hinnebusch BJ, Preston JF III, Romeo T (2005) Depolymerization of β-1, 6-N-acetyl-D-glucosamine disrupts the integrity of diverse bacterial biofilms. J Bacteriol 187:382–387

Iwase T, Uehara Y, Shinji H, Tajima A, Seo H, Takada K, Agata T, Mizunoe Y (2010) *Staphylococcus epidermidis* Esp inhibits *Staphylococcus aureus* biofilm formation and nasal colonization. Nature 465:346–351

Jabra-Rizk MA, Meiller TF, James C, Shirtliff ME (2006) Effect of farnesol on *Staphylococcus aureus* biofilm formation and antimicrobial susceptibility. Antimicrob Agents Chemother 50:1463–1469

Jain AGY, Agrawal R, Khare P, Jain SK (2007) Biofilms a microbial life perspective: a critical review. Crit Rev Ther Drug Carrier Syst 24:393–443

Janek T, Łukaszewicz M, Krasowska A (2012) Antiadhesive activity of the biosurfactant pseudofactin II secreted by the Arctic bacterium *Pseudomonas fluorescens* BD5. BMC Microbiol 12:24

Jayaraman A, Wood TK (2008) Bacterial quorum sensing: signals, circuits, and implications for biofilms and disease. Annu Rev Biomed Eng 10:145–167

Jiang P, Li J, Han F, Duan G, Lu X, Gu Y, Yu W (2011) Antibiofilm activity of an exopolysaccharide from marine bacterium *Vibrio* sp. QY101. PLoS ONE 6:e18514

Junker LM, Clardy J (2007) High-throughput screens for small-molecule inhibitors of *Pseudomonas aeruginosa* biofilm development. Antimicrob Agents Chemother 51:3582–3590

Kalishwaralal K, BarathManiKanth S, Pandian SR, Deepak V, Gurunathan S (2010) Silver nanoparticles impede the biofilm formation by *Pseudomonas aeruginosa* and *Staphylococcus epidermidis*. Colloids Surf B Biointerfaces 79:340–344

Kaplan JB (2009) Therapeutic potential of biofilm-dispersing enzymes. Int J Artif Organs 32:545–554

Kaplan JB (2010) Biofilm dispersal: mechanisms, clinical implications and potential therapeutic uses. J Dent Res 89:205–218

Karlsson AJ, Pomerantz WC, Neilsen KJ, Gellman SH, Palecek SP (2009) Effect of sequence and structural properties on 14-helical betapeptide activity against *Candida albicans* planktonic cells and biofilms. ACS Chem Biol 4:567–579

Karlsson AJ, Flessner RM, Gellman SH, Lynn DM, Palecek SP (2010) Polyelectrolyte multilayers fabricated from antifungal β-peptides: design of surfaces that exhibit antifungal activity against *Candida albicans*. Biomacromolecules 11:2321–2328

Khan MSA, Ahmad I (2012a) Biofilm inhibition by *Cymbopogon citratus* and *Syzygium aromaticum* essential oils in the strains of *Candida albicans*. J Ethnopharmacol 140:416–423

Khan MSA, Ahmad I (2012b) Antibiofilm activity of certain phytocompounds and their synergy with fluconazole against *Candida albicans* biofilms. J Antimicrob Chemother 67:618–621

Kim JS, Kuk E, Yu KN, Kim JH, Park SJ, Lee HJ, Kim SH, Park YK, Park YH, Hwang CY, Kim YK, Lee YS, Jeong DH, Cho MH (2007) Antimicrobial effects of silver nanoparticles. Nanomedicine 3:95–101

Kishen A, Upadya M, Tegos GP, Hamblin MR (2010) Efflux pump inhibitor potentiates antimicrobial photodynamic inactivation of *Enterococcus faecalis* biofilm. Photochem Photobiol 86:1343–1349

Kiran GS, Sabarathnam B, Selvin J (2010) Biofilm disruption potential of a glycolipid biosurfactant from marine *Brevibacterium casei*. FEMS Immunol Med Microbiol 59:432–438

Kiran S, Sharma P, Harjai K, Capalash N (2011) Enzymatic quorum quenching increases antibiotic susceptibility of multidrug resistant *Pseudomonas aeruginosa*. Iran J Microbiol 3:1–12

Knowles JR, Roller S, Murray DB, Naidu AS (2005) Antimicrobial action of carvacrol at different stages of dual-species biofilm development by *Staphylococcus aureus* and *Salmonella enterica* Serovar Typhimurium. Appl Environ Microbiol 71:797–803

Kokare CR, Chakraborty S, Khopade AN, Mahadik KR (2009) Biolfilm: importance and applications. Indian J Biotechnol 8:159–168

Konopka K, Dorocka-Bobkowska B, Gebremedhin S, Duzguneş N (2010) Susceptibility of *Candida* biofilms to histatin 5 and fluconazole. Antonie Van Leeuwenhoek 97:413–417

Ku TSN, Palanisamy SKA, Lee SA (2010) Susceptibility of *Candida albicans* biofilms to azithromycin, tigecycline and vancomycin and the interaction between tigecycline and antifungals. Int J Antimicrob Agents 36:441–446

Kudva IT, Jelacic S, Tarr PI, Youderian P, Hovde CJ (1999) Biocontrol of *Escherichia coli* O157 with O157-specific bacteriophages. Appl Environ Microbiol 65:3767–3773

Kuramitsu HK, He X, Lux R, Anderson MH, Shi W (2007) Interspecies interactions within oral microbial communities. Microbiol Mol Biol Rev 71:653–670

Kuzma L, Ozalski MR, Walencka E, Ozalska BR, Wysokinska H (2007) Antimicrobial activity of diterpenoids from hairy roots of *Salvia sclarea* L.: salvipisone as a potential anti-biofilm agent active against antibiotic resistant *Staphylococci*. Phytomedicine 14:31–35

Lazar V (2011) Quorum sensing in biofilms: how to destroy the bacterial citadels or their cohesion/power? Anaerobe 17:280e285

Lazzell AL, Chaturvedi AK, Pierce CG, Prasad D, Uppuluri P, Lopez-Ribot JL (2009) Treatment and prevention of *Candida albicans* biofilms with caspofungin in a novel central venous cathetermurine model of candidiasis. J Antimicrob Chemother 64:567–570

Lee J, Jayaraman A, Wood TK (2007) Indole is an inter-species biofilm signal mediated by SdiA. BMC Microbiol 7:42

Lewis K (2005) Persister cells and the riddle of biofilm survival. Biokhimiya 70:327–336

Lonn-Stensrud J, Landin MA, Benneche T, Petersen FC, Scheie AA (2009) Furanones, potential agents for preventing *Staphylococcus epidermidis* biofilm infections? J Antimicrob Chemother 63:309–331

Mardh PA, Rodrigues AG, Genc M, Novikova N, Martinez-de-Oliveira J, Guaschino S (2002) Facts and myths on recurrent vulvovaginal candidosis—a review on epidemiology, clinical manifestations, diagnosis, pathogenesis and therapy. Int J STD AIDS 13:522–539

Martinez LR, Mihu MR, Tar M, Cordero RJ, Han G, Friedman AJ, Friedman JM, Nosanchuk JD (2010) Demonstration of antibiofilm and antifungal efficacy of chitosan against candidal biofilms, using an in vivo central venous catheter model. J Infect Dis 201:1436–1440

Martins M, Henriques M, Lopez-Ribot JL, Oliveira R (2012) Addition of DNase improves the in vitro activity of antifungal drugs against *Candida albicans* biofilms. Mycoses 55:80–85

Matejuk A, Leng Q, Begum MD, Woodle MC, Scaria P, Chou ST, Mixson AJ (2010) Peptide-based antifungal therapies against emerging infections. Drugs Future 35:197–217

Meire MA, Coenye T, Nelis HJ, De Moor RC (2012) Evaluation of Nd:YAG and Er:YAG irradiation, antibacterial photodynamic therapy and sodium hypochlorite treatment on *Enterococcus faecalis* biofilms. Int Endod J 45:482–491

Mermel LA, Allon M, DE Bouza C, Flynn P, O'Grady NP, Raad II, Rijnders BJ, Sherertz RJ, Warren DK (2009) Clinical practice guidelines for the diagnosis andmanagement of intravascular catheter-related infection: 2009 update by the infectious diseases society of America. Clin Infect Dis 49:1–45

Miceli MH, Bernardo SM, Lee SA (2009) In vitro analyses of the combination of high-dose doxycycline and antifungal agents against *Candida albicans* biofilms. Int J Antimicrob Agents 34:326–332

Mireles JR, Toguchi A, Harshey RM (2001) *Salmonella enterica* serovar *Typhimurium* swarming mutants with altered biofilm-forming abilities: surfactin inhibits biofilm formation. J Bacteriol 183:5848–5854

Molin S, Tolker-Nielson T (2003) Gene transfer occurs with enhanced efficiency in biofilms and induces enhanced stabilisation of the biofilm structure. Curr Opin Biotechnol 14:255–261

Monteiro AS, Miranda TT, Lula I, Denadai AML, Sinisterra RD, Santoro MM, Santos VL (2011a) Inhibition of *Candida albicans* CC biofilms formation in polystyrene plate surfaces by biosurfactant produced by *Trichosporon montevideense* CLOA72. Colloids Surf B Biointerfaces 84:467–476

Monteiro DR, Gorup LF, Silva S, Negri M, de Camargo ER, Oliveira R, Barbosa DB, Henriques M (2011b) Silver colloidal nanoparticles: antifungal effect against adhered cells and biofilms of *Candida albicans* and *Candida glabrata*. Biofouling 27:711–719

Muller FM, Seidler M, Beauvais A (2011) *Aspergillus fumigatus* biofilms in the clinical setting. Med Mycol 49(Suppl 1):S96–S100

Musk DJ, Banko DA, Hergenrother PJ (2005) Iron salts perturb biofilm formation and disrupt existing biofilms of *Pseudomonas aeruginosa*. Chem Biol 12:789–796

Musthafa KS, Saroja V, Pandian SK, Ravi AV (2011) Antipathogenic potential of marine *Bacillus* sp. SS4 on N-acyl-homoserine-lactonemediated virulence factors production in *Pseudomonas aeruginosa* (PAO1). J Biosci 36:55–67

Namasivayam SKR, Preethi M, Bharani ARS, Robin G, Latha B (2012) Biofilm inhibitory effect of silver nanoparticles coated catheter against *Staphylococcus aureus* and evaluation of its synergistic effects with antibiotics. Int J Biol Pharm Res 3:259–265

Nava-Ortiz CAB, Burillo G, Concheiro A et al (2010) Cyclodextrin-functionalized biomaterials loaded with miconazole prevent *Candida albicans* biofilm formation in vitro. Acta Biomater 6:1398–1404

Nikolaev YA, Plankunov VK (2007) Biofilm- "city of microbes" or an analogue of multicellular organisms? Microbiology 76:125–138

Nithyanand P, Thenmozhi R, Rathna J, Pandian SK (2010) Inhibition of biofilm formation in *Streptococcus pyogenes* by coral associated Actinomycetes. Curr Microbiol 60:454–460

Niu C, Gilbert ES (2004) Colorimetric method for identifying plant essential oil components that affect biofilm formation and structure. Appl Environ Microbiol 70:6951–6956

Njoroge J, Sperandio V (2009) Jamming bacterial communication: new approaches for the treatment of infectious diseases. EMBO Mol Med 1:201–210

Nostro A, Roccaro AS, Bisignano G, Marino A, Cannatelli MA, Pizzimenti FC, Cioni PL, Procopio F, Blanco AR (2007) Effects of oregano, carvacrol and thymol on *Staphylococcus aureus* and *Staphylococcus epidermidis* biofilms. J Med Microbiol 56:519–523

Oosterhof JJH, Buijssen KJDA, Busscher HJ, van der Laan BFAM, van der Mei HC (2006) Effects of quaternary ammonium silane coatings on mixed fungal and bacterial biofilms on tracheoesophageal shunt prostheses. Appl Environ Microbiol 72:3673–3677

Padmapriya B, Suganthi S (2013) Antimicrobial and anti adhesive activity of purified biosurfactants produced by *Candida* species. Middle East J Sci Res 14:1359–1369

Peters BM, Jabra-Rizk MA, Scheper MA, Leid JG, Costerton JW, Shirtliff ME (2010) Microbial interactions and differential protein expression in *Staphylococcus aureus*–*Candida albicans* dual-species biofilms. FEMS Immunol Med Microbiol 59:493–503

Prince AS (2002) Biofilms, antimicrobial resistance, and airway infection. N Engl J Med 347:1110–1111

Rabea EI, Badawy ME, Stevens CV, Smagghe G, Steurbaut W (2003) Chitosan as antimicrobial agent: applications and mode of action. Biomacromolecules 4:1457–1465

Radford DR, Sweet SP, Challacombe SJ, Walter JD (1998) Adherence of *Candida albicans* to denture-base materials with different surface finishes. J Dent 26:577–583

Raffa RB, Iannuzzo JR, Levine DR, Saeid KK, Schwartz RC, Sucic NT, Terleckyj OD, Jeffrey M (2005) Young bacterial communication ("Quorum Sensing") via ligands and receptors: a novel pharmacologic target for the design of antibiotic drugs. J Pharmacol Exp Ther 312:417–423

Ramage G, Saville DO, Lopez-Ribot JL (2005) *Candida* biofilm: an update. Eukaryot Cell 4:633–638

Rasmussen TB, Givskov M (2006) Quorum sensing inhibitors: a bargain of effects. Microbiology 152:895–904

Rasmussen TB, Skindersoe ME, Bjarnsholt T, Phipps RK, Christensen KB, Jensen PO, Andersen JB, Koch B, Larsen TO, Hentzer M, Eberl L, Hoiby N, Givskov M (2005) Identity and effects of quorum-sensing inhibitors produced by *Penicillium* species. Microbiology 151:1325–1340

Ravichandiran V, Shanmugam K, Anupama K, Fomas S, Princy A (2012) Structure-based virtual screening for plant-derived SdiA-selective ligands as potential antivirulent agents against uropathogenic *Escherichia coli*. Eur J Med Chem 48:200–205

Redding S, Bhatt HR, Rawls G, Siegel K, Scott Lopez-Ribot J (2009) Inhibition of *Candida albicans* biofilm formation on denture material. Oral Surg Oral Med Oral Pathol Oral Radiol Endod 107:669–672

Ren D, Bedzyk LA, Ye RW, Thomas SM, Wood TK (2004) Stationary-phase quorum-sensing signals affect autoinducer-2 and gene expression in *Escherichia coli*. Appl Environ Microbiol 70:2038–2043

Ren DC, Zuo RJ, Barrios AF, Bedzyk LA, Eldridge GR, Pasmore ME, Wood TK (2005) Differential gene expression for investigation of *Escherichia coli* biofilm inhibition by plant extract ursolic acid. Appl Environ Microbiol 71:4022–4034

Repentigny D, Lewandowski D, Jolicoeur P (2004) Immunopathogenesis of oropharyngeal candidiasis in human immunodeficiency virus infection. Clin Microbiol Rev 17:729–759

Ribeiro APD, Andrade MC, Bagnato VS, Vergani CE, Primo FL, Tedesco AC, Pavarina AC (2013) Antimicrobial photodynamic therapy against pathogenic bacterial suspensions and biofilms using chloro-aluminum phthalocyanine encapsulated in nanoemulsions. Lasers Med Sci. doi:10.1007/s10103-013-1354-x

Rivardo F, Turner RJ, Allegrone G, Ceri H, Martinotti MG (2009) Anti-adhesion activity of two biosurfactants produced by *Bacillus* spp. prevents biofilm formation of human bacterial pathogens. Appl Microbiol Biotechnol 83:541–553

Rodrigues LR, van der Mei HC, Teixeira JA, Oliveira R (2004) Biosurfactant from *Lactococcus lactis* 53 inhibits microbial adhesion on silicone rubber. Appl Microbiol Biotechnol 66:306–311

Rodrigues L, Banat IM, Teixeira J, Oliveira R (2006) Biosurfactants: potential applications in medicine. J Antimicrob Chemother 57:609–618

Rohde H, Burandt EC, Siemssen N, Frommelt L, Burdlski C, Wurster S, Scherpe S, Davies AP, Harris LG, Horstkotte MA, Knobloch JKM, Ragunath C, Kaplan JB, Mack D (2007) Polysaccharide intercellular adhesin or protein factors in biofilm accumulation of *Staphylococcus epidermidis* and *Staphylococcus aureus* isolated from prosthetic hip and knee joint infections. Biomaterials 28:1711–1720

Rossland E, Langsrud T, Granum PE, Sorhaug T (2005) Production of antimicrobial metabolites by strains of *Lactobacillus* or *Lactococcus* co-cultured with *Bacillus cereus* in milk. Int J Food Microbiol 98:193–200

Rufino RD, Luna JM, Sarubbo LA, Rodrigues LRM, Teixeira JAC, Campos-Takaki GM (2011) Antimicrobial and anti-adhesive potential of a biosurfactant rufisan produced by *Candida lipolytica* UCP0988. Colloids Surf B Biointerfaces 84:1–5

Rukayadi Y, Han S, Yong D, Hwang JK (2011) In vitro activity of xanthorrhizol against *Candida glabrata*, *C. guilliermondii*, and *C. parapsilosis* biofilms. Med Mycol 49:1–9

Ryjenkov DA, Tarutina M, Moskvin OV, Gomelsky M (2005) Cyclic diguanylate is a ubiquitous signaling molecule in bacteria: insights into biochemistry of the GGDEF protein domain. Bacteriology 187:1792–1798

Sachachter B (2003) Slimy business-the biotechnology of biofilms. Nat Biotechnol 21:361–365

Saino E, Sbarra MS, Arciola CR, Scavone M, Bloise N, Nikolov P, Ricchelli F, Visai L (2010) Photodynamic action of Tri-meso (N-methyl-pyridyl), meso (N-tetradecyl-pyridyl) porphine on *Staphylococcus epidermidis* biofilms grown on Ti6Al4V alloy. Int J Artif Organs 33:636–645

Sambanthamoorthy K, Gokhale AA, Lao W, Parashar V, Neiditch MB, Semmelhack MF, Lee I, Waters CM (2011) Identification of a novel benzimidazole that inhibits bacterial biofilm formation in a broad-spectrum manner. Antimicrob Agents Chemother 55:4369–4378

Sarabhai S, Sharma P, Capalash N (2013) Ellagic acid derivatives from *Terminalia chebula* Retz. downregulate the expression of quorum sensing genes to attenuate *Pseudomonas aeruginosa* PAO1 virulence. PLoS ONE 8:e53441

Schinabeck MK, Long LA, Hossain MA, Chandra J, Mukherjee PK, Mohamed S, Ghannoum MA (2004) Rabbit model of *Candida albicans* biofilm infection: liposomal amphotericin B antifungal lock therapy. Antimicrob Agents Chemother 48:1727–1732

Secinti KD, Ozalp H, Attar A, Sargon MF (2011) Nanoparticle silver ion coatings inhibit biofilm formation on titanium implants. J Clin Neurosci 18:391–395

Shen Y, Koller T, Kreikemeyer B, Nelson DC (2013) Rapid degradation of *Streptococcus pyogenes* biofilms by PlyC, a bacteriophage-encoded endolysin. J Antimicrob Chemother. doi:10.1093/jac/dkt104

Shirtliff ME, Krom BP, Meijering RA, Peters BM, Zhu J, Scheper MA, Harris ML, Jabra-Rizk MA (2009) Farnesol-induced apoptosis in *Candida albicans*. Antimicrob Agents Chemother 53:392–2401

Sillankorva S, Oliveira DR, Vieira MJ, Sutherland IW, Azeredo J (2004) Bacteriophage V S1 infection of *Pseudomonas fluorescens* planktonic cells versus biofilms. Biofouling 20:133–138

Sillankorva S, Neubauer P, Azeredo J (2010) Phage control of dual species biofilms of *Pseudomonas fluorescens* and *Staphylococcus lentus*. Biofouling 26:567–575

Simoes M, Sillankorva S, Pereira MO, Azeredo J, Vieira MJ (2007) The effect of hydrodynamic conditions on the phenotype of *Pseudomonas fluorescens* biofilms. Biofouling 24:249–258

Singh P, Cameotra S (2004) Potential applications of microbial surfactants in biomedical sciences. Trends Biotechnol 22:142–146

Singh N, Pemmaraju SC, Pruthi PA, Cameotra SS, Pruthi V (2013) *Candida* biofilm disrupting ability of di-rhamnolipid (RL-2) produced from *Pseudomonas aeruginosa* DSVP20. Appl Biochem Biotechnol 169:2374–2391

Sintim HO, Smith JA, Wang J, Nakayama S, Yan L (2010) Paradigm shift in discovering next-generation anti-infective agents: targeting quorum sensing, c-di-GMP signaling and biofilm formation in bacteria with small molecules. Future Med Chem 2:1005–1035

Smith AW (2005) Biofilms and antibiotic therapy: is there a role for combating bacterial resistance by the use of novel drug delivery systems. Adv Drug Deliv Rev 57:1539–1550

Smith KM, Bu YG, Suga H (2003a) Library screening for synthetic agonists and antagonists of a *Pseudomonas aeruginosa* autoinducer. Chem Biol 10:563–571

Smith KM, Bu YG, Suga H (2003b) Induction and inhibition of *Pseudomonas aeruginosa* quorum sensing by synthetic autoinducer analogs. Chem Biol 10:81–89

Soderling EM, Marttinen AM, Haukioja AL (2011) Probiotic lactobacilli interfere with *Streptococcus mutans* biofilm formation in vitro. Curr Microbiol 62:618–622

Splendiani A, Livingston AG, Nicolella C (2006) Control membrane-attached biofilms using surfactants. Biotechnol Bioeng 94:15–23

Stark M, Liu LP, Deber CM (2002) Cationic hydrophobic peptides with antimicrobial activity. Antimicrob Agents Chemother 46:3585–3590

Stewart PS (2003) New ways to stop biofilm infections. Lancet 361:97

Suzuki N, Yoneda M, Hatano Y, Iwamoto T, Masuo Y, Hirofuji T (2011) *Enterococcus faecium* WB2000 inhibits biofilm formation by oral cariogenic *Streptococci*. Int J Dent. doi:10.1155/2011/834151

Tahmourespour A, Kermanshahi RK (2011) The effect of a probiotic strain (*Lactobacillus acidophilus*) on the plaque formation of oral *Streptococci*. Bosn J Basic Med Sci 11:37–40

Takano E, Chakraburtty R, Nihira T, Yamada Y, Bibb MJ (2001) A complex role for the γ-butyrolactone SCB1 in regulating antibiotic production in *Streptomyces coelicolor* A3(2). Mol Microbiol 41:1015–1028

Tamayo R, Pratt JT, Camilli A (2007) Roles of cyclic diguanylate in the regulation of bacterial pathogenesis. Annu Rev Microbiol 61:131–148

Taraszkiewicz A, Fila G, Grinholc M, Nakonieczna J (2013) Innovative strategies to overcome biofilm resistance. Bio Med Res Int. doi:10.1155/2013/150653

Taweechaisupapong S, Singhara S, Lertsatitthanakorn P, Khunkitti W (2010) Antimicrobial effects of *Boesenbergia pandurata* and *Piper sarmentosum* leaf extracts on planktonic cells and biofilm of oral pathogens. Pak J Pharm Sci 23:224–231

Tetz GV, Artemenko NK, Tetz VV (2009) Effect of DNase and antibiotics on biofilm characteristics. Antimicrob Agents Chemother 53:1204–1209

Teughels W, Van Assche N, Sliepen I, Quirynen M (2006) Effect of material characteristics and/or surface topography on biofilm development. Clin Oral Implants Res 17:68–81

Tew GN, Liu D, Chen B, Doerksen RJ, Kaplan J, Carroll PJ, Klein ML, DeGrado WF (2002) De novo design of biomimetic antimicrobial polymers. Proc Natl Acad Sci USA 99:5110–5114

Thaweboon S, Thaweboon B (2009) In vitro antimicrobial activity of *Ocimum americanum* L. essential oil against oral microorganisms. Southeast Asian J Trop Med Public Health 40:1025–1033

Thenmozhi R, Nithyanand P, Rathna J, Pandian SK (2009) Antibiofilm activity of coral associated bacteria against different clinical M serotypes of *Streptococcus pyogenes*. FEMS Immunol Med Microbiol 57:284–294

Toulet D, Debarre C, Imbert C (2012) Could liposomal amphotericin B (L-AMB) lock solutions be useful to inhibit *Candida* spp. biofilms on silicone biomaterials? J Antimicrob Chemother 67:430–432

Tournu H, Dijck PV (2012) *Candida* biofilms and the host: models and new concepts for eradication. Int J Microbiol. doi:10.1155/2012/845352

Trampuz A, Widmer AF (2006) Infections associated with orthopedic implants. Curr Opin Infect Dis 19:349–356

Tsai PW, Yang CY, Chang HT, Lan CY (2011) Human antimicrobial peptide LL-37 inhibits adhesion of *Candida albicans* by interacting with yeast cell-wall carbohydrates. PLoS ONE 6: e17755. doi:10.1371/journal.pone.0017755

Valle J, Re DS, Henry N, Fontaine T, Balestrino D, Latour-Lambert P, Ghigo JM (2006) Broad-spectrum biofilm inhibition by a secreted bacterial polysaccharide. Proc Natl Acad Sci USA 103:12558–12563

van Hamme JD, Singh A, Ward OP (2006) Physiological aspects Part 1 in a series of papers devoted to surfactants in microbiology and biotechnology. Biotechnol Adv 24:604–620

Venkatesh M, Rong L, Raad I, Versalovic J (2009) Novel synergistic antibiofilm combinations for salvage of infected catheters. J Med Microbiol 58:936–944

Villa F, Cappitelli F (2013) Plant-derived bioactive compounds at sub-lethal concentrations: towards smart biocide-free antibiofilm strategies. Phytochem Rev 12:245–254

Wainwright M (2003) The use of dyes in modern biomedicine. Biotech Histochem 78:147–155

Walencka E, Sadowska B, Rozalska S, Hryniewicz W, Rozalska B (2005) Lysostaphin as a potential therapeutic agent for staphylococcal biofilm eradication. Pol J Microbiol 54:191–200

Walencka E, Rozalska S, Sadowska B, Rozalska B (2008) The influence of *Lactobacillus acidophilus*-derived surfactants on staphylococcal adhesion and biofilm formation. Folia Microbiol 53:61–66

Waters CM, Bassler BL (2005) Quorum sensing: cell-to-cell communication in bacteria. Annu Rev Cell Dev Biol 21:319–346

Weiss EI, Lev-Dor R, Kashamn Y, Goldhar J, Sharon N, Ofek I (1998) Inhibiting interspecies coaggregation of plaque bacteria with a craneberry juice constituent. J Am Dent Assoc 129:1719–1723

Wertheim HF, Melles DC, Vos MC, van Leeuwen W, van Belkum A, Verbrugh HA, Nouwen JL (2005) The role of nasal carriage in *Staphylococcus aureus* infections. Lancet Infect Dis 5:751–762

Wilson M (2001) Bacterial biofilms and human disease. Sci Prog 84:235–254

Worthington RJ, Justin J, Richards MC (2012) Small molecule control of bacterial biofilms. Org Biomol Chem 10:7457–7474

Wuertz SOS, Hausner M (2004) Microbial communities and their interactions in biofilm systems: an overview. Water Sci Technol 49:327–336

Xavier JB, Picioreanu C, Rani SA, van Loosdrecht MCM, Stewart PS (2005) Biofilm-control strategies based on enzymic disruption of the extracellular polymeric substance matrix: a modelling study. Microbiology 151:3817–3832

Xiong Y, Liu Y (2010) Biological control of microbial attachment: a promising alternative for mitigating membrane biofouling. Appl Microbiol Biotechnol 86:825–837

Yan H, Chen W (2010) 3′,5′-cyclic diguanylic acid: a small nucleotide that makes big impacts. Chem Soc Rev 39:2914–2924

Yang L, Rybtke MT, Jakobsen TH, Hentzer M, Bjarnsholt T, Givskov M, Tolker-Nielsen T (2009) Computer-aided identification of recognized drugs as *Pseudomonas aeruginosa* quorum-sensing inhibitors. Antimicrob Agents Chemother 53:2432–2443

Zumbuehl A, Ferreira L, Kuhn D, Astashkina A, Long L, Yeo Y, Iaconis T, Ghannoum M, Fink GR, Langer R, Kohane DS (2007) Antifungal hydrogels. Proc Natl Acad Sci USA 104:12994–12998

The Effect of Plasmids and Other Biomolecules on the Effectiveness of Antibiofilm Agents

L.C. Gomes, P.A. Araújo, J.S. Teodósio, M. Simões, and F.J. Mergulhão

Abstract This chapter describes the impact of cell transformation with a recombinant plasmid and the effects of the presence of selected biomolecules (bovine serum albumin—BSA, alginate, yeast extract and humic acids) on biofilm resistance to quaternary ammonium compounds (QACs), which are often used in medical applications to prevent microbial contamination. Two case studies are presented, the first concerning cell transformation with recombinant plasmids and the second addressing potential interfering substances. In the first case study, the pET28 and pUC8 plasmids were used to transform *Escherichia coli* JM109(DE3), and biofilm formation, removal and antimicrobial susceptibility to the cationic biocide benzyldimethyldodecylammonium chloride (BDMDAC) were assessed. Plasmid-bearing cells formed biofilms with higher cell densities, whereas non-transformed cells had higher viabilities. It was found that biocide treatment was not efficient for biofilm removal and that the thickness of the biofilms formed by non-transformed cells is less affected by the treatment, a fact that can be associated with a higher protein content of the biofilm matrix. Despite being unsuccessful at removing the biofilms, BDMDAC was very effective at killing the cells since complete inactivation was attained for transformed and non-transformed strains. In the second case study, it was possible to conclude that BSA, alginate and yeast extract resulted in mild interferences in the antibacterial activity of benzalkonium chloride (BAC) and cetyltrimethyl ammonium bromide (CTAB) against *Bacillus cereus* and *Pseudomonas fluorescens*. Humic acids have a severe impact on the activity of these QACs and can even trigger metabolic activation in some circumstances. These observations suggest that the presence of the tested biomolecules should be taken into account when using QACs as disinfection agents.

L.C. Gomes • P.A. Araújo • J.S. Teodósio • M. Simões • F.J. Mergulhão (✉)
LEPABE, Department of Chemical Engineering, Faculty of Engineering, University of Porto, Rua Dr. Roberto Frias, s/n, 4200-465, Porto, Portugal
e-mail: filipem@fe.up.pt

1 Introduction

Biofilms cause serious problems in environmental, industrial and biomedical fields. Nevertheless, the biofilms with the worst reputation are those found in the health sector (Bryers 2008), since more than 50 % of all microbial infections in humans are believed to be linked to the formation of biofilms (Costerton et al. 1999). This chapter focuses on biofilms made by *Bacillus cereus*, *Escherichia coli* and *Pseudomonas fluorescens*, which can be hugely problematic in both medical and industrial environments.

With over 250 serotypes, the Gram-negative bacterium *E. coli* is a highly versatile organism ranging from harmless gut commensal to a dangerous pathogen (Beloin et al. 2008). Its frequent community lifestyle and the availability of a wide array of genetic tools have contributed to establish *E. coli* as an excellent model organism for biofilm studies (Beloin et al. 2008; Wood 2009). In the health sector, pathogenic strains of *E. coli* are responsible for 70–95 % of urinary tract infections, one of the most common bacterial diseases (Dorel et al. 2005; Jacobsen et al. 2008). These infections are especially frequent in cases of catheterisation (due to biofilm development on the indwelling catheters) where the incidence of infection increases 5–10 % per day (Dorel et al. 2005). *B. cereus* also exists in hospital environments and can attach to the surface of catheters and cause persistent and chronic infections, especially among immunosuppressed patients (Kuroki et al. 2009; Bottone 2010). *P. fluorescens* is an unusual agent for disease in humans. Nonetheless, this bacterium demonstrates haemolytic activity and it has been known to infect donated blood (Gibb et al. 1995). Other strains from the *Pseudomonas* genus are notorious for their impact in medical settings. For example, *Pseudomonas aeruginosa* has been shown to form biofilms on the tissues of the cystic fibrosis lung (Govan and Deretic 1996) and on abiotic surfaces such as contact lenses and catheter lines (Miller and Ahearn 1987; Nickel et al. 1985).

The biofilm mode of growth leads to a large increase in resistance to antimicrobial agents, including antibiotics, biocides, and preservatives, compared with cultures grown in suspension (Stewart and Costerton 2001; Brown and Smith 2003). In fact, when cells exist in a biofilm, they become 10 to 1,000 times less susceptible to the effects of antimicrobial agents. Some factors that contribute to biofilm resistance to antibiotics and biocides include physical or chemical diffusion barriers to agent penetration within the biofilm matrix, slow growth rate of biofilm cells due to nutrient limitation, activation of the general stress response, and the presence of "persister" cells or antibiotic-resistant small-colony variants (Mah and O'Toole 2001). This high level resistance makes most device-related infections difficult or impossible to eradicate by conventional antimicrobial chemotherapy. Therefore, there is now a perceived need to elucidate resistance mechanisms and develop effective antibiofilm strategies.

In the first case study presented in this chapter, the effects of a QAC—benzyldimethyldodecylammonium chloride (BDMDAC)—on transformed and non-transformed *E. coli* cells were assessed regarding their biofilm removal capacity and bacterial inactivation. Some biocides, such as the QACs, are used externally on the skin to prevent or limit microbial infection (antiseptics and topical antimicrobials) and for preoperative skin disinfection and are incorporated into pharmaceutical products (preservatives) to avoid microbial contamination (Russell 2003). The cationic biocide BDMDAC has already demonstrated strong antibacterial activity, causing damage of the cytoplasmic membrane and the consequent release of essential intracellular components (Ferreira et al. 2011). In the second case-study presented on this chapter, the influence of potential interfering substances (bovine serum albumin—BSA, alginate, yeast extract and humic acids) was studied on the antimicrobial activity of other two QACs—benzalkonium chloride (BAC) and cetyltrimethyl ammonium bromide (CTAB)—against *B. cereus* and *P. fluorescens*, as it was previously shown that the presence of organic matter can affect the efficiency of biocides (Russell 2003). The biomolecules used throughout this experiment are recognised as potentially interfering agents in the European Standard EN-1276 (1997) and biofilm matrix components that have an important role in biofilm resistance against antimicrobial agents (Cloete 2003).

1.1 The Influence of Plasmids on Biofilm Formation and Resistance

A plasmid is an extrachromosomal, circular and double-stranded DNA molecule that carries its own origin of replication. Under natural conditions, many plasmids are transmitted to new hosts by conjugation, a procedure by which donor cells can transfer genes to recipient cells. However, the plasmids used in this work—pET28 and pUC8—are incapable of directing their own conjugation because they lack the *tra* gene and therefore they are non-conjugative plasmids (Ehlers 2000).

The pET28 vector (Fig. 1) is a 5.4 kb plasmid harbouring a kanamycin resistance gene and a pMB1 origin of replication, and is usually present in about 20–60 copies per cell (Prather et al. 2003). This vector is used for recombinant protein expression using the transcriptional promoter and termination sequences from phage T7. Recombinant proteins are expressed as fusions with histidine residues to enable purification with a nickel affinity column. Expression of the T7 polymerase that will transcribe the cloned gene is achieved through the inducible *lacUV5* promoter that is present in the host cell chromosome via a lysogenic insertion. To prevent "leaky" expression from this promoter, a copy of the *lacI* repressor is also present on the plasmid to reduce toxicity effects related with the expression of the cloned gene prior to induction.

The pUC8 vector (Fig. 2) is a 2.7 kb plasmid containing the bla_{TEM-1} ampicillin-resistance gene (Paterson and Bonomo 2005) and a mutated pMB1 origin of replication, which is responsible for its high copy number (500–700 copies per cell) (Prather et al. 2003). It also contains the β-lactamase promoter to transcribe the

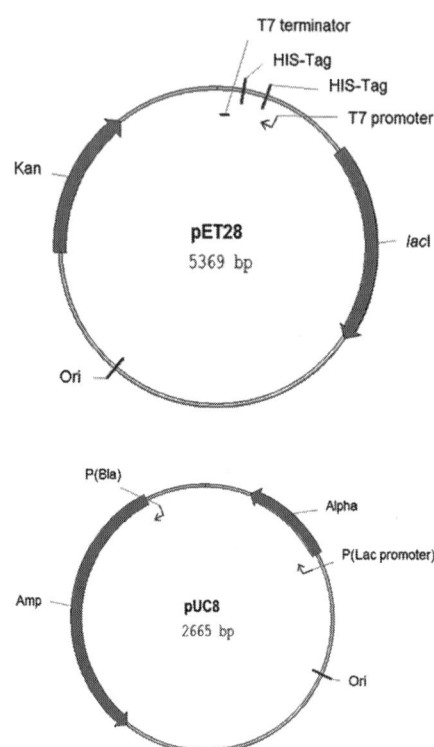

Fig. 1 Plasmid pET28 map. This plasmid harbours (1) a pMB1 origin of replication (Ori), (2) a repressor for the *lac* promoter (*lac*I), (3) a transcriptional promoter from the T7 phage (T7 promoter), (4) an affinity purification tag (HIS-Tag), (5) a T7 transcriptional terminator (T7 terminator), and (6) a kanamycin resistance gene (Kan)

Fig. 2 Plasmid pUC8 map. This plasmid harbours (1) a mutated pMB1 origin of replication (Ori), (2) a lactose promoter [P(Lac promoter)], (3) an α-complementation site for blue/white screening, (4) a β-lactamase promoter [P(Bla)], and (5) an ampicillin resistance gene (Amp)

resistance gene and a fragment of the *lacZ* gene for α-complementation. When a gene is cloned in this plasmid using the existing multiple cloning site, it disrupts the *lacZ* fragment. When this recombinant plasmid containing the cloned gene is transformed into suitable host cells, the colonies can be readily selected through a procedure called "blue/white" screening.

1.2 Physiological Effects of the Presence of Plasmids

Plasmid presence can have a variety of effects on host physiology. Some investigations have considered plasmids as "cellular parasites" (Diaz Ricci and Hernández 2000) since it was recognised that the introduction and expression of foreign DNA in a host organism often changes the metabolism of that organism as a consequence of the "metabolic burden" (Glick 1995). The term "metabolic burden" (sometimes also called "metabolic load" or "metabolic drain") is defined as the amount of host cell resources (raw materials and energy) that is required to maintain and express foreign DNA (Glick 1995). Concerning the physiological alterations at culture level, several studies with *E. coli* have shown that plasmid-bearing cells exhibited lower specific growth rates than plasmid-free cells (Cheah et al. 1987; Ryan et al. 1989; Khosravi et al. 1990; Seo and Bailey 1985; Birnbaum and Bailey 1991; Ow et al. 2006; Flores et al. 2004), resulting in lower biomass yields at the end of fermentation (Ow et al. 2006). The larger the plasmid size (Khosravi et al. 1990; Ryan et al. 1989; Cheah et al. 1987) or higher the copy number (Seo and Bailey 1985; Birnbaum and Bailey 1991), the more severe will be the impact on cell growth. Cheah et al. (1987) compared the behaviour of plasmid pUC8 and four recombinant derivatives containing inserts of different sizes. Although growth in log phase was unaffected by plasmid size, maximum cell density decreased as plasmid size increased and, with the largest plasmid, cell death was accelerated once the stationary phase of cell growth was reached (Cheah et al. 1987). Growth retardation in plasmid-bearing cells is possibly caused by the redirection of intracellular resources such as amino acids, nucleotides and metabolic energy to support plasmid-related activities and by inhibitory mechanisms on host cell metabolism (Simões et al. 2005b; Andersson et al. 1996). At the cellular level, alterations in the amount of *E. coli* cell proteins and ribosomal components have been shown to occur after the introduction of multicopy plasmids (Birnbaum and Bailey 1991). The levels of stress proteins were higher for recombinant strains, while metabolic enzymes showed lower values. Cell filamentation can also occur during plasmid DNA production in *E. coli*, causing a decrease in growth rate or, eventually, no further cell division, leading to low biomass and plasmid DNA productivity (Teodósio et al. 2012). Another detrimental effect of the metabolic burden is a reduced cellular viability in plasmid-bearing cells (Diaz Ricci and Hernández 2000), possibly as a result of the increased stress suffered by these cells. Nonetheless, it is clear that the metabolic burden associated with the plasmid alone is small when compared to the effect of recombinant protein expression (Andersson et al. 1996; Bentley et al. 1990; Da Silva and Bailey 1986).

We may think that plasmids will always affect hosts negatively because they rarely encode functions that are absolutely necessary for their growth, but that is not necessarily true. In nature, plasmids usually provide cells with a growth advantage, showing that under certain culture conditions plasmids can positively affect host performance (Diaz Ricci and Hernández 2000). Rhee et al. (1994) observed that *E. coli* JM109 displays a better growth and a higher metabolic activity in a minimal media when carrying three plasmids than without plasmids. Diaz Ricci and Hernández (2000) have confirmed those results using different culture conditions and other plasmids. An additional report suggested that the presence of a low copy number plasmid, encoding multiple antibiotic resistance, did not affect the maximum growth rate (Klemperer et al. 1979).

The close proximity of bacterial cells in biofilms provides an excellent environment for the exchange of genetic material (Ong et al. 2009) carried by the plasmid. Thereby, the effects of *E. coli* plasmids on biofilm formation have been described on numerous studies, the vast majority of which have used conjugative plasmids (Ghigo 2001; Reisner et al. 2003, 2006; May and Okabe 2008; Yang et al. 2008; Król et al. 2011; Norman et al. 2008). Ghigo (2001) provided the first evidence that natural conjugative plasmids induce the biofilm formation of different *E. coli* K-12 strains, despite that most laboratory *E. coli* K-12 strains are poor biofilm formers. Interestingly, Ghigo's results suggest that the conjugative pili responsible for the horizontal transfer of the plasmid may also act as cell adhesins, which connect the cells and stabilise biofilm structures in hydrodynamic biofilm systems (Ghigo 2001). These results were supported by Reisner et al. (2003) who noted that biofilm formation was most common for natural *E. coli* isolates that harboured conjugative plasmids. Their data suggested that initial attachment to the surface was not improved in the presence of a derepressed IncF plasmid; however, the difference in biofilm development between plasmid-free and plasmid-carrying strains became pronounced after 20 h. Plasmid-carrying strains continued to accumulate biomass and give rise to a dramatically different biofilm architecture (Reisner et al. 2003). Later, May and Okabe (2008) reported that the conjugative factor of F plasmid was involved in colonic acid and curli production during biofilm formation, which promoted cell–surface adherence.

Few studies have addressed the effect of non-conjugative plasmids on biofilm formation. *E. coli* O157:H7 carries a 92-kb virulent and non-conjugative plasmid (pO157) (Burland et al. 1998; Lim et al. 2010b) that influences biofilm formation and architecture (Lim et al. 2010a). Under smooth flow conditions, pO157 enabled biofilm development through increased production of extracellular polymeric substances (EPS) and generation of hyperadherent variants (Lim et al. 2010a). It has been reported (Huang et al. 1993, 1994) that when a plasmid containing a mutated pMB1 origin (the same as in pUC8) was transformed into *E. coli* DH5α, the plasmid-bearing cells formed biofilms with a higher cell density when compared to non-transformed cells. Conversely, Gallant et al. (2005) revealed that strains of *E. coli* carrying TEM-1-encoding plasmid vectors (like that used in this work) grew normally but showed reduced adhesion and biofilm formation.

1.3 The Effects of Potential Interfering Substances on Biofilm Resistance

It is assumed that organic material can potentially interfere with the microbiocidal activity of disinfectants and other antimicrobial compounds (Otzen 2011; Aal et al. 2008; Russell 2008; Stringfellow et al. 2009). This interference is usually due to the reaction between the biocide and the organic matter, reducing the concentration of antimicrobial agent for attacking microorganisms. Another possibility is that organic material protects microorganisms from biocide action (Russell 2008). Consequently, longer contact times and/or higher disinfectant dosages are needed to maintain biocide effectiveness when organic matter is present (Ruano et al. 2001).

Many references related to the study of interfering substances can be found in literature; however, most refer to the effects caused by BSA and water hardness. Aal et al. (2008) evaluated the bactericidal activity of disinfectants referred in the

German Veterinary Society guidelines as references for testing disinfectants used in the dairy and food industries. In order to simulate the conditions found in real life, they used low fat milk as an organic load and reported the significance of choosing an appropriate disinfectant because the inclusion of a challenging substance (organic material) is important to access the proper bactericidal activity. Bessems (1998) demonstrated that a QAC tested on three microorganisms (a Gram-positive, a Gram-negative bacterium and a yeast) had a similar killing rate in the absence of interfering substances. After the inclusion of 17dH water hardness, a strong reduction of the killing activity was found for the Gram-negative bacteria; however, the same behaviour was not observed for the other two microorganisms. Jono et al. (1986) assessed the effect of dried yeast extract and human serum on the activity of BAC, concluding that the bactericidal activity of the QAC was inhibited by solutions of 2.5 % dried yeast extract and 10 % human serum. The inhibition by yeast extract was more pronounced than human serum at the given concentrations. They also concluded that the presence of dried yeast increased the concentration of biocide necessary to kill bacteria.

In the treatment of oral infections, one potential factor reducing the activity of the disinfecting agents is also the chemical environment of the root canal. The root canal system contains a complex mixture of organic and inorganic compounds. Portenier et al. (2001) showed that a number of organic compounds, including BSA, reduced the antimicrobial effectiveness of root canal medicaments (calcium hydroxide, chlorhexidine and iodine potassium iodide). Later, Pappen et al. (2010) revealed that high concentrations of BSA significantly decreased the antimicrobial activity of sodium hypochlorite against oral microorganisms.

2 Case Studies

2.1 Case Study 1: The Influence of Plasmids pET28 and pUC8 on Biofilm Formation and Resistance

The influence of the presence of a non-conjugative plasmid on the biofilm-forming and resistance capacity of *E. coli* cells was studied in a flow cell system (Fig. 3), as described by Teodósio et al. (2012). The planktonic cell concentration was similar for transformed and non-transformed strains, but the number of sessile cells was higher for the plasmid-bearing strains when compared to the non-transformed strain (Teodósio et al. 2012). This observation has also been reported by other authors (Huang et al. 1993, 1994). Diaz Ricci and Hernández (2000) showed that, in certain conditions, plasmids can positively affect cell growth of *E. coli* JM109, as it did here in regard to biofilm. The percentage of viable biofilm-associated cells was about 2-fold lower for the transformed strains (Teodósio et al. 2012). Previous studies suggest that cells bearing a plasmid may suffer from a metabolic burden resulting from plasmid maintenance, replication and/or protein expression (Glick

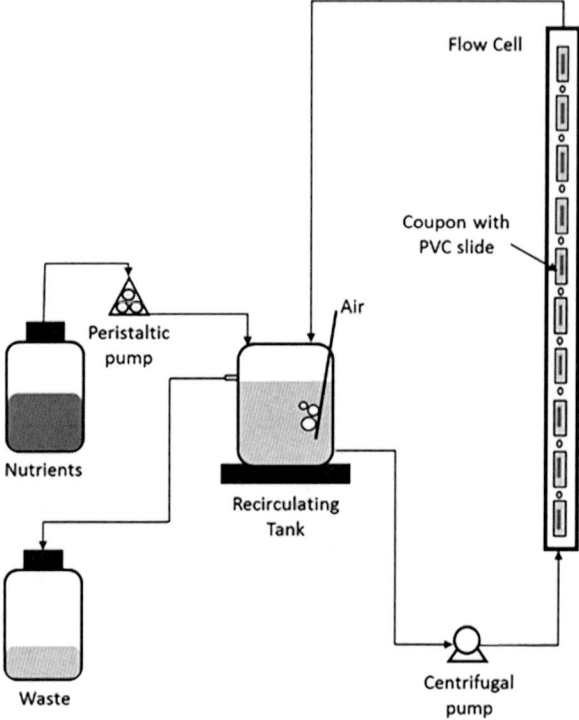

Fig. 3 Schematic representation of the flow cell system used to study the effect of non-conjugative plasmids on the biofilm-forming and resistance capacity of *E. coli* strains

1995; Ow et al. 2006; Wang et al. 2006). A detrimental effect of the metabolic burden was a reduced cellular viability in plasmid-bearing cells, an outcome previously mentioned by Diaz Ricci and Hernández (2000), possibly as a result of the stress suffered by these cells. An increase in glucose metabolism in transformed cells may have been accompanied by the increase in ATP and fermentation by-products, and this accumulation of end-products caused metabolic stress (Diaz Ricci et al. 1991, 1992). As it is known that several stress conditions favour biofilm formation, the conjugation of the above-mentioned stress factors may have stimulated biofilm formation by the transformed strains as thicker biofilms were obtained in these cases (Teodósio et al. 2012).

Biofilm susceptibility experiments were performed using a quaternary ammonium compound—BDMDAC—that was circulated in the system (Fig. 3) at the same flow conditions used for biofilm formation (Teodósio et al. 2012). The tested biocide treatment was not sufficient for complete biofilm removal. It seems that the biofilms formed by the non-transformed strain were more resistant to this treatment, although the total number of cells was relatively constant for all tested strains during the treatment (Teodósio et al. 2012). In a previous study (Simões et al. 2005a), the same concentration of a quaternary ammonium compound was also ineffective in removal of *Pseudomonas* biofilms. Simões et al. (2005a) demonstrated that the presence of BAC increased biofilm mechanical stability. On the other hand, it is known that the antimicrobial mode of action of certain cationic surfactants as QACs has been attributed to their positive charge, which promotes an electrostatic interaction with negatively charged sites on cell membrane (Cloete et al. 1998). So, it is believable that electrostatic interactions could increase cell-to-cell cross-linking by preventing cell removal, resulting in more compact biofilms. Possibly, the QAC cations may also cross-link the anionic groups of biofilm polymers (such as polysaccharides), providing greater binding force in a developed biofilm. The content of the main EPS components in bacterial biofilms (proteins and polysaccharides) was analysed, and a higher protein content and similar polysaccharide percentage was determined for the non-transformed strain (Teodósio et al. 2012). The results showed that protein content is probably a better indicator for biofilm resistance than the polysaccharide composition, as suggested by Pereira and Vieira (2001).

Despite being unsuccessful at removing the biofilms, BDMDAC was very effective on killing the cells since complete inactivation was attained for all the strains. However, cells transformed with pUC8 were the most resistant to inactivation with this biocide, indicating that antimicrobial susceptibility of biofilms can be plasmid dependent (Teodósio et al. 2012).

2.2 Case Study 2: The Effects of the Presence of BSA, Alginate, Yeast Extract and Humic Acids on Biofilm Resistance

The antibacterial activity of BAC, CTAB and their combination was determined by respirometry (Simões et al. 2005b) in the absence and presence of four potentially interfering substances (Araújo et al. 2013). Higher inactivation rates were observed for *B. cereus* when compared to *P. fluorescens* at the same QAC concentration. *B. cereus* is more susceptible due to the fact that it is a Gram-positive bacterium that lacks an outer membrane, which typically provides increased protection to Gram-negative bacteria (Araújo et al. 2013). This result was also reported by Lawrence (1950), who has shown that Gram-positive bacteria are more affected by cationic surfactants because of their higher ratio between acidic and basic groups, e.g. nucleus and enzyme systems, than Gram-negative bacteria.

The selected interfering substances influenced the antimicrobial activity of the QACs to some extent. The inactivation of *B. cereus* was not significantly affected by the presence of any interfering substance, except that the presence of humic acids increased concentrations necessary for total inactivation (Araújo et al. 2013). The antimicrobial action of the QACs against *P. fluorescens* was not significantly influenced by the presence of most potential interfering substances, except for humic acids, and the antimicrobial activity of the QACs against the bacterial consortium was significantly affected by the presence of the tested interfering substances (Araújo et al. 2013). Humic acids were the interfering substances that had the most notorious effect in reducing the surfactant activity. Ishiguro et al. (2007) reported that cationic surfactants bind intensely to humic substances and Koopal et al. (2004) also verified the formation of complexes of humic acid-cationic surfactant.

Respiratory activity potentiation happened when the QACs were used on *P. fluorescens* and the bacterial consortium in the presence of humic acids and yeast extract (Araújo et al. 2013). Humic acids were proposed to replace synthetic surfactants in industrial applications (Visser 1985), so it is possible that their inclusion in a solution of QACs may interfere with the chemical characteristics of the solution, leading to an apparent reduced antimicrobial efficacy. Since QACs are membrane active agents, their use at sub-lethal concentrations could improve membrane permeability and consequently the nutrient influx. Humic acids might be broken down to smaller molecules that could be utilised by cells as a carbon (Camper 2004) or nutrient source (Salati et al. 2011). In fact, it was found that the growth rates of many anaerobic and aerobic microorganisms are increased by humic substances that stimulate enzyme activity (Hartung 1992; Pouneva 2005). In a similar way, yeast extract is a nitrogen source widely used as component of growth media (Hakobyan et al. 2012).

3 Conclusions

The main aim of this work was to analyse the impacts of cell transformation with a recombinant plasmid and also the presence of organic compounds on the antimicrobial activity of different biocides. In the case of plasmids, biocide treatment results indicate that transformed cells are more resistant to short exposure periods and that BDMDAC is ineffective for *E. coli* biofilm removal but effective for biofilm inactivation. This result can be explained by a high protein content of the biofilm formed by the plasmid-free strain. From the tested interfering substances, humic acids have the most severe impact on the biocide activity, even causing metabolic activation in some circumstances. These results highlight the importance of specific biomolecules on the efficacy of biocides that are currently used in medical applications.

Acknowledgements The authors acknowledge the financial support provided by Operational Programme for Competitiveness Factors—COMPETE, European Fund for Regional Development—FEDER and by the Portuguese Foundation for Science and Technology—FCT, through Projects PTDC/EBB-BIO/102863/2008 and PTDC/EBB-EBI/105085/2008. Luciana Gomes acknowledges the receipt of a Ph.D. grant from FCT (SFRH/BD/80400/2011).

References

Aal S, Hunsinger B, Böhm R (2008) Determination of the bactericidal activity of chemical disinfectants bacteria in dairies according to the DVG-guidelines. Hyg Med 33:463–471

Andersson L, Yang S, Neubauer P, Enfors S-o (1996) Impact of plasmid presence and induction on cellular responses in fed batch cultures of *Escherichia coli*. J Biotechnol 46:255–263

Araújo P, Lemos M, Mergulhão F, Melo L, Simões M (2013) The influence of interfering substances on the antimicrobial activity of selected quaternary ammonium compounds. Int J Food Sci. doi:10.1155/2013/237581

Beloin C, Roux A, Ghigo J-M (2008) *Escherichia coli* biofilms. In: Romeo T (ed) Current topics in microbiology and immunology, bacterial biofilms. Springer, Heidelberg, pp 249–289

Bentley WE, Mirjalili N, Andersen DC, Davis RH, Kompala DS (1990) Plasmid-encoded protein: the principal factor in the "metabolic burden" associated with recombinant bacteria. Biotechnol Bioeng 35:668–681

Bessems E (1998) The effect of practical conditions on the efficacy of disinfectants. Int Biodeter Biodegr 41:177–183

Birnbaum S, Bailey JE (1991) Plasmid presence changes the relative levels of many host cell proteins and ribosome components in recombinant *Escherichia coli*. Biotechnol Bioeng 37:736–745

Bottone EJ (2010) *Bacillus cereus*, a volatile human pathogen. Clin Microbiol Rev 23:382–398

Brown MRW, Smith AW (2003) Antimicrobial agents and biofilms. In: Wilson M, Devine D (eds) Medical implications of biofilms. Cambridge University Press, Cambridge, pp 36–55

Bryers JD (2008) Medical biofilms. Biotechnol Bioeng 100:1–18

Burland V, Shao Y, Perna NT, Plunkett G, Blattner FR, Sofia HJ (1998) The complete DNA sequence and analysis of the large virulence plasmid of *Escherichia coli* O157:H7. Nucleic Acids Res 26:4196–4204

Camper AK (2004) Involvement of humic substances in regrowth. Int J Food Microbiol 92:355–364

Cheah UE, Weigand WA, Stark BC (1987) Effects of recombinant plasmid size on cellular processes in *Escherichia coli*. Plasmid 18:127–134

Cloete TE (2003) Resistance mechanisms of bacteria to antimicrobial compounds. Int Biodeter Biodegr 51:277–282

Cloete TE, Jacobs L, Brözel VS (1998) The chemical control of biofouling in industrial water systems. Biodegradation 9:23–37

Costerton JW, Stewart PS, Greenberg EP (1999) Bacterial biofilms: a common cause of persistent infections. Science 284:1318–1322

Da Silva NA, Bailey JE (1986) Theoretical growth yield estimates for recombinant cells. Biotechnol Bioeng 28:741–746

Diaz Ricci JC, Hernández ME (2000) Plasmid effects on *Escherichia coli* metabolism. Crit Rev Biotechnol 20:79–108

Diaz Ricci JC, Hitzmann B, Bailey JE (1991) In vivo NMR analysis of the influence of pyruvate decarboxylase and alcohol dehydrogenase of *Zymomonas mobilis* on the anaerobic metabolism of *Escherichia coli*. Biotechnol Progr 7:305–310

Diaz Ricci JC, Tsu M, Bailey JE (1992) Influence of expression of the pet operon on intracellular metabolic fluxes of *Escherichia coli*. Biotechnol Bioeng 39:59–65

Dorel C, Lejeune P, Jubelin G (2005) Role of biofilms in infections caused by *Escherichia coli*. In: Pace JL, Rupp ME, Finch R (eds) Biofilms, infection, and antimicrobial therapy. CRC, Boca Raton, FL, pp 73–80

Ehlers LJ (2000) Gene transfer in biofilms. In: Allison D, Gilbert P, Lappin-Scott H, Wilson M (eds) Community structure and co-operation in biofilms. Cambridge University Press, Cambridge, pp 215–254

EN-1276 European Standard (1997) Chemical disinfectants and antiseptics—Quantitative suspension test for the evaluation of bactericidal activity of chemical disinfectants and antiseptics used in food, industrial, domestic, and institutional areas—Test method and requirements (phase 2, step 1)

Ferreira C, Pereira AM, Pereira MC, Melo LF, Simões M (2011) Physiological changes induced by the quaternary ammonium compound benzyldimethyldodecylammonium chloride on *Pseudomonas fluorescens*. J Antimicrob Chemother 66:1036–1043

Flores S, de Anda-Herrera R, Gosset G, Bolívar FG (2004) Growth-rate recovery of *Escherichia coli* cultures carrying a multicopy plasmid, by engineering of the pentose-phosphate pathway. Biotechnol Bioeng 87:485–494

Gallant CV, Daniels C, Leung JM, Ghosh AS, Young KD, Kotra LP, Burrows LL (2005) Common β-lactamases inhibit bacterial biofilm formation. Mol Microbiol 58:1012–1024

Ghigo J-M (2001) Natural conjugative plasmids induce bacterial biofilm development. Nature 412:442–445

Gibb AP, Martin KM, Davidson GA, Walker B, Murphy WG (1995) Rate of growth of *Pseudomonas fluorescens* in donated blood. J Clin Pathol 48:717–718

Glick BR (1995) Metabolic load and heterologous gene expression. Biotechnol Adv 13:247–261

Govan JR, Deretic V (1996) Microbial pathogenesis in cystic fibrosis: mucoid *Pseudomonas aeruginosa* and *Burkholderia cepacia*. Microbiol Rev 60:539–574

Hakobyan L, Gabrielyan L, Trchounian A (2012) Yeast extract as an effective nitrogen source stimulating cell growth and enhancing hydrogen photoproduction by *Rhodobacter sphaeroides* strains from mineral springs. Int J Hydrogen Energy 37:6519–6526

Hartung HA (1992) Stimulation of anaerobic digestion with peat humic substance. Sci Total Environ 113:17–33

Huang C-T, Peretti SW, Bryers JD (1993) Plasmid retention and gene expression in suspended and biofilm cultures of recombinant *Escherichia coli* DH5α(pMJR1750). Biotechnol Bioeng 41:211–220

Huang C-T, Peretti SW, Bryers JD (1994) Effects of medium carbon-to-nitrogen ratio on biofilm formation and plasmid stability. Biotechnol Bioeng 44:329–336

Ishiguro M, Tan W, Koopal LK (2007) Binding of cationic surfactants to humic substances. Colloids Surf A 306:29–39

Jacobsen SM, Stickler DJ, Mobley HLT, Shirtliff ME (2008) Complicated catheter-associated urinary tract infections due to *Escherichia coli* and *Proteus mirabilis*. Clin Microbiol Rev 21:26–59

Jono K, Takayama T, Kuno M, Higashide E (1986) Effect of alkyl chain length of benzalkonium chloride on the bactericidal activity and binding to organic materials. Chem Pharm Bull 34:4215–4224

Khosravi M, Ryan W, Webster DA, Stark BC (1990) Variation of oxygen requirement with plasmid size in recombinant *Escherichia coli*. Plasmid 23:138–143

Klemperer RMM, Ismail NTAJ, Brown MRW (1979) Effect of R plasmid RP1 on the nutritional requirements of *Escherichia coli* in batch culture. J Gen Appl Microbiol 115:325–331

Koopal LK, Goloub TP, Davis TA (2004) Binding of ionic surfactants to purified humic acid. J Colloid Interface Sci 275:360–367

Król JE, Nguyen HD, Rogers LM, Beyenal H, Krone SM, Top EM (2011) Increased transfer of a multidrug resistance plasmid in *Escherichia coli* biofilms at the air-liquid interface. Appl Environ Microbiol 77:5079–5088

Kuroki R, Kawakami K, Qin L, Kaji C, Watanabe K, Kimura Y, Ishiguro C, Tanimura S, Tsuchiya Y, Hamaguchi I, Sakakura M, Sakabe S, Tsuji K, Inoue M, Watanabe H (2009) Nosocomial bacteremia caused by biofilm-forming *Bacillus cereus* and *Bacillus thuringiensis*. Intern Med 48:791–796

Lawrence CA (1950) Mechanism of action and neutralizing agents for surface-active materials upon microorganisms. Ann N Y Acad Sci 53:66–75

Lim JY, La HJ, Sheng H, Forney LJ, Hovde CJ (2010a) Influence of plasmid pO157 on *Escherichia coli* O157:H7 Sakai biofilm formation. Appl Environ Microbiol 76:963–966

Lim JY, Yoon J, Hovde CJ (2010b) A brief overview of *Escherichia coli* O157:H7 and its plasmid O157. J Microbiol Biotechnol 20:5–14

Mah T-FC, O'Toole GA (2001) Mechanisms of biofilm resistance to antimicrobial agents. Trends Microbiol 9:34–39

May T, Okabe S (2008) *Escherichia coli* harboring a natural IncF conjugative F plasmid develops complex mature biofilms by stimulating synthesis of colanic acid and curli. J Bacteriol 190:7479–7490

Miller MJ, Ahearn DG (1987) Adherence of *Pseudomonas aeruginosa* to hydrophilic contact lenses and other substrata. J Clin Microbiol 25:1392–1397

Nickel JC, Ruseska I, Wright JB, Costerton JW (1985) Tobramycin resistance of *Pseudomonas aeruginosa* cells growing as a biofilm on urinary catheter material. Antimicrob Agents Chemother 27:619–624

Norman A, Hansen LH, She Q, Sørensen SJ (2008) Nucleotide sequence of pOLA52: a conjugative IncX1 plasmid from *Escherichia coli* which enables biofilm formation and multidrug efflux. Plasmid 60:59–74

Ong C-LY, Beatson SA, McEwan AG, Schembri MA (2009) Conjugative plasmid transfer and adhesion dynamics in an *Escherichia coli* biofilm. Appl Environ Microbiol 75:6783–6791

Otzen D (2011) Protein–surfactant interactions: a tale of many states. Biochim Biophys Acta 1814:562–591

Ow D, Nissom P, Philp R, Oh S, Yap M (2006) Global transcriptional analysis of metabolic burden due to plasmid maintenance in *Escherichia coli* DH5 alpha during batch fermentation. Enzyme Microb Technol 39:391–398

Pappen FG, Qian W, Aleksejūnienė J, de Toledo Leonardo R, Leonardo MR, Haapasalo M (2010) Inhibition of sodium hypochlorite antimicrobial activity in the presence of bovine serum albumin. J Endod 36:268–271

Paterson DL, Bonomo RA (2005) Extended-spectrum β-lactamases: a clinical update. Clin Microbiol Rev 18:657–686

Pereira MO, Vieira MJ (2001) Effects of the interactions between glutaraldehyde and the polymeric matrix on the efficacy of the biocide against *Pseudomonas fluorescens* biofilms. Biofouling 17:93–101

Portenier I, Haapasalo H, Rye A, Waltimo T, Ørstavik D, Haapasalo M (2001) Inactivation of root canal medicaments by dentine, hydroxylapatite and bovine serum albumin. Int Endod J 34:184–188

Pouneva ID (2005) Effect of humic substances on the growth of microalgal cultures. Russ J Plant Physiol 52:410–413

Prather KJ, Sagar S, Murphy J, Chartrain M (2003) Industrial scale production of plasmid DNA for vaccine and gene therapy: plasmid design, production, and purification. Enzyme Microb Technol 33:865–883

Reisner A, Haagensen JAJ, Schembri MA, Zechner EL, Molin S (2003) Development and maturation of *Escherichia coli* K-12 biofilms. Mol Microbiol 48:933–946

Reisner A, Höller BM, Molin S, Zechner EL (2006) Synergistic effects in mixed *Escherichia coli* biofilms: conjugative plasmid transfer drives biofilm expansion. J Bacteriol 188:3582–3588

Rhee J, Diaz Ricci JC, Bode J, Schügerl K (1994) Metabolic enhancement due to plasmid maintenance. Biotechnol Lett 16:881–884

Ruano M, El-Attrache J, Villegas P (2001) Efficacy comparisons of disinfectants used by the commercial poultry industry. Avian Dis 45:972–977

Russell AD (2003) Biocide use and antibiotic resistance: the relevance of laboratory findings to clinical and environmental situations. Lancet 3:794–803

Russell AD (2008) Factors influencing the efficacy of antimicrobial agents. In: Russell AD, Hugo WB, Ayliffe GAJ (eds) Principles and practice of disinfection, preservation and sterilization. Blackwell, London, pp 98–127

Ryan W, Parulekar SJ, Stark BC (1989) Expression of β-lactamase by recombinant *Escherichia coli* strains containing plasmids of different sizes—effects of pH, phosphate, and dissolved oxygen. Biotechnol Bioeng 34:309–319

Salati S, Papa G, Adani F (2011) Perspective on the use of humic acids from biomass as natural surfactants for industrial applications. Biotechnol Adv 29:913–922

Seo J-H, Bailey JE (1985) Effects of recombinant plasmid content on growth properties and cloned gene product formation in *Escherichia coli*. Biotechnol Bioeng 27:1668–1674

Simões M, Pereira MO, Vieira MJ (2005a) Effect of mechanical stress on biofilms challenged by different chemicals. Water Res 39:5142–5152

Simões M, Pereira MO, Vieira MJ (2005b) Validation of respirometry as a short-term method to assess the efficacy of biocides. Biofouling 21:9–17

Stewart PS, Costerton JW (2001) Antibiotic resistance of bacteria in biofilms. Lancet 358:135–138

Stringfellow K, Anderson P, Caldwell D, Lee J, Byrd J, McReynolds J, Carey J, Nisbet D, Farnell M (2009) Evaluation of disinfectants commonly used by the commercial poultry industry under simulated field conditions. Poultry Sci 88:1151–1155

Teodósio JS, Simões M, Mergulhão FJ (2012) The influence of nonconjugative *Escherichia coli* plasmids on biofilm formation and resistance. J Appl Microbiol 113:373–382

Visser SA (1985) Physiological action of humic substances on microbial cells. Soil Biol Biochem 17:457–462

Wang Z, Xiang L, Shao J, Wegrzyn A, Wegrzyn G (2006) Effects of the presence of ColE1 plasmid DNA in *Escherichia coli* on the host cell metabolism. Microb Cell Fact 5:34

Wood TK (2009) Insights on *Escherichia coli* biofilm formation and inhibition from whole-transcriptome profiling. Environ Microbiol 11:1–15

Yang X, Ma Q, Wood TK (2008) The R1 conjugative plasmid increases *Escherichia coli* biofilm formation through an envelope stress response. Appl Environ Microbiol 74:2690–2699

Antimicrobial Coatings to Prevent Biofilm Formation on Medical Devices

Phat L. Tran, Abdul N. Hamood, and Ted W. Reid

Abstract Under different environmental conditions, bacteria colonize and develop biofilms on diverse surfaces including those of medical devices. The development of biofilms on medical devices is one of the most serious challenges that the healthcare systems face. In response, various methods have been developed to prevent biofilm formation on such devices. In this chapter, we discuss different strategies designed to prevent biofilm formation on three medical devices: central venous catheters, urinary tract catheters, and contact lenses. These strategies are based on modifying the surface of these devices by either coating or impregnating them with a variety of antimicrobial agents. For central venous catheters, we describe coating with silver, chlorhexidine silver sulfadiazine, or organoselenium. For urinary tract catheters, we describe coating with hydrogel, silver, triclosan, gendine, nitric oxide, and antibiotics. We also describe novel approaches to prevent biofilm development on urinary tract catheters including the utilization of quorum-sensing inhibitors and biological coatings (bacteria or bacteriophages). For contact lenses, we discussed coating with either a non-covalent coating (furanones, silver, or polyquaternium compounds) or a covalent coating (furanones, polyquaternium compounds, cationic peptides, or organoselenium). We review the mechanism (s) through which each agent inhibits biofilm development and the influence of

---

P.L. Tran (✉)
Department of Ophthalmology and Visual Sciences, Texas Tech University Health Sciences Center, Lubbock, TX, USA
e-mail: phat.tran@ttuhsc.edu

A.N. Hamood
Department of Immunology and Molecular Microbiology, Texas Tech University Health Sciences Center, Lubbock, TX, USA

T.W. Reid
Department of Ophthalmology and Visual Sciences, Texas Tech University Health Sciences Center, Lubbock, TX, USA

Department of Immunology and Molecular Microbiology, Texas Tech University Health Sciences Center, Lubbock, TX, USA

the material from which the medical device was made on the quality of coating formed by different agents on these devices. Additionally, we review different in vitro assays, animal models for biofilm development, and clinical trials used to assess the effectiveness of each agent and the rate of success of each coating based on these assessments. Finally, we summarize any reported toxicity associated with these coatings.

1 Introduction

Biofilms are microbially derived sessile communities in which the cells are irreversibly attached to a substratum, interface, or to each other (Donlan and Costerton 2002). Within the biofilm, bacteria are embedded within an extracellular polysaccharide matrix (EPS) (Donlan and Costerton 2002). The growth as well as the expression of different bacterial genes within the biofilm is different from those of their free-living counterpart (Donlan and Costerton 2002). Available evidence, from studies performed on *Pseudomonas aeruginosa*, suggests that biofilms develop in four stages: reversible attachment, irreversible attachment, maturation, and dispersion (Sauer et al. 2002). In the reversible attachment, *P. aeruginosa* attaches to a substrate using its polar flagellum (Sauer et al. 2002). During irreversible attachment *P. aeruginosa* becomes nonmotile, and this transition involves the development of bacterial clusters (Sauer et al. 2002). These bacterial clusters will develop and their thickness increases during the maturation stage (Sauer et al. 2002). During maturation, pores and channels develop within the biofilm. During the dispersion stage, some bacteria detach from their biofilm structure. Within the biofilm, microorganisms are highly resistant to different antimicrobial agents including antibiotics, germicides, and disinfectants (Sauer et al. 2002). Among the different mechanisms that contribute to their resistance are the ability of the EPS to reduce the penetration of antimicrobial agents; the alternation in the growth of microorganisms within the biofilms; and the physiological changes that are induced by the alternation in the growth mode (Suci et al. 1994; Duguid et al. 1992; Desai et al. 1998).

Biofilm develops on numerous medical devices causing extensive medical and economical losses. Among these medical devices are the following: urinary catheters, central venous catheters, prosthetic heart valves, contact lenses (CLs), contact lens cases, dental unit water lines, and intrauterine devices. Therefore, medical devices that prevent or significantly reduce biofilm development are urgently needed. The most successful approach so far has been coating the medical device surface with antimicrobial agents. In this chapter, we describe different strategies that were developed to coat these medical devices (urinary catheters, central venous catheters, contact lenses, and contact lens cases) with different antimicrobial agents.

2 Central Venous Catheters

2.1 Catheter-Related Blood Stream Infections

Central venous catheters (CVCs) are used in patient management including chronically ill patients and patients requiring acute or long-term medical care (Novikov et al. 2012). They are used to deliver medications and nutritional support that cannot be provided solely through peripheral venous catheters and to measure different hemodynamic parameters (Maaskant et al. 2009). CVCs were primarily used in intensive care units (Hewlett and Rupp 2012). However, currently CVCs are used in other healthcare settings, as well as long-term care facilities, home health care, and outpatient hemodialysis centers (Hewlett and Rupp 2012). Catheter-related blood stream infections (CRBSIs) result in significant morbidity and mortality and a tremendous increase in healthcare cost (Hewlett and Rupp 2012). A recent epidemiology study indicated that the mortality rate associated with central venous catheter-related infections may reach as high as 36 % (Akoh 2011). It is estimated that in the USA, 80,000 CRBSIs occur in intensive care units each year (Mermel 2000), and there are a total of 250,000 cases/year if entire hospitals are considered (Maki et al. 2006). Microbial pathogens may access and colonize the external surface of short-term intravenous catheter through patients' skin at the insertion sites (Mermel 2011). The pathogens may later colonize the luminal surfaces of the catheters and cause blood stream infections (Hewlett and Rupp 2012). Although other routes of inoculation are thought to be involved less frequently, CRBSIs can also be caused by hematogenous seeding of the catheter from a distant site or by infusion of a contaminated substance (Hewlett and Rupp 2012).

The surface of the central venous catheter is very suitable for the colonization of different microbial pathogens (Casey et al. 2008). Once inserted, the catheter surface is covered by a film of different host proteins including fibrin, collagen, fibrinogen, elastin, laminin, and fibronectin (Vaudaux et al. 1994). All these factors facilitate the colonization of microbial pathogens and the development of biofilm on both the external and internal surfaces of the catheter.

The most commonly isolated microorganism from the blood of patients with catheter-related blood stream infections are the following: coagulation negative *staphylococci*, *Staphylococcus aureus*, *enterococci*, and other bacteria that are commonly found on skin (Betjes 2011; O'Grady et al. 2011). The mortality rate among patients on dialysis who become infected with *S. aureus* CRBSIs may reach 20 % (Betjes 2011). Gram-negative pathogens including *P. aeruginosa* and *Acinetobacter baumannii* may be associated with CRBSIs but to a lesser degree than Gram-positive species (Prospero et al. 2006; Miller and O'Grady 2012).

Different antimicrobial agents have been either coated or incorporated into the polymer of CVCs, and the efficacy of these agents in inhibiting either microbial colonization or the incidence of catheter-related blood stream infections was evaluated, as is discussed below.

2.2 Antimicrobial Coatings of CVCs

2.2.1 Silver

Several clinical trials demonstrated that silver impregnated central venous catheters did not significantly reduce the rate of catheter colonization and the incidence of CRBSIs (Dunser et al. 2005; Bach et al. 1999; Moretti et al. 2005). Hagau et al. demonstrated that in critically ill patients, silver impregnated venous catheters had no effect on the bacterial colonization of the catheters (Hagau et al. 2009). Whereas standard catheters were first colonized 3 days after the insertion, silver-integrated catheters were first colonized 5 days after insertion (Hagau et al. 2009). However, over time, the incidence of colonization and infection between silver impregnated and standard polyurethane catheters may be significantly different (Hagau et al. 2009). Corral et al. conducted a randomized controlled trial to evaluate the effectiveness of Oligon vantex silver catheter (OVSC) in reducing the rate of catheter colonization in critically ill patients (Corral et al. 2003). OVSC was produced from a thermoplastic polyurethane elastomer by incorporating silver particles at 0.5–1 % (w/w). As such, the continuous release of silver particles from the inserted catheter would produce a maximum antimicrobial effect (Hentschel and Munstedt 1999). In comparison with standard polyurethane CVCs, OVSC reduced the incidence of catheter colonization by only 16 % more than the controls (Corral et al. 2003). Silver iontophoretic catheters are utilized in the USA for long-term catheter-related catheterization. These catheters consist of two silver wires connected to an electrical power source that is disposed in a parallel and helical manner around the proximal subcutaneous segment of the catheter (Hentschel and Munstedt 1999). The iontophoretic reaction allows a controlled and sustained release of silver ions (Raad et al. 1996). However, a randomized controlled trial demonstrated no significant difference between silver-iontophoretic catheters and control catheters in the incidence of catheter colonization and CRBSIs (Bong et al. 2003).

2.2.2 Chlorhexidine Silver Sulfadiazine

Chlorhexidine [1, 6-bis(4′-chlorophenylbiguanide)hexane] is a divalent cationic biguanide agent (Kampf and Kramer 2004). As a cation, it binds to the negatively charged bacterial cell wall, displaces the cations that stabilize the cell membrane, and cause it to be leaky (Russell 1986). At high concentrations, chlorhexidine binding destroys the structural integrity of the membrane which leads to cell death (Russell 1986). When used together, chlorhexidine and silver sulfadiazine function in a synergistic way. Chlorhexidine is a very effective disinfectant with a broad spectrum bactericidal activity against Gram-positive and Gram-negative bacteria. However, it is most effective against Gram-positive bacteria. As a cationic agent, it disrupts bacterial cell membranes and increases the uptake of silver salts

(Bong et al. 2003). First generation chlorhexidine silver sulfadiazine (CH-SS)-coated CVCs were coated with CH-SS on the external luminal surface only (Bach et al. 1996). Several clinical trials demonstrated the effectiveness of these catheters in reducing the rate of catheter colonization and CRBSIs in comparison with the standard non-coated catheters (Veenstra et al. 1999; Tennenberg et al. 1997; Brun-Buisson et al. 2004). Later on, a second generation catheter was developed. The internal surface of these catheters is coated with chlorhexidine only. The coating extends into the extension set and hubs (Miller and O'Grady 2012). The external surface, however, is coated with chlorhexidine and silver sulfadiazine (Miller and O'Grady 2012). In addition, the amount of chlorhexidine in the outer surface is three times that on the inner surface which allows for an extended release of CH-SS (Miller and O'Grady 2012). Randomized, prospective, controlled clinical trials revealed that the use of these second generation CH-SS catheters reduced the overall risk of catheter colonization and CRBSIs (in comparison with uncoated catheters) (Rupp et al. 2005; Brun-Buisson et al. 2004; Ostendorf et al. 2005). However, coated catheter did not reduce the incidence of catheter-related bacteremia (Brun-Buisson et al. 2004; Ostendorf et al. 2005). Due to the extensive utilization of CH-SS catheters, the threat of chlorhexidine-resistant bacteria from these devices may occur. The emergence of chlorhexidine-resistant mutants among Gram-negative bacteria including; *Proteus* species, *P. aeruginosa*, and *Serratia* species has been reported (Stickler et al. 1993a; Marrie and Costerton 1981). Reduced susceptibility of staphylococci to chlorhexidine has also been reported (Horner et al. 2012) and was found to involve efflux pumps (Horner et al. 2012). CH-SS are largely safe and nontoxic; however, some patients developed anaphylactic reaction from using them (Oda et al. 1997; Pittaway and Ford 2002).

2.2.3 Minocycline/Rifampicin

In addition to antiseptics, CVCs have also been coated with antibiotics, with the most extensively analyzed being those coated with minocycline and rifampicin. Minocycline (7-dimethylamino-6-deoxytetracycline) is a second generation, semi-synthetic tetracycline that has been used as an antibiotic against Gram-negative and Gram-positive bacteria (Garrido-Mesa et al. 2013). Minocycline, which binds to the 30S subunit ribosome and inhibits protein synthesis, has a longer half-life than tetracycline (Garrido-Mesa et al. 2013; Klein and Cunha 1995). Rifampicin, which is a synthetic bactericidal antibiotic, is used to treat infections caused by different bacteria including *Mycobacterium* and methicillin-resistant *S. aureus* (MRSA), and also used as a prophylactic therapy against *Neisseria meningitis* infections. By inhibiting the DNA-dependent RNA polymerase, rifampicin inhibits bacterial DNA-dependent RNA synthesis (Calvori et al. 1965). Mutations within the RNA polymerase that alter the rifampicin binding sites lead to rifampicin-resistant bacterial strains (Feklistov et al. 2008). Minocycline/rifampicin were impregnated on the external and internal surfaces of CVCs. Multicenter, randomized, prospective trials showed that minocycline/rifampicin impregnated catheters significantly reduced catheter colonization and the incidence of CRBSIs (Darouiche et al. 1999; Raad et al. 1997). In comparison with the first generation CH-SS catheters, minocycline/rifampicin catheters reduced the rate of catheter colonization by threefold and the rate of CRBSIs by 12-fold (Raad et al. 1997). Among critically ill patients, minocycline/rifampicin CVCs significantly reduced nosocomial bloodstream infections and the length of stays in intensive care units (Hanna et al. 2003). The catheters also reduced the risk of catheter-related infections in patients with acute renal failure (Chatzinikolaou et al. 2003). In vivo analysis revealed that following 21 days of sequential exposure to different Gram-positive pathogens, including methicillin-resistant *S. aureus*, methicillin-resistant *Staphylococcus epidermidis* (MRSE), and vancomycin-resistant enterococci (VRE), minocycline/ rifampicin catheters retained their antibacterial activity (Aslam and Darouiche 2007). In vitro and in vivo studies using a rat subcutaneous implantation model showed that when challenged with rifampicin-resistant *S. aureus*, antiseptic-coated catheters were less susceptible to colonization than the minocycline/rifampicin catheter (Sampath et al. 2001).

2.2.4 Organoselenium Coating of Hemodialysis Catheters

Selenium catalyzes the formation of superoxide radicals ($O_2^{\cdot-}$) which inhibits bacterial attachment to solid surfaces. The detailed mechanism of function of selenium compounds is described in the "Antimicrobial mechanism of selenium" section. Covalent binding of organoselenium to solid surfaces prevents bacterial colonization of those surfaces (Tran et al. 2009).

Cellulose discs coated with organoselenium methacrylate polymers inhibited colonization and biofilm formation by *S. aureus* and *P. aeruginosa* (Tran et al. 2009). Using in vitro and in vivo biofilm assays, Tran et al. demonstrated the effectiveness of hemodialysis catheters coated with the organoselenium agent (selenocyanato diacetic acid, SCAA) in preventing biofilm development by *S. aureus* (Tran et al. 2012b). Small pieces of uncoated polyurethane decathlon high flow, long-term catheters were coated internally and externally by covalent attachment of SCAA (Tran et al. 2012b). To visualize the biofilm by confocal laser scanning microscopy (CLSM), Tran et al. utilized the *S. aureus* strain AH133 (Malone et al. 2009; Tran et al. 2012b). This strain carried plasmid pMC11 which contains the gene that codes for green fluorescent protein (Malone et al. 2009). The inhibitory effect of SCAA-coated catheter pieces in vitro was examined using two systems: the static biofilm system and the flow-through continuous culture system (Schaber et al. 2007). In the static biofilm system, catheter pieces were incubated with AH133 in Tryptic Soy broth (TSB) for 24 h using 24-well microtiter plates. The pieces were then rinsed and vigorously vortexed to detach bacterial cells, and the amount of the biofilm was assessed using a crystal violet assay. In addition, the number of *S. aureus* AH133 cells within the biofilm was determined. In comparison with uncoated catheter pieces, SCAA-coated pieces reduced AH133 biofilm by over 5 logs (Tran et al. 2012b). To assess the stability of the SCAA coating, catheter pieces were stored in PBS at room temperature for 6–8 weeks before utilizing them in the static biofilm system. Regardless of the storage time, the SCAA-coated pieces still significantly inhibited the development of AH133 biofilms (Tran et al. 2012b). In the flow-through continuous-culture system, catheter pieces were colonized with AH133 for 1 h and subjected to a continuous stream of TSB for 5 days using a peristaltic pump (Tran et al. 2012b). Image analysis revealed the AH133 formed well-developed biofilms on the outer and inner surfaces of the uncoated catheters (Tran et al. 2012b). However, no biofilm was detected on either surface of the SCAA-coated catheter pieces (Tran et al. 2012b).

Tran et al. utilized the murine model of biofilm development to determine if SCAA-coated catheter inhibits the development of AH133 biofilm in vivo (Tran et al. 2012b). Catheter pieces were subcutaneously inserted in the back of adult mice (Tran et al. 2012b). Strain AH133 was inoculated within the vicinity of the inserted catheter pieces (Tran et al. 2012b). After 3 days, the catheter pieces were extracted, and biofilms were analyzed (Tran et al. 2012b). Strain AH133 formed mature biofilm on the inner and outer surfaces of the uncoated catheter pieces (Fig. 1) (Tran et al. 2012b). Structural aspects of the biofilm were examined using the COMSTAT program (Heydorn et al. 2000). On uncoated catheter pieces, AH133 produced biofilm with marked biomass, average thickness, surface area, and surface area/biomass ratio (Tran et al. 2012b). In contrast, no biofilm developed on either the outer or the inner surfaces of the SCAA-coated catheter pieces (Tran et al. 2012b).

3 Urinary Tract Catheters

3.1 Catheter-Associated Urinary Tract Infections

Catheter-associated urinary tract infections (UTIs) are the most common type of healthcare-related infections, accounting for about 40 % of healthcare associated infections in the USA (NNIS 2004; Tambyah 2004). In the USA, more than 30 million bladder catheters are placed annually (Tambyah 2004). As a result, thousands of cases of catheter-associated UTIs occur (Feng et al. 2000). It is estimated that 10–50 % of patients undergoing short-term catheterization (up to 7 days) develop catheter-associated urinary tract infections whereas almost all patients undergoing long-term catheterization (greater than 28 days) will develop catheter-associated urinary tract infections (Stickler 1996).

The urinary tract catheter system may either be closed or open. Whereas in the closed system, the catheter empties into a plastic bag, in the open system, the catheter drains into an open collection container (Kaye and Hessen 1994). Therefore, open system catheters become contaminated quickly and patients develop catheter-associated urinary tract infections (Stickler 1996). In vitro and in vivo studies demonstrated the development of bacterial biofilm on urinary catheters.

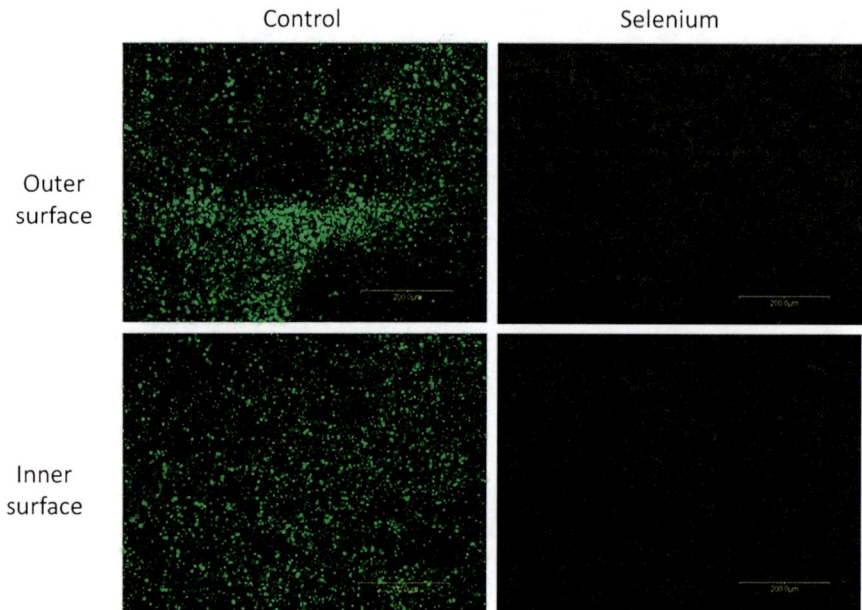

Fig. 1 Selenium coating inhibited the in vivo development of *S. aureus* biofilm on inner and outer surfaces of the hemodialysis catheter. This was done using the murine model of biofilm development (described in detail in the published article: Tran et al., An organo-selenium inhibits *S. aureus* biofilms on hemodialysis catheters in vivo. *Antimicrob Agents Chemother.* 56:972–978. 2012). Catheter pieces were examined by confocal laser scanning microscopy. Control, mouse with uncoated catheter pieces implanted; Selenium, mouse with selenium-coated catheter pieces implanted

Several studies documented the development of biofilm on urinary catheters obtained from patients (Nickel et al. 1989; Stickler et al. 1993b). A buildup of a conditional film that is formed from electrolytes, organic molecules, and host proteins, on the surface of the urinary catheters, provides a suitable surface for attachment and colonization of bacterial pathogens (Donlan and Costerton 2002; Siddiq and Darouiche 2012). Single species bacteria including *Staphylococcus epidermidis*, *Enterococcus faecalis*, *Escherichia coli*, and *Proteus mirabilis* initially colonize the urinary tract catheter (Donlan and Costerton 2002). If the catheter remains in place, biofilms formed by mixed species including *Pseudomonas aeruginosa*, *Proteus mirabilis*, and *Klebsiella pneumoniae* develop (Stickler 1996). Compared with biofilms developed on other devices, many urinary catheter biofilms are associated with the precipitation of minerals such as calcium phosphate (hydroxyapatite) and magnesium ammonium phosphate (sturvite) (Tunney et al. 1999). Certain microorganisms within the urinary catheter biofilm produce urease enzyme which hydrolyzes the urea of the urine producing ammonia. The ammonia produced increases the local pH and causes the precipitation of these minerals (Donlan and Costerton 2002; Tunney et al. 1999). These mineral containing biofilm, termed encrustations, are primarily found by urease producing bacteria including *Proteus mirabilis, Klebsiella pneumoniae, Pseudomonas aeruginosa, Proteus vulgaris*, and *Morganella morganii* (Stickler et al. 1993a, 1998).

3.2 Antimicrobial Coatings of Urinary Tract Catheters

3.2.1 Hydrogels

Hydrogels are cross-linked, hydrophilic polymers that decrease microbial adherence and encrustation (Saint et al. 1998). Due to the potential difference in the catheter material or the type of the hydrogel, the efficacy of the hydrogel-coated catheter in preventing catheter-associated urinary tract infections is not universally accepted (Siddiq and Darouiche 2012). Using an in vitro model of catheterized bladder and human urine that was inoculated with *Proteus mirabilis* obtained from an encrusted catheter, Morris and Stickler (Morris and Stickler 1998) assessed the degree of encrustation on different catheters by determining the amount of calcium and magnesium deposited on the bladder. Although all tested catheters were eventually blocked, the mean time to blockage for hydrogel gated latex catheters was longer than that of uncoated silicone (34 h vs. 47 h) (Morris and Stickler 1998) (Beiko et al. 2004).

3.2.2 Silver

The antibacterial activity of silver has been appreciated since ancient times. Silver ions produce a broad spectrum bactericidal effect (Clement and Jarrett 1994) by interacting with thiol groups and subsequently inactivating several vital enzymes (Flemming et al. 1990; Liau et al. 1997). Silver ions also enhance pyrimidine dimerization within the DNA molecule through photodynamic reactions and interfere with DNA replication (Fox and Modak 1974; Russell and Hugo 1994). Using electron microscopy and X-ray microanalysis, Feng et al. demonstrated that silver causes several morphological changes in *Escherichia coli* and *Staphylococcus aureus* by detaching the cytoplasmic membrane from the cell wall (Feng et al. 2000). In addition, small, electron-dense molecules were deposited around the cell wall or within the bacteria (Feng et al. 2000). Furthermore, silver was detected within the electron dense molecules, the cytoplasm, and the DNA molecules (Feng et al. 2000).

Meta analysis showed that urinary catheters coated with silver alloy were more effective in preventing urinary tract infection than those coated with silver oxide (Saint et al. 1998; Davenport and Keeley 2005). Silver alloy-coated catheters reduced catheter-associated urinary tract infection by 45 % (Davenport and Keeley 2005). The greatest reduction was detected in postoperative patients, intensive care patients, and burn patients (Davenport and Keeley 2005). Schumm and Lam analyzed the results of different clinical trials that were conducted to determine the effectiveness of indwelling catheters in preventing urinary tract infections in adults who underwent short-term urinary catheterization (Schumm and Lam 2008). Less than a week after catheterization, silver alloy catheters significantly reduced the incidence of asymptomatic bacteriuria (Schumm and Lam 2008).

3.2.3 Triclosan

Triclosan (2, 4, 4′-triclora 2′-hydroxydiphenyl ether) is a nonionic broad spectrum antimicrobial agent that has been utilized in numerous medical and personal care products (Bhargava and Leonard 1996). Triclosan activity is concentration and formulation dependent (Bhargava and Leonard 1996) and specifically targets the enoyl-acyl carrier protein reductase (FabI) component of the type II fatty acid synthesis system, inhibiting fatty acid synthesis (Heath et al. 1998, 2002). Triclosan is a slow binding inhibition of FabI in Gram-negative and Gram-positive bacteria (Heath et al. 1999). *Pseudomonas aeruginosa* is resistant to triclosan due to the presence of FABI, a triclosan insensitive isofunctional FAD-dependent enoyl acyl carrier protein, as well as a multidrug efflux pump (Schweizer 1998).

In vitro studies showed that triclosan-eluting urethral stents inhibited the growth of common bacterial uropathogens including *Enterococcus faecalis*, *Klebsiella pneumoniae*, *Staphylococcus aureus*, and *Proteus mirabilis* in a dose-dependent way but had no effect on *Pseudomonas aeruginosa* (Chew et al. 2006). Sublethal concentrations of triclosan did not affect the adhesion and internalization of uropathogenic *Escherichia coli* to kidney or bladder cells in vitro, but significantly reduced the amount of TNF-α secreted by a bladder cell line (Elwood et al. 2007). Cadieux et al. utilized the rabbit urinary tract infection model to examine the effectiveness of triclosan impregnated stents on the growth and survival of *Proteus mirabilis* (Cadieux et al. 2006). Triclosan impregnated stents significantly reduced the number of viable *Proteus mirabilis* recovered in urine (Cadieux et al. 2006). However, there was no significant difference in the encrustation between the control and triclosan impregnated stents (Cadieux et al. 2006). Based on the results of a clinical study, Cadieux et al. suggested that a triclosan eluting stent alone is not sufficient to reduce device-associated infections in patients who needed long-term urethral stents (Cadieux et al. 2009). In this study, patients received a control stent for 3 months (Cadieux et al. 2009). The control stent was then removed and replaced with a triclosan stent for an additional 3 months (Cadieux et al. 2009). Overall, similar types of microorganisms were isolated from urine culture and the stent during each indwell period (Cadieux et al. 2009). Another model to prevent catheter encrustation is the catheterized bladder model in which triclosan is directly delivered into the residual urine in the catheterized bladder (Stickler et al. 2003; Jones et al. 2005). Instead of water, the retention balloon was inflated with triclosan (Stickler et al. 2003; Jones et al. 2005). Triclosan, which diffused through the balloon membrane, attacked planktonic cells and prevented the colonization of the catheter surface (Stickler et al. 2003; Jones et al. 2005). In addition, the reduction in bacterial activity would prevent the increase in the urinary pH (Stickler et al. 2003; Jones et al. 2005). Analysis of *Proteus mirabilis* biofilm development using this model showed that in comparison with a control catheter, a latex catheter inflated with triclosan had a controlled urinary pH, reduced the number of bacteria in urine, and showed no surge of encrustation (Williams and Stickler 2008). However, different results were obtained when the model was utilized to determine if triclosan prevented encrustation by microflora of uropathogens that commonly infect patients undergoing long-term catheterization (Williams and Stickler 2008). While *Proteus mirabilis, Escherichia coli*, and *Klebsiella pneumoniae* were eliminated from the residual urine, there was no effect on *Enterococcus faecalis* and *Pseudomonas aeruginosa* (Williams and Stickler 2008).

3.2.4 Gendine and Nitric Oxide

Hachem et al. described Gendine as a novel antimicrobial catheter coating (Hachem et al. 2009). In comparison with uncoated and silver hydrogel-coated catheters, Gendine-coated catheters significantly reduced biofilm produced by different pathogens including *Pseudomonas aeruginosa, Enterococcus faecalis, Klebsiella pneumoniae*, and *Candida* (Hachem et al. 2009). Gendine-coated catheters were more efficient than silver hydrogel-coated ones in preventing *Escherichia coli* colonization in a rabbit model (Hachem et al. 2009).

Regev-Shoshani et al. impregnated Foley urinary catheters with nitric oxide (NO) (Regev-Shoshani et al. 2010). NO, which is a small naturally produced, hydrophobic, free radical gas, is bacteriostatic and bacteriocidal (Fang 1997; McMullin et al. 2005). Coated catheters slowly released NO over 14 days and prevented bacterial colonization and biofilm formation on the luminal and exterior surfaces of the catheters (Regev-Shoshani et al. 2010).

3.2.5 Antibiotics

Antibiotic-coated urethral catheters have also been developed. Norfloxacin, a fluroquinolone synthetic antibiotic, was impregnated into a coating layer on the outer and inner surfaces of a urethral catheter (Park et al. 2003). Norfloxacin-coated catheters generated a considerable zone of inhibition against *Escherichia coli, Klebsiella pneumoniae*, and *Proteus vulgaris* for 6 days (Park et al. 2003). A similar approach was used to generate gentamicin-coated catheters (Cho et al. 2003). Kowalczuk et al. (Kowalczuk et al. 2012) described covalent and non-covalent attachment of sparfloxacin to the surface of heparin-coated urinary catheters. The catheters prevented colonization and biofilm development of *Escherichia coli, Staphylococcus aureus*, and *Staphylococcus epidermidis* for about 3 days (Kowalczuk et al. 2012), and inhibition assays confirmed their antimicrobial activity (Kowalczuk et al. 2012). Pugach et al. (1999) described silicon Foley catheters that were externally coated with ciprofloxacin liposome. The average time to positive *E. coli* urine cultures improved from 3.5 to 5.3 days when ciprofloxacin liposome-coated catheters were compared to uncoated catheters in a rabbit model (Pugach et al. 1999). Coated catheters also decreased the bacteremia rate by 30 % (Pugach et al. 1999). The risk associated with the antibiotic-coated urinary catheters is the emergence of antibiotic-resistant strains (Siddiq and Darouiche 2012). Due to the higher concentration of bacteria in urine than on skin, such a risk is greater with urinary than vascular catheters (Siddiq and Darouiche 2012).

3.2.6 Quorum-Sensing Inhibitors

The cell-to-cell communication system (quorum sensing, QS) influences biofilm development and the production of different virulence factors in Gram-negative and Gram-positive bacteria (Njoroge and Sperandio 2009). One of the *Staphylococcus aureus* QS systems consists of RNAIII activating protein (RAP) which phosphorylates its target protein, the 21 Kda TRAP (Njoroge and Sperandio 2009). The QS inhibitor RNAIII-inhibiting peptide (RIP) inhibits TRAP phosphorylation (Gov et al. 2001). RIP has been shown to inhibit biofilm formation and staphylococcal infection (Balaban et al. 2005). Cirioni et al. (2007) implanted RIP-coated urethral stents in rat bladders and examined the effectiveness of RIP-coated catheters in inhibiting *Staphylococcus aureus* biofilm. RIP suppressed the formation of *Staphylococcus aureus* on urethral stents and significantly reduced the number of *Staphylococcus aureus* microorganisms (CFU/ml) in urine cultures (Cirioni et al. 2007). When the RIP-coated urethral stents were utilized in conjunction with the antibiotic teicoplanin (intraperitoneally injected), no bacteria were recovered either on the stent or in urine cultures (Cirioni et al. 2007).

3.2.7 Biological Coatings

A novel approach to prevent the development of catheter-associated biofilm by pathogenic bacteria in patients who require a long-term urinary catheter is coating the surfaces of the catheter with benign bacteria (Sunden et al. 2006). Studies were conducted using the nonpathogenic *Escherichia coli* strain 83972 which persistently colonized a person without symptomatic infection during 3 years of observation (Andersson et al. 1991). An in vivo study showed that the non-p fimbriae PapG deletion mutant of 83972 (Hu2117) reduced the colonization of urinary catheters by different uropathogens (Trautner et al. 2002). An initial clinical study showed that the introduction of a Hu2117-coated indwelling catheter in a human bladder reduced the incidence of symptomatic urinary tract infections (Trautner et al. 2007). However, both in vitro and in vivo studies demonstrated that in comparison with the absence of uropathogens, the adherence of 83920 or Hu2117 to an unmodified silicone urinary catheter is low (Trautner et al. 2007, 2008). Mannose is the ligand for FimH which is the adhesive component of type I fimbriae in 83920 (Bouckaert et al. 2005). Therefore, Trautner et al. (2012) covalently immobilized mannose on silicone substrates. Pre-exposure of the mannose-modified surface to *Escherichia coli* 83920 produced a protective biofilm that reduced the adherence of *Enterococcus faecalis* by 80-fold (Trautner et al. 2012). However, in clinical trial, *Pseudomonas aeruginosa* overgrew *Escherichia coli* Hu2117 on urinary catheters (Prasad et al. 2009). Therefore, Liao et al. (2012) attempted to add a *P. aeruginosa* lytic phage to the Hu2117 catheter. Fu et al. (2010) had previously shown that pretreatment of catheter pieces with a cocktail of *P. aeruginosa* lytic phages successfully reduced the 48-h mean *Pseudomonas aeruginosa* biofilm cell density by 99.9 %. Treatment of *Escherichia coli* Hu2117-coated catheter segments with *P. aeruginosa* phages prevented *Pseudomonas aeruginosa* colonization (Liao et al. 2012).

4 Contact Lenses

4.1 Contact Lens-Related Bacterial Infection

More than 250 million people in the world wear contact lenses. One of the early problems for contact lenses was poor oxygenation of the cornea, caused by the fact that the contacts were made of an impermeable hard acrylic polymer. This problem was overcome by the advent of the use of a silicone hydrogel polymer for the manufacture of the lenses. While these lenses are safer for the cornea from an oxygenation standpoint, they still cause acute red eye infections as well as cases of corneal ulceration. In fact the advent of silicone hydrogel lenses has not reduced the incidence of these events (Willcox 2013). In a recent report (Yildiz et al. 2012), it was found that out of 507 cases of corneal ulcers, at a single institution, 223 (43.9 %) were contact lens related. In addition, the investigators observed a significant increase in the number of cases of presumed bacterial keratitis associated with soft contact lens wear over the 3-year period of their study.

Microbial adhesion to contact lenses is believed to be one of the initiating events in the formation of many corneal infiltrations, including microbial keratitis, that occur during contact lens wear (Willcox 2013). In earlier days, patients were told to take their lenses out daily for removal of protein and lipids, which accumulated from the tear film, as well as for sterilization. Now, newer lenses can be worn for 30 days with FDA approval. This extended wear is potentially a situation where a biofilm can form and cause grave damage to the eye (Poggio et al. 1989; Holden et al. 1996; Sankaridurg et al. 1996b, 1999, 2000; Jalbert et al. 2000; Keay et al. 2000; Corrigan et al. 2001). In a 1989 epidemiological study (Poggio et al. 1989), the investigators found almost a five times higher incidence of ulcerative keratitis among extended-wear soft contact lens users compared with daily wear soft contact lens users. Later studies with extended wear contacts showed 4–7 times the risk of ulcerative keratitis (Dart et al. 1991; Schein et al. 1989). When daily disposable CLs became available (1995–1999), it was supposed that frequent replacement of the lenses would reduce the risk of microbial keratitis. However, an epidemiological study from the Netherlands reported that the expected reduction in cases did not happen (Cheng et al. 1999). This is also seen in the more recent study by (Yildiz et al. 2012), where they also saw an increase in the number of cases with time.

Bacterial colonization of CLs has also been implicated in CL-induced inflammation. Specifically, CL acute red eye (CLARE), CL peripheral ulcer (CLPU), and infiltrative keratitis have all been associated with adherence of bacteria to hydrogel CLs. In particular, many CLARE cases have been associated with *Haemophilus influenzae* (Sankaridurg et al. 1996a), *Acinetobacter* sp. (Corrigan et al. 2001), *Pseudomonas aeruginosa* (Holden et al. 1996; Sankaridurg et al. 1996b), *Aeromonas hydrophila* (Sankaridurg et al. 1996b), *Serratia liquefaciens* (Sankaridurg et al. 1996b), *Serratia marcescens* (Holden et al. 1996), and *Pseudomonas putida* (Holden et al. 1996). Infiltrative keratitis and CLPU have been associated with *Staphylococcus aureus* (Jalbert et al. 2000), *Streptococcus pneumoniae* (Sankaridurg et al. 1999), *Abiotrophia defective* (Keay et al. 2000), *Acinetobacter* sp. (Corrigan et al. 2001).

Fungal biofilms are also associated with contact lens-related infections. *Fusarium keratoplasticum* sp. nov. and *Fusarium petroliphilum* stat. nov. are two phylogenetic species that are among the most frequently isolated fusaria in outbreaks of contact lens-associated keratitis (Short et al. 2013). Other in vitro studies showed that *Acanthamoeba castellanii* trophozoites (Beattie et al. 2011) and *Candida* spp. (*albicans, parapsilosis, tropicalis, glabrata,* and *krusei*) (Estivill et al. 2011) are also capable of forming biofilms on different medical devices. In addition, bacteria are thought to have a role in fungal attachment. For instance, *Pseudomonas* had been shown to enhance the absorption of *Acanthamoeba* to contact lenses (Simmons et al. 1998).

While bacterial contaminates that colonize contact lenses have been studied for many years (Hovding 1981), and biofilms themselves have been studied for almost 20 years (Costerton et al. 1978), until recently most of the studies on contact lens biofilms were devoted to looking at methods of sterilization (Szczotka-Flynn et al. 2010). However, it is known that the biofilms on lenses protect bacteria and fungi from disinfectants. It has been shown that fungal hyphae can penetrate the surface of most types of CLs (Willcox 2013). Also, *Acanthamoeba* adhere in greater numbers to first-generation silicone hydrogel lenses compared with the second-generation or hydroxyethyl methacrylate-based soft lenses (Willcox 2013).

Bacteria that are known to form biofilms on contact lenses include: *Pseudomonas aeruginosa* (Burnham et al. 2012; Henriques et al. 2005); *Staphylococcus epidermidis* and *S. aureus* (Catalanotti et al. 2005; Henriques et al. 2005); and *Serratia marcescens* (Hume et al. 2003). Also, in a study of 28 contact lens patients with keratitis, the bacterial genera, *Achromobacter, Stenotrophomonas,* and *Delftia,* were found in all clinical groups (Wiley et al. 2012).

Table 1 Coatings to block biofilm formation on contact lenses

Active component	Mechanism of Action	Advantages/disadvantages
(a) Coatings non-covalently attached		
Furanones	Inhibition of quorum sensing	Toxic to murine fibroblasts; showed enhancement of biofilms at low conc.[a]
Silver	Formation of silver thiol bonds with enzymes and membrane proteins, possibly intercalates with DNA	Toxic to mammalian cells; limited activity; turns tissue black; expensive; causes less killing with Gram-positive bacteria[b]
Polyquaternium compounds	Chelation of bacterial components	Requires direct contact with bacteria; causes cytotoxicity and inflammation[c]
(b) Coatings covalently attached		
Furanones (fimbrolides)	Quorum-sensing inhibition	Only causes about 1 log killing[d]
Polyquaternium compounds	Rupture of bacterial cell membrane	Only causes about 1 log killing[e]
Cationic peptides (melimine)	Unknown	Causes 1–4 logs of killing; stability unknown from proteolysis[f]
Organoselenium (polymerized throughout the contact lens)	Catalysis of superoxide formation	Causes 5–7 logs of killing for Gram negatives and Gram positives; stable for the life of the lens (if incorporated into the polymer); inexpensive[g]

[a]Kuehl et al. (2009)
[b]Atiyeh et al. (2007), Trop et al. (2006), Poon and Burd (2004), Hidalgo et al. (1998), Lee and Moon (2003), Willcox et al. (2010)
[c]Paimela et al. (2012)
[d]Zhu et al. (2008)
[e]Tiller et al. (2001, 2002)
[f]Willcox et al. (2008), Cole et al. (2010), Chen et al. (2012), Dutta et al. (2013)
[g]Mathews et al. (2006), Tran et al. (2012a, 2013)

4.2 Antimicrobial Coating of Contact Lenses

In an attempt to improve the safety of the extended wear lenses, different materials have been developed in order to reduce the possibility of biofilm formation which could lead to bacterial keratitis and corneal ulcers. These compounds fall into two main classes: (1) Compounds that must leach off the lens in order to kill bacteria or inhibit their ability to form biofilms and (2) those compounds that are covalently attached to the lens and yet can either kill bacteria or inhibit their ability to form biofilms. These different coating materials are discussed in detail below and are listed in Table 1 along with their advantages and disadvantages.

4.2.1 Compounds That Are Released from Lenses

Furanones

These are quorum-sensing blockers. These compounds showed an ability to inhibit biofilm on medical device polymers (Baveja et al. 2004) but yielded equivocal results with contact lenses for their ability to block biofilm formation (George et al. 2005). However, a more recent paper (Kuehl et al. 2009) showed that these compounds (1) will inhibit biofilm formation; (2) need to be free to inhibit biofilms; however, (3) were toxic for murine fibroblasts; and (4) at sub-inhibitory concentrations, showed an enhancement of *S. aureus* biofilm formation (Table 1). This would appear to make them poor candidates for addition to contact lenses.

Polyquaternium Compounds (Polyquats)

These compounds have only been used in contact lens case solutions (Alcon, Inc.). They are thought to kill by chelation of bacterial components (Weisbarth et al. 2007). If these compounds do kill by chelation then they would have a limited range of activity. Also, they have been shown to increase cytoxicity and inflammation in human corneal epithelial cells (Paimela et al. 2012) (Table 1). They may also kill by cell wall rupture (see below under "Covalent attachment to the Lens"). They generally consist of long-chain molecules with cationic ends and can be attached to surfaces. It is felt that the cationic end pierces components of the bacterial cell wall causing it to rupture and die. This is thought to be an advantage since it would not lead to the development of resistance in microorganisms. However, since these long-chain compounds need to have direct contact with the microorganisms in order to penetrate their cell walls, this would restrict the amount of killing that could take place and the surface could become overwhelmed by bacteria with time.

Silver

Silver is widely used since it has antimicrobial activity against a broad spectrum of bacteria (Yin et al. 1999) and fungi (Wright et al. 1999). It is thought that silver atoms bind to thiol groups in essential enzymes and subsequently cause their deactivation. Silver also forms stable sulfur–silver bonds with proteins in the cell membrane that are involved in ion transport (Klueh et al. 2000). It was also proposed that Ag^+ enters the cell and intercalates between the purine and pyrimidine base pairs disrupting the hydrogen bonding between the two antiparallel strands and denaturing DNA molecules (Klueh et al. 2000). Many antimicrobial medical devices use silver as their active agent. However, silver has been shown to be cytotoxic to fibroblasts and keratinocytes (Atiyeh et al. 2007; Trop et al. 2006;

Poon and Burd 2004; Hidalgo et al. 1998; Lee and Moon 2003) (Table 1). Also, with bandages, enough silver to inhibit biofilm formation has been shown to stain tissue black. Those characteristics would tend to rule it out for contact lens use; however, silver-impregnated contact lens cases are on the market (CIBA Vision). In a study with silver impregnated etafilcon A lenses, Willcox et al. showed complete inhibition at 20 ppm silver against *P. aeruginosa* and approximately 6 logs of inhibition against *S. aureus* at 20 ppm silver (Willcox et al. 2010). At 20 ppm silver, they only showed 1 log of inhibition against *Acanthamoeba*.

4.2.2 Compounds That Can Be Covalently Attached to a Contact Lens and Inhibit Biofilm Formation

Fimbrolides (Furanones)

These are also quorum-sensing inhibitors. While these furanones were shown to have toxicity to mammalian cells (see above under furanones), the safety of the fimbrolide class of furanones attached to a contact lens was studied by a short-term clinical assessment (Zhu et al. 2008) (Table 1). They found that the fimbrolide-coated lenses reduced biofilm formation by 67–92 % for different bacteria and 70 % for *Acanthamoeba*. They also saw no significant ocular response, by slit lamp, after 1 month in an animal model or overnight in humans.

Polyquaternium Compounds (Polyquats) and Polymeric Pyridinium Compounds

These compounds generally consist of long-chain molecules with cationic ends and can be attached to surfaces. The mode of action of these materials as well as the limitation of this type of compound has been discussed above. Also, studies on these compounds attached to surfaces show only about one log of biofilm inhibition on contact lenses (Tiller et al. 2001, 2002) (Table 1).

Cationic Peptides (Melimine)

A peptide was synthesized that contained portions of the sequences of the antimicrobial cationic peptides mellitin and protamine. The peptide was named melimine. This peptide was covalently attached to contact lenses and was able to reduce approximately 80 % of *S. aureus* and *P. aeruginosa* adhesion in vitro (Willcox et al. 2008; Cole et al. 2010; Chen et al. 2012) (Table 1). A rabbit study showed a reduction in contact lens-induced acute red eye and contact lens-induced peripheral ulcers (Cole et al. 2010). A later study found that attachment of melimine to etafilcon A lenses inhibited *P. aeruginosa, S. aureus, A. castellanii*, and *Fusarium solani* by 3.1, 3.9, 1.2, and 1.0 logs, respectively (Dutta et al. 2013). However, there is no data on stability of these peptides on the lens and no data on the ability of the lens to still demonstrate antimicrobial activity after time in an eye.

Organoselenium

The use of selenium-coated lenses to block bacterial attachment has been reported (Mathews et al. 2006). This study was based upon the ability of organo-selenium compounds covalently attached to a polymer matrix to generate superoxide radicals. In this case an organo-selenium compound was covalently attached to the surface of a contact lens, and the lens was studied after two months in a rabbit eye. No effects were seen on the rabbit eye after two months of wear. The lenses from the eye were then tested against *P. aeruginosa*. The lenses were studied by scanning electron microscopy and no biofilm was present on the lens, while control lenses showed extensive biofilm growth. Similar lenses were shown to completely inhibit biofilm formation by *S. aureus*, *P. aeruginosa*, and *S. marcescens* (control lenses showed over 6 logs of growth under the same conditions) (Tran et al. 2012a). In addition, it was shown that human transformed corneal epithelial cells could grow (in vitro) under the lens with no toxic effect, and the lenses were fully active after soaking for over 90 days in PBS (Tran et al. 2013). More recently, a selenomethacrylate compound was copolymerized in a hydrogel lens and displayed no loss of superoxide generating activity (Tran et al. 2013).

4.3 Antimicrobial Mechanism of Selenium

Selenium is best known for its nutritional essentiality as the catalytic trace element component of enzymes, for example, glutathione (GSH) peroxidases and thioredoxin reductase. In these selenium enzymes the selenium atom catalyzes the oxidation of glutathione and the reduction of H_2O_2 to water. However, the organo-selenium molecule has recently been shown to function as a catalyst for the formation of superoxide radicals ($O_2^{\bullet-}$) (Fig. 2) from the oxidation of thiols (Seko and Imura 1997). A possible mechanism by which the organoselenium molecule serves as a catalytic generator of superoxide radicals ($O_2^{\bullet-}$) from the oxidation of thiols as was reported by Chaudiere et al. in 1992. This catalytic ability of selenium has been known for over 60 years (Feigl and West 1947), but the pro-oxidative characteristics of several selenium compounds were elucidated much later (Seko et al. 1989).

Superoxide radical formation appears to account for most of the observed toxicity of selenium towards different bacteria such as *Staphylococcus epidermidis*, *Staphylococcus aureus*, *Listeria monocytogenes*, *Salmonella typhimurium*, and *Escherichia coli* in vitro (Babior et al. 1975; Bortolussi et al. 1987; Hoepelman et al. 1990; Kramer and Ames 1988; Rosen and Klebanoff 1981). It has also been shown that organo-selenium can be covalently attached to different biomaterials

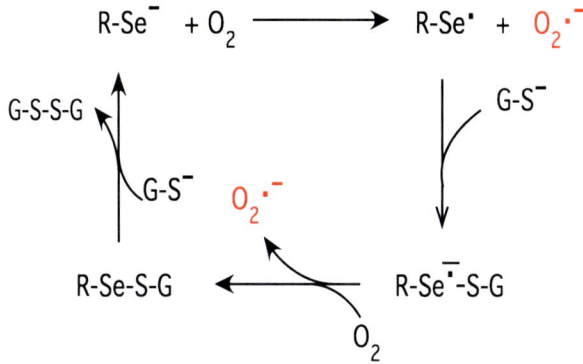

Fig. 2 Organoselenium catalyzes the formation of superoxide radicals. Selenium promotes bacterial toxicity through the catalytic production of the short-lived superoxide radical on the surface of any material to which it is attached. As seen in the reaction above, the molecule R-Se$^-$ is regenerated. Thus, RSe$^-$ is catalytic and is not changed by the reaction. The other important component in the reaction is free sulfhydryls. Sulfhydryl species are abundant in all body fluids. Many bacterial membranes also contain sulfhydryl groups

and medical devices such as intravenous catheter, contact lenses, and block the formation of *Staphylococcus aureus* and *Pseudomonas aeruginosa* biofilms (Low et al. 2011; Mathews et al. 2006; Reid et al. 2010; Tran et al. 2009, 2012b) (Table 1). Concentrations of organo-selenium as low as 0.1 % or 0.2 % were sufficient to inhibit bacterial attachment to these materials. As can be seen in Fig. 2, an organo-selenium molecule can serve as a reducing agent and donate electrons to oxygen. This oxidized selenium molecule can then become reduced back to its original state by obtaining electrons from sulfhydral compounds. In vivo a large source of sulfhydral compounds is glutathione, which is found in body fluids at around 150 μM.

It is important to remember that although selenium has the ability to catalyze the formation of superoxide radicals, selenium is essential for life (Thomson 2004) with a recommended dietary allowance of 55 micrograms per day for both men and women (Institute for Medicine, Food and Nutrition Board 2000). This is because selenium is incorporated into 25 different proteins in the body (Kryukov et al. 2003). Also, superoxide radicals are utilized as second messengers for normal cell growth mechanisms. Superoxide radicals in body fluids have an estimated half-life of 400 ns or less and a diffusion pathway of 55–3,000 nm (Saran and Bors 1989). This is probably why no toxicity was observed in corneas from eyes containing contact lenses coated with organoselenium (Mathews et al. 2006). In addition, cancer patients given milligram quantities of organo-selenium for a year showed no observed toxicity (Reid et al. 2004).

5 Contact Lens Cases

5.1 Bacterial Contamination of Contact Lens Cases

A recent review found that 24–81 % of lens storage cases are contaminated with biofilms, with the frequency of contamination increasing in wearers suffering from microbial keratitis (reviewed by Szczotka-Flynn et al. 2010). In another study, more than 70 % of the cases used in daily wear for 1 month were contaminated regardless of the case type. Contamination of contact lenses is less frequent and associated with fewer organisms than lens cases, but correlates more closely with organisms that cause corneal infections (Das et al. 2007; Martins et al. 2002; McLaughlin-Borlace et al. 1998). Inadequate case cleaning is the most common cause occurring in 72 % of contact lens wearers (Radford et al. 1993).

5.2 Antimicrobial Coating of Contact Lens Cases

There is a need for an effective means of maintaining storage cases free from bacterial and fungal contamination. The only two types of compounds that have been tested for this purpose are silver compounds and organoselenium compounds.

5.2.1 Silver

The use of silver to keep lens cases free from contamination has been reviewed recently (Dantam et al. 2011). They found that a silver impregnated case, MicroBlock (CIBA Vision), was the best at reduction of *P. aeruginosa* (2.4 logs), *Serratia marcescens* (3.3 logs), *D. acidovorans* (2.8 logs), *Fusarium solani* (0.5 logs), in solution. The i-clean case (Sauflon Pharmaceuticals) was the best at reduction of *S. aureus* (5.4 logs), and Nano-case (Marietta Vision) was most effective at *Stenotrophomonas maltophilia* (0.2 logs). All of these studies only measured the reduction of bacteria growing in solution. Oddly, only the MicroBlock demonstrated release of silver into solution and did this over a 28-day period. More recently it was found in a patient study, where the cases themselves were monitored, more than 70 % of the storage cases used in daily wear CL care for a month were contaminated irrespective of whether they contained silver impregnation or not (Dantam et al. 2012). However, silver-impregnated cases were colonized by reduced levels of Gram-negative bacteria. If the silver cases did not remain wet, they showed 94 % contamination.

5.2.2 Organoselenium

While reduction of bacterial number in solution is important, cleaning of the case is the main problem. Most patients replace the cleaning solution every day; however, many don't clean the case. Thus, in order to prevent bacterial colonization of lenses and solutions, it is important to reduce the ability of bacteria to form biofilms on lens cases. Recently, it was shown that a case made with organo-selenium incorporated into the polymer of a polypropylene case eliminated biofilm formation on the case material (Reid et al. 2012, 2013). It was found that organoselenium completely inhibited biofilm formation by *S. aureus*, *P. aeruginosa*, *Stenotrophomonas maltophilia*, and *E. coli* in vitro. In addition it was found that this complete inhibition was still present after 8 weeks of soaking the case, and the same results were obtained when the case was maintained dry. Thus, if the lens case solution was changed daily, no contamination should occur for a contact lens stored in the case.

6 Summary

It is now well recognized that biofilm development is a serious problem for medical devices. Solutions for this problem have progressed over the last 10 years. Initially, devices were impregnated with materials such as silver and other antimicrobials that would leach out from the device in order to control bacterial attachment. These earlier attempts had toxicity problems and led to the development of covalent attachment of materials such as quaternary amines, furanones, and organoselenium compounds. These resulted in a more stable device coating with less toxicity. Recently, it was found that the organoselenium compounds could be incorporated during the polymerization process and still inhibit microbial attachment. This could provide an antimicrobial that is uniformly distributed throughout the device. Thus, the device could degrade or wear, with time, and still have an antimicrobial surface.

Acknowledgments Part of the research was supported by NIH Grant 5R44DK074187-03.

References

Akoh JA (2011) Vascular access infections: epidemiology, diagnosis, and management. Curr Infect Dis Rep 13:324–332

Andersson P, Engberg I, Lidin-Janson G, Lincoln K, Hull R, Hull S, Svanborg C (1991) Persistence of Escherichia coli bacteriuria is not determined by bacterial adherence. Infect Immun 59:2915–2921

Aslam S, Darouiche RO (2007) Prolonged bacterial exposure to minocycline/rifampicin-impregnated vascular catheters does not affect antimicrobial activity of catheters. J Antimicrob Chemother 60:148–151

Atiyeh BS, Costagliola M, Hayek SN, Dibo SA (2007) Effect of silver on burn wound infection control and healing: review of the literature. Burns 33:139–148

Babior BM, Curnutte JT, Kipnes RS (1975) Biological defense mechanisms. Evidence for the participation of superoxide in bacterial killing by xanthine oxidase. J Lab Clin Med 85: 235–244

Bach A, Schmidt H, Bottiger B, Schreiber B, Bohrer H, Motsch J, Martin E, Sonntag HG (1996) Retention of antibacterial activity and bacterial colonization of antiseptic-bonded central venous catheters. J Antimicrob Chemother 37:315–322

Bach A, Eberhardt H, Frick A, Schmidt H, Bottiger BW, Martin E (1999) Efficacy of silver-coating central venous catheters in reducing bacterial colonization. Crit Care Med 27:515–521

Balaban N, Stoodley P, Fux CA, Wilson S, Costerton JW, Dell'Acqua G (2005) Prevention of staphylococcal biofilm-associated infections by the quorum sensing inhibitor RIP. Clin Orthop Relat Res 48–54

Baveja JK, Willcox MD, Hume EB, Kumar N, Odell R, Poole-Warren LA (2004) Furanones as potential anti-bacterial coatings on biomaterials. Biomaterials 25:5003–5012

Beattie TK, Tomlinson A, Seal DV, McFadyen AK (2011) Salicylate inhibition of acanthamoebal attachment to contact lenses. Optom Vis Sci 88:1422–1432

Beiko DT, Knudsen BE, Watterson JD, Cadieux PA, Reid G, Denstedt JD (2004) Urinary tract biomaterials. J Urol 171:2438–2444

Betjes MG (2011) Prevention of catheter-related bloodstream infection in patients on hemodialysis. Nat Rev Nephrol 7:257–265

Bhargava HN, Leonard PA (1996) Triclosan: applications and safety. Am J Infect Control 24: 209–218

Bong JJ, Kite P, Wilco MH, McMahon MJ (2003) Prevention of catheter related bloodstream infection by silver iontophoretic central venous catheters: a randomised controlled trial. J Clin Pathol 56:731–735

Bortolussi R, Vandenbroucke-Grauls CM, van Asbeck BS, Verhoef J (1987) Relationship of bacterial growth phase to killing of Listeria monocytogenes by oxidative agents generated by neutrophils and enzyme systems. Infect Immun 55:3197–3203

Bouckaert J, Berglund J, Schembri M et al (2005) Receptor binding studies disclose a novel class of high-affinity inhibitors of the Escherichia coli FimH adhesin. Mol Microbiol 55:441–455

Brun-Buisson C, Doyon F, Sollet JP, Cochard JF, Cohen Y, Nitenberg G (2004) Prevention of intravascular catheter-related infection with newer chlorhexidine-silver sulfadiazine-coated catheters: a randomized controlled trial. Intensive Care Med 30:837–843

Burnham GW, Cavanagh HD, Robertson DM (2012) The impact of cellular debris on Pseudomonas aeruginosa adherence to silicone hydrogel contact lenses and contact lens storage cases. Eye Contact Lens 38:7–15

Cadieux PA, Chew BH, Knudsen BE, Dejong K, Rowe E, Reid G, Denstedt JD (2006) Triclosan loaded ureteral stents decrease Proteus mirabilis 296 infection in a rabbit urinary tract infection model. J Urol 175:2331–2335

Cadieux PA, Chew BH, Nott L, Seney S, Elwood CN, Wignall GR, Goneau LW, Denstedt JD (2009) Use of triclosan-eluting ureteral stents in patients with long-term stents. J Endourol 23:1187–1194

Calvori C, Frontali L, Leoni L, Tecce G (1965) Effect of rifamycin on protein synthesis. Nature 207:417–418

Casey AL, Mermel LA, Nightingale P, Elliott TS (2008) Antimicrobial central venous catheters in adults: a systematic review and meta-analysis. Lancet Infect Dis 8:763–776

Catalanotti P, Lanza M, Del Prete A, Lucido M, Catania MR, Galle F, Boggia D, Perfetto B, Rossano F (2005) Slime-producing Staphylococcus epidermidis and S. aureus in acute bacterial conjunctivitis in soft contact lens wearers. New Microbiol 28:345–354

Chatzinikolaou I, Finkel K, Hanna H, Boktour M, Foringer J, Ho T, Raad I (2003) Antibiotic-coated hemodialysis catheters for the prevention of vascular catheter-related infections: a prospective, randomized study. Am J Med 115:352–357

Chaudiere J, Courtin O, Leclaire J (1992) Glutathione oxidase activity of selenocystamine: a mechanistic study. Arch Biochem Biophys 296:328–336

Chen R, Willcox MD, Cole N, Ho KK, Rasul R, Denman JA, Kumar N (2012) Characterization of chemoselective surface attachment of the cationic peptide melimine and its effects on antimicrobial activity. Acta Biomater 8:4371–4379

Cheng KH, Leung SL, Hoekman HW, Beekhuis WH, Mulder PG, Geerards AJ, Kijlstra A (1999) Incidence of contact-lens-associated microbial keratitis and its related morbidity. Lancet 354: 181–185

Chew BH, Cadieux PA, Reid G, Denstedt JD (2006) In-vitro activity of triclosan-eluting ureteral stents against common bacterial uropathogens. J Endourol 20:949–958

Cho YW, Park JH, Kim SH et al (2003) Gentamicin-releasing urethral catheter for short-term catheterization. J Biomater Sci Polym Ed 14:963–972

Cirioni O, Ghiselli R, Minardi D et al (2007) RNAIII-inhibiting peptide affects biofilm formation in a rat model of staphylococcal ureteral stent infection. Antimicrob Agents Chemother 51:4518–4520

Clement JL, Jarrett PS (1994) Antibacterial silver. Met Based Drugs 1:467–482

Cole N, Hume EB, Vijay AK, Sankaridurg P, Kumar N, Willcox MD (2010) In vivo performance of melimine as an antimicrobial coating for contact lenses in models of CLARE and CLPU. Invest Ophthalmol Vis Sci 51:390–395

Corral L, Nolla-Salas M, Ibanez-Nolla J, Leon MA, Diaz RM, Cruz Martin M, Iglesia R, Catalan R (2003) A prospective, randomized study in critically ill patients using the Oligon Vantex catheter. J Hosp Infect 55:212–219

Corrigan KM, Harmis NY, Willcox MD (2001) Association of Acinetobacter species with contact lens-induced adverse responses. Cornea 20:463–466

Costerton JW, Geesey GG, Cheng KJ (1978) How bacteria stick. Sci Am 238:86–95

Dantam J, Zhu H, Stapleton F (2011) Biocidal efficacy of silver-impregnated contact lens storage cases in vitro. Invest Ophthalmol Vis Sci 52:51–57

Dantam J, Zhu H, Willcox M, Ozkan J, Naduvilath T, Thomas V, Stapleton F (2012) In vivo assessment of antimicrobial efficacy of silver-impregnated contact lens storage cases. Invest Ophthalmol Vis Sci 53:1641–1648

Darouiche RO, Raad II, Heard SO et al (1999) A comparison of two antimicrobial-impregnated central venous catheters. Catheter Study Group. N Engl J Med 340:1–8

Dart JK, Stapleton F, Minassian D (1991) Contact lenses and other risk factors in microbial keratitis. Lancet 338:650–653

Das S, Sheorey H, Taylor HR, Vajpayee RB (2007) Association between cultures of contact lens and corneal scraping in contact lens related microbial keratitis. Arch Ophthalmol 125: 1182–1185

Davenport K, Keeley FX (2005) Evidence for the use of silver-alloy-coated urethral catheters. J Hosp Infect 60:298–303

Desai M, Buhler T, Weller PH, Brown MR (1998) Increasing resistance of planktonic and biofilm cultures of Burkholderia cepacia to ciprofloxacin and ceftazidime during exponential growth. J Antimicrob Chemother 42:153–160

Donlan RM, Costerton JW (2002) Biofilms: survival mechanisms of clinically relevant microorganisms. Clin Microbiol Rev 15:167–193

Duguid IG, Evans E, Brown MR, Gilbert P (1992) Effect of biofilm culture upon the susceptibility of Staphylococcus epidermidis to tobramycin. J Antimicrob Chemother 30:803–810

Dunser MW, Mayr AJ, Hinterberger G, Florl CL, Ulmer H, Schmid S, Friesenecker B, Lorenz I, Hasibeder WR (2005) Central venous catheter colonization in critically ill patients: a prospective, randomized, controlled study comparing standard with two antiseptic-impregnated catheters. Anesth Analg 101:1778–1784

Dutta D, Cole N, Kumar N, Willcox MD (2013) Broad spectrum antimicrobial activity of melimine covalently bound to contact lenses. Invest Ophthalmol Vis Sci 54:175–182

Elwood CN, Chew BH, Seney S, Jass J, Denstedt JD, Cadieux PA (2007) Triclosan inhibits uropathogenic Escherichia coli-stimulated tumor necrosis factor-alpha secretion in T24 bladder cells in vitro. J Endourol 21:1217–1222

Estivill D, Arias A, Torres-Lana A, Carrillo-Munoz AJ, Arevalo MP (2011) Biofilm formation by five species of Candida on three clinical materials. J Microbiol Methods 86:238–242

Fang FC (1997) Perspectives series: host/pathogen interactions. Mechanisms of nitric oxide-related antimicrobial activity. J Clin Invest 99:2818–2825

Feigl F, West PW (1947) Test for selenium based on catalytic effect. Anal Chem 19:351–353

Feklistov A, Mekler V, Jiang Q, Westblade LF, Irschik H, Jansen R, Mustaev A, Darst SA, Ebright RH (2008) Rifamycins do not function by allosteric modulation of binding of Mg^{2+} to the RNA polymerase active center. Proc Natl Acad Sci 105:14820–14825

Feng QL, Wu J, Chen GQ, Cui FZ, Kim TN, Kim JO (2000) A mechanistic study of the antibacterial effect of silver ions on Escherichia coli and Staphylococcus aureus. J Biomed Mater Res 52:662–668

Flemming CA, Ferris FG, Beveridge TJ, Bailey GW (1990) Remobilization of toxic heavy metals adsorbed to bacterial wall-clay composites. Appl Environ Microbiol 56:3191–3203

Fox CL Jr, Modak SM (1974) Mechanism of silver sulfadiazine action on burn wound infections. Antimicrob Agents Chemother 5:582–588

Fu W, Forster T, Mayer O, Curtin JJ, Lehman SM, Donlan RM (2010) Bacteriophage cocktail for the prevention of biofilm formation by Pseudomonas aeruginosa on catheters in an in vitro model system. Antimicrob Agents Chemother 54:397–404

Garrido-Mesa N, Zarzuelo A, Galvez J (2013) What is behind the non-antibiotic properties of minocycline? Pharmacol Res 67:18–30

George M, Pierce G, Gabriel M, Morris C, Ahearn D (2005) Effects of quorum sensing molecules of Pseudomonas aeruginosa on organism growth, elastase B production, and primary adhesion to hydrogel contact lenses. Eye Contact Lens 31:54–61

Gov Y, Bitler A, Dell'Acqua G, Torres JV, Balaban N (2001) RNAIII inhibiting peptide (RIP), a global inhibitor of Staphylococcus aureus pathogenesis: structure and function analysis. Peptides 22:1609–1620

Hachem R, Reitzel R, Borne A, Jiang Y, Tinkey P, Uthamanthil R, Chandra J, Ghannoum M, Raad I (2009) Novel antiseptic urinary catheters for prevention of urinary tract infections: correlation of in vivo and in vitro test results. Antimicrob Agents Chemother 53:5145–5149

Hagau N, Studnicska D, Gavrus RL, Csipak G, Hagau R, Slavcovici AV (2009) Central venous catheter colonization and catheter-related bloodstream infections in critically ill patients: a comparison between standard and silver-integrated catheters. Eur J Anaesthesiol 26:752–758

Hanna HA, Raad II, Hackett B, Wallace SK, Price KJ, Coyle DE, Parmley CL (2003) Antibiotic-impregnated catheters associated with significant decrease in nosocomial and multidrug-resistant bacteremias in critically ill patients. Chest 124:1030–1038

Heath RJ, Yu YT, Shapiro MA, Olson E, Rock CO (1998) Broad spectrum antimicrobial biocides target the FabI component of fatty acid synthesis. J Biol Chem 273:30316–30320

Heath RJ, Rubin JR, Holland DR, Zhang E, Snow ME, Rock CO (1999) Mechanism of triclosan inhibition of bacterial fatty acid synthesis. J Biol Chem 274:11110–11114

Heath RJ, White SW, Rock CO (2002) Inhibitors of fatty acid synthesis as antimicrobial chemotherapeutics. Appl Microbiol Biotechnol 58:695–703

Henriques M, Sousa C, Lira M, Elisabete M, Oliveira R, Oliveira R, Azeredo J (2005) Adhesion of Pseudomonas aeruginosa and Staphylococcus epidermidis to silicone-hydrogel contact lenses. Optom Vis Sci 82:446–450

Hentschel T, Munstedt H (1999) Thermoplastic polyurethane–the material used for the Erlanger silver catheter. Infection 27(Suppl 1):S43–S45

Hewlett AL, Rupp ME (2012) New developments in the prevention of intravascular catheter associated infections. Infect Dis Clin North Am 26:1–11

Heydorn A, Nielsen AT, Hentzer M, Sternberg C, Givskov M, Ersboll BK, Molin S (2000) Quantification of biofilm structures by the novel computer program COMSTAT. Microbiology 146(Pt 10):2395–2407

Hidalgo E, Bartolome R, Barroso C, Moreno A, Dominguez C (1998) Silver nitrate: antimicrobial activity related to cytotoxicity in cultured human fibroblasts. Skin Pharmacol Appl Skin Physiol 11:140–151

Hoepelman IM, Bezemer WA, Vandenbroucke-Grauls CM, Marx JJ, Verhoef J (1990) Bacterial iron enhances oxygen radical-mediated killing of Staphylococcus aureus by phagocytes. Infect Immun 58:26–31

Holden BA, La Hood D, Grant T, Newton-Howes J, Baleriola-Lucas C, Willcox MD, Sweeney DF (1996) Gram-negative bacteria can induce contact lens related acute red eye (CLARE) responses. CLAO J 22:47–52

Horner C, Mawer D, Wilcox M (2012) Reduced susceptibility to chlorhexidine in staphylococci: is it increasing and does it matter? J Antimicrob Chemother 67:2547–2559

Hovding G (1981) The conjunctival and contact lens bacterial flora during lens wear. Acta Ophthalmol 59:387–401

Hume EB, Stapleton F, Willcox MD (2003) Evasion of cellular ocular defenses by contact lens isolates of Serratia marcescens. Eye Contact Lens 29:108–112

Institute of Medicine, Food and Nutrition Board (2000) Dietary reference intakes: vitamin C, vitamin E, selenium, and carotenoids. National Academy Press, Washington, DC

Jalbert I, Willcox MD, Sweeney DF (2000) Isolation of Staphylococcus aureus from a contact lens at the time of a contact lens-induced peripheral ulcer: case report. Cornea 19:116–120

Jones GL, Russell AD, Caliskan Z, Stickler DJ (2005) A strategy for the control of catheter blockage by crystalline Proteus mirabilis biofilm using the antibacterial agent triclosan. Eur Urol 48:838–845

Kampf G, Kramer A (2004) Epidemiologic background of hand hygiene and evaluation of the most important agents for scrubs and rubs. Clin Microbiol Rev 17:863–893, table of contents

Kaye D, Hessen MT (1994) Infections associated with foreign bodies in the urinary tract. In: Bisno AL, Waldovogel FA (eds) Infections associated with indwelling medical devices, 2nd edn. American Society for Microbiology, Washington, DC, pp 291–307

Keay L, Harmis N, Corrigan K, Sweeney D, Willcox M (2000) Infiltrative keratitis associated with extended wear of hydrogel lenses and Abiotrophia defectiva. Cornea 19:864–869

Klein NC, Cunha BA (1995) Tetracyclines. Med Clin North Am 79:789–801

Klueh U, Wagner V, Kelly S, Johnson A, Bryers JD (2000) Efficacy of silver-coated fabric to prevent bacterial colonization and subsequent device-based biofilm formation. J Biomed Mater Res 53:621–631

Kowalczuk D, Ginalska G, Piersiak T, Miazga-Karska M (2012) Prevention of biofilm formation on urinary catheters: comparison of the sparfloxacin-treated long-term antimicrobial catheters with silver-coated ones. J Biomed Mater Res B Appl Biomater 100B:1874–1882

Kramer GF, Ames BN (1988) Mechanisms of mutagenicity and toxicity of sodium selenite (Na2SeO3) in Salmonella typhimurium. Mutat Res 201:169–180

Kryukov GV, Castellano S, Novoselov SV, Lobanov AV, Zehtab O, Guigo R, Gladyshev VN (2003) Characterization of mammalian selenoproteomes. Science 300:1439–1443

Kuehl R, Al-Bataineh S, Gordon O, Luginbuehl R, Otto M, Textor M, Landmann R (2009) Furanone at subinhibitory concentrations enhances staphylococcal biofilm formation by luxS repression. Antimicrob Agents Chemother 53:4159–4166

Lee AR, Moon HK (2003) Effect of topically applied silver sulfadiazine on fibroblast cell proliferation and biomechanical properties of the wound. Arch Pharm Res 26:855–860

Liao KS, Lehman SM, Tweardy DJ, Donlan RM, Trautner BW (2012) Bacteriophages are synergistic with bacterial interference for the prevention of Pseudomonas aeruginosa biofilm formation on urinary catheters. J Appl Microbiol 113:1530–1539

Liau SY, Read DC, Pugh WJ, Furr JR, Russell AD (1997) Interaction of silver nitrate with readily identifiable groups: relationship to the antibacterial action of silver ions. Lett Appl Microbiol 25:279–283

Low D, Hamood A, Reid T, Mosley T, Tran P, Song L, Morse A (2011) Attachment of selenium to a reverse osmosis membrane to inhibit biofilm formation of S. aureus. J Membr Sci 378: 171–178

Maaskant JM, De Boer JP, Dalesio O, Holtkamp MJ, Lucas C (2009) The effectiveness of chlorhexidine-silver sulfadiazine impregnated central venous catheters in patients receiving high-dose chemotherapy followed by peripheral stem cell transplantation. Eur J Cancer Care (Engl) 18:477–482

Maki DG, Kluger DM, Crnich CJ (2006) The risk of bloodstream infection in adults with different intravascular devices: a systematic review of 200 published prospective studies. Mayo Clinic Proc Mayo Clinic 81:1159–1171

Malone CL, Boles BR, Lauderdale KJ, Thoendel M, Kavanaugh JS, Horswill AR (2009) Fluorescent reporters for Staphylococcus aureus. J Microbiol Methods 77:251–260

Marrie TJ, Costerton JW (1981) Prolonged survival of Serratia marcescens in chlorhexidine. Appl Environ Microbiol 42:1093–1102

Martins EN, Farah ME, Alvarenga LS, Yu MC, Hoflin-Lima AL (2002) Infectious keratitis: correlation between corneal and contact lens cultures. CLAO J 28:146–148

Mathews SM, Spallholz JE, Grimson MJ, Dubielzig RR, Gray T, Reid TW (2006) Prevention of bacterial colonization of contact lenses with covalently attached selenium and effects on the rabbit cornea. Cornea 25:806–814

McLaughlin-Borlace L, Stapleton F, Matheson M, Dart JK (1998) Bacterial biofilm on contact lenses and lens storage cases in wearers with microbial keratitis. J Appl Microbiol 84:827–838

McMullin BB, Chittock DR, Roscoe DL, Garcha H, Wang L, Miller CC (2005) The antimicrobial effect of nitric oxide on the bacteria that cause nosocomial pneumonia in mechanically ventilated patients in the intensive care unit. Respir Care 50:1451–1456

Mermel LA (2000) Prevention of intravascular catheter-related infections. Ann Intern Med 132:391–402

Mermel LA (2011) What is the predominant source of intravascular catheter infections? Clin Infect Dis 52:211–212

Miller DL, O'Grady NP (2012) Guidelines for the prevention of intravascular catheter-related infections: recommendations relevant to interventional radiology for venous catheter placement and maintenance. J Vasc Interv Radiol 23:997–1007

Moretti EW, Ofstead CL, Kristy RM, Wetzler HP (2005) Impact of central venous catheter type and methods on catheter-related colonization and bacteraemia. J Hosp Infect 61:139–145

Morris NS, Stickler DJ (1998) Encrustation of indwelling urethral catheters by Proteus mirabilis biofilms growing in human urine. J Hosp Infect 39:227–234

National Nosocomial Infections Surveillance System (2004) National Nosocomial Infections Surveillance (NNIS) System Report, data summary from January 1992 through June 2004, issued October 2004. Am J Infect Control 32:470–485

Nickel JC, Downey JA, Costerton JW (1989) Ultrastructural study of microbiologic colonization of urinary catheters. Urology 34:284–291

Njoroge J, Sperandio V (2009) Jamming bacterial communication: new approaches for the treatment of infectious diseases. EMBO Mol Med 1:201–210

Novikov A, Lam MY, Mermel LA, Casey AL, Elliott TS, Nightingale P (2012) Impact of catheter antimicrobial coating on species-specific risk of catheter colonization: a meta-analysis. Antimicrob Resist Infect Control 1:40

Oda T, Hamasaki J, Kanda N, Mikami K (1997) Anaphylactic shock induced by an antiseptic-coated central venous [correction of nervous] catheter. Anesthesiology 87:1242–1244

O'Grady NP, Alexander M, Burns LA et al (2011) Guidelines for the prevention of intravascular catheter-related infections. Clin Infect Dis 52:e162–e193

Ostendorf T, Meinhold A, Harter C, Salwender H, Egerer G, Geiss HK, Ho AD, Goldschmidt H (2005) Chlorhexidine and silver-sulfadiazine coated central venous catheters in haematological patients–a double-blind, randomised, prospective, controlled trial. Support Care Cancer 13:993–1000

Paimela T, Ryhanen T, Kauppinen A, Marttila L, Salminen A, Kaarniranta K (2012) The preservative polyquaternium-1 increases cytoxicity and NF-kappaB linked inflammation in human corneal epithelial cells. Mol Vis 18:1189–1196

Park JH, Cho YW, Cho YH, Choi JM, Shin HJ, Bae YH, Chung H, Jeong SY, Kwon IC (2003) Norfloxacin-releasing urethral catheter for long-term catheterization. J Biomater Sci Polym Ed 14:951–962

Pittaway A, Ford S (2002) Allergy to chlorhexidine-coated central venous catheters revisited. Br J Anaesth 88:304–305, Author reply 305

Poggio EC, Glynn RJ, Schein OD, Seddon JM, Shannon MJ, Scardino VA, Kenyon KR (1989) The incidence of ulcerative keratitis among users of daily-wear and extended-wear soft contact lenses. N Engl J Med 321:779–783

Poon VK, Burd A (2004) In vitro cytotoxity of silver: implication for clinical wound care. Burns 30:140–147

Prasad A, Cevallos ME, Riosa S, Darouiche RO, Trautner BW (2009) A bacterial interference strategy for prevention of UTI in persons practicing intermittent catheterization. Spinal Cord 47:565–569

Prospero E, Barbadoro P, Savini S, Manso E, Annino I, D'Errico MM (2006) Cluster of Pseudomonas aeruginosa catheter-related bloodstream infections traced to contaminated multidose heparinized saline solutions in a medical ward. Int J Hyg Environ Health 209:553–556

Pugach JL, DiTizio V, Mittelman MW, Bruce AW, DiCosmo F, Khoury AE (1999) Antibiotic hydrogel coated Foley catheters for prevention of urinary tract infection in a rabbit model. J Urol 162:883–887

Raad I, Hachem R, Zermeno A, Dumo M, Bodey GP (1996) In vitro antimicrobial efficacy of silver iontophoretic catheter. Biomaterials 17:1055–1059

Raad I, Darouiche R, Dupuis J et al (1997) Central venous catheters coated with minocycline and rifampin for the prevention of catheter-related colonization and bloodstream infections. A randomized, double-blind trial. The Texas Medical Center Catheter Study Group. Ann Intern Med 127:267–274

Radford CF, Woodward EG, Stapleton F (1993) Contact lens hygiene compliance in a university population. J Br Contact Lens Assoc 16:105–111

Regev-Shoshani G, Ko M, Miller C, Av-Gay Y (2010) Slow release of nitric oxide from charged catheters and its effect on biofilm formation by Escherichia coli. Antimicrob Agents Chemother 54:273–279

Reid ME, Stratton MS, Lillico AJ, Fakih M, Natarajan R, Clark LC, Marshall JR (2004) A report of high-dose selenium supplementation: response and toxicities. J Trace Elem Med Biol 18: 69–74

Reid T, Tran P, Cortez J, Mosley T, Shashtri M, Spallholz J, Pot S, Hamood A (2010) Medical devices coated with organo-selenium inhibit bacterial and cellular attachment. Int Med Dev 5:23–31

Reid TW, Tran P, Mosley T, Jarvis C, Thomas J, Tran K, Hanes R, Hamood A (2012) Antimicrobial properties of selenium covalently incorporated into the polymer of contact lens case material. Association for Research in Vision and Ophthalmology, Fort Lauderdale, FL, 3 May. http://www.iovs.org/am

Reid TW, Tran P, Mosley T, Jarvis C, Webster D, Hanes R, Hamood A (2013) Selenium covalently incorporated into the polymer of contact lens case material inhibits bacterial biofilm formation. Association for Research in Vision and Ophthalmology, Seattle, WA, 5 May. http://www.iovs.org/am

Rosen H, Klebanoff SJ (1981) Role of iron and ethylenediaminetetraacetic acid in the bactericidal activity of a superoxide anion-generating system. Arch Biochem Biophys 208:512–519

Rupp ME, Lisco SJ, Lipsett PA et al (2005) Effect of a second-generation venous catheter impregnated with chlorhexidine and silver sulfadiazine on central catheter-related infections: a randomized, controlled trial. Ann Intern Med 143:570–580

Russell AD (1986) Chlorhexidine: antibacterial action and bacterial resistance. Infection 14: 212–215

Russell AD, Hugo WB (1994) Antimicrobial activity and action of silver. Prog Med Chem 31: 351–370

Saint S, Elmore JG, Sullivan SD, Emerson SS, Koepsell TD (1998) The efficacy of silver alloy-coated urinary catheters in preventing urinary tract infection: a meta-analysis. Am J Med 105:236–241

Sampath LA, Tambe SM, Modak SM (2001) In vitro and in vivo efficacy of catheters impregnated with antiseptics or antibiotics: evaluation of the risk of bacterial resistance to the antimicrobials in the catheters. Infect Control Hosp Epidemiol 22:640–646

Sankaridurg PR, Vuppala N, Sreedharan A, Vadlamudi J, Rao GN (1996a) Gram negative bacteria and contact lens induced acute red eye. Indian J Ophthalmol 44:29–32

Sankaridurg PR, Willcox MD, Sharma S et al (1996b) Haemophilus influenzae adherent to contact lenses associated with production of acute ocular inflammation. J Clin Microbiol 34:2426–2431

Sankaridurg PR, Sharma S, Willcox M, Sweeney DF, Naduvilath TJ, Holden BA, Rao GN (1999) Colonization of hydrogel lenses with Streptococcus pneumoniae: risk of development of corneal infiltrates. Cornea 18:289–295

Sankaridurg PR, Sharma S, Willcox M, Naduvilath TJ, Sweeney DF, Holden BA, Rao GN (2000) Bacterial colonization of disposable soft contact lenses is greater during corneal infiltrative events than during asymptomatic extended lens wear. J Clin Microbiol 38:4420–4424

Saran M, Bors W (1989) Oxygen radicals acting as chemical messengers: a hypothesis. Free Radic Res Commun 7:213–220

Sauer K, Camper AK, Ehrlich GD, Costerton JW, Davies DG (2002) Pseudomonas aeruginosa displays multiple phenotypes during development as a biofilm. J Bacteriol 184:1140–1154

Schaber JA, Hammond A, Carty NL, Williams SC, Colmer-Hamood JA, Burrowes BH, Dhevan V, Griswold JA, Hamood AN (2007) Diversity of biofilms produced by quorum-sensing-deficient clinical isolates of Pseudomonas aeruginosa. J Med Microbiol 56:738–748

Schein OD, Glynn RJ, Poggio EC, Seddon JM, Kenyon KR (1989) The relative risk of ulcerative keratitis among users of daily-wear and extended-wear soft contact lenses. A case-control study. Microbial Keratitis Study Group. N Engl J Med 321:773–778

Schumm K, Lam TB (2008) Types of urethral catheters for management of short-term voiding problems in hospitalized adults: a short version Cochrane review. Neurourol Urodyn 27: 738–746

Schweizer HP (1998) Intrinsic resistance to inhibitors of fatty acid biosynthesis in Pseudomonas aeruginosa is due to efflux: application of a novel technique for generation of unmarked chromosomal mutations for the study of efflux systems. Antimicrob Agents Chemother 42:394–398

Seko Y, Imura N (1997) Active oxygen generation as a possible mechanism of selenium toxicity. Biomed Environ Sci 10:333–339

Seko Y, Saito Y, Kitahara J, Imura N (1989) Active oxygen generation by the reaction of selenite with reduced glutathione in vitro. In: Wendel A (ed) Selenium in biology and medicine. Springer, Berlin, pp 70–73

Short DP, O'Donnell K, Thrane U, Nielsen KF, Zhang N, Juba JH, Geiser DM (2013) Phylogenetic relationships among members of the Fusarium solani species complex in human infections and the descriptions of F. keratoplasticum sp. nov. and F. petroliphilum stat. nov. Fungal Genet Biol 53:59–70

Siddiq DM, Darouiche RO (2012) New strategies to prevent catheter-associated urinary tract infections. Nat Rev Urol 9:305–314

Simmons PA, Tomlinson A, Seal DV (1998) The role of Pseudomonas aeruginosa biofilm in the attachment of Acanthamoeba to four types of hydrogel contact lens materials. Optom Vis Sci 75:860–866

Stickler DJ (1996) Bacterial biofilms and the encrustation of urethral catheters. Biofouling 9:293–305

Stickler D, Ganderton L, King J, Nettleton J, Winters C (1993a) Proteus mirabilis biofilms and the encrustation of urethral catheters. Urol Res 21:407–411

Stickler D, King J, Nettleton J, Winters C (1993b) The structure of urinary catheter encrusting bacterial biofilms. Cells Mater 3:315–319

Stickler D, Morris N, Moreno MC, Sabbuba N (1998) Studies on the formation of crystalline bacterial biofilms on urethral catheters. Eur J Clin Microbiol Infect Dis 17:649–652

Stickler DJ, Jones GL, Russell AD (2003) Control of encrustation and blockage of Foley catheters. Lancet 361:1435–1437

Suci PA, Mittelman MW, Yu FP, Geesey GG (1994) Investigation of ciprofloxacin penetration into Pseudomonas aeruginosa biofilms. Antimicrob Agents Chemother 38:2125–2133

Sunden F, Hakansson L, Ljunggren E, Wullt B (2006) Bacterial interference–is deliberate colonization with Escherichia coli 83972 an alternative treatment for patients with recurrent urinary tract infection? Int J Antimicrob Agents 28(Suppl 1):S26–S29

Szczotka-Flynn LB, Pearlman E, Ghannoum M (2010) Microbial contamination of contact lenses, lens care solutions, and their accessories: a literature review. Eye Contact Lens 36:116–129

Tambyah PA (2004) Catheter-associated urinary tract infections: diagnosis and prophylaxis. Int J Antimicrob Agents 24(Suppl 1):S44–S48

Tennenberg S, Lieser M, McCurdy B, Boomer G, Howington E, Newman C, Wolf I (1997) A prospective randomized trial of an antibiotic- and antiseptic-coated central venous catheter in the prevention of catheter-related infections. Arch Surg 132:1348–1351

Thomson CD (2004) Assessment of requirements for selenium and adequacy of selenium status: a review. Eur J Clin Nutr 58:391–402

Tiller JC, Liao CJ, Lewis K, Klibanov AM (2001) Designing surfaces that kill bacteria on contact. Proc Natl Acad Sci USA 98:5981–5985

Tiller JC, Lee SB, Lewis K, Klibanov AM (2002) Polymer surfaces derivatized with poly(vinyl-N-hexylpyridinium) kill airborne and waterborne bacteria. Biotechnol Bioeng 79:465–471

Tran PL, Hammond AA, Mosley T et al (2009) Organoselenium coating on cellulose inhibits the formation of biofilms by Pseudomonas aeruginosa and Staphylococcus aureus. Appl Environ Microbiol 75:3586–3592

Tran P, Hamood A, Mosley T, Jarvis C, Thomas J, Lackey B, Reid TW (2012a) Organo-selenium coated contact lenses: effect upon bacterial biofilm attachment. Association for Research in Vision and Ophthalmology, Fort Lauderdale, FL, 3 May. http://www.iovs.org/am

Tran PL, Lowry N, Campbell T et al (2012b) An organoselenium compound inhibits Staphylococcus aureus biofilms on hemodialysis catheters in vivo. Antimicrob Agents Chemother 56:972–978

Tran P, Hamood A, Webster D, Jarvis C, Hanes R, Reid TW (2013) Selenium contact lens hydrogel polymer: inhibition of bacterial biofilm formation. Association for Research in Vision and Ophthalmology, Seattle, WA, 5 May. http://www.iovs.org/am

Trautner BW, Darouiche RO, Hull RA, Hull S, Thornby JI (2002) Pre-inoculation of urinary catheters with Escherichia coli 83972 inhibits catheter colonization by Enterococcus faecalis. J Urol 167:375–379

Trautner BW, Hull RA, Thornby JI, Darouiche RO (2007) Coating urinary catheters with an avirulent strain of Escherichia coli as a means to establish asymptomatic colonization. Infect Control Hosp Epidemiol 28:92–94

Trautner BW, Cevallos ME, Li H, Riosa S, Hull RA, Hull SI, Tweardy DJ, Darouiche RO (2008) Increased expression of type-1 fimbriae by nonpathogenic Escherichia coli 83972 results in an increased capacity for catheter adherence and bacterial interference. J Infect Dis 198:899–906

Trautner BW, Lopez AI, Kumar A, Siddiq DM, Liao KS, Li Y, Tweardy DJ, Cai C (2012) Nanoscale surface modification favors benign biofilm formation and impedes adherence by pathogens. Nanomedicine 8:261–270

Trop M, Novak M, Rodl S, Hellbom B, Kroell W, Goessler W (2006) Silver-coated dressing acticoat caused raised liver enzymes and argyria-like symptoms in burn patient. J Trauma 60:648–652

Tunney MM, Jones DS, Gorman SP (1999) Biofilm and biofilm-related encrustation of urinary tract devices. Methods Enzymol 310:558–566

Vaudaux PE, Lew DP, Waldvogel FA (1994) Host factors predisposing to and influencing therapy of foreign body infections. In: Bisno AL, Waldvogel FA (eds) Infections associated with indwelling medical devices. American Society for Microbiology, Washington, DC, pp 1–29

Veenstra DL, Saint S, Sullivan SD (1999) Cost-effectiveness of antiseptic-impregnated central venous catheters for the prevention of catheter-related bloodstream infection. JAMA 282: 554–560

Weisbarth RE, Gabriel MM, George M, Rappon J, Miller M, Chalmers R, Winterton L (2007) Creating antimicrobial surfaces and materials for contact lenses and lens cases. Eye Contact Lens 33:426–429, discussion 434

Wiley L, Bridge DR, Wiley LA, Odom JV, Elliott T, Olson JC (2012) Bacterial biofilm diversity in contact lens-related disease: emerging role of Achromobacter, Stenotrophomonas, and Delftia. Invest Ophthalmol Vis Sci 53:3896–3905

Willcox MD (2013) Microbial adhesion to silicone hydrogel lenses: a review. Eye Contact Lens 39:61–66

Willcox MD, Hume EB, Aliwarga Y, Kumar N, Cole N (2008) A novel cationic-peptide coating for the prevention of microbial colonization on contact lenses. J Appl Microbiol 105: 1817–1825

Willcox MDP, Hume EBH, Vijay AK, Petcavich R (2010) Ability of silver-impregnated contact lenses to control microbial growth and colonisation. J Optometry 3:143–148

Williams GJ, Stickler DJ (2008) Effect of triclosan on the formation of crystalline biofilms by mixed communities of urinary tract pathogens on urinary catheters. J Med Microbiol 57: 1135–1140

Wright JB, Lam K, Hansen D, Burrell RE (1999) Efficacy of topical silver against fungal burn wound pathogens. Am J Infect Control 27:344–350

Yildiz EH, Airiani S, Hammersmith KM, Rapuano CJ, Laibson PR, Virdi AS, Hongyok T, Cohen EJ (2012) Trends in contact lens-related corneal ulcers at a tertiary referral center. Cornea 31:1097–1102

Yin HQ, Langford R, Burrell RE (1999) Comparative evaluation of the antimicrobial activity of ACTICOAT antimicrobial barrier dressing. J Burn Care Rehabil 20:195–200

Zhu H, Kumar A, Ozkan J et al (2008) Fimbrolide-coated antimicrobial lenses: their in vitro and in vivo effects. Optom Vis Sci 85:292–300

Medicinal Plants and Phytocompounds: A Potential Source of Novel Antibiofilm Agents

Iqbal Ahmad, Fohad Mabood Husain, Meenu Maheshwari, and Maryam Zahin

Abstract Medicinal plants and plant-derived bioactive compounds are well known for their contribution to primary health care as a source novel drug discovery for various ailments. Emergence and spread of microbial drug resistance due to various mechanisms has impacted the efficacy of almost all old and new antibacterial drugs. The biofilm mode of microbial growth has significantly increased the survival strategies and resistance levels of microbes to drugs, making the treatment of infections more difficult. Currently, efforts are going on to develop novel strategies including targeting biofilms to treat infections. Various natural products are known to inhibit biofilm formation or preformed biofilms. In recent years medicinal plants and phytocompounds were reported with promising antibiofilm activity in vitro from different parts of the world. In this chapter we have reviewed the current literature on antibiofilm agents derived from medicinal plants and/or plant-derived compounds. Plant extracts and phytocompounds of various classes have been found effective against bacterial or fungal biofilms, with some compounds showing activity against both. Such compounds are expected to be effective against mixed biofilms. Interestingly, certain quorum-sensing inhibiting plant extracts or compounds can also inhibit biofilms made by bacteria such as *Pseudomonas aeruginosa*. The majority of the antibiofilm phytocompounds identified so far have been tested in vitro; however, only a few compounds have been reported effective under in vivo condition. This could be due to the lack of access of the investigators to suitable animal models for different diseases to assess the therapeutic efficacy of these antibiofilm agents. The results of this chapter indicated that the phytocompounds may be effective alone or in combination with antibiotics, as in the treatment of systemic infection. However, further investigation on their mode

I. Ahmad (✉) · F.M. Husain · M. Maheshwari
Department of Agricultural Microbiology, Aligarh Muslim University, Aligarh, UP, 202002, India
e-mail: ahmadiqbal8@yahoo.co.in; fahadamu@gmail.com

M. Zahin
James Graham Brown Cancer Center, University of Louisville, Louisville, KY, 40202, USA

of action and in vivo efficacy are prerequisites to obtain broad-spectrum antifungal agents of clinical value.

1 Introduction

Infectious diseases are one of the leading causes of morbidity and mortality globally and their magnitude is higher in developing and underdeveloped countries. The antibiotic era during the twentieth century had reduced the threat of infectious diseases. Nevertheless, over the years, there has been an increase in drug resistance among pathogenic bacteria. The increased prevalence of antibiotic resistance has led to the introduction of combination therapy which has increased treatment efficacy and contained drug resistance to some extent (Athamna et al. 2005). Although combination therapy provided the answer to antibiotic resistance for a while, there have been reports of emerging resistance to drugs in combination and multidrug resistance in common pathogenic bacteria (so called "superbugs") (Rodas-Suárez et al. 2006). The major groups of problematic MDR bacteria include *Mycobacterium tuberculosis*, methicillin-resistant *Staphylococcus aureus* (MRSA), and ESβL producing MDR enteric bacteria and others (Ahmad et al. 2009). The problem of drug resistance in fungi, especially *Candida albicans*, has also become a problematic issue in the treatment of fungal infections.

Various mechanisms of bacterial drug resistance are known, which arise as a result of mutations and/or acquisition of new resistance genes by genetic exchange mechanisms (Tenover 2006). Among the other factors contributing to the ability of microbes to combat antimicrobials is their ability to exist in biofilms that allow them to withstand harsh environmental conditions. These biofilms have been implicated in a wide range of hospital infections and food spoilages, thus posing a serious concern in both the food and medical industries. Biofilm formation and antibiotic resistance among bacterial pathogens represents a major hurdle in human health (Rogers et al. 2010). A report from the U.S. National Institutes of Health states that >80 % of microbial infections are biofilm based (Davies et al. 1998). Biofilms are structural communities encased in a self-secreted exopolymeric substance (EPS). It has been reported that bacteria living within biofilm are 1,000-fold more tolerant to antibiotics and are inherently insensitive to the host immune response (Caraher et al. 2007).

Due to the increase in complexity of most microbial infections and the resistance to conventional therapy, researchers have been compelled to identify alternatives for the treatment of infections (Ahmad et al. 2009). In the last few years, efforts have been directed towards developing preventive strategies that can be used to disarm microorganisms without killing them (Cegelski et al. 2008; Rasko and Sperandio 2010). An innovative approach is the use of antibiofilm agents that are effective at inhibiting biofilm formation and destroying preformed biofilms (Roman et al. 2013). In addition, as these substances do not exert their action by killing cells, they theoretically do not impose a selective pressure to cause the development of resistance (Rasko and Sperandio 2010). Observing the processes of biofilm formation, it is reasonable to expect that interfering with the key steps that orchestrate genesis of virtually every biofilm could be a way for new preventive strategies that do not necessarily exert lethal effects on cells but rather sabotage their propensity for a sessile lifestyle. Various strategies and agents have been found useful in preventing biofilm formation on various surfaces including medical devices. Many of these compounds are useful in destroying biofilms in medical settings as well as preventing biofouling. In search of broad-spectrum antibiofilm agents, medicinal plant extracts and natural products have attracted the attention of scientists (Villa and Cappitelli 2013). Plant extracts, and other biologically active compounds isolated from plants, have gained widespread interest in this regard as they have been known to cure diseases and illnesses since ancient times (Ahmad and Beg 2001). Plant extracts/compounds are widely accepted due to the perception that they are safe and they have a long history of use in folk medicine as immune boosters and for the prevention and treatment of several diseases (Ahmad et al. 2006).

Over the years, the use of medicinal plants, which form the backbone of traditional medicine, has grown to an estimated 80 % of the population, of mostly developing countries that rely on traditional medicines for their primary health care (Ahmad et al. 2006). Modern science and technological advances are accelerating the discovery and development of innovative, plant-derived pharmaceuticals that have improved therapeutic activity and reduced side effects. Plant-derived substances under intensive research for possible applications include crude extracts of leaves, roots, stems, and individual compounds isolated from these essential oils and oil components. Although a considerable amount of research on plants and the active constituents is currently underway, the focus is mainly on their antimicrobial properties against planktonic bacteria. Resistant biofilms remain largely unexplored even though they have been shown to be more tolerant to antimicrobial agents than their planktonic counterparts (Costerton et al. 2003; Caraher et al. 2007; Taraszkiewicz et al. 2013).

To overcome the problem of multidrug resistance in pathogenic microorganisms, various strategies have been suggested and some have been implemented in chemotherapy such as the combinational approach (Ahmad et al. 2009). With increased understanding of the role of biofilms in pathogenesis and drug resistance, disruption of biofilms has been considered a novel drug target. Thus, antibiofilm agents with broad-spectrum activities and proven safety could be ideal candidates for drug development to combat infection caused by majority of biofilm-forming pathogens. There has been an ongoing effort to obtain such agents from various natural products including phytocompounds. Reports have appeared in the last few years indicating a promising potential of medicinal extracts and phytocompound to be exploited in chemotherapy alone or in combination with classical antibiotics. In this chapter we have reviewed the scientific reports published in the last decade to provide the current state of knowledge on the possible role of medicinal plant extracts or phytocompounds as antibiofilm agents.

2 Antibiofilm Compounds from Plants

The role of natural products, including medicinal plants and phytocompounds, as anti-infective agents is well established and has been discussed in several excellent review articles (Cowan 1999; Gibbons 2005; Savoia 2012). Considering the vast diversity in the chemical structures of phtyocompounds and their known bioactivities (Harborne et al. 1999), it is expected that medicinal plant extracts and phytocompounds will hopefully provide promising antibiofilm agents if screened systematically.

Many plant extracts and compounds are already known to change the hydrophobicity and adhesion of bacteria to attachment sites (Ahmad and Aqil 2007). In the last few years an increasing number of reports have been published showing a possible role of phytocompounds in interfering with biofilm processes (Rezanska et al. 2012) and in few cases mechanism has been proposed.

Biofilm formation involves adhesion, maturation, and differentiation steps. Molecular mechanisms have also been explored in *C. albicans* and other pathogens (Khan and Ahmad 2013). The role of microscopic techniques like scanning electron microscopy (SEM), transmission electron microscopy (TEM), confocal laser scanning microscopy (CLSM), and atomic force microscopy (AFM) has made the study of biofilm inhibition easier (Khan and Ahmad 2013). Some of the plants and their derived bioactive compounds acting on different stages of biofilm formation are listed in Table 1.

3 Inhibitors of Bacterial Biofilm

In recent years various studies have indicated roles for natural products in inhibiting biofilm and associated functions. Sandasi et al. (2010) demonstrated that seven culinary herbs reduced biofilm adhesion of both the clinical and the type strains of *Listeria monocytogenes* by at least 50 % but only three (*Rosmarinus officinalis, Mentha piperita,* and *Melaleuca alternifolia*) inhibited preformed biofilms. Al-Bakri et al. (2010) studied the antibiofilm activity of seven *Salvia* species and found that both, plant extract and volatile oil of *S. triloba*, demonstrated an antibiofilm activity against MRSA clinical strains. *S. triloba* extract, at a concentration of 0.78 mg/mL, exhibited an 86.2 % and 83.4 % reduction against MRSA strains and a 98.3 % reduction against *S. aureus*. On the other hand, *S. triloba* volatile oil at 12.5 % concentration demonstrated 99.8 % and 94.3 % biofilm reduction in MRSA strains which was comparable to that of *S. aureus* (98.7 %). The antibiofilm activity of the folkloric medicinal plant *Andrographis paniculata* against biofilm forming *P. aeruginosa* isolated from cystic fibrosis sputum was studied, and it was found that six extracts of *A. paniculata* showed significant antibiofilm activity, with methanolic extract inhibiting biofilm growth maximally (Murugan et al. 2011). *Streptococcus mutans,* a gram-positive oral bacterium, has

Table 1 Inhibitory action of selected plants and their derived bioactive compounds at different stages of biofilm formation

Inhibitor	Organism	Stage of action	References
Ursolic acid	E. coli, V. harveyi, P. aeruginosa	Adherence	Ren et al. (2005)
Garlic extract	C. albicans	Adherence, maturation	Shuford et al. (2005)
R. officinalis, E. angustifoli, T. vulgaris, M. piperita extract	L. monocytogenes	Adherence	Sandasi et al. (2010)
Boesenbergia pandurata oil	C. albicans	Maturation	Taweechaisupapong et al. (2010)
Naringenin, kaemferol, quercitin, apigenin	V. harveyi	Maturation	Vikram et al. (2010)
Proanthocyadins A1, Curcuma longa (oil), Kaurenoic acid	S. mutans	Adherence	Daglia et al. (2010), Lee et al. (2011a, b), Jeong et al. (2013)
T. catappa, C. spinosa, C. cyminum extract	P. aeruginosa	Maturation	Taganna et al. (2011), Issac Abraham et al. (2011, 2012)
C. leptophloeos, B. acuruana, P. moniliformis extract, Allicin	S. epidermidis	Adherence	Trentin et al. (2011), Cruz-Villalón and Pérez-Giraldo (2011)
Eugenol, carvone, caeveol, carvacrol, thymol	P. aeruginosa	Adherence	Soumya et al. (2011a, b)
Muscari comosum extract	C. albicans	Dispersion	Villa et al. (2012)
R. Ulmifolus extract	S. aureus	Adherence	Quave et al. (2012)
Chelerythrine, sanguinarine, DHBF, proAc	S. aureus, S. epidermidis	Maturation	Artini et al. (2012)
Cinnamaldehyde, carvacrol, thymol, eugenol	L. monocytogenes	Maturation	Upadhyay et al. (2013)

long been implicated as a primary causative agent of dental caries. The biofilm forming potential of these bacteria has been targeted using natural products. *Embilica officinalis* fruit extract (benzene fraction), *Trachyspermum ammi* seed extract (petrol ether fraction), and *Salvadora persica* have been found to be effective inhibitors in vitro (Hasan et al. 2012; Khan et al. 2012; Al-Sohaibani and Murugan 2012). Extensive studies of the anti-*Staphylococcus epidermidis* biofilm activity of 45 aqueous extracts were published by Trentin et al. (2011). At 4 mg/mL, the most effective were extracts derived from *Bauhinia acuruana* branches (orchidtree), *Chamaecrista desvauxii* fruits, *B. acuruana* fruits, and *Pityrocarpa moniliformis* leaves, which decreased biofilm formation by 81.7, 87.4, 77.8, and 77 %, respectively. When applied at tenfold lower concentration, noteworthy biofilm inhibition was observed only in the presence of *Commiphora leptophloeos* stem bark (corkwood) and *Senna macranthera* fruit extracts (reductions of 67.3 and 66.7 %, respectively). The extract 220D-F2 from the root of *Rubus ulmifolius* was used to inhibit *S. aureus* biofilm formation to a degree that can be correlated with increased antibiotic susceptibility without limiting bacterial growth (Quave et al. 2012). *Achyranthes aspera*, an ethanomedicinal herb, was evaluated for its potential to inhibit growth and biofilm formation by a cariogenic *S. mutans*isolate. The biofilm inhibition percentage obtained for methanol, benzene, petroleum ether, and aqueous extracts (125 μg/mL) were <94, <74, <62, and <42 %, respectively (Murugan et al. 2013). In addition to the plant extracts, various polysaccharides isolated from plants including okra fruit (Lengsfeld et al. 2004; Wittschier et al. 2007), aloe vera (Xu et al. 2010), liquorice root (Wittschier et al. 2007, 2009), ginseng (Lee et al. 2004, 2006, 2009), and blackcurrant (Wittschier et al. 2007) have been shown to inhibit binding of *Helicobacter pylori* to gastric cells and mucin in vitro.

The essential oil of *Curcuma longa* inhibited the formation of *S. mutans* biofilms by interfering with its adherence at concentrations higher than 0.5 mg/mL (Lee et al. 2011a, b). Essential oil and hydrosol of *Satureja thymbra* and polytoxinol, a compound based on essential oil, were shown to be effective against biofilms formed by *Salmonella, Listeria, Pseudomonas, Staphylococcus*, and *Lactobacillus* spp. (Al-Shuneigat et al. 2005; Chorianopoulos et al. 2008). Essential oil components, viz., eugenol (structure 1), carvone, caeveol, carvacrol, and thymol (structure 2), interfered with adherence phenomena and inhibited biofilm formation by *P. aeruginosa* strains (Soumya et al. 2011a, b). Kavanaugh and Ribbeck (2012) demonstrated that certain essential oils can eradicate bacteria within biofilms with higher efficiency than certain important antibiotics, making them interesting candidates for the treatment of biofilms.

H_3CO
HO

1

CH_3

OH

H_3C CH_3

2

Phytocompounds have also been evaluated for their biofilm inhibitory potential. Zeng et al. (2008) carried out an analysis of 51 active compounds used in traditional Chinese medicine. Five of them had a proven ability to inhibit biofilm formation, with the flavonoid baicalein (structure 3) being the most effective. This substance is contained, for example, in *Oroxylum indicum* or in the roots of *Scutellaria baicalensis*. Baicalin, the glucuronide of baicalein, has significant antibiofilm activity against *Burkholderia cenocepacia* or *B. multivorans* (Brackman et al. 2009). Jeon et al. (2009) studied the antibiofilm activity of myricetin (structure 4) (flavonol) and *tt*-farnesol (structure 5), compounds ubiquitously found in fruits (cranberries and red wine grapes), and propolis (a resinous mixture collected from tree buds, sap flows, or other botanical sources by honey bees), against *S. mutans* causing dental caries. They showed that the mixture of the natural products, in combination with fluoride, disrupted the accumulation and structural organization of EPS and bacterial cells in the matrix, which affected the biochemical and physiological properties of the biofilms.

3

4

5

Cranberry (*Vaccinium macrocarpon*) contains less frequent A-type linked proanthocyanidins. In contrast to the more frequent proanthocyanidins containing B-type linkages, the A-type showed far greater anti-adhesive activity to uropathogenic *E. coli* (Howell et al. 2005). Daglia et al. (2010) found that dealcoholised red wine was able to inhibit *S. mutans* biofilm formation on human teeth. Proanthocyanidins, for example, proanthocyanidin A1 (structure 6), were the components most involved in the anti-adhesion and antibiofilm activity. Aesculetin (structure 7), present in horse chestnut or *Aesculus hippocastanum*, was proved to be efficient in preventing biofilm formation by *S. aureus* (Durig et al. 2010). Hancock et al. (2010) studied the biofilm inhibitory activity of ellagic acid (EA) (structure 8) and tannic acid (TA) (structure 9) against two *E. coli* strains VR50, a urinary tract strain and F18, a commensal isolate. Both compounds reduced biofilm formation in VR50 and F18 significantly. TA and EA reduced biofilm formation by 44–80 and 22–26 %, respectively. However, no synergistic effect of the two compounds was observed.

6

7

8

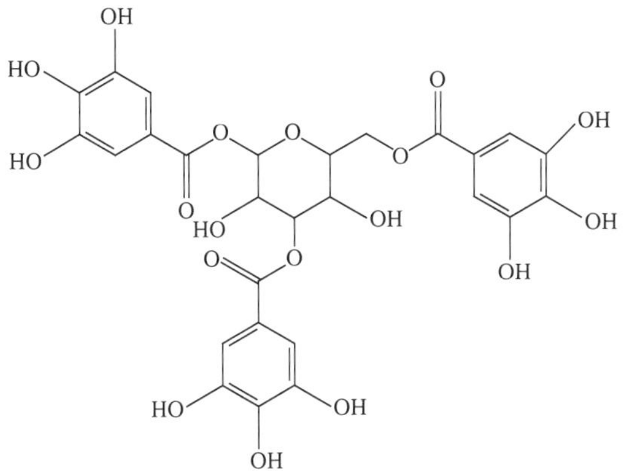

9

Plant auxin 3-indolylacetonitrile (IAN) was found to inhibit the biofilm formation of both *E. coli* O157:H7 and *P. aeruginosa* without affecting its growth. IAN more effectively inhibited biofilms than indole for the two pathogenic bacteria. Additionally, IAN decreased the production of virulence factors including 2-heptyl-3-hydroxy-4(1H)-quinolone (PQS), pyocyanin, and pyoverdine in *P. aeruginosa*. DNA microarray analysis indicated that IAN repressed genes involved in curli formation and glycerol metabolism, whereas IAN induced indole-related genes and prophage genes in *E. coli* O157:H7. It appeared that IAN inhibited the biofilm formation of *E. coli* by reducing curli formation and inducing indole production (Lee et al. 2011a, b). Boswellic acids are pentacyclic triterpenes, which are produced in plants belonging to the genus *Boswellia*. One of the acid acetyl-11-keto-b-boswellic acid (AKBA) (structure 10) inhibited the formation of biofilms by *S. aureus* and *Staphylococcus epidermidis* and also reduced the preformed biofilms generated by these bacteria (Raja et al. 2011).

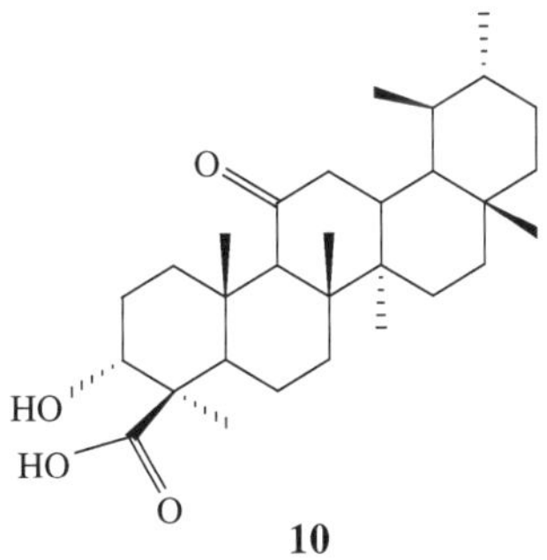

10

Coenye et al. (2012) investigated five plant extracts with antibiofilm activity. Sub-MIC concentrations of *Rhodiola crenulata* (arctic root), *Epimedium brevicornum* (rowdy lamb herb), and *Polygonum cuspidatum* (Japanese knotweed) extracts inhibited *Propionibacterium acnes* biofilm formation by 64.8, 98.5, and

99.2 %, respectively. Moreover, active compounds (resveratrol, icariin, and salidroside) within the extracts were identified and tested against three *P. acnes* strains. The most effective compound was resveratrol from *P. cuspidatum*, which reduced biofilm formation by 80 % for each strain at a concentration of 0.32 % (w/v). Icariin extracted from *E. brevicornum* reduced biofilm formation by 40–70 % at concentrations of 0.01–0.08 % (w/v). The antibiofilm activity of salidroside (0.02–0.25 % concentration) extracted from *R. crenulata* was strain dependent and yielded a biofilm reduction of 40 % for *P. acnes* LMG 16711 and less than 20 % for other tested strains. Importantly, the antibiofilm activity was detected at sub-inhibitory concentrations. Two phenolics ferulic and gallic acids demonstrated preventive action on biofilm formation and showed a higher potential to reduce the mass of biofilms formed by the Gram-negative bacteria (Borges et al. 2012).

Plant-derived compounds Chelerythrine (CH) (structure 11), Sanguinarine (SA) (structure 12), DiHydroxyBenzoFuran (DHBF) (structure 13), and proAnthocyanidin A2-phosphatidylCholine (proAc) (structure 14) were evaluated for biofilm formation inhibition and mature biofilm disruption in *S. aureus* and *S. epidermidis*. All four compounds affected biofilm formation, with comparable efficacy; in fact the inhibition ranged between 1.3 and 5.5-fold of the strongest inhibitory effect (SA on *S. epidermidis*). On *S. aureus*, SA and CH showed a similar inhibitory action with EC50 values of 24.5 and 15.2 µM, respectively, while DHBF and proAc were more effective, EC50 = 8.2 and EC50 = 6.9 µM, respectively. *S. epidermidis* RP62A was more sensitive to the compounds, with CH, SA, and proAc inhibiting at EC50 = 8.6, 4.4, and 7.6 µM, respectively, while the DHBF was less effective with EC50 = 23.5 µM (Artini et al. 2012).

11

12

13

14

The antibiofilm abilities of 522 plant extracts against *P. aeruginosa* PA14 was examined (Cho et al. 2013). Three species of *Carex* plant extracts inhibited *P. aeruginosa* biofilm formation by more than 80 % without affecting planktonic cell grow that a concentration of 200 μg/mL. The most active extract of *Carex pumila,* the resveratrol dimer ε-viniferin (structure 15), was one of the main antibiofilm compounds effective against *P. aeruginosa*. The compounds *trans*-resveratrol (structure 16) and ε-viniferin dose-dependently inhibited the biofilm formation of two *P. aeruginosa* strains, PAO1 and PA14. Specifically, *trans*-resveratrol inhibited *P. aeruginosa* PAO1 biofilm formation by 92 % at 50 μg/mL, and ε-viniferin inhibited *P. aeruginosa* PA14 biofilm formation by 82 % at 50 μg/mL without affecting planktonic cell growth. Interestingly, ε-viniferin inhibited the biofilm formation of enterohemorrhagic *Escherichia coli* O157:H7 by 98 % at 10 μg/mL (Cho et al. 2013).

15

16

Kumar et al. (2013) evaluated the inhibition of biofilm formation in the presence of zingerone alone and its ability to increase the susceptibility of the pathogen to ciprofloxacin. SEM of catheter surfaces showed thinner *P. aeruginosa* biofilms in the presence of zingerone. Further, biofilm was inhibited and eradicated in the presence of zingerone (structure 17) alone and in combination with ciprofloxacin. Highly significant inhibition ($p \leq 0.001$) was observed when the phytocompound and antibiotic were used as adjunct therapy. Brandenburg et al. (2013) showed that both the D and L isoforms of tryptophan inhibited *P. aeruginosa* biofilm formation on tissue culture plates, with an equimolar ratio of D and L isoforms producing the greatest inhibitory effect. Addition of D-/L-tryptophan to existing biofilms inhibited further biofilm growth and caused partial biofilm disassembly.

17

In a study conducted by Upadhyay et al. (2013), sub-MICS of plant-derived compounds like cinnamaldehyde (structure 18), carvacrol, thymol (structure 2), and eugenol (structure 1) were investigated for inhibiting *Listeria monocytogenes* biofilm formation and inactivating mature biofilms at 37, 25, and 4 °C on polystyrene plates and stainless-steel coupons. All compounds inhibited biofilm synthesis and inactivated fully formed *Listeria monocytogenes* biofilms on both matrices at all temperatures tested ($P < 0.05$). Real-time quantitative PCR data revealed that all compounds tested downregulated critical *Listeria monocytogenes* biofilm-associated genes ($P < 0.05$). Kaurenoic acid (KA), a single chemical compound from *A. continentalis*, was investigated for its inhibitory effect on the ability of *S. mutans* to adhere to saliva-coated hydroxyapatite beads (S-HAs) and biofilm formation. The adherence of *S. mutans* was significantly inhibited in a dose-dependent manner in the presence of KA. In addition, the adherence to S-HAs was obviously inhibited at 3–4 μg/mL of KA. Biofilm formation was significantly inhibited at 3 μg/mL of KA and completely inhibited at 4 μg/mL of KA. Biofilm formation on the surface of resin teeth was also significantly inhibited when treated with 3 μg/mL of KA and completely inhibited at 4 μg/mL (Jeong et al. 2013).

18

4 Inhibitors of Fungal Biofilm

Xanthorrhizol isolated from *Curcuma xanthorrhiza* (Rukayadi et al. 2011) and the oil of *Ocimum americanum* (Thaweboon and Thaweboon 2009) showed potent in vitro activity against *Candida* biofilms. In 2008, 30 plant oils including 10 terpenic derivatives and corresponding to the major components of essential oils were tested for their activity against *C. albicans* biofilms (Agarwal et al. 2008). Almost all the studied terpenic derivatives showed antibiofilm activity; however, carvacrol, geraniol, and thymol exhibited the strongest activity. Moreover, these compounds also proved to be efficient against biofilms of *C. glabrata* and *C. parapsilosis* (Dalleau et al. 2008). Polyphenols, extracted from green tea, also showed effects against *C. albicans* biofilms. Epigallocatechin-3-gallate (structure 19), the most abundant polyphenol in green tea extract, reduced the *C. albicans* biofilm metabolic activity by 80 % (Xie and Lou 2008). (R)-goniothalamin, the most abundant styryl lactone in the *Goniothalamus* genus (*Annonaceae* family), was active against *C. albicans* biofilms (Martins et al. 2009). Peppermint, eucalyptus, ginger grass, and clove oils resulted in a reduction in *C. albicans* biofilm formation. The main component of eucalyptus oil, 1,8- cineole, showed potent antibiofilm activity against *C. albicans* biofilms (Hendry et al. 2009). *Candida* biofilm formation was also inhibited more effectively by *Boesenbergia pandurata* (finger root) oil; biofilms were reduced by 63–98 % when sub-MIC volumes (from 4 to 32 μL/mL) were used. Moreover, a significant disruption of mature biofilms was observed when similar volumes of the tested oils were applied (Taweechaisupapong et al. 2010). Additionally, the antifungal activity of tea tree oil has been studied by De Prijck and coworkers against *C. albicans* biofilms. The tea tree oil was released from modified polydimethyl siloxane disks as a model for incorporating antifungals into medical devices to prevent biofilm formation by *Candida* spp. (De Prijck et al. 2010). The efficacy of sub-lethal concentrations of *Muscari comosum* bulb extract in modulating yeast adhesion and subsequent biofilm development on abiotic surfaces and its role as extracellular signal responsible for biofilm dispersion was reported (Villa et al. 2012).

19

Purpurin (structure 20) (1,2,4-trihydroxy-9,10-anthraquinone), a natural red anthraquinone pigment commonly found in madder root (*Rubia tinctorum* L.) at sub-lethal concentrations (3 μg/mL), inhibited *C. albicans* biofilm formation and reduced the metabolic activity of mature biofilms in a concentration-dependent manner. SEM images showed that purpurin-treated *C. albicans* biofilms were scanty and exclusively consisted of aggregates of blastospores (Tsang et al. 2012).

20

Khan and Ahmad (2012a) reported that essential oil components eugenol and cinnamaldehyde were more active against preformed biofilms than amphotericin B and fluconazole against both clinical and reference strains of *C. albicans* (*C. albicans* 04 and *C. albicans* SC5314, respectively). At 0.5× MIC, eugenol and cinnamaldehyde were the most inhibitory compounds against biofilm formation. Light and electron microscopic studies revealed the deformity of three-dimensional structures of biofilms formed in the presence of sub-MICs of eugenol and cinnamaldehyde. Combination studies showed that synergy was highest between eugenol and fluconazole (fractional inhibitory concentration index = 0.14) against preformed biofilms of *C. albicans* SC5314. In another study, promising in vitro antibiofilm activity by *Cymbopogon citratus* and *Syzygium aromaticum* oil was demonstrated against *C. albicans* strains that displayed formation of moderate to strong biofilms. Tested oils were more active against preformed biofilms compared to amphotericin B and fluconazole. At 0.5× MIC, *Cymbopogon citratus* followed by *Syzygium aromaticum* were inhibitory against biofilm formation. Light and electron microscopic studies revealed the deformity of three-dimensional structures of biofilms formed in the presence of sub-MICs of *Cymbopogon citratus* (Khan and Ahmad 2012b).

5 Broad Spectrum Inhibitors of Biofilm

Huber et al. (2003) investigated biofilm inhibition by three compounds (−)-epigallocatechin (EGCG), ellagic acid, and tannic acid against *Burkholderia cepacia*. While treatment with tannic acid showed no real reduction in film thickness, both EGCG and ellagic acid produced a marked reduction, with ellagic acid being the most effective. Antibiofilm activity of green tea polyphenols was also demonstrated on the attached pathogenic yeast, *C. albicans* (Evensen and Braun 2009), with EGCG being more effective than epigallocatechin or epicatechin-3-gallate (structure 19). This study suggests that the metabolic instability produced by the catechin-induced proteasome inactivation was a contributor to the decrease in the growth rate constant as well as biofilm formation and maintenance.

Shuford et al. (2005) have reported the effect of fresh garlic extract on both the adherence and mature phases of *C. albicans* biofilms. Reduction assays of the biofilms using XXT (2,3-bis-(2-methoxy-4-nitro-5-sulfophenyl)-5-[(phenylamino) carbonyl]-2H tetrazoliumhydroxide) showed reduction at both phases. However, this study revealed nothing as to the nature of this inhibition. Rasmussen et al. (2005) identified garlic extract as having specificity for quorum-sensing-controlled virulence genes in *P. aeruginosa*. In vitro analysis of *P. aeruginosa* biofilms showed considerable destruction of the biofilm when exposed to a combination of garlic extract and tobramycin. Exposure to either compound alone had little to no effect on the biofilm. Allicin (structure 21), applied at sub-inhibitory concentration, was involved in specific enzymatic inhibition of polysaccharide intracellular adhesin (PIA) synthesis. Suppression of PIA production, the main substance in *S. epidermidis* agglutination, led to the prevention of biofilm formation by this pathogen (Cruz-Villalón and Pérez-Giraldo 2011).

21

Carneiro et al. (2011) tested sub-MIC concentrations of casbane diterpene (structure 22) (CS) extracted from *Croton nepetaefolius* bark against two Gram-positive bacteria (*S. aureus* and *S. epidermidis*), five Gram-negative bacteria (*Pseudomonas fluorescens, P. aeruginosa, Klebsiella oxytoca, K. pneumoniae,* and *E. coli*), and three yeasts (*Candida tropicalis, C. albicans,* and *C. glabrata*). *S. aureus* and *S. epidermidis* biofilms were significantly disrupted when CS was applied (125 and 250 µg/mL, respectively). Among Gram-negative bacteria, *K. oxytoca* biofilm formation was not affected by CS, and *K. pneumoniae* biofilms were reduced by 45 %. Administration of CS at a concentration of 125 µg/mL caused complete inhibition of *P. fluorescens* biofilms (by 80 %). However, lower concentrations of CS supported *P. aeruginosa* biofilm formation. Similar results were obtained for *E. coli*. The authors explained the observed phenomena by the enhanced production of exopolysaccharides due to the stress induced by the presence of CS in the culture. Further, casbane diterpene activity against *C. albicans* and *C. tropicalis* was observed, reducing biofilm formation by 50 % (at concentrations of 62.5 and 15.6 µg/mL, respectively). Weak antibiofilm activity has been also found in peppermint (*Mentha piperita*) extract against biofilms of *P. aeruginosa* and *C. albicans* (Sandasi et al. 2011).

22

Studies with cinnamaldehyde and 2-nitrocinnamaldehdye showed an inhibitory effect on the biofilms of two *Vibrio* mutants, *V. anguillarum* 4411 and *V. vulnificus* LMG (Brackman et al. 2008). The authors observed that cinnamaldehyde affected the total mass of the biofilm but not the number of viable cells. These results, in addition to further analysis, led to the conclusion that these compounds were affecting the production/accumulation of the exopolysacccharide matrix (Brackman et al. 2008). As discussed in the previous section, cinnamaldehyde is also a potent inhibitor of *Candida* biofilms (Khan and Ahmad 2012a), thus this compound could potentially be a broad spectrum biofilm inhibitor.

6 Quorum-Sensing Inhibitors as Antibiofilm Agent

Quorum sensing (QS) plays a vital role in biofilm formation and virulence factor production in several bacterial species (De Kievit et al. 2001). Consequently, compounds that interfere with QS systems are expected to interfere with biofilm formation also. Extracts of the south Florida plants, *Conocarpus erectus, Bucida buceras*, and *Callistemon viminalis*, showed considerable inhibition of QS-regulated LasA protease, LasB elastase, pyoverdin production, and biofilm production in *P. aeruginosa* (Adonizio et al. 2008). Aqueous extracts of edible plants and fruits such as *Ananas comosus, Musa paradiciaca, Manilkara zapota*, and *Ocimum sanctum* also demonstrated a significant reduction in the biofilm formation abilities of *P. aeruginosa* strain PAO1 (Musthafa et al. 2010). Taganna et al. (2011) found that a tannin-rich component of *Terminalia catappa* leaves (TCF12) was able to inhibit the maturation of biofilms of *P. aeruginosa* to significant levels. The methanolic extract obtained from *Cuminum cyminum*, a traditional food ingredient in South Indian dishes, was shown to act as quorum-sensing inhibitor (QSI). By interfering with the acyl-homoserine lactone activity, it inhibited biofilm formation in several bacterial pathogens (Issac Abraham et al. 2012). The extract of *Capparis spinosa* also showed a high degree of anti-quorum-sensing activity in a dose-dependent manner without affecting the bacterial growth of *Serratia marcescens, P. aeruginosa, E. coli*, and *Proteus mirabilis*. At a concentration of 2 mg/mL, an inhibition of *E. coli* biofilm formation by 73 % was observed. For the pathogens *Serratia marcescens, P. aeruginosa,* and *P. mirabilis*, biofilm biomass was reduced by 79, 75, and 70 %, respectively. Moreover, the mature biofilm structure was disrupted for all of the studied pathogens (Issac Abraham et al. 2011). Similarly, 83 % *P. aeruginosa* biofilm inhibition was achieved with *Lagerstroemia speciosa* (giant crape myrtle) extract, a concentration of 10 mg/mL. Application of the extract to *P. aeruginosa* PAO1 biofilms increased bacterial susceptibility to tobramycin. Significant inhibition of QS-regulated virulence factors: LasA protease, LasB elastase, and pyoverdin production, was also recorded (Singh et al. 2012). *Melia dubia* (bead tree) bark extracts reduced *E. coli* biofilm formation by 84 % at a concentration of 30 mg/mL. Bacterial swarming, regulated by QS, was inhibited by 75 %, resulting in decreased biofilm expansion (Ravichandiran et al. 2012).

The biofilm inhibitor ursolic acid (structure 23) was identified from 13,000 samples of compounds purified from whole plants and separated parts such as fruits, leafs, roots, and stems. Ursolic acid from the tree *Diospyros dendo* added at the rate of 10 µg/mL decreased biofilm formation in *E. coli, V. harveyi*, and *P. aeruginosa* PAO1. Transcriptome analyses showed the induction of chemotaxis and motility genes in *E. coli* treated with the plant-derived compound, suggesting that ursolic acid may function as a signal that tells cells to remain motile, hindering cell adhesion or destabilizing already formed biofilms (Ren et al. 2005). Hamamelitannin (structure 24) extracted from the bark of *Hamamelis virginiana* (witch hazel) did not affect the growth of *Staphylococcus* spp., but it did prevent biofilm formation and cell attachment in vitro. Further evidence was provided by implantation of grafts (soaked in hamamelitannin) into an animal model, which drastically decreased the bacterial load in comparison to the controls (Kiran et al. 2008). Two furocoumarins, bergamottin (structure 25) and dihydroxybergamottin (structure 26), isolated from grape fruit juice, were shown to have strong AI-1 and AI-2 inhibitory activities at concentrations as low as 1 µg/mL. Further investigation showed that both compounds inhibited biofilms made by *E. coli* O157:H7, *Salmonella typhimurium*, and *P. aeruginosa*, without inhibition of bacterial growth (Girennavar et al. 2008). Vanillin (structure 27) (4-hydroxy-3-methoxybenzaldehyde), a well-known food flavoring agent, was studied for its QSI properties against different individual acylhomoserine lactone (AHL) molecules using bioindicator strains. Vanillin showed significant inhibition in short-chain [C4-HSL (69 %) and 3-Oxo-C8-HSL (59.8 %)] and long-chain AHL molecules. Biofilm formation by *A. hydrophila* on a polystyrene surface was also inhibited up to 46.3 %. Results suggested that vanillin could be used as a potential QSI compound that reduces biofilm formation on RO membranes (Ponnusamy et al. 2009).

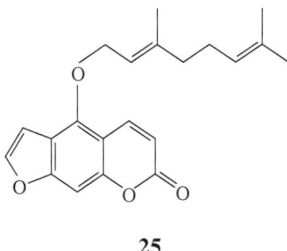

25

26

27

A number of flavonoids found in citrus species, including naringenin (structure 28), kaempferol (structure 29), apigenin (structure 30), and quercetin (structure 31), which are antagonists of homoserine lactones and AI-2-mediated cell–cell signaling in *V. harveyi*, were able to inhibit biofilm formation by *V. harveyi* BB120 and *E. coli* O157:H7 in a dose-dependent manner (Vikram et al. 2010). Flavan-3-ol catechin, one of the flavonoids from the bark of *Combretum albiflorum*, reduced biofilm formation by *P. aeruginosa* PAO1 (Vandeputte et al. 2010).

28

29

30

31

The antibiofilm activity of curcumin (structure 32) against uropathogens using a standard quantitative biofilm assay revealed a concentration-dependent reduction in biofilm biomass of uropathogens when treated with curcumin. Curcumin, at 100 μg/mL, efficiently dislodged the biofilm biomass by 52 %, 89 %, 52 %, and 76 % in *E. coli*, *P. aeruginosa* PAO1, *P. mirabilis*, and *S. marcescens*, respectively. In addition, curcumin was found to be very effective in disrupting the mature (preformed) biofilms of uropathogens. The CLSM images indicated major disruption in the biofilm architecture as well as reduced thickness in curcumin-treated mature biofilms of uropathogens. In *E. coli*, the thickness of the biofilm was reduced from 16 to 10 μm, whereas in *P. mirabilis* it was 11 μm in the control and 6.36 μm in curcumin-treated biofilm. Likewise, *S. marcescens* also displayed a higher reduction from 12 to 3.78 μm in biofilm thickness (Packiavathy et al. 2012). Six sesquiterpene lactones (SLs) of the goyazensolide and isogoyazensolide type isolated from the Argentine herb *Centratherum punctatum* were found to alter biofilm formation, elastase activity, and production of N-acyl-homoserinelactones (AHLs) at lower concentrations. Compounds 2, 3, and 5 displayed significant inhibitory effects on *P. aeruginosa* biofilm formation at 0.5 μg/mL, compound 3 being (1.32 μM) the most potent (42 %) (Amaya et al. 2012). Lastly, ellagic acid derivatives from *Terminalia chebula* showed a significant reduction ($p < 0.001$) in QS-regulated production of extracellular virulence factors in *P. aeruginosa* PAO1. Biofilm formation and alginate were significantly ($p < 0.05$) reduced with enhanced (20 %) susceptibility to tobramycin (Sarabhai et al. 2013).

32

7 In Vivo Efficacy of Antibiofilm Agents

Various phytocompounds have been reported to possess antibiofilm activity in vitro; however, their in vivo efficacy has not yet been fully explored. The in vivo use of garlic extracts as a potential therapy for lung infections was reported by Bjarnsholt et al. (2005). Mice infected with *P. aeruginosa* and treated with a garlic/tobramycin combination showed significantly improved clearing of their bacterial infections as compared to a placebo control group. Highly effective antibiofilm activity was observed for fresh *Allium sativum* extract (fresh garlic extract, FGE). Fourfold treatment of a *P. aeruginosa* biofilm with FGE (at 24 h intervals) resulted in biofilm reduction by 6 log10 units. Moreover, in vivo prophylactic treatment of a mouse model of kidney infection with FGE (35 mg/mL) for

14 days resulted in a 3 log10 unit decrease in the bacterial load on the 5th day after infection compared to untreated animals. In addition, FGE protected renal tissue from bacterial adherence and resulted in a milder inflammatory response and histopathological changes of infected tissues (Harjai et al. 2010).

Cady et al. (2012) demonstrated that S-phenyl-L-cysteine sulfoxide and its breakdown product, diphenyl disulfide, significantly reduced the amount of biofilm formation by *P. aeruginosa*. These organosulfur compounds did not reduce planktonic cell growth and only affected biofilm formation. In a *Drosophila*-based infection model, both S-phenyl-L-cysteine sulfoxide and diphenyl disulfide significantly reduced the *P. aeruginosa* recovered 18 h postinfection (relative to the control) and were nonlethal to the fly hosts.

To determine the effects of the crude and petroleum ether fraction of *T. ammi* on the oral colonization and cariogenic potential of *S. mutans* in vivo, a mouse model was utilized. The caries score was found to be reduced in the treated groups, reducing the total smooth surface as well as the sulcal surface caries. The percentage of total smooth surface caries and sulcal surface caries was 84.83 % and 87.61 % for crude extract and 53.93 % and 73.11 % for the petroleum ether fraction of *T. ammi*, respectively (Khan et al. 2012).

8 Conclusions

Plants represent a sustainable source of antibiofilm agents which have unique modes of action and properties. Present screening and evaluation results of medicinal plants and phytocompounds are promising. It is expected that sub-MICs of plant-derived compounds might offer an elegant way to interfere with various steps involved in biofilm formation. These antibiofilm compounds might also be effective in enhancing antibacterial drug efficacy through improved penetration. Isolation of active compounds from extracts and exploring the mode of action of phytocompounds is further needed in order to identify the most ideal broad-spectrum agents. Many compounds derived from plants or active fractions of extracts have shown promising activity in vitro. Such compounds are expected to exhibit activity in in vivo models, e.g., activity of garlic and its derived compounds have shown encouraging biofilm inhibitory activity in vivo in different animal models. Other compounds have to be tested in vivo to uncover their therapeutic potential.

Acknowledgments FMH is grateful to the Department of Science and Technology (DST), New Delhi, India, for the financial assistance in the form of an INSPIRE fellowship. The authors are also thankful to Mr. Mohammad Shavez Khan (Research Scholar, AMU, Aligarh) for his kind help in the preparation of structures of phytocompounds.

References

Adonizio A, Kong KF, Mathee K (2008) Inhibition of quorum sensing controlled virulence factor production in *Pseudomonas aeruginosa* by south Florida plant extracts. Antimicrob Agents Chemother 52:198–203

Agarwal V, Lal P, Pruthi V (2008) Prevention of *Candida albicans* biofilm by plant oils. Mycopathologia 165(1):13–19

Ahmad I, Aqil F (2007) *In vitro* efficacy of bioactive extracts of 15 medicinal plants against ESβL-producing multidrug-resistant enteric bacteria. Microbiol Res 162(3):264–275

Ahmad I, Beg AZ (2001) Antimicrobial and phytochemical studies on 45 Indian medicinal plants against multi-drug resistant human pathogens. J Ethnopharmacol 74(2):113–123

Ahmad I, Aqil F, Ahmad F, Owais M (2006) Herbal medicines: prospects and constraints. In: Ahmad I, Aqil F, Owais M (eds) Modern phytomedicine: turning medicinal plants into drugs. Wiley-Blackwell, Germany, pp 59–78

Ahmad I, Zahin M, Aqil F, Khan MSA, Ahmad S (2009) Novel approaches to combating drug-resistant bacteria. In: Ahmad I, Aqil F (eds) New strategies combating bacterial infections. Wiley-Blackwell, Germany, pp 47–70

Al-Bakri AG, Othman G, Afifi FU (2010) Determination of the antibiofilm, anti-adhesive, and anti-MRSA activities of seven *Salvia* species. Pharmacogn Mag 6:264–270

Al-Shuneigat J, Cox SD, Markham JL (2005) Effects of a topical essential oil-containing formulation on bio-film-forming coagulase-negative Staphylococci. Lett Appl Microbiol 41:52–55

Al-Sohaibani S, Murugan K (2012) Anti-biofilm activity of *Salvadora persica* on cariogenic isolates of *Streptococcus mutans*: in vitro and molecular docking studies. Biofouling 28:29–38

Amaya S, Pereira JA, Borkosky SA, Valdez JC, Bardón A, Arena ME (2012) Inhibition of quorum sensing in *Pseudomonas aeruginosa* by sesquiterpene lactones. Phytomedicine 19(13):1173–1177

Artini M, Papa R, Barbato G, Scoarughi GL, Cellini A, Morazzoni P, Bombardelli E, Selan L (2012) Bacterial biofilm formation inhibitory activity revealed for plant derived natural compounds. Bioorg Med Chem 20:920–926

Athamna A, Athamna M, Nura A, Shlyakov E, Bast DJ, Farrell D, Rubinstein E (2005) Is *in vitro* antibiotic combination more effective than single-drug therapy against anthrax? Antimicrob Agents Chemother 49:1323–1325

Bjarnsholt T, Jensen P, Rasmussen TB, Christophersen L, Calum H, Hentzer M, Hougen HP, Rygaard J, Moser C, Eberl L, Iby NH, Givskov M (2005) Garlic blocks quorum sensing and promotes rapid clearing of pulmonary *Pseudomonas aeruginosa* infections. Microbiology 151:3873–3880

Borges A, Saavedra MJ, Simões M (2012) The activity of ferulic and gallic acids in biofilm prevention and control of pathogenic bacteria. Biofouling 28(7):755–767

Brackman G, Defoirdt T, Miyamoto C, Bossier P, Van Calenbergh S, Nelis H, Coenye T (2008) Cinnamaldehyde and cinnamaldehyde derivatives reduce virulence in *Vibrio* spp. by decreasing the DNA-binding activity of the quorum sensing response regulator LuxR. BMC Microbiol 8:149

Brackman G, Hillaert U, Van Calenbergh S, Nelis HJ, Coenye T (2009) Use of quorum sensing inhibitors to interfere with biofilm formation and development in *Burkholderia multivorans* and *Burkholderia cenocepacia*. Res Microbiol 160:144–151

Brandenburg KS, Rodriguez KJ, Michael JF, Schurr J, McAnulty CJC, Murphy CJ, Abbott NL (2013) Trytophan inhibits biofilm formation by *Pseudomonas aeruginosa*. Antimicrob Agents Chemother 57(4):1921–1925

Cady NC, McKean KA, Behnke J, Kubec R, Mosier AP, Kasper SH, Burz DS, Musah RA (2012) Inhibition of biofilm formation, quorum sensing and infection in *Pseudomonas aeruginosa* by natural products-inspired organosulfur compounds. PLoS One 7(6):e38492

Caraher E, Reynolds G, Murphy P, McClean S, Callaghan M (2007) Comparison of antibiotic susceptibility of *Burkholderia cepacia* complex organisms when grown planktonically or as biofilm *in vitro*. Eur J Clin Microbiol Infect Dis 26:213–221

Carneiro VA, Dos Santos HS, Arruda FVS, Bandeira PN, Albuquerque MRJR, Pereira MO, Henriques M, Cavada BS, Teixeira EH (2011) Casbane diterpene as a promising natural antimicrobial agent against biofilm-associated infections. Molecules 16(1):190–201

Cegelski L, Marshall GR, Eldridge GR, Hultgren SJ (2008) The biology and future prospects of antivirulence therapies. Nat Rev Microbiol 6:17–27

Cho HS, Lee J, Ryu SY, Joo SW, Cho MH, Lee J (2013) Inhibition of *Pseudomonas aeruginosa* and *Escherichia coli* O157:H7 biofilm formation by plant metabolite ε-Viniferin. J Agric Food Chem 61(29):7120–7126

Chorianopoulos NG, Giaouris FD, Skendamis PN, Haroutounian SA, Nychas GJE (2008) Disinfectant test against monoculture andmixed-culture biofilms composed of technological, spoilage and pathogenic bacteria: bactericidal effect of essential oil and hydrosol of *Satureja thymbra* and comparison with standard acid–base sanitizers. J Appl Microbiol 104:1586–1596

Coenye T, Brackman G, Rigole P, Witte ED, Honraet K, Rossel B, Nelis HJ (2012) Eradication of *Propionibacterium acnes* biofilms by plant extracts and putative identification of icariin, resveratrol and salidroside as active compounds. Phytomedicine 19:409–412

Costerton W, Veeh R, Shirtliff M, Pasmore M, Post C, Ehrlich G (2003) The application of biofilm science to the study and control of chronicbacterial infections. J Clin Invest 112:1466–1477

Cowan MM (1999) Plant products as antimicrobial agents. Clin Microbiol Rev 12:564–582

Cruz-Villalón G, Pérez-Giraldo C (2011) Effect of allicin on the production of polysaccharide intercellular adhesin in *Staphylococcus epidermidis*. J Appl Microbiol 110:723–728

Daglia M, Stauder M, Papetti A, Signoretto C, Giusto G, Canepari P, Pruzzo C, Gazzani G (2010) Antiadhesion and antibiofilm activities of high molecular weight coffee components against *Streptococcus mutans*. Food Chem 119:1182–1188

Dalleau S, Cateau E, Bergès T, Berjeaud JM, Imbert C (2008) *In vitro* activity of terpenes against *Candida* biofilms. Int J Antimicrob Agents 31(6):572–576

Davies DG, Parsek MR, Pearson JP, Iglewski BH, Costerton JW, Greenberg EP (1998) The involvement of cell-to-cell signals in the development of a bacterial biofilm. Science 280:295–298

De Kievit TR, Gillis R, Marx S, Brown C, Iglewski BH (2001) Quorum-sensing genes in *Pseudomonas aeruginosa* biofilms: their role and expression patterns. Appl Environ Microbiol 67:1865–1873

De Prijck K, De Smet N, Coenye T, Schacht E, Nelis HJ (2010) Prevention of *Candida albicans* biofilm formation by covalently bound dimethylaminoethylmethacrylate and polyethylenimine. Mycopathologia 170(4):213–221

Durig A, Kouskoumvekaki I, Vejborg RM, Klemm P (2010) Chemoinformatics-assisted development of new anti-biofilm compounds. Appl Microbiol Biotechnol 87:309–317

Evensen NA, Braun PC (2009) The effects of tea polyphenols on *Candida albicans*: inhibition of biofilm formation and proteasome inactivation. Can J Microbiol 55:1033–1039

Gibbons S (2005) Plants as a source of bacterial resistance modulators and anti-infective agents. Phytochem Rev 4:63–78

Girennavar B, Cepeda ML, Soni KA, Vikram A, Jesudhasan P, Jayaprakasha GK, Pillai SD, Patil BS (2008) Grapefruit juice and its furocoumarins inhibits autoinducer signaling and biofilm formation in bacteria. Int J Food Microbiol 125(2):204–208

Hancock V, Dahl M, Vejborg RM, Klemm P (2010) Dietary plant components ellagic acid and tannic acid inhibit Escherichia coli biofilm formation. J Med Microbiol 59(4):496–498

Harborne JB, Baxter H, Moss GP (1999) Phytochemical dictionary: a handbook of bioactive compounds from plants. Taylor and Francis, London

Harjai K, Kumar R, Singh S (2010) Garlic blocks quorum sensing and attenuates the virulence of *Pseudomonas aeruginosa*. FEMS Immunol Med Microbiol 58:161–168

Hasan S, Danishuddin M, Adil M, Singh K, Verma PK (2012) Efficacy of *E. officinalis* on the cariogenic properties of *Streptococcus mutans*: a novel and alternative approach to suppress quorum-sensing mechanism. PLoS One 7(7):e40319

Hendry ER, Worthington T, Conway BR, Lambert PA (2009) Antimicrobial efficacy of eucalyptus oil and 1,8-cineole alone and in combination with chlorhexidine digluconate against microorganisms grown in planktonic and biofilm cultures. J Antimicrob Chemother 64(6):1219–1225

Howell AB, Reed JD, Krueger CG, Winterbottom R, Cunningham DG, Leahy M (2005) A-type cranberry proanthocyanidins and uropathogenic bacterial anti-adhesion activity. Phytochemistry 66:2281–2291

Huber B, Eberl L, Feucht W, Polster J (2003) Influence of polyphenols on bacterial biofilm formation and quorum-sensing. Z Naturforsch C 58:879–884

Issac Abraham SV, Palani A, Ramaswamy BR, Shunmugiah KP, Arumugam VR (2011) Antiquorum sensing and antibiofilm potential of *Capparis spinosa*. Arch Med Res 42:658–668

Issac Abraham SV, Palani A, Khadar Syed M, Shunmugiah KP, Arumugam VR (2012) Antibiofilm and quorum sensing inhibitory potential of *Cuminum cyminum* and its secondary metabolite methyl eugenol against Gram negative bacterial pathogens. Food Res Int 45:85–92

Jeon JG, Klein MI, Xiao J, Gregoire S, Rosalen PL, Koo H (2009) Influences of naturally occurring agents in combination with fluoride on gene expression and structural organization of *Streptococcus mutans* in biofilms. BMC Microbiol 9:228–237

Jeong S, Kim B, Keum K, Lee K, Kang S, Park B, Lee Y, You Y (2013) Kaurenoic acid from *Aralia continentalis* inhibits biofilm formation of *Streptococcus mutans*. Evid Based Complement Alternat Med 160592:9

Kavanaugh NL, Ribbeck K (2012) Selected antimicrobial essential oils eradicate *Pseudomonas* spp. and *Staphylococcus aureus* biofilms. Appl Environ Microbiol 78:4057–4061

Khan MSA, Ahmad I (2012a) Antibiofilm activity of certain phytocompounds and their synergy with fluconazole against *Candida albicans* biofilms. J Antimicrob Chemother 67:618–621

Khan MSA, Ahmad I (2012b) Biofilm inhibition by *Cymbopogon citratus* and *Syzygium aromaticum* essential oils in the strains of *Candida albicans*. J Ethnopharmacol 140:416–423

Khan MSA, Ahmad I (2013) Microscopy in mycological research with especial reference to ultrastructures and biofilm studies. In: Mendez-Vilas A (ed) Current microscopy contributions to advances in sciences and technology microscopy. FormatexSpain, Spain, pp 646–659

Khan R, Adil M, Danishuddin M, Verma PK, Khan AU (2012) In vitro and in vivo inhibition of *Streptococcus mutans* biofilm by *Trachyspermum ammi* seeds: an approach of alternative medicine. Phytomedicine 19(8–9):747–755

Kiran MD, Adikesavan NV, Cirioni O, Giacometti A, Silvestri C, Scalise G, Ghiselli R, Saba V, Orlando F, Shoham M, Balaban N (2008) Discovery of a quorum-sensing inhibitor of drug-resistant Staphylococcal infections by structure-based virtual screening. Mol Pharmacol 73:1578–1586

Kumar L, Chibber S, Harjai K (2013) Zingerone inhibit biofilm formation and improve antibiofilm efficacy of ciprofloxacin against *Pseudomonas aeruginosa* PAO1. Fitoterapia 90:73–78, http://dx.doi.org/10.1016/j.fitote.2013.06.017

Lee JH, Lee JS, Chung MS, Kim KH (2004) In vitro anti-adhesive activity of an acidic polysaccharide from *Panax ginseng* on *Porphyromonas gingivalis* binding to erythrocytes. Planta Med 70:566–568

Lee JH, Shim JS, Lee JS, Kim MK, Chung MS, Kim KH (2006) Pectin-like acidic polysaccharide from *Panax ginseng* with selective antiadhesive activity against pathogenic bacteria. Carbohydr Res 341:1154–1163

Lee JH, Shim JS, Chung MS, Lim ST, Kim KH (2009) In vitro anti-adhesive activity of green tea extract against pathogen adhesion. Phytother Res 23:460–466

Lee JH, Cho MH, Lee J (2011a) Indole production promotes *Escherichia coli* mixed culture growth with *Pseudomonas aeruginosa* by inhibiting quorum signalling. Environ Microbiol 13:62–73

Lee KH, Kim BS, Keum KS, Yu HH, Kim YH, Chang BS, Ra JY, Moon HD, Seo BR, Choi NY, You YO (2011b) Essential oil of *Curcuma longa* inhibits *Streptococcus mutans* biofilm formation. J Food Sci 76:H226–H230

Lengsfeld C, Titgemeyer F, Faller G, Hensel A (2004) Glycosylated compounds from okra inhibit adhesion of Helicobacter pylori to human gastric mucosa. J Agric Food Chem 52:1495–1503

Martins CV, de Resende MA, da Silva DL (2009) *In vitro* studies of anticandidal activity of goniothalamin enantiomers. J Appl Microbiol 107(4):1279–1286

Murugan K, Selvanayaki K, Al-Sohaibani S (2011) Antibiofilm activity of *Andrographis paniculata* against cystic fibrosis clinical isolate. *Pseudomonas aeruginosa*. World J Microbiol Biotechnol 27:1661–1668

Murugan K, Sekar K, Sangeetha S, Ranjitha S, Sohaibani SA (2013) Antibiofilm and quorum sensing inhibitory activity of *Achyranthes aspera* on cariogenic *Streptococcus mutans*: an *in vitro* and *in silico* study. Pharm Biol 51:728–736

Musthafa KS, Ravi AV, Annapoorani A, Packiavathy ISV, Pandian SK (2010) Evaluation of anti-quorum-sensing activity of edible plants and fruits through inhibition of the N-acyl-homoserine lactone system in *Chromobacterium violaceum* and *Pseudomonas aeruginosa*. Chemotherapy 56:333–339

Packiavathy IASV, Priya S, Pandian SK, Ravi AV (2012) Inhibition of biofilm development of uropathogens by curcumin—an anti-quorum sensing agent from *Curcuma longa*. Food Chem 148:453–460, http://dx.doi.org/10.1016/j.foodchem.2012.08.002

Ponnusamy K, Paul D, Kweon JH (2009) Inhibition of quorum sensing mechanism and *Aeromonas hydrophila* biofilm formation by vanillin. Environ Eng Sci 26:1359–1363

Quave CL, Estévez-Carmona M, Compadre CM, Hobby G, Hendrickson H, Beenken KE, Smeltzer MS (2012) Ellagic acid derivatives from *Rubus ulmifolius* inhibit *Staphylococcus aureus* biofilm formation and improve response to antibiotics. PLoS One 7:e28737

Raja AF, Ali F, Khan IA, Shawl AS, Arora DS, Shah BA, Taneja SC (2011) Acetyl-11-keto-β-boswellic acid (AKBA); targeting oral cavity pathogens. BMC Microbiol 11:54–62

Rasko DA, Sperandio V (2010) Anti-virulence strategies to combat bacteria-mediated disease. Nat Rev Drug Discov 9:117–128

Rasmussen TB, Bjarnsholt T, Skindersoe ME, Hentzer M, Kristoffersen P, Köte M (2005) Screening for quorum-sensing inhibitors (QSI) by use of a novel genetic system, the QSI selector. J Bacteriol 187:1799–1814

Ravichandiran V, Shanmugam K, Anupama K, Thomas S, Princy A (2012) Structure-based virtual screening for plant-derived SdiA-selective ligands as potential antivirulent agents against uropathogenic *Escherichia coli*. Eur J Med Chem 48:200–205

Ren D, Zuo R, González Barrios AF, Bedzyk LA, Eldridge GR, Pasmore ME, Wood TK (2005) Differential gene expression for investigation of *Escherichia coli* biofilm inhibition by plant extract ursolic acid. Appl Environ Microbiol 71(7):4022–4034

Rezanska T, Cejkova A, Masak J (2012) Natural products: strategic tools for modulation of biofilm formation. In: Atta-ur-Rahman FRS (ed) Studies in natural products chemistry. Elsevier, Amsterdam, pp 269–303

Rodas-Suárez OR, Flores-Pedroche JF, Betancourt-Rule JM, Quiñones-Ramírez EI, Vázquez-Salinas C (2006) Occurrence and antibiotic sensitivity of *Listeria monocytogenes* strains isolated from oysters, fish, and estuarine water. Appl Environ Microbiol 72:7410–7412

Rogers SA, Huigens RW, Cavanagh J, Melander C (2010) Synergistic effects between conventional antibiotics and 2-aminoimidazole-derived antibiofilm agents. J Antimicrob Chemother 54:2112–2118

Roman S, Ines J, Stefanie W, Alexander T (2013) New approaches to control infections: anti-biofilm strategies against gram-negative bacteria. Chimia Int J Chem 67:286–290

Rukayadi Y, Han S, Yong D, Hwang JK (2011) *In vitro* activity of xanthorrhizol against *Candida glabrata*, *C. guilliermondii*, and *C. parapsilosis* biofilms. Med Mycol 49:1–9

Sandasi M, Leonard CM, Viljoen AM (2010) The *in vitro* antibiofilm activity of selected culinary herbs and medicinal plants against *Listeria monocytogenes*. Lett Appl Microbiol 50:30–35

Sandasi M, Leonard CM, Van Vuuren SF, Viljoen AM (2011) Peppermint (*Mentha piperita*) inhibits microbial biofilms in vitro. South Afr J Bot 77(1):6

Sarabhai S, Sharma P, Capalash N (2013) Ellagic acid derivatives from *Terminalia chebula* Retz. downregulate the expression of quorum sensing genes to attenuate *Pseudomonas aeruginosa* PAO1 virulence. PLoS One 8(1):e53441

Savoia D (2012) Plant-derived antimicrobials compounds: alternatives to antibiotics. Future Microbiol 7(8):979–990

Shuford JA, Steckelberg JM, Patel R (2005) Effects of fresh garlic extract on *Candida albicans* biofilms. Antimicrob Agents Chemother 49(1):473

Singh BN, Singh HB, Singh A, Singh BR, Mishra A, Nautiyal CS (2012) *Lagerstroemia speciosa* fruit extract modulates quorum sensing-controlled virulence factor production and biofilm formation in *Pseudomonas aeruginosa*. Microbiology 158(2):529–538

Soumya EA, Houari A, Hassan L, Remmal A, Koraichi SI (2011a) *In vitro* activity of four common essential oil components against biofilm-producing *Pseudomonas aeruginosa*. Res J Microbiol 6:394–401

Soumya EA, Koraichi SI, Hassan L, Ghizlane Z, Hind M, Remmal A (2011b) Carvacrol and thymol components inhibiting *Pseudomonas aeruginosa* adherence and biofilm formation. Afr J Microbiol Res 5:3229–3232

Taganna JC, Quanico JP, Perono RMG, Amor EC, Rivera WL (2011) Tannin–rich fraction from *Terminalia catappa* inhibits quorum sensing (QS) in *Chromobacterium violaceum* and the QS-controlled biofilm maturation and LasA staphylolytic activity in *Pseudomonas aeruginosa*. J Ethnopharmacol 134:865–871

Taraszkiewicz A, Fila G, Grinholc M, Nakonieczna J (2013) Innovative strategies to overcome biofilm resistance. Biomed Res Int 2013:150653, http://dx.doi.org/10.1155/2013/150653

Taweechaisupapong S, Singhara S, Lertsatitthanakorn P, Khunkitti W (2010) Antimicrobial effects of *Boesenbergia pandurata* and *Pipersarmentosum* leaf extracts on planktonic cells and biofilm of oral pathogens. Pak J Pharm Sci 23(2):224–231

Tenover FC (2006) Mechanisms of antimicrobial resistance in bacteria. Am J Med 119:S3–S10

Thaweboon S, Thaweboon B (2009) *In vitro* antimicrobial activity of *Ocimum americanum* L. essential oil against oral microorganisms. Southeast Asian J Trop Med Public Health 40(5):1025–1033

Trentin DS, Giordani RB, Zimmer KR, da Silva AG, da Silva MV, Correia MT, Baumvol IJ, Macedo AJ (2011) Potential of medicinal plants from the Brazilian semi-arid region (Caatinga) against *Staphylococcus epidermidis* planktonic and biofilm lifestyles. J Ethnopharmacol 137:327–335

Tsang PW-K, Bandara HMHN, Fong W-P (2012) Purpurin suppresses *Candida albicans* biofilm formation and hyphal development. PLoS One 7(11):e50866

Upadhyay A, Upadhyaya I, Kollanoor-Johny A, Venkitanarayanan K (2013) Antibiofilm effect of plant derived antimicrobials on *Listeria monocytogenes*. Food Microbiol 36(1):79–89, http://dx.doi.org/10.1016/j.fm.2013.04.010

Vandeputte OM, Kiendrebeogo M, Rajaonson S, Diallo B, Mol A, Jaziri ME, Baucher M (2010) Identification of catechin as one of the flavonoids from *Combretum albiflorum* bark extract that reduces the production of quorum-sensing-controlled virulence factors in *Pseudomonas aeruginosa* PAO1. Appl Environ Microbiol 71:243–253

Vikram A, Jayaprakasha GK, Jesudhasan PR, Pillai SD, Patil BS (2010) Suppression of bacterial cell–cell signaling, biofilm formation and type III secretion system by citrus flavonoids. J Appl Microbiol 109:515–527

Villa F, Cappitelli F (2013) Plant-derived bioactive compounds at sub-lethal concentrations: towards smart biocide-free antibiofilm strategies. Phytochem Rev 12:245–254

Villa F, Borgonovo G, Cappitelli F, Giussani B, Bassoli A (2012) Sub-lethal concentrations of *Muscari comosum* bulb extract suppress adhesion and induce detachment of sessile yeast cells. Biofouling 28:1107–1117

Wittschier N, Lengsfeld C, Vorthems S, Stratmann U, Ernst JF, Verspohl EJ, Hensel A (2007) Large molecules as anti-adhesive compounds against pathogens. J Pharm Pharmacol 59:777–786

Wittschier N, Faller G, Hensel A (2009) Aqueous extracts and polysaccharides from liquorice roots (*Glycyrrhiza glabra* L.) inhibit adhesion of *Helicobacter pylori* to human gastric mucosa. J Ethnopharmacol 125:218–223

Xie CF, Lou HX (2008) Chemical constituents from the Chinese bryophytes and their reversal of fungal resistance. Curr Org Chem 12(8):619–628

Xu C, Ruan XM, Li HS, Guo BX, Ren XD, Shuang JL, Zhang Z (2010) Anti-adhesive effect of an acidic polysaccharide from *Aloe vera* L. var. chinensis (Haw.) Berger on the binding of *Helicobacter pylori* to the MKN-45 cell line. J Pharm Pharmacol 62:1753–1759

Zeng Z, Qian L, Cao L, Tan H, Huang Y, Xue X, Shen Y, Zhou S (2008) Virtual screening for novel quorum sensing inhibitors to eradicate biofilm formation of *Pseudomonas aeruginosa*. Appl Microbiol Biotechnol 79:119–126

Staphylococcus aureus Biofilm Formation and Inhibition

Carolyn B. Rosenthal, Joe M. Mootz, and Alexander R. Horswill

Abstract *Staphylococcus aureus* is a prominent cause of chronic infections. Generally these infections are considered to be communities of bacteria that are matrix-encased and attached to a surface, which is frequently referred to as a biofilm. These infections can occur on host tissue, such as on bone in osteomyelitis and heart valves in infective endocarditis, or they can occur on foreign implanted materials. In this review, we summarize the latest knowledge in the basic principles of *S. aureus* biofilm formation, and we outline the current understanding of biofilm matrix components and the impact of quorum sensing in modulating biofilm structure. A strong emphasis is placed on biofilm inhibition through an examination of the latest literature on exogenous enzyme approaches or small-molecule treatments for inhibiting biofilms. These small molecules include a number of recently reported natural products with bioactivity against *S. aureus* biofilms and some limited examples of anti-biofilm synthetic compounds. Overall the goal is to provide readers with a basic understanding of *S. aureus* biofilm development and give a fresh look at the ever-growing array of new treatment options that may lead to innovative therapies for these challenging chronic infections.

1 Introduction

Staphylococcus aureus is a notorious pathogen capable of causing a spectrum of acute and chronic infections. A tremendous amount of effort has been placed on understanding the acute nature of disease caused by this pathogen and there is continual concern about the growing levels of antibiotic resistance (DeLeo and Chambers 2009; Chambers and Deleo 2009; Gordon and Lowy 2008; Spellberg et al. 2013). At the same time, *S. aureus* is one of the most common etiological

---

C.B. Rosenthal • J.M. Mootz • A.R. Horswill (✉)
Department of Microbiology, Roy J. and Lucille A. Carver College of Medicine, University of Iowa, 540F EMRB, Iowa City, IA, 52242, USA
e-mail: alex-horswill@uiowa.edu

agents of chronic infections. It is the leading cause of infective endocarditis and osteomyelitis (Fowler et al. 2005; Lew and Waldvogel 2004), a common cause of foreign body infections (Zimmerli et al. 2004), and a frequent invader in chronic lung disease (Valenza et al. 2008). It is generally believed that the ability of *S. aureus* to attach to surfaces, or to itself, and develop a matrix-encased community of cells called a "biofilm" is a factor in the progression of chronic disease (Kiedrowski and Horswill 2011).

S. aureus biofilm development has been a focal point of research over the last decade. Numerous studies have investigated surface adhesins, matrix components, and transcriptional regulators, all with the goal of better understanding how *S. aureus* forms a biofilm and with the eventual goal of improving treatment options. The challenge presented by biofilm infections is that these structures are characterized by resistance to chemotherapies and host defenses, properties that promote bacterial persistence in the host (Parsek and Singh 2003; del Pozo and Patel 2007; Patel 2005; Brady et al. 2008; Costerton 2005; Otto 2013). However, when a *S. aureus* biofilm is dispersed, it regains susceptibility to antimicrobials (Lauderdale et al. 2010; Boles and Horswill 2008), suggesting that an improved understanding of formation and dispersal mechanisms could aid the development of more effective treatments for chronic infections.

How significant are *S. aureus* biofilm infections to our healthcare system? For purposes of this review, we will cover implant-associated infections, which are a growing problem for healthcare systems worldwide. Annually, there are two million nonvascular indwelling devices implanted in the USA, nearly half of which will become infected (Darouiche 2004; Zimmerli et al. 2004). *S. aureus* and coagulase negative staphylococci are the most common isolates (Blot et al. 2005; Darouiche 2004; Mermel et al. 2009; Warren et al. 2006; Wu et al. 2003), accounting for nearly two-thirds of this type of infection. For treating implant infections, long courses of antibiotics are required and additional surgeries are often necessary (Darouiche 2001), sometimes leading to the removal of the infected device (Darouiche 2004; Zimmerli et al. 2004). All of these additional procedures worsen patient outcomes. Both catheter-related bloodstream infections and implant-associated infections add significant burden to the healthcare system (Blot et al. 2005; Darouiche 2004), which manifests as an increase in the length of a hospital stay and raised total costs (Dimick et al. 2001). In order to improve patient outcomes and reduce the burden on our healthcare systems, a better understanding of how *S. aureus* forms and disassembles a biofilm is needed.

In this review, we will closely examine advances in our knowledge of *S. aureus* biofilm development. We will summarize *S. aureus* adhesins and matrix components that are important for a biofilm to form, signaling mechanisms that modulate biofilm integrity, and enzymatic mechanisms of biofilm dispersal. As an emphasis in this review, we will focus on the growing field of small-molecule biofilm inhibitors that have been identified from natural products and synthetic libraries. Since there is more literature on *S. aureus* biofilms than can be covered here, the interested reader is referred to other recent reviews to obtain additional perspective

(Gotz 2002; O'gara 2007; Kiedrowski and Horswill 2011; Boles and Horswill 2011; Kaplan 2010; Otto 2013).

2 *S. aureus* Biofilm Development

A bacterial biofilm can be defined as a community of cells encased in an extracellular matrix. In the staphylococci, biofilm development is thought to occur in four stages: (1) attachment; (2) microcolony formation; (3) maturation; and (4) detachment (see Fig. 1). During the first stage, free-floating cells attach to an abiotic or biotic substratum such as a foreign body or host tissue. The mechanisms used to facilitate attachment are largely dependent upon substrate surface chemistry, with electrostatic or hydrophobic interactions facilitating bacterial attachment to abiotic surfaces and various non-covalent interactions mediating adhesin attachment to biotic surfaces. Following this phase, bacteria multiply to form microcolonies (Stage 2), which are defined as small aggregates of cells that contain some matrix material. It is often considered an intermediate stage of biofilm formation that links the attachment step with the mature biofilm, and it is sometimes not considered a separate stage. However, it has been repeatedly observed through in vitro studies (Shanks et al. 2005; Yarwood et al. 2004; Bateman et al. 2001) and in clinical samples (Stoodley et al. 2008), and for these reasons we will consider the microcolony an independent stage. The continued growth of microcolonies and production of biofilm matrix components result in the significant accumulation of biomass and development of a mature biofilm (Stage 3). This stage has the characteristic surface structure often associated with bacterial biofilms, such as tower formation and water channels, and the cells display the maximal level of resistance to antimicrobials. Finally, mechanical and active mechanisms can trigger cellular detachment from the biofilm (Stage 4). During this stage, the biofilm matrix is typically targeted for degradation resulting in bacterial dissemination, which allows free-floating cells to reinitiate the biofilm development process at new sites. Detachment of the biofilm restores bacterial susceptibility to chemotherapies and is an active area of research interest (Lauderdale et al. 2010; Boles and Horswill 2008, 2011; Kaplan 2010).

3 *S. aureus* Biofilm Matrix

One of the most important components of bacterial biofilms is the extracellular matrix material that is essential for cellular encasement and community function. The biofilm matrix provides protection against both mechanical and chemical environmental stresses, including resistance against antimicrobial peptides, antibiotics, and uptake by phagocytes. A great deal of effort has been spent on identifying the components of the *S. aureus* biofilm matrix, and what has become clear is that

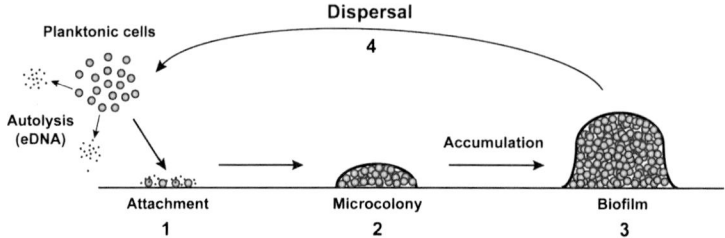

Fig. 1 Schematic of *S. aureus* biofilm development. Stage 1, a subpopulation of planktonic cells lyse, release eDNA, and adhere to a conditioned surface using a combination of surface adhesins, eDNA, and PIA. Stage 2, the attached cells grow to a microcolony that begins to display some morphological features of a biofilm. Stage 3, the biomass accumulates and the structure matures into an established biofilm that displays the expected characteristics, such as tower formation and antimicrobial resistance. Stage 4, active (quorum-sensing) or mechanical mechanisms lead to biofilm detachment and a return to the planktonic state. Multiple factors are involved in the detachment phase, such as proteases and PSMs

the matrix composition varies across strain types and is highly dependent on the environmental conditions promoting biofilm development. Much of the current knowledge of the biofilm matrix is based on studies of enzymes or molecules that destroy this cohesive material, and many of these matrix-degrading agents are summarized in Table 1.

The first extensively studied matrix component is the exopolysaccharide termed PIA (*P*olysaccharide *I*ntercellular *A*dhesin) or PNAG (*P*oly *N*-*A*cetyl*G*lucosamine). PIA is primarily composed of a β1-6 acetylglucosamine homopolymer, is partially de-acetylated (15–20 %), and is negatively charged (Mack et al. 1996). It is produced and secreted by the proteins encoded in the *ica* (*i*ntercellular *a*dhesion) gene locus, *icaADBC* (Cramton et al. 1999), which include a *N*-acetylglucosamine transferase (IcaA and IcaB) (Gerke et al. 1998), a predicted exporter (IcaC) (Gerke et al. 1998), and a deacetylase (IcaD) (Vuong et al. 2004b). This collection of proteins builds the PIA polymer from UDP-*N*-acetylglucosamine to a structure that is over 100 subunits in length. The *ica* locus is important for biofilm formation in many *S. aureus* strains and expression is induced by a variety of environmental conditions including low oxygen, glucose, osmolarity, temperature, and in the presence of sub-inhibitory concentrations of antibiotics (Fitzpatrick et al. 2005; Cramton et al. 2001). However, in a number of studies, *S. aureus* strains have been identified that do not require the *ica* locus to generate a robust biofilm, and many of these strains are clinical MRSA isolates (Beenken et al. 2003; Lauderdale et al. 2010; Boles et al. 2010; Boles and Horswill 2008; O'neill et al. 2007).

The PIA-independent *S. aureus* strains rely on proteins and extracellular DNA (eDNA) as the important components of the biofilm matrix. In the host, biofilm development initiates with attachment to extracellular matrix material including fibrinogen, fibronectin, and collagen, which coat foreign bodies (Francois et al. 1998, 2000). *S. aureus* possesses numerous surface-exposed MSCRAMMs (*M*icrobial *S*urface *C*omponents *R*ecognizing *A*dhesive *M*atrix *M*olecules) as well

Table 1 Natural signaling and enzymatic mechanisms of modulating *S. aureus* biofilms

Process or agent	Mechanism	References
Native mechanisms that modulate *S. aureus* biofilm integrity		
Autoinducing peptide (AIP)	*agr* quorum-sensing signal disperses biofilms	Boles and Horswill (2008), Lauderdale et al. (2010)
PSMs	Surfactant properties promote dispersal of Staphylococcal biofilms	Periasamy et al. (2012), Vuong et al. (2000)
Nuclease	Degradation of eDNA in biofilm matrix	Kiedrowski et al. (2011), Mann et al. (2009)
V8 protease	Cleavage of fibronectin-binding proteins (FnbpAB)	McGavin et al. (1997), Marti et al. (2010), O'neill et al. (2008)
Staphopains (cysteine proteases)	Cleavage of unknown surface or biofilm matrix proteins	Mootz et al. (2013)
D-amino acids	Improper incorporation of D-amino acids into peptidoglycan, reduced production of surface proteins	Hochbaum et al. (2011), Kolodkin-Gal et al. (2010)
Hyaluronate lyase	Enzyme that cleaves hyaluronic acid and may prevent biofilms	Pecharki et al. (2008)
Additional enzyme mechanisms		
Dispersin B	*N*-acetylglucosaminidase that cleaves PIA/PNAG	Kaplan et al. (2004)
Lysostaphin	Glycylglycine endopeptidase that cleaves the pentaglycine cross-bridge of staphylococcal peptidoglycan	Wu et al. (2003)
Esp protease	*S. epidermidis* protease able to inhibit *S. aureus* nasal colonization and biofilm formation	Sugimoto et al. (2013), Iwase et al. (2010)
LasB protease	*Pseudomonas aeruginosa* protease capable of inhibiting *S. aureus* biofilm formation	Park et al. (2012)

as secreted proteins that contain binding domains for these matrix proteins (Foster and Hook 1998). Many *S. aureus* MSCRAMMs have an important role in biofilm formation including SasC (Schroeder et al. 2009), SasG (Conrady et al. 2008; Corrigan et al. 2007), FnbpAB (O'neill et al. 2008), Protein A (Merino et al. 2009), and ClfB (Abraham and Jefferson 2012). These adhesins are particularly important in the initiation of endovascular infections, bone and joint infections, and prosthetic device infections (Gordon and Lowy 2008). In addition to MSCRAMMs, other surface proteins, such as Bap (Trotonda et al. 2005), have been identified with important biofilm roles, but the mechanisms through which these additional proteins contribute to attachment and/or cell–cell adhesion are still under investigation.

As outlined above, eDNA is an important matrix material that is thought to be released into the surrounding milieu by the carefully regulated autolysis of a subpopulation of cells (Mann et al. 2009; Rice et al. 2007). The eDNA provides structural support for the biofilm and may be involved in cell-to-cell or cell-to-surface adhesion. β-toxin is the only protein identified in the *S. aureus* biofilm matrix that can hold onto eDNA and provide bridging support (Huseby et al. 2010). β-toxin has a three-dimensional structure that resembles Nuc (Huseby et al. 2007), and it is capable of binding eDNA and oligomerizing to form higher ordered states. The multimer is protease susceptible, providing the first link between eDNA and proteins in forming the biofilm framework, and *S. aureus* mutants in β-toxin are defective in biofilm formation in vitro and in vivo (Huseby et al. 2010). However, many clinical strains of *S. aureus* do not produce β-toxin due to the presence of a converting prophage (van Wamel et al. 2006), suggesting that other eDNA-binding proteins await identification in the biofilm matrix.

4 Quorum Sensing in *S. aureus* Biofilms

In *S. aureus*, biofilm formation and detachment are regulated by the *agr* (*a*ccessory *g*ene *r*egulator) quorum-sensing system (see Fig. 2). Quorum sensing is a common mechanism utilized by most bacteria to respond to their environment and coordinate a group response. In *S. aureus*, this self-population monitoring leads to global changes in gene expression that influence biofilm formation, and the signal that controls these events is an autoinducing peptide (AIP). The *agr* quorum-sensing system is a chromosomal locus that encodes the proteins that produce and respond to the AIP signal (reviewed in Thoendel et al. 2011; Novick and Geisinger 2008). Under low *agr* expression conditions, cell surface protein expression is high while secreted enzyme expression is low, making *S. aureus* cells more adherent and sessile. When a critical threshold of AIP is reached, either due to growth of the cellular community or the accumulation of a high local signal concentration, a regulatory change occurs that leads to increased expression of the RNAIII transcript. RNAIII is a 514-bp transcript that is major *agr* effector, and high levels of this transcript induce production of extracellular virulence factors that include toxins, superantigens, and exo-enzymes. Multiple studies have linked the induction of the *agr* system with the inhibition of *S. aureus* biofilms (Yarwood et al. 2004; Boles and Horswill 2008; Lauderdale et al. 2010; Periasamy et al. 2012), and currently the primary inhibitory factors are thought to be exo-enzymes and phenol-soluble modulins (PSMs). These studies have demonstrated that there is an inverse correlation between *agr* expression and levels of biofilm biomass, resulting in the characteristic waves of biofilm growth and detachment seen throughout *S. aureus* biofilm development (Yarwood et al. 2004).

The PSMs are surfactant molecules that have been identified in both *S. aureus* and *S. epidermidis* and are under direct control by the response regulator AgrA, which induces their transcription (Wang et al. 2007; Vuong et al. 2004a; Queck et al. 2008). In *S. aureus*, genetic deletion of the *psmα* and *psmβ* operons results in increased biofilm biomass and induction of PSMs is able to detach established biofilms (Periasamy et al. 2012). In the related pathogen *Staphylococcus*

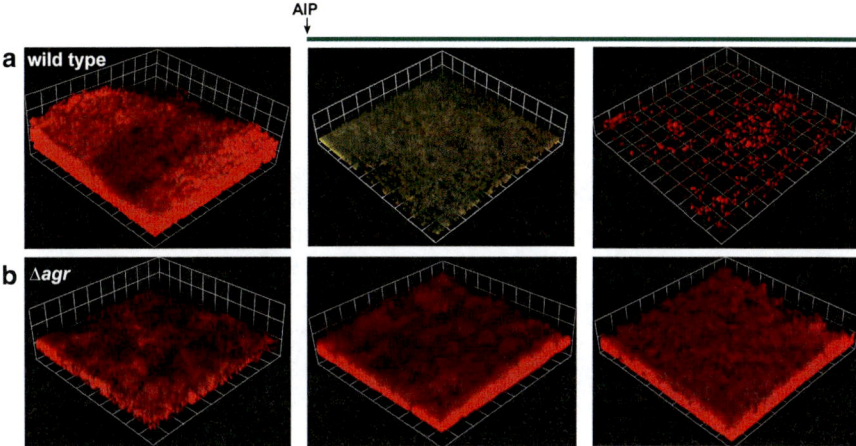

Fig. 2 Quorum-sensing-mediated detachment of *S. aureus* biofilms. *S. aureus* biofilms were grown for 2 days with a constitutive RFP expressing plasmid and an *agr*-controlled GFP plasmid. AIP signal was added to the growth media and biofilm integrity and fluorescence were monitored for 2 days (for further details, see Boles and Horswill 2008). In panel (**a**), AIP was added to a *S. aureus* wild-type strain, and in panel (**b**) it was added to a Δ*agr* mutant strain. The images are reconstructions of confocal microscopy pictures, and they show that the wild-type strain will detach from the biofilm during quorum-sensing induction. Reprinted with permission from Boles and Horswill (2008)

epidermidis, the prevention of PSMβ expression limited dissemination using a mouse model of foreign body infection (Wang et al. 2011b). Altogether, these results emphasize the significance of PSMs as modulators of biofilm development.

The other major class of molecules positively regulated by the *agr* system and impacting biofilm development are the extracellular proteases. Numerous studies have demonstrated that various regulatory conditions that lead to high extracellular protease levels negatively impact *S. aureus* biofilm formation (Mootz et al. 2013; Boles and Horswill 2011; Lauderdale et al. 2009, 2010; O'neill et al. 2008; Marti et al. 2010; Tsang et al. 2008; Zielinska et al. 2012). Most *S. aureus* strains secrete at least ten proteases including a metalloprotease (Aur), seven serine proteases (SspA and SplA-F), and two cysteine proteases (Staphopains ScpA and SspB). All of these proteases are induced when RNAIII levels accumulate (Thoendel et al. 2011), and not surprisingly, the deletion of all these proteases results in increased abundance of secreted and surface-associated virulence factors (Kolar et al. 2013). Several proteins with biofilm roles have been identified that are cleaved by these core proteases. For instance, Aur metalloprotease cleaves surface-exposed clumping factor ClfB (McAleese et al. 2001) and PSMs (Zielinska et al. 2011). The increased stability of PSMs in an *aur* mutant leads to increased osteoblast cell death and bone destruction in a murine model of osteomyelitis (Cassat et al. 2013). SspA (V8) serine protease degrades the fibronectin-binding MSCRAMMs (McGavin et al. 1997) as well as Protein A (Karlsson et al. 2001), and there is evidence that SspA might be important in biofilm remodeling (McGavin et al. 1997; Marti et al. 2010; O'neill et al. 2008). The Staphopains have been implicated as inhibitors of biofilm formation on both biotic and abiotic surfaces (Mootz et al. 2013). Both ScpA and SspB cleave the host proteins fibrinogen and collagen (Ohbayashi et al. 2011), and in addition, ScpA is capable of degrading elastin (Potempa et al. 1988). Therefore, the Staphopains may prevent *S. aureus* biofilm formation, or disassemble established biofilms, through a complex mechanism of cleaving both *S. aureus*-specific proteins and host proteins. However, the protein targets of Staphopains have yet to be elucidated within a biofilm. The other major family of secreted proteases are the Spl proteases, which were initially linked to biofilm dispersal (Boles and Horswill 2008). However, these enzymes have very specific cleavage sites and their targets and role in biofilm development are not yet clear.

5 Enzymatic Mechanisms of Biofilm Inhibition

In this section, we will cover enzymatic treatments that can inhibit *S. aureus* biofilm formation. While *S. aureus*-secreted proteases have demonstrated self-cleavage capability, other bacterial proteases can tap into these pathways and we present some examples that lead to biofilm inhibition. Some of the other enzymes presented, such as nuclease and hyaluronate lyase, are produced by *S. aureus* but are not part of the quorum-sensing dispersal pathway discussed above. Finally, we cover commercially produced enzymes that have demonstrated successful inhibition of *S. aureus* biofilms in vitro and in vivo.

5.1 Proteases

Recently, the serine protease Esp produced by *S. epidermidis* has gained attention as a cross-species inhibitor of *S. aureus* nasal colonization (Iwase et al. 2010). Esp is a homolog of the *S. aureus* V8 (SspA) protease that is known to have anti-biofilm properties (Marti et al. 2010; O'neill et al. 2008). Follow-up studies have demonstrated that Esp cleaves surface and secreted proteins in *S. aureus* that influence biofilm formation (Sugimoto et al. 2013), with one of the most important of these being the major autolysin Atl (Chen et al. 2013). Other bacterial pathogens produce cross-acting proteases with activity against *S. aureus* biofilms. As one example, *Pseudomonas aeruginosa* secretes the elastase LasB, which prevents *S. aureus* biofilm formation and disperses established biofilms (Park et al. 2012). Finally, there have been many demonstrations that commercial proteases, such as proteinase K and trypsin, have anti-biofilm properties against many *S. aureus* biofilm forming strains (Lauderdale et al. 2010; Boles and Horswill 2008; O'neill et al. 2007; Chaignon et al. 2007).

5.2 Nuclease and Other DNases

The first identified and probably best studied *S. aureus* enzyme is nuclease (Nuc). Nuc is a potent, secreted endonuclease that can degrade single- and double-stranded DNA, as well as RNA, in a calcium-dependent manner (Cunningham et al. 1956; Cuatrecasas et al. 1967). Although *nuc* gene expression was originally thought to be part of the *agr* regulon, recent studies have shown it is controlled by the SaeRS two-component system (Olson et al. 2013b). Exogenous treatment of Nuc enzyme, or controlled expression of *nuc*, prevents *S. aureus* biofilm formation (Kiedrowski et al. 2011). *S. aureus* mutants lacking *nuc* produce larger biofilms compared to wild-type strains due to accumulation of high MW eDNA, and this phenotype is conserved across *S. aureus* strain types (Kiedrowski et al. 2011). In addition to the *S. aureus* Nuc, other DNases have been shown to prevent *S. aureus* biofilm formation on diverse abiotic surfaces, such as glass, titanium, and plastic (Lauderdale et al. 2010; Izano et al. 2008; Mann et al. 2009).

5.3 Hyaluronate Lyase

Hyaluronate lyases are secreted bacterial enzymes that cleave the β-1,4 glycosidic bond of hyaluronic acid, a host matrix polymer (Hynes and Walton 2000). The *S. aureus* hyaluronate lyase, encoded by the gene *hysA*, was initially described as a "spreading factor" for its ability to promote infection dissemination in a murine skin infection model (Duran-Reynals 1933). More recently, it has been demonstrated that many *S. aureus* strains contain multiple forms of this enzyme (Hart et al. 2013). The abundance of hyaluronic acid in the mammalian host, particularly at in vivo biofilm infection sites (Jiang et al. 2011; Laurent and Fraser 1992), makes it a potential component of the biofilm matrix in vivo. Recent studies in our laboratory have found that there is increased biofilm formation in the presence of hyaluronic acid in a MRSA *hysA* mutant, and exogenous addition of purified HysA reduced biofilm formation (Rosenthal and Horswill, unpublished observations). Along these lines, the *Streptococcus intermedius* hyaluronate lyase was shown to be important for biofilm dispersal (Pecharki et al. 2008). These findings suggest that the role of hyaluronate lyase in *S. aureus* biofilm dispersal warrants further investigation.

5.4 Dispersin B

Dispersin B is an enzyme produced by the dental pathogen *Actinobacillus actinomycetemcomitans* that has been shown to function as a *N*-acetyl-glucosaminidase. When added exogenously to staphylococcal biofilms, the enzyme can cleave PIA in the biofilm matrix and rapidly degrade the established biofilm (Donelli et al. 2007; Kaplan et al. 2004). Kaplan et al. demonstrated that concentrations as low as 40 ng/mL resulted in greater than a 50 % decrease in an established *S. epidermidis* biomass after 9 min, and 4.8 μg/mL completely abolished biomass in 2 min. Dispersin B is not bacteriocidal, demonstrating that the rapid effect on biofilms is a result of PIA digestion in the matrix and destabilization of the structure. However, many lineages of *S. aureus* form PIA-independent biofilms that do not respond to Dispersin B treatment and instead are sensitive to DNases or proteases (Izano et al. 2008). To date, no endogenous staphylococcal PIA-degrading enzymes have been identified, but it is possible that they remain to be discovered.

5.5 Lysostaphin

Lysostaphin is a glycylglycine endopeptidase produced by *Staphylococcus simulans*, and this enzyme cleaves the pentaglycine cross-bridge of staphylococcal peptidoglycan (Schindler and Schuhardt 1964). While this enzyme is known for its ability to lyse *S. aureus* cells at low concentrations, it has also been found to be effective at inhibiting both *S. aureus* and *S. epidermidis* biofilms. At a low concentration, lysostaphin will kill *S. aureus* cells in a biofilm and disrupt the extracellular matrix in vitro on polystyrene and polycarbonate surfaces (Wu et al. 2003). When administered in combination with nafcillin, lysostaphin was able to eradicate established *S. aureus* biofilms in catheters implanted into the jugular veins of mice (Kokai-Kun et al. 2009). Additionally, when catheters were pretreated with lysostaphin, the mice were completely protected from MRSA infection of the indwelling catheters. While this is an intriguing new method of enzymatic biofilm dispersal, the exact mechanism by which this occurs is still unknown. One possibility is that there is cell wall debris in the matrix, which is targeted by lysostaphin resulting in biofilm disruption. Alternatively, the destruction of the biofilm could be due to rapid lysis of cells followed by matrix destabilization (Wu et al. 2003). Despite our limited understanding of the lysostaphin anti-biofilm mechanism, the success of using this enzyme to treat staphylococcal biofilm infections suggests that it could be attractive for further development.

6 Small-Molecule Inhibitors of *S. aureus* Biofilms

In recent years, there have been increasing reports of naturally occurring and synthetic small molecules that inhibit *S. aureus* biofilm formation (listed in Table 2). In many cases for these molecules, the anti-biofilm mechanism is not known in detail, nor has the agent been tested in animal models of infection. We will present the highlights from a few of the better characterized examples (see Fig. 3).

Table 2 Synthetic and natural product small molecules that inhibit *S. aureus* biofilms

Process or agent	Anti-biofilm mechanism	References
Natural products		
Cis-2-decanoic acid	Unknown	Davies and Marques (2009)
Tannic acid	Induction of IsdA and potential modulation of cell wall	Lee et al. (2013), Payne et al. (2013)
Ellagic acid	Unknown	Quave et al. (2012)
Magnolol	Inhibition of autolysis and eDNA release	Wang et al. (2011a)
4,5-Disubstituted-2-aminoimidazole-triazole conjugates	Zinc-chelation	Su et al. (2011)
N-acetyl-L-cysteine	Unknown	Drago et al. (2013)
Synthetic		
Chelators	Metal chelation	Venkatesh et al. (2009)
Aryl rhodanines	Inhibition of attachment	Opperman et al. (2009)

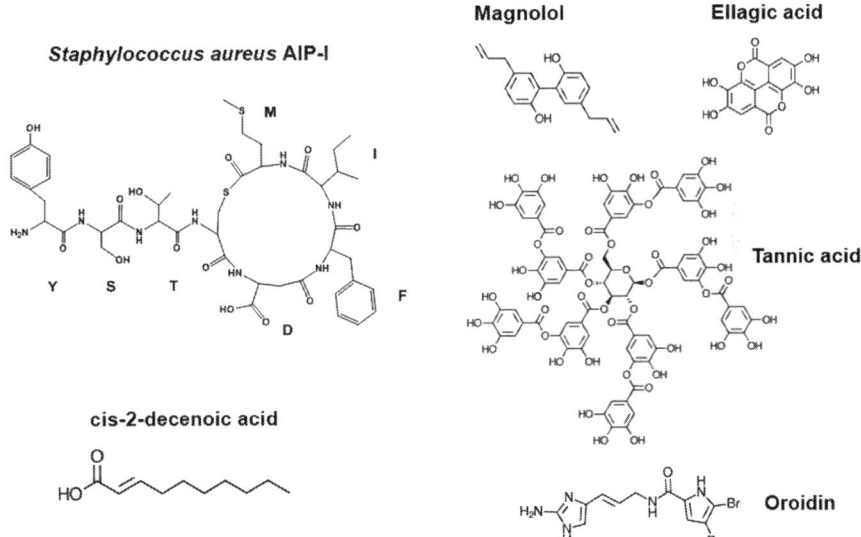

Fig. 3 Structures of natural product inhibitors of *S. aureus* biofilm formation. Representative examples of bacterial-derived compounds are shown along with plant- or marine-derived compounds

6.1 Natural Product Inhibitors

As outlined in the preceding sections, many bacteria can produce agents that prevent biofilm formation. In one recent example, D-tyrosine, D-phenylalanine, and D-proline, as well as a mixture of D-amino acids, were shown to be effective at inhibiting *S. aureus* biofilm formation (Kolodkin-Gal et al. 2010; Hochbaum et al. 2011). One mechanism could be the improper incorporation of D-amino acids in the peptidoglycan peptide side chains in the place of the terminal D-alanine. In support of this proposal, the addition of D-alanine restored biofilm formation in the presence of both D-tyrosine and the D-amino acid mixture. Additionally, biofilms that were treated with D-amino acids were found to have fewer surface proteins when compared to biofilms treated with L-amino acids by confocal microscopy (Hochbaum et al. 2011). The anti-biofilm characteristics of D-amino acids are not limited to *S. aureus*, as the authors found that D-amino acids also prevented biofilm formation in other bacteria, such as *Bacillus subtilis* and *Escherichia coli*. Considering D-amino acids are produced in late biofilm cultures by *B. subtilis*, there may be some general conserved nature to this biofilm dispersal strategy across bacterial species.

In another recent example of a bacterial derived agent, the fatty acid messenger, *cis*-2-decanoic acid, was identified from culture supernatants of *P. aeruginosa* as being broadly biofilm inhibitory (Davies and Marques 2009). The fatty acid disrupted established biofilms across a large variety of bacterial species, including *S. aureus*, and the authors postulated that it might function as a signaling molecule in multispecies biofilms to induce dispersal in a synchronized manner. More recently, *cis*-2-decanoic acid has been shown to enhance the effectiveness of antibiotics against *S. aureus* biofilms (Jennings et al. 2012). However, more work is required to determine the specific mechanism through which biofilm degradation occurs.

Plants are an abundant source of natural products and increasingly compounds are being identified that are bioactive against *S. aureus*. Recently, ellagic acid derivatives were identified from *Rubus ulmifolius* (Elmleaf blackberry) as being anti-biofilm in nature (Quave et al. 2012). Ellagic acid prevented biofilm formation in all lineages of *S. aureus* and enhanced the susceptibility of these biofilms to antibiotics. Additionally, the ellagic acid bioactivity was effective on human plasma conditioned surfaces to provide a more relevant mimic of the in vivo disease state. At this time, preliminary studies suggest the anti-biofilm mechanism is not through alteration of PIA production or inhibition of *sarA* expression (Quave et al. 2012). More studies are needed to further define the ellagic acid mode of action against *S. aureus* biofilms.

Tannic acid is another example of a plant-derived natural product that has recently been shown to be anti-biofilm in nature (Lee et al. 2013; Payne et al. 2013). In looking at the tannin-induced change in *S. aureus* extracellular proteins, the immunodominant surface protein A (IsdA) increased in abundance. In follow-up analysis, induced expression of the *isdA* gene was found to inhibit biofilm formation and an *isdA* mutant was not sensitive to treatment, further suggesting the tannic acid triggered increase in IsdA levels is the reason for the biofilm phenotype. Tea is abundant in tannins and the exposure of *S. aureus* biofilms to tea replicated these findings in vitro. Importantly, the authors also demonstrated that tannins were anti-biofilm in vivo using a rat model of throat colonization (Payne et al. 2013). IsdA is a lytic transglycosylase and could be modifying the *S. aureus* peptidoglycan and altering biofilm development. Supporting this proposal, a catalytically inactive form of IsdA had no effect on the biofilm capacity. However, the exact mechanism of IsdA on biofilm limitation awaits further elucidation.

As another recent example of plant-derived compounds, magnolol, a major component isolated from the stem bark of *Magnolia* species, has been demonstrated to have anti-biofilm properties against *S. aureus* (Wang et al. 2011a). Through transcriptional profiling, it was shown that magnolol decreased *cidA* expression and increased *lrgAB* expression, both loci with known connections to autolysis (Rice and Bayles 2008). Supporting the profiling, magnolol treatment reduced *S. aureus* autolysin activity in a dose-dependent manner (Wang et al. 2011a), suggesting that this natural product controls lysis and eDNA release, which is essential for *S. aureus* biofilm formation (Rice et al. 2007).

Finally, there has been success from marine-derived natural products in identifying anti-biofilm agents. Multiple reports have demonstrated that the 2-aminoimidazole motif found in the sponge-derived natural products oroidin and related compounds are effective at preventing *S. aureus* and other bacterial pathogens from forming biofilms (Su et al. 2011; Stowe et al. 2011). These compounds can synergize with antibiotics to increase the potency against MRSA biofilm (Rogers et al. 2010). Interestingly, the 2-aminoimidazole scaffold might be functioning as a biofilm inhibitor through a zinc chelation mechanism (Rogers et al. 2009). There are critical *S. aureus* surface proteins, like SasG, that are important for biofilm formation and require zinc to build a proper surface structure (Gruszka et al. 2012; Conrady et al. 2008; Corrigan et al. 2007). Whether or not the 2-aminoimidazole containing compounds function through SasG inhibition mechanism or through another zinc-dependent process remains to be determined.

6.2 Synthetic Inhibitors

Some success has been achieved by screening combinatorial libraries for anti-biofilm compounds. The aryl rhodamines are an interesting class of compounds identified initially by screening against *S. epidermidis* biofilms and successfully tested against other bacterial pathogens like *S. aureus* (Opperman et al. 2009). These compounds are broadly anti-biofilm against a diverse array of clinical *S. aureus* isolates in the low micromolar range, and they retain some activity against other Gram-positive pathogens. The aryl rhodamines inhibit attachment of *S. aureus* to surfaces, although the exact mechanism of action is not yet clear.

Multiple chelators are known to inhibit staphylococcal biofilm formation. The chelator EDTA was found to be effective at decreasing *S. aureus* and *S. epidermidis* adherence and biofilm formation (Kadry et al. 2009; Venkatesh et al. 2009; Shanks et al. 2006), and additionally, the calcium chelator EGTA was found to be effective at preventing biofilm formation of *S. epidermidis*. Polysaccharide production and surface hydrophobicity were significantly reduced when *S. epidermidis* strains were treated with both EDTA and EGTA (Kadry et al. 2009). Whether some of these EGTA anti-biofilm properties will function against *S. aureus* are unclear. While the exact mechanism of action for these chelators against *S. aureus* biofilms is still unknown, it is tempting to speculate that they could be inhibiting the function of metal-dependent surface proteins that are important for biofilm formation (Conrady et al. 2008).

7 Conclusions and Future Perspectives

Significant advances in the *S. aureus* biofilm field have been made in the past decade. Through ongoing studies on biofilm formation mechanisms, a tremendous number of surface factors and transcriptional regulators have been identified that contribute to biofilm development. In this review, we focused on the basics of *S. aureus* biofilm development, provided an overview on matrix components, and described the *agr* quorum-sensing system as one example of a prominent biofilm regulator. The rest of the article summarized the many recent examples of enzyme and small-molecule-based anti-biofilm mechanisms. Due to space limitations, many other adhesins were not covered in detail herein, such as Atl (Bose et al. 2012; Houston et al. 2011), FnbpAB (Geoghegan et al. 2013), and Bap (Valle et al. 2012), and there are recent exciting advances in our knowledge of these structures. There are also a number of characterized regulators that have been demonstrated to have biofilm roles, such as SarA (Beenken et al. 2010; Tsang et al. 2008; Beenken et al. 2003), MgrA (Trotonda et al. 2008), ArlRS (Fournier and Hooper 2000), and CcpA (Seidl et al. 2008), to name a few. The many complexities of autolysis and eDNA release during *S. aureus* biofilm development were also only briefly covered (Mann et al. 2009; Rice and Bayles 2008; Rice et al. 2007). Due to the rapid growth in knowledge on the *S. aureus* biofilm development pathways, much of this information extends beyond this review, and the interested reader is encouraged to examine the recent literature. All these advances are important as our understanding of biofilm formation and dispersal mechanisms continues to grow, and the hope is this new insight will aid our ability to pioneer approaches to preventing *S. aureus* biofilms.

Going forward, there are many factors that will have to be considered as we continue investigating developmental pathways and inhibitors of *S. aureus* biofilm formation. It is increasingly being appreciated that *S. aureus* preferentially binds to matrix proteins in vivo rather than directly to abiotic material (Walker and Horswill 2012; Beenken et al. 2010; Mootz et al. 2013). As research advances, adapting biofilm assays to take this into account will be important to properly assess the efficacy of new anti-biofilm therapies. Additionally, there should be consideration of other surfaces for *S. aureus* biofilm formation. This pathogen has the ability to bind human mucin proteins that are abundant in many host environments (Shuter et al. 1996), and mucin has been shown to enhance biofilm capacity and antimicrobial tolerance of other lung pathogens (Landry et al. 2006). *S. aureus* can even grow on some of the sugars that would be liberated from glycosylated mucins in such an environment (Olson et al. 2013a), and growth on sugars often favors biofilm formation due to acid secretion (Boles and Horswill 2008). There also needs to be consideration for the tremendous differences across *S. aureus* strain types in terms of biofilm capacity, and there should be attempts to assess effectiveness across various biofilm techniques, such as flow cells or the newer microfluidic-based methods (Moormeier et al. 2013). The simple microtiter plastic attachment assays are useful as an investigation starting point, but they do not always have the robustness and consistency to make broad conclusions about *S. aureus* biofilms.

Beyond assays, there are other factors that should be taken into account as the *S. aureus* biofilm field advances. For instance, there are few definitions of what constitutes a *S. aureus* biofilm. Currently, a structure has to "look like a biofilm" by some type of microscopy method and display enhanced resistance to antimicrobials. There are no uniform dimensions (size, thickness) or other standards that can be relied upon as a general biofilm definition. As the field progresses, there should be consideration for more standardization in *S. aureus* biofilm research and defining these structures in a rigorous manner for study comparisons across the field. If researchers had key biomarkers of biofilms to track, such as a gene or secreted product that is only induced in a biofilm in a conserved manner, there would be tremendous benefit to monitoring this biomarker during treatment tests in vitro or in vivo.

In this review, natural mechanisms of *S. aureus* biofilm dispersal were covered, along with many enzyme and small-molecule treatment approaches. The field is evolving quickly and many alternative and creative strategies to prevent *S. aureus* biofilms are under investigation. Phage therapy is one example that is currently being explored to prevent or eliminate established biofilms (Kelly et al. 2012). Electrical currents have shown promise in preliminary studies to prevent *S. aureus* biofilms and those of other pathogens (del Pozo et al. 2009), and additional mechanism studies suggest that hypochlorous acid (bleach) produced from media salts could be the reason for the anti-biofilm activity (Sandvik et al. 2013). The semimetal gallium has shown promise in preventing *S. aureus* biofilms in preliminary in vitro studies (Baldoni et al. 2010). There are also many examples of alterations of surface chemistry as an anti-biofilm strategy, and as one representative, Slippery Liquid-Infused Porous Surfaces (SLIPS) were recently shown to be effective at preventing *S. aureus* attachment and biofilm development (Epstein et al. 2012).

The field of *S. aureus* biofilms has made tremendous strides in recent years as our understanding of surface adhesins, regulatory networks, enzyme treatments, natural product and synthetic inhibitors, and quantitative assays continues to improve. Despite these successes, many areas outlined herein are still in need of further improvement as the field continues to mature. Perhaps most importantly, only a select few of the current in vitro biofilm-related discoveries have transitioned to animal model confirmation. Due to the high level of variability in biofilm results, it will be critically important to continue testing new treatment approaches in relevant animal models of biofilm infection. Altogether, the recent advances in biofilm knowledge hold great promise for developing effective approaches for treating some of the devastating *S. aureus* chronic infections.

Acknowledgments C.B.R. was supported by NIH Training Grants GM008365 and DE023520. J.M.M was supported by NIH T32 Training Grant No. AI07511. This work was supported by grant AI083211 (Project 3 to ARH) from the National Institute of Allergy and Infectious Diseases.

References

Abraham NM, Jefferson KK (2012) Staphylococcus aureus clumping factor B mediates biofilm formation in the absence of calcium. Microbiology 158:1504–1512

Baldoni D, Steinhuber A, Zimmerli W, Trampuz A (2010) In vitro activity of gallium maltolate against Staphylococci in logarithmic, stationary, and biofilm growth phases: comparison of conventional and calorimetric susceptibility testing methods. Antimicrob Agents Chemother 54:157–163

Bateman BT, Donegan NP, Jarry TM, Palma M, Cheung AL (2001) Evaluation of a tetracycline-inducible promoter in *Staphylococcus aureus* in vitro and in vivo and its application in demonstrating the role of *sigB* in microcolony formation. Infect Immun 69:7851–7857

Beenken KE, Blevins JS, Smeltzer MS (2003) Mutation of sarA in Staphylococcus aureus limits biofilm formation. Infect Immun 71:4206–4211

Beenken KE, Mrak LN, Griffin LM, Zielinska AK, Shaw LN, Rice KC, Horswill AR, Bayles KW, Smeltzer MS (2010) Epistatic relationships between sarA and agr in Staphylococcus aureus biofilm formation. PLoS One 5:e10790

Blot S, Depuydt P, Annemans L, Benoit D, Hoste E, De Waele JJ, Decruyenaere J, Vogelaers D, Colardy F, Vandewoude KH (2005) Clinical and economic outcomes in critically ill patients with nosocomial catheter-related bloodstream infections. Clin Infect Dis 41:1591–1598

Boles BR, Horswill AR (2008) *agr*-mediated dispersal of *Staphylococcus aureus* biofilms. PLoS Pathog 4:e1000053

Boles BR, Horswill AR (2011) Staphylococcal biofilm disassembly. Trends Microbiol 19:449–455

Boles BR, Thoendel M, Roth AJ, Horswill AR (2010) Identification of genes involved in polysaccharide-independent Staphylococcus aureus biofilm formation. PLoS One 5:e10146

Bose JL, Lehman MK, Fey PD, Bayles KW (2012) Contribution of the Staphylococcus aureus Atl AM and GL murein hydrolase activities in cell division, autolysis, and biofilm formation. PLoS One 7:e42244

Brady RA, Leid JG, Calhoun JH, Costerton JW, Shirtliff ME (2008) Osteomyelitis and the role of biofilms in chronic infection. FEMS Immunol Med Microbiol 52:13–22

Cassat JE, Hammer ND, Campbell JP, Benson MA, Perrien DS, Mrak LN, Smeltzer MS, Torres VJ, Skaar EP (2013) A secreted bacterial protease tailors the Staphylococcus aureus virulence repertoire to modulate bone remodeling during osteomyelitis. Cell Host Microbe 13:759–772

Chaignon P, Sadovskaya I, Ragunah C, Ramasubbu N, Kaplan JB, Jabbouri S (2007) Susceptibility of Staphylococcal biofilms to enzymatic treatments depends on their chemical composition. Appl Microbiol Biotechnol 75:125–132

Chambers HF, Deleo FR (2009) Waves of resistance: *Staphylococcus aureus* in the antibiotic era. Nat Rev Microbiol 7:629–641

Chen C, Krishnan V, Macon K, Manne K, Narayana SV, Schneewind O (2013) Secreted proteases control autolysin-mediated biofilm growth of Staphylococcus aureus. J Biol Chem 288:29440–29452

Conrady DG, Brescia CC, Horii K, Weiss AA, Hassett DJ, Herr AB (2008) A zinc-dependent adhesion module is responsible for intercellular adhesion in Staphylococcal biofilms. Proc Natl Acad Sci USA 105:19456–19461

Corrigan RM, Rigby D, Handley P, Foster TJ (2007) The role of Staphylococcus aureus surface protein SasG in adherence and biofilm formation. Microbiology 153:2435–2446

Costerton JW (2005) Biofilm theory can guide the treatment of device-related orthopaedic infections. Clin Orthop Relat Res: 7–11

Cramton SE, Gerke C, Schnell NF, Nichols WW, Gotz F (1999) The intercellular adhesion (ica) locus is present in Staphylococcus aureus and is required for biofilm formation. Infect Immun 67:5427–5433

Cramton SE, Ulrich M, Gotz F, Doring G (2001) Anaerobic conditions induce expression of polysaccharide intercellular adhesin in Staphylococcus aureus and Staphylococcus epidermidis. Infect Immun 69:4079–4085

Cuatrecasas P, Fuchs S, Anfinsen CB (1967) The binding of nucleotides and calcium to the extracellular nuclease of Staphylococcus aureus. Studies by gel filtration. J Biol Chem 242:3063–3067

Cunningham L, Catlin BW, De Garile MP (1956) A deoxyribonuclease of *Micrococcus pyogenes*. J Am Chem Soc 78:4642–4644

Darouiche R (2001) Device-associated infections: a macroproblem that starts with microadherence. Clin Infect Dis 33:1567–1572

Darouiche R (2004) Treatment of infections associated with surgical implants. N Engl J Med 350:1422–1429

Davies DG, Marques CN (2009) A fatty acid messenger is responsible for inducing dispersion in microbial biofilms. J Bacteriol 191:1393–1403

Del Pozo JL, Patel R (2007) The challenge of treating biofilm-associated bacterial infections. Clin Pharmacol Ther 82:204–209

Del Pozo JL, Rouse MS, Mandrekar JN, Steckelberg JM, Patel R (2009) The electricidal effect: reduction of Staphylococcus and pseudomonas biofilms by prolonged exposure to low-intensity electrical current. Antimicrob Agents Chemother 53:41–45

Deleo FR, Chambers HF (2009) Reemergence of antibiotic-resistant *Staphylococcus aureus* in the genomics era. J Clin Invest 119:2464–2474

Dimick JP, Pelz RK, Consuji R, Swoboda SM, Hendrix CW, Lipsett PA (2001) Increased resource use associated with catheter-related bloodstream infection in the surgical intensive care unit. Arch Surg 136:229–234

Donelli G, Francolini I, Romoli D, Guaglianone E, Piozzi A, Ragunath C, Kaplan JB (2007) Synergistic activity of dispersin B and cefamandole nafate in inhibition of Staphylococcal biofilm growth on polyurethanes. Antimicrob Agents Chemother 51:2733–2740

Drago L, De Vecchi E, Mattina R, Romano CL (2013) Activity of N-acetyl-L-cysteine against biofilm of Staphylococcus aureus and Pseudomonas aeruginosa on orthopedic prosthetic materials. Int J Artif Organs 36:39–46

Duran-Reynals F (1933) Studies on a certain spreading factor existing in bacteria and its significance for bacterial invasiveness. J Exp Med 58:161–181

Epstein AK, Wong TS, Belisle RA, Boggs EM, Aizenberg J (2012) Liquid-infused structured surfaces with exceptional anti-biofouling performance. Proc Natl Acad Sci USA 109:13182–13187

Fitzpatrick F, Humphreys H, O'gara JP (2005) The genetics of Staphylococcal biofilm formation–will a greater understanding of pathogenesis lead to better management of device-related infection? Clin Microbiol Infect 11:967–973

Foster TJ, Hook M (1998) Surface protein adhesins of Staphylococcus aureus. Trends Microbiol 6:484–488

Fournier B, Hooper DC (2000) A new two-component regulatory system involved in adhesion, autolysis, and extracellular proteolytic activity of Staphylococcus aureus. J Bacteriol 182:3955–3964

Fowler VG Jr, Miro JM, Hoen B, Cabell CH, Abrutyn E, Rubinstein E, Corey GR, Spelman D, Bradley SF, Barsic B, Pappas PA, Anstrom KJ, Wray D, Fortes CQ, Anguera I, Athan E, Jones P, Van Der Meer JT, Elliott TS, Levine DP, Bayer AS (2005) *Staphylococcus aureus* endocarditis: a consequence of medical progress. JAMA 293:3012–3021

Francois P, Vaudaux P, Lew PD (1998) Role of plasma and extracellular matrix proteins in the physiopathology of foreign body infections. Ann Vasc Surg 12:34–40

Francois P, Schrenzel J, Stoerman-Chopard C, Favre H, Herrmann M, Foster TJ, Lew DP, Vaudaux P (2000) Identification of plasma proteins adsorbed on hemodialysis tubing that promote Staphylococcus aureus adhesion. J Lab Clin Med 135:32–42

Geoghegan JA, Monk IR, O'gara JP, Foster TJ (2013) Subdomains N2N3 of fibronectin binding protein A mediate Staphylococcus aureus biofilm formation and adherence to fibrinogen using distinct mechanisms. J Bacteriol 195:2675–2683

Gerke C, Kraft A, Sussmuth R, Schweitzer O, Gotz F (1998) Characterization of the N-acetylglucosaminyltransferase activity involved in the biosynthesis of the Staphylococcus epidermidis polysaccharide intercellular adhesin. J Biol Chem 273:18586–18593

Gordon RJ, Lowy FD (2008) Pathogenesis of methicillin-resistant Staphylococcus aureus infection. Clin Infect Dis 46(Suppl 5):S350–S359

Gotz F (2002) Staphylococcus and biofilms. Mol Microbiol 43:1367–1378

Gruszka DT, Wojdyla JA, Bingham RJ, Turkenburg JP, Manfield IW, Steward A, Leech AP, Geoghegan JA, Foster TJ, Clarke J, Potts JR (2012) Staphylococcal biofilm-forming protein has a contiguous rod-like structure. Proc Natl Acad Sci USA 109:E1011–E1018

Hart ME, Tsang LH, Deck J, Daily ST, Jones RC, Liu H, Hu H, Hart MJ, Smeltzer MS (2013) Hyaluronidase expression and biofilm involvement in Staphylococcus aureus UAMS-1 and its sarA, agr and sarA agr regulatory mutants. Microbiology 159:782–791

Hochbaum AI, Kolodkin-Gal I, Foulston L, Kolter R, Aizenberg J, Losick R (2011) Inhibitory effects of D-amino acids on Staphylococcus aureus biofilm development. J Bacteriol 193:5616–5622

Houston P, Rowe SE, Pozzi C, Waters EM, O'gara JP (2011) Essential role for the major autolysin in the fibronectin-binding protein-mediated Staphylococcus aureus biofilm phenotype. Infect Immun 79:1153–1165

Huseby M, Shi K, Brown CK, Digre J, Mengistu F, Seo KS, Bohach GA, Schlievert PM, Ohlendorf DH, Earhart CA (2007) Structure and biological activities of beta toxin from Staphylococcus aureus. J Bacteriol 189:8719–8726

Huseby MJ, Kruse AC, Digre J, Kohler PL, Vocke JA, Mann EE, Bayles KW, Bohach GA, Schlievert PM, Ohlendorf DH, Earhart CA (2010) Beta toxin catalyzes formation of nucleoprotein matrix in Staphylococcal biofilms. Proc Natl Acad Sci USA 107:14407–14412

Hynes WL, Walton SL (2000) Hyaluronidases of Gram-positive bacteria. FEMS Microbiol Lett 183:201–207

Iwase T, Uehara Y, Shinji H, Tajima A, Seo H, Takada K, Agata T, Mizunoe Y (2010) Staphylococcus epidermidis Esp inhibits Staphylococcus aureus biofilm formation and nasal colonization. Nature 465:346–349

Izano EA, Amarante MA, Kher WB, Kaplan JB (2008) Differential roles of poly-N-acetylglucosamine surface polysaccharide and extracellular DNA in Staphylococcus aureus and Staphylococcus epidermidis biofilms. Appl Environ Microbiol 74:470–476

Jennings JA, Courtney HS, Haggard WO (2012) Cis-2-decenoic acid inhibits S. aureus growth and biofilm in vitro: a pilot study. Clin Orthop Relat Res 470:2663–2670

Jiang D, Liang J, Noble PW (2011) Hyaluronan as an immune regulator in human diseases. Physiol Rev 91:221–264

Kadry AA, Fouda SI, Shibl AM, Abu El-Asrar AA (2009) Impact of slime dispersants and anti-adhesives on in vitro biofilm formation of Staphylococcus epidermidis on intraocular lenses and on antibiotic activities. J Antimicrob Chemother 63:480–484

Kaplan JB (2010) Biofilm dispersal: mechanisms, clinical implications, and potential therapeutic uses. J Dent Res 89:205–218

Kaplan JB, Ragunath C, Velliyagounder K, Fine DH, Ramasubbu N (2004) Enzymatic detachment of Staphylococcus epidermidis biofilms. Antimicrob Agents Chemother 48:2633–2636

Karlsson A, Saravia-Otten P, Tegmark K, Morfeldt E, Arvidson S (2001) Decreased amounts of cell wall-associated protein A and fibronectin-binding proteins in Staphylococcus aureus sarA mutants due to up-regulation of extracellular proteases. Infect Immun 69:4742–4748

Kelly D, Mcauliffe O, Ross RP, Coffey A (2012) Prevention of Staphylococcus aureus biofilm formation and reduction in established biofilm density using a combination of phage K and modified derivatives. Lett Appl Microbiol 54:286–291

Kiedrowski MR, Horswill AR (2011) New approaches for treating Staphylococcal biofilm infections. Ann N Y Acad Sci 1241:104–121

Kiedrowski MR, Kavanaugh JS, Malone CL, Mootz JM, Voyich JM, Smeltzer MS, Bayles KW, Horswill AR (2011) Nuclease modulates biofilm formation in community-associated methicillin-resistant Staphylococcus aureus. PLoS One 6:e26714

Kokai-Kun JF, Chanturiya T, Mond JJ (2009) Lysostaphin eradicates established Staphylococcus aureus biofilms in jugular vein catheterized mice. J Antimicrob Chemother 64:94–100

Kolar SL, Ibarra JA, Rivera FE, Mootz JM, Davenport JE, Stevens SM, Horswill AR, Shaw LN (2013) Extracellular proteases are key mediators of Staphylococcus aureus virulence via the global modulation of virulence-determinant stability. Microbiologyopen 2:18–34

Kolodkin-Gal I, Romero D, Cao S, Clardy J, Kolter R, Losick R (2010) D-amino acids trigger biofilm disassembly. Science 328:627–629

Landry RM, An D, Hupp JT, Singh PK, Parsek MR (2006) Mucin-Pseudomonas aeruginosa interactions promote biofilm formation and antibiotic resistance. Mol Microbiol 59:142–151

Lauderdale KJ, Boles BR, Cheung AL, Horswill AR (2009) Interconnections between Sigma B, agr, and proteolytic activity in Staphylococcus aureus biofilm maturation. Infect Immun 77:1623–1635

Lauderdale KJ, Malone CL, Boles BR, Morcuende J, Horswill AR (2010) Biofilm dispersal of community-associated methicillin-resistant Staphylococcus aureus on orthopedic implant material. J Orthop Res 28:55–61

Laurent TC, Fraser JR (1992) Hyaluronan. FASEB J 6:2397–2404

Lee JH, Park JH, Cho HS, Joo SW, Cho MH, Lee J (2013) Anti-biofilm activities of quercetin and tannic acid against Staphylococcus aureus. Biofouling 29:491–499

Lew DP, Waldvogel FA (2004) Osteomyelitis. Lancet 364:369–379

Mack D, Fischer W, Krokotsch A, Leopold K, Hartmann R, Egge H, Laufs R (1996) The intercellular adhesin involved in biofilm accumulation of Staphylococcus epidermidis is a linear beta-1,6-linked glucosaminoglycan: purification and structural analysis. J Bacteriol 178:175–183

Mann EE, Rice KC, Boles BR, Endres JL, Ranjit D, Chandramohan L, Tsang LH, Smeltzer MS, Horswill AR, Bayles KW (2009) Modulation of eDNA release and degradation affects Staphylococcus aureus biofilm maturation. PLoS One 4:e5822

Marti M, Trotonda MP, Tormo-Mas MA, Vergara-Irigaray M, Cheung AL, Lasa I, Penades JR (2010) Extracellular proteases inhibit protein-dependent biofilm formation in Staphylococcus aureus. Microbes Infect 12:55–64

Mcaleese FM, Walsh EJ, Sieprawska M, Potempa J, Foster TJ (2001) Loss of clumping factor B fibrinogen binding activity by Staphylococcus aureus involves cessation of transcription, shedding and cleavage by metalloprotease. J Biol Chem 276:29969–29978

Mcgavin MJ, Zahradka C, Rice K, Scott JE (1997) Modification of the Staphylococcus aureus fibronectin binding phenotype by V8 protease. Infect Immun 65:2621–2628

Merino N, Toledo-Arana A, Vergara-Irigaray M, Valle J, Solano C, Calvo E, Lopez JA, Foster TJ, Penades JR, Lasa I (2009) Protein A-mediated multicellular behavior in *Staphylococcus aureus*. J Bacteriol 191:832–843

Mermel LA, Allon M, Bouza E, Craven DE, Flynn P, O'grady NP, Raad I, Rijnders BJ, Sherertz RJ, Warren DK (2009) Clinical practice guidelines for the diagnosis and management of intravascular catheter-related infection: 2009 update by the Infectious Diseases Society of America. Clin Infect Dis 49:1–45

Moormeier DE, Endres JL, Mann EE, Sadykov MR, Horswill AR, Rice KC, Fey PD, Bayles KW (2013) Use of microfluidic technology to analyze gene expression during Staphylococcus aureus biofilm formation reveals distinct physiological niches. Appl Environ Microbiol 79:3413–3424

Mootz JM, Malone CL, Shaw LN, Horswill AR (2013) Staphopains modulate Staphylococcus aureus biofilm integrity. Infect Immun 81:3227–3238

Novick RP, Geisinger E (2008) Quorum sensing in Staphylococci. Annu Rev Genet 42:541–564

O'gara JP (2007) ica and beyond: biofilm mechanisms and regulation in Staphylococcus epidermidis and Staphylococcus aureus. FEMS Microbiol Lett 270:179–188

O'neill E, Pozzi C, Houston P, Smyth D, Humphreys H, Robinson DA, O'gara JP (2007) Association between methicillin susceptibility and biofilm regulation in *Staphylococcus aureus* isolates from device-related infections. J Clin Microbiol 45:1379–1388

O'neill E, Pozzi C, Houston P, Humphreys H, Robinson DA, Loughman A, Foster TJ, O'gara JP (2008) A novel Staphylococcus aureus biofilm phenotype mediated by the fibronectin-binding proteins, FnBPA and FnBPB. J Bacteriol 190:3835–3850

Ohbayashi T, Irie A, Murakami Y, Nowak M, Potempa J, Nishimura Y, Shinohara M, Imamura T (2011) Degradation of fibrinogen and collagen by staphopains, cysteine proteases released from Staphylococcus aureus. Microbiology 157:786–792

Olson ME, King JM, Yahr TL, Horswill AR (2013a) Sialic acid catabolism in Staphylococcus aureus. J Bacteriol 195:1779–1788

Olson ME, Nygaard TK, Ackermann L, Watkins RL, Zurek OW, Pallister KB, Griffith S, Kiedrowski MR, Flack CE, Kavanaugh JS, Kreiswirth BN, Horswill AR, Voyich JM (2013b) Staphylococcus aureus nuclease is an SaeRS-dependent virulence factor. Infect Immun 81:1316–1324

Opperman TJ, Kwasny SM, Williams JD, Khan AR, Peet NP, Moir DT, Bowlin TL (2009) Aryl rhodanines specifically inhibit staphylococcal and enterococcal biofilm formation. Antimicrob Agents Chemother 53:4357–4367

Otto M (2013) Staphylococcal infections: mechanisms of biofilm maturation and detachment as critical determinants of pathogenicity. Annu Rev Med 64:175–188

Park JH, Lee JH, Cho MH, Herzberg M, Lee J (2012) Acceleration of protease effect on Staphylococcus aureus biofilm dispersal. FEMS Microbiol Lett 335:31–38

Parsek MR, Singh PK (2003) Bacterial biofilms: an emerging link to disease pathogenesis. Annu Rev Microbiol 57:677–701

Patel R (2005) Biofilms and antimicrobial resistance. Clin Orthop Relat Rese: 41–7

Payne DE, Martin NR, Parzych KR, Rickard AH, Underwood A, Boles BR (2013) Tannic acid inhibits Staphylococcus aureus surface colonization in an IsaA-dependent manner. Infect Immun 81:496–504

Pecharki D, Petersen FC, Scheie AA (2008) Role of hyaluronidase in Streptococcus intermedius biofilm. Microbiology 154:932–938

Periasamy S, Joo HS, Duong AC, Bach TH, Tan VY, Chatterjee SS, Cheung GY, Otto M (2012) How Staphylococcus aureus biofilms develop their characteristic structure. Proc Natl Acad Sci USA 109:1281–1286

Potempa J, Dubin A, Korzus G, Travis J (1988) Degradation of elastin by a cysteine proteinase from Staphylococcus aureus. J Biol Chem 263:2664–2667

Quave CL, Estevez-Carmona M, Compadre CM, Hobby G, Hendrickson H, Beenken KE, Smeltzer MS (2012) Ellagic acid derivatives from Rubus ulmifolius inhibit Staphylococcus aureus biofilm formation and improve response to antibiotics. PLoS One 7:e28737

Queck SY, Jameson-Lee M, Villaruz AE, Bach TH, Khan BA, Sturdevant DE, Ricklefs SM, Li M, Otto M (2008) RNAIII-independent target gene control by the agr quorum-sensing system: insight into the evolution of virulence regulation in Staphylococcus aureus. Mol Cell 32:150–158

Rice KC, Bayles KW (2008) Molecular control of bacterial death and lysis. Microbiol Mol Biol Rev 72:85–109, table of contents

Rice KC, Mann EE, Endres JL, Weiss EC, Cassat JE, Smeltzer MS, Bayles KW (2007) The *cidA* murein hydrolase regulator contributes to DNA release and biofilm development in *Staphylococcus aureus*. Proc Natl Acad Sci USA 104:8113–8118

Rogers SA, Huigens RW 3rd, Melander C (2009) A 2-aminobenzimidazole that inhibits and disperses gram-positive biofilms through a zinc-dependent mechanism. J Am Chem Soc 131:9868–9869

Rogers SA, Huigens RW III, Cavanagh J, Melander C (2010) Synergistic effects between conventional antibiotics and 2-aminoimidazole-derived antibiofilm agents. Antimicrob Agents Chemother 54:2112–2118

Sandvik EL, Mcleod BR, Parker AE, Stewart PS (2013) Direct electric current treatment under physiologic saline conditions kills Staphylococcus epidermidis biofilms via electrolytic generation of hypochlorous acid. PLoS One 8:e55118

Schindler C, Schuhardt V (1964) Lysostaphin: a new bacteriolytic agent for the Staphylococcus. Proc Natl Acad Sci USA 51:414–421

Schroeder K, Jularic M, Horsburgh SM, Hirschhausen N, Neumann C, Bertling A, Schulte A, Foster S, Kehrel BE, Peters G, Heilmann C (2009) Molecular characterization of a novel Staphylococcus aureus surface protein (SasC) involved in cell aggregation and biofilm accumulation. PLoS One 4:e7567

Seidl K, Goerke C, Wolz C, Mack D, Berger-Bachi B, Bischoff M (2008) Staphylococcus aureus CcpA affects biofilm formation. Infect Immun 76:2044–2050

Shanks RM, Donegan NP, Graber ML, Buckingham SE, Zegans ME, Cheung AL, O'toole GA (2005) Heparin stimulates Staphylococcus aureus biofilm formation. Infect Immun 73:4596–4606

Shanks RM, Sargent JL, Martinez RM, Graber ML, O'toole GA (2006) Catheter lock solutions influence staphylococcal biofilm formation on abiotic surfaces. Nephrol Dial Transplant 21:2247–2255

Shuter J, Hatcher VB, Lowy FD (1996) Staphylococcus aureus binding to human nasal mucin. Infect Immun 64:310–318

Spellberg B, Bartlett JG, Gilbert DN (2013) The future of antibiotics and resistance. N Engl J Med 368:299–302

Stoodley P, Nistico L, Johnson S, Lasko LA, Baratz M, Gahlot V, Ehrlich GD, Kathju S (2008) Direct demonstration of viable Staphylococcus aureus biofilms in an infected total joint arthroplasty. A case report. J Bone Joint Surg Am 90:1751–1758

Stowe SD, Richards JJ, Tucker AT, Thompson R, Melander C, Cavanagh J (2011) Anti-biofilm compounds derived from marine sponges. Mar Drugs 9:2010–2035

Su Z, Peng L, Worthington RJ, Melander C (2011) Evaluation of 4,5-disubstituted-2-aminoimidazole-triazole conjugates for antibiofilm/antibiotic resensitization activity against MRSA and Acinetobacter baumannii. ChemMedChem 6:2243–2251

Sugimoto S, Iwamoto T, Takada K, Okuda K, Tajima A, Iwase T, Mizunoe Y (2013) Staphylococcus epidermidis Esp degrades specific proteins associated with Staphylococcus aureus biofilm formation and host-pathogen interaction. J Bacteriol 195:1645–1655

Thoendel M, Kavanaugh JS, Flack CE, Horswill AR (2011) Peptide signaling in the Staphylococci. Chem Rev 111:117–151

Trotonda MP, Manna AC, Cheung AL, Lasa I, Penades JR (2005) SarA positively controls bap-dependent biofilm formation in Staphylococcus aureus. J Bacteriol 187:5790–5798

Trotonda MP, Tamber S, Memmi G, Cheung AL (2008) MgrA represses biofilm formation in Staphylococcus aureus. Infect Immun 76:5645–5654

Tsang LH, Cassat JE, Shaw LN, Beenken KE, Smeltzer MS (2008) Factors contributing to the biofilm-deficient phenotype of Staphylococcus aureus sarA mutants. PLoS One 3:e3361

Valenza G, Tappe D, Turnwald D, Frosch M, Konig C, Hebestreit H, Abele-Horn M (2008) Prevalence and antimicrobial susceptibility of microorganisms isolated from sputa of patients with cystic fibrosis. J Cyst Fibros 7:123–127

Valle J, Latasa C, Gil C, Toledo-Arana A, Solano C, Penades JR, Lasa I (2012) Bap, a biofilm matrix protein of Staphylococcus aureus prevents cellular internalization through binding to GP96 host receptor. PLoS Pathog 8:e1002843

Van Wamel WJ, Rooijakkers SH, Ruyken M, Van Kessel KP, Van Strijp JA (2006) The innate immune modulators Staphylococcal complement inhibitor and chemotaxis inhibitory protein of *Staphylococcus aureus* are located on beta-hemolysin-converting bacteriophages. J Bacteriol 188:1310–1315

Venkatesh M, Rong L, Raad I, Versalovic J (2009) Novel synergistic antibiofilm combinations for salvage of infected catheters. J Med Microbiol 58:936–944

Vuong C, Saenz HL, Gotz F, Otto M (2000) Impact of the agr quorum-sensing system on adherence to polystyrene in Staphylococcus aureus. J Infect Dis 182:1688–1693

Vuong C, Durr M, Carmody AB, Peschel A, Klebanoff SJ, Otto M (2004a) Regulated expression of pathogen-associated molecular pattern molecules in Staphylococcus epidermidis: quorum-sensing determines pro-inflammatory capacity and production of phenol-soluble modulins. Cell Microbiol 6:753–759

Vuong C, Kocianova S, Voyich JM, Yao Y, Fischer ER, Deleo FR, Otto M (2004b) A crucial role for exopolysaccharide modification in bacterial biofilm formation, immune evasion, and virulence. J Biol Chem 279:54881–54886

Walker JN, Horswill AR (2012) A coverslip-based technique for evaluating Staphylococcus aureus biofilm formation on human plasma. Front Cell Infect Microbiol 2:39

Wang R, Braughton KR, Kretschmer D, Bach TH, Queck SY, Li M, Kennedy AD, Dorward DW, Klebanoff SJ, Peschel A, Deleo FR, Otto M (2007) Identification of novel cytolytic peptides as key virulence determinants for community-associated MRSA. Nat Med 13:1510–1514

Wang D, Jin Q, Xiang H, Wang W, Guo N, Zhang K, Tang X, Meng R, Feng H, Liu L, Wang X, Liang J, Shen F, Xing M, Deng X, Yu L (2011a) Transcriptional and functional analysis of the effects of magnolol: inhibition of autolysis and biofilms in Staphylococcus aureus. PLoS One 6:e26833

Wang R, Khan BA, Cheung GY, Bach TH, Jameson-Lee M, Kong KF, Queck SY, Otto M (2011b) Staphylococcus epidermidis surfactant peptides promote biofilm maturation and dissemination of biofilm-associated infection in mice. J Clin Invest 121:238–248

Warren DK, Quadir WW, Hollenbeak CS, Elward AM, Cox MJ, Fraser VJ (2006) Attributable cost of catheter-associated bloodstream infections among intensive care patients in a nonteaching hospital. Crit Care Med 34:2084–2089

Wu JA, Kusuma C, Mond JJ, Kokai-Kun JF (2003) Lysostaphin disrupts Staphylococcus aureus and Staphylococcus epidermidis biofilms on artificial surfaces. Antimicrob Agents Chemother 47:3407–3414

Yarwood JM, Bartels DJ, Volper EM, Greenberg EP (2004) Quorum sensing in Staphylococcus aureus biofilms. J Bacteriol 186:1838–1850

Zielinska AK, Beenken KE, Joo HS, Mrak LN, Griffin LM, Luong TT, Lee CY, Otto M, Shaw LN, Smeltzer MS (2011) Defining the strain-dependent impact of the Staphylococcal accessory regulator (sarA) on the alpha-toxin phenotype of Staphylococcus aureus. J Bacteriol 193:2948–2958

Zielinska AK, Beenken KE, Mrak LN, Spencer HJ, Post GR, Skinner RA, Tackett AJ, Horswill AR, Smeltzer MS (2012) sarA-mediated repression of protease production plays a key role in the pathogenesis of Staphylococcus aureus USA300 isolates. Mol Microbiol 86:1183–1196

Zimmerli W, Trampuz A, Ochsner PE (2004) Prosthetic-joint infections. N Engl J Med 351:1645–1654

Novel Targets for Treatment of *Pseudomonas aeruginosa* Biofilms

Morten Alhede, Maria Alhede, and Thomas Bjarnsholt

Abstract *Pseudomonas aeruginosa* causes infection in all parts of the human body. The bacterium is naturally resistant to a wide range of antibiotics. In addition to resistance mechanisms such as efflux pumps, the ability to form aggregates, known as biofilm, further reduces *Pseudomonas aeruginosa*'s susceptibility to antibiotics. The presence of such biofilms is acknowledged to equal a persistent infection due to their inherent high tolerance to all antimicrobials and immune cells. In this chapter we discuss the mechanisms of biofilm tolerance. The latest biofilm research is reviewed and future treatment strategies such as quorum sensing inhibitors, silver, and antibodies are thoroughly evaluated.

M. Alhede
Department of International Health, Immunology and Microbiology, University of Copenhagen, 2200 Copenhagen, Denmark

Department of Clinical Microbiology, Rigshospitalet, 2100 Copenhagen, Denmark
e-mail: malhede@sund.ku.dk

M. Alhede
Department of International Health, Immunology and Microbiology, University of Copenhagen, 2200 Copenhagen, Denmark
e-mail: mvg@sund.ku.dk

T. Bjarnsholt (✉)
Department of International Health, Immunology and Microbiology, University of Copenhagen, 2200 Copenhagen, Denmark

Department of Clinical Microbiology, Rigshospitalet, 2100 Copenhagen, Denmark

Institute of International Health, Immunology and Microbiology, University of Copenhagen, Blegdamsvej 3B, 2200 Copenhagen, Denmark
e-mail: tbjarnsholt@sund.ku.dk

1 *Pseudomonas aeruginosa*

P. aeruginosa is a Gram-negative bacterium belonging to the γ-proteobacteria. It is found in environments such as soil, water, plants, common food, and mammalian tissues but are not a part of the normal human flora (Hardalo and Edberg 1997). The complete sequence of the genome of *P. aeruginosa* strain PAO1 was published in Nature in the year 2000 and was noted for its large size of 6.3 million base pairs and 5,570 open reading frames (30 % larger than *Escherichia coli* K12) (Stover et al. 2000).

Depending on the habitat, *P. aeruginosa* holds the potential to express an impressive arsenal of virulence factors, which are regulated by 468 transcriptional regulators (Stover et al. 2000). In spite of its huge arsenal of toxins, the bacterium primarily infects hospitalized and immunocompromised humans where it causes chronic infections in tissues such as heart (endocarditis) (Reyes and Lerner 1983), respiratory tracts of cystic fibrosis (CF) patients (Bjarnsholt et al. 2009), paranasal sinuses (rhinosinusitis) (Oncel et al. 2010), chronic wounds (Fazli et al. 2009), caries (El-Solh et al. 2004), osteomyelitis (Sapico 1996), and intravenous catheters and stents (Tacconelli et al. 2009). Chronic *P. aeruginosa* infections are particularly common in patients at intensive care units, and it is the most frequent Gram-negative etiologic agent associated with infections of indwelling catheters and foreign body implants (Brouqui et al. 1995).

The bacterium causes chronic pulmonary infections in 80 % of adults with the genetic disorder CF and is also a common cause of bacterial pneumonia in patients with HIV (FitzSimmons 1993; Afessa and Green 2000; Emerson et al. 2002; Schaedel et al. 2002). In otherwise healthy individuals persistent *P. aeruginosa* infections can be found in relation to periodontitis, keratitis, otitis media, and burn wounds (Barbosa et al. 2001; Post 2001; Fleiszig and Evans 2002).

2 Biofilms of *P. aeruginosa*

In chronic infections the bacteria persists despite the host defense and antibiotic treatment. The ability for *P. aeruginosa,* and most other bacteria, to grow as aggregated and sessile communities is one important factor involved in its persistence in chronic infections. The term biofilm is applied to bacterial cells that live as aggregates embedded in a self-produced matrix, which can be on a surface or in suspension. Biofilm is a life-form that increases the bacterial survival of environmental insults including otherwise detrimental effects of antibiotics and immune cells (Stewart and Costerton 2001; Aaron et al. 2002; Jesaitis et al. 2003).

The importance of the bacterial biofilm phenotype is becoming increasingly renowned as improved methods to study aggregated bacteria have become available. Therefore, a dramatic accumulation of evidence for its widespread presence has occurred. Especially within chronic infections, biofilms have been found to play a detrimental role. In this respect, the increased tolerance of biofilms has strengthened the belief that a chronic infection equals the biofilm state of growth (Burmølle et al. 2010; Høiby et al. 2010). Once a mature biofilm has formed, it is almost impossible to eradicate it with antimicrobials and a chronic inflammation will occur. The best option is to remove or debride the infected tissue or implant, and if that is not an option chronic suppressive therapy seems to be the only alternative (Høiby et al. 2010).

The tolerance of biofilms has been linked to its slow growth since both in vitro aggregates in suspension and flow-cell biofilms have the same slow growth rate as stationary phase shaking cultures (Alhede et al. 2011). Interestingly, a recent study showed that the growth rate of biofilms is independent of age, but that the tolerance to antibiotics increases with age. It was found that the tolerance towards antibiotics was reversible by physical disruption, suggesting that internal structures of the matrix components plays the major role in surviving otherwise lethal treatments with antibiotics and resistance to phagocytes (Alhede et al. 2011).

In addition to the inherent tolerance of the biofilm, traditional resistance mechanisms (e.g., efflux pumps and other adaptive resistance systems) are also prominent players in biofilm infections (Ciofu 2003; Haagensen et al. 2007; Pamp et al. 2008). The implication of adaptive resistance in metabolically active biofilm cells has led to the effective combination therapy for early eradication of *P. aeruginosa* in cystic fibrosis patients (Hansen et al. 2008). However, due to the rise in multi-resistant strains, and the fact that mature biofilms are close to impossible to eradicate, new and alternative targets are needed in order to treat chronic biofilm infections.

3 Novel Treatments

As stated by (Høiby et al. 2010), the first and preferred strategy against biofilm infection would be to prevent invading bacteria from forming aggregates. Since the aggregates show increased tolerance to both antibiotics and the immune system, development of drugs that impede surface attachment or other specific events in the early stages of aggregation may keep infecting bacteria in a planktonic, susceptible state (Bjarnsholt et al. 2005a, b). Killing infecting bacteria has long been the preferred strategy. This has been achieved by conventional antimicrobials targeting basal life processes of the bacteria. But the dissemination of resistance and the lack of new antimicrobials have initiated the search for new strategies. It is generally accepted that the application of lethal or growth inhibiting compounds will impose a selective pressure upon the bacteria resulting in resistance genes and hence a purification in the population (Nnis System 2004; Clatworthy et al. 2007; Hawkey 2008; Spellberg et al. 2008; Boucher et al. 2009). Therefore, a constant development of new drugs is essential, as drugs already in use become obsolete.

3.1 Weakening of Biofilm

Most pathogenic bacteria including *P. aeruginosa* produce compounds that impair the immune system, e.g., inhibition of antimicrobial production, antimicrobial degradation, inhibition of chemotaxis, and induction of apoptosis and necrosis (Kharazmi et al. 1984a; Bortolussi et al. 1987; Kharazmi 1991; Allen et al. 2005; Bjarnsholt et al. 2005a, b; Jensen et al. 2007; Alhede et al. 2009). Once a chronic infection has been established, the most obvious alternative to antibiotic-mediated killing would be to attenuate the bacteria with respect to pathogenicity in order to enable the immune system to clear the biofilm infection (Bjarnsholt and Givskov 2007). The novel treatment strategies explained in the following sections all target the biofilm in such a way that it becomes susceptible to antibiotic treatment or the immune system.

4 Quorum Sensing Inhibitors

Probably, the most studied novel strategy in antimicrobials is the development of quorum sensing inhibitors (QSIs). This strategy targets the regulation of virulence expression since bacteria, including *P. aeruginosa*, regulate a range of social behaviors (e.g., metabolism, virulence, and motility) to exploit their survival potential. Cooperative behaviors are maintained through inter- and extracellular chemical crosstalk comparable to higher organisms (Shapiro 1998). Gram-negative bacteria execute their cross talk by means of signal molecules such as N-acyl homoserine lactones (AHLs) (Withers et al. 2001). Among those synchronized activities is the expression of virulence factors (Davies et al. 1998; Smith and Iglewski 2003). This type of bacterial communication was termed QS by (Fuqua et al. 1994). QS systems allow bacteria to "sense" bacterial density in the environment and respond by gross changes in gene expression. It has been proposed that this mechanism enables arrest in the production of virulence factors until enough bacteria have been amassed to defeat the host defense (Waters and Bassler 2005).

One of the most important virulence factors produced by the model organism *P. aeruginosa* is rhamnolipid. Jensen and colleagues found that *P. aeruginosa* produces the compound in a QS regulated manner and proved that the bacterium kills PMNs with this substance in liquid culture and biofilms (Jensen et al. 2007). Later, Alhede et al. demonstrated that *P. aeruginosa* biofilms growing in vitro in flow cells initiate rhamnolipid production upon contact with human neutrophils (PMNs). Hence, the bacterium is able to detect the presence of these immune cells and react by producing rhamnolipid in a QS-dependent manner. Due to the molecules bipolar structure, the rhamnolipids were found to stick to the biofilm surface and thus create a shield that kills immune cells (Alhede et al. 2009). The effect of rhamnolipids in vivo has been shown in several studies and clearly demonstrates how potent this compound is. Mutants not able to produce rhamnolipids are cleared at a faster rate than corresponding wild-type bacteria (Alhede et al. 2009; van Gennip et al. 2009). Further stressing the importance of rhamnolipids is the association between the production of rhamnolipids by colonizing *P. aeruginosa* isolates and the development of Ventilator Associated Pneumonia (VAP). The authors of the study showed that VAP occurred more frequently in patients colonized during the entire observation period by isolates producing high levels of rhamnolipids (Kohler et al. 2010).

In addition to rhamnolipids, QS regulates a range of other virulence factors such as proteases, elastases, and lipases (Kharazmi et al. 1984a, b; Doring et al. 1986; Kharazmi et al. 1986, 1989; Kharazmi 1991). Attenuating bacteria by targeting the regulation of virulence will assist the immune system and consequently facilitate eradication (Hentzer et al. 2003b; Bjarnsholt et al. 2005a, b). It has been put forward (yet not proven) that this strategy imposes a weaker selective pressure with respect to development of resistance compared with conventional antibiotics (Hentzer et al. 2003a). However, even though QSIs target non-vital functions, the fitness of the bacteria could be reduced as a consequence of lost virulence and the presence of immune cells and thus impose selection (Defoirdt et al. 2010).

In addition to controlling the production of virulence factors, QS has also been shown to control biofilm tolerance to antibiotics such as tobramycin, ciprofloxacin, and ceftazidime. QS-deficient biofilms are more prone to killing by these antibiotics (Bjarnsholt et al. 2005a, b; Bjarnsholt and Givskov 2007) and are less tolerant to PMNs (Jesaitis et al. 2003; Bjarnsholt et al. 2005a, b) than a QS proficient biofilm. Since a large number of virulence factors are controlled by QS, blockage will likely result in many beneficial effects.

Numerous researchers have searched for compounds that could block the QS system and thereby enable biofilm eradication (Bjarnsholt and Givskov 2008). Several proof of concept studies have been published, but the first promising compounds were the synthetic furanones C-30 and C-56 (Hentzer et al. 2003b; Wu et al. 2004). QSIs do not kill or detach the biofilm directly but they render the biofilm more susceptible to antibiotics, as was the case with these furanones. In vitro, *P. aeruginosa* biofilms were significantly less tolerant to 100 µg/mL tobramycin when treated with furanone C-30 (Hentzer et al. 2003b). In addition, in vivo studies in a pulmonary mouse model confirmed the potential of the furanones by demonstrating that bacteria were cleared faster in furanone-treated vs. untreated mice (Hentzer et al. 2003b; Wu et al. 2004). Recently, two QSIs from natural sources have been isolated: Iberin from horseradish and Ajoene from garlic (Jakobsen et al. 2012a, b).

QS deficiency leads to reduced tolerance to a variety of conventional antibiotics, and QSI compounds that block production of the rhamnolipid shield should make the biofilm more prone to eradication by the immune system (Bjarnsholt et al. 2005a, b; Rasmussen et al. 2005; Alhede et al. 2009). Consequently, prophylactic administration of QSIs or administration of QSIs in combination with antibiotics or other antimicrobials may become a useful strategy in the treatment of biofilm infections. Recently, an interesting paper from Christensen et al. showed a synergistic effect of combining tobramycin with a QSI in a murine implant

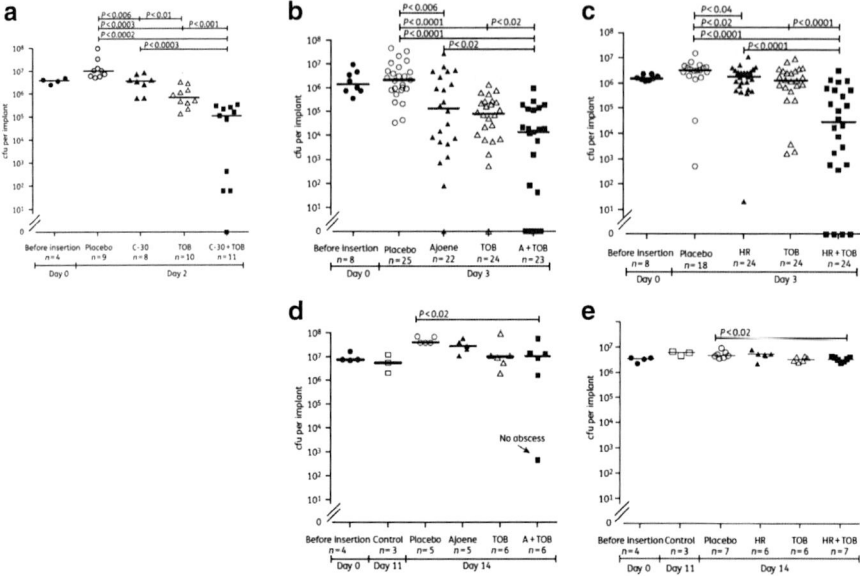

Fig. 1 Synergistic antibacterial efficacy of combination treatment with tobramycin and quorum-sensing inhibitors against *Pseudomonas aeruginosa* in an intraperitoneal foreign-body infection mouse model. Clearance of implants pre-colonized with wild-type *P. aeruginosa* inserted in the peritoneal cavity of BALB/c mice treated with either placebo (*open circles*), QSI (*filled triangles*), tobramycin (TOB) (*open triangles*), or a combination of TOB and QSI (QSI + TOB) (*filled squares*). *Squares, triangles,* and *circles* represent cfu/implant in individual mice and *horizontal bars* represent the medians. The QSIs depicted are furanone C-30 (**a**), Ajoene (**b** + **d**), and Horseradish extract (**c** + **e**). Adopted from Christensen et al. (2012)

infection model. Significant clearance was found with all tested QSIs (C-30, Ajoene, and horseradish juice), and the authors stressed the point that the best application of this treatment was on early initiation (Christensen et al. 2012) (see Fig. 1). The authors speculated that the effect seen was due to a reduction in extracellular DNA (eDNA). eDNA has been shown to reduce the effect of aminoglycoside via cation chelation (Mulcahy et al. 2008; Chiang et al. 2013), and the release of eDNA has been shown to be controlled by QS (Allesen-Holm et al. 2006). Further, QS controlled virulence factors (e.g., rhamnolipid) are known to lyse immune cells which leads to release of the host DNA (Jensen et al. 2007; Alhede et al. 2009). The presence of eDNA from both bacteria and host seems to be suppressed by QSIs and could thus explain the higher susceptibility to tobramycin.

5 DNase

As stated earlier in this chapter, the matrix of the biofilm plays a very important role in the tolerance to antimicrobials. In addition to its vital stabilizing effect (Montanaro et al. 2011), eDNA has been shown to chelate cations in the biofilm (i.e., aminoglycosides). Hence, targeting eDNA seems to be an important antibiofilm target.

In 2002, cleaving DNA with DNases was demonstrated to be effective in preventing biofilm formation of *P. aeruginosa* in vitro in flow cells (Whitchurch et al. 2002). However, DNase treatment of already existing biofilms only had an effect on immature biofilms younger than 84 h. Older biofilms seemed to be independent of the stability offered by DNA (e.g., more polysaccharide) or able to inactivate the DNase (Whitchurch et al. 2002). Similar results were shown by (Tetz et al. 2009), who also showed that coadministration of DNase together with β-lactam antibiotics to a 24-h-old *P. aeruginosa* biofilm, significantly reduced the biomass as compared to the control (Tetz et al. 2009).

DNases are thus promising drugs against biofilm formation and are already administered to chronically infected CF patients with significant result (Frederiksen et al. 2006; Alipour et al. 2009; Kaplan 2009). It has been shown that necrotic PMNs release F-actin and DNA that via filament bundles enhance *P. aeruginosa* biofilm formation in vitro (Walker et al. 2005; Parks et al. 2009). As in the case without the presence of PMNs, inhibition of biofilm formation in the presence of DNase was observed, but only in young biofilms. Interestingly, it was found that the mature biofilms could be disrupted with a combination of the DNase and polyvalent anion polyaspartat (Tang et al. 2005; Parks et al. 2009). It was hypothesized that the bundles of F-actin and DNA are stabilized by multivalent cations (i.e., histones and antimicrobial peptides) and are dissolved by multivalent anions such as polyaspartate. The dissociation of the bundles increases the access of DNase to cleavage sites and thus facilitates biofilm disruption (Tang et al. 2005; Tolker-Nielsen and Hoiby 2009).

These finding might explain the efficacy of inhaled DNase in CF, which is associated with a reduction in infectious burden and incidences of pulmonary exacerbations (Robinson 2002; Frederiksen et al. 2006). The potential of DNase treatment can be enhanced by the addition of anionic polymers to disrupt biofilms in vivo. However, Tolker-Nielsen and Høiby foresee several problems with polyvalent anions that need to be addressed before CF patients can be treated. First, they speculate that the anions will either bind to or struggle to diffuse into the sputum. Second, they point out that biofilms that give rise to PMN accumulation and lung tissue damage are located in the respiratory part of the airways, where inhalation therapy (i.e., *DNase and* polyaspartate) is out of reach (Tolker-Nielsen and Hoiby 2009). If the above problems can be solved, this treatment will help thousands of patients with CF, but it also seems to be promising for other biofilm infections.

6 Silver

The antibacterial effect of silver has been known and utilized since the nineteenth century but was "forgotten" with the introduction of antibiotics. With the emergence of multiple resistant bacteria, silver is facing its revival (Chopra 2007). The main focus of silver as an antimicrobial agent has been in the topical treatment of infected wounds. The mode of action of silver is multiple, unlike most antibiotics. Silver interferes with several components of bacterial cell structures and functions, including cell membrane integrity, respiratory chains, transmembranous energy and electrolyte transport, and enzyme activities (Lansdown 2002).

In general it is very difficult to compare the antimicrobial efficiency of silver containing products (i.e., dressings) because of a complete lack of standardized test methods, and the fact that silver is used in many different formulations (e.g., silver nitrate, silver sulfadiazine (SSD), and nano crystalline silver). Most silver dressings exploit the highly reactive silver cation to achieve their antimicrobial effect. Manufacturers are then distinguished by how the silver is incorporated into the dressing and the amount of silver that is released (Toy and Macera 2011).

Silver's multiple of modes of action are proposed to be less affected by the microenvironmental variations found in biofilms than are antibiotics (Bjarnsholt et al. 2007). Furthermore, silver is known to decrease bacterial adhesion and destabilize the biofilm matrix (Klueh et al. 2000; Chaw et al. 2005). Hence, silver could prove to be an efficient antibiofilm drug. Many silver-containing wound dressings have shown very promising results against *P. aeruginosa* when they are growing in dilute solutions (Parsons et al. 2005; Castellano et al. 2007); however, few studies have examined its efficacy against biofilms (Kostenko et al. 2010; Bowler et al. 2012). As is the case with all other antimicrobials, we have found that silver containing dressings loose their effect as the biofilm matures (unpublished data). Bjarnsholt et al. demonstrated that to eradicate a mature in vitro biofilm (4 day old) with silver sulfadiazine, concentrations as high as 5–10 µg/mL were needed. This concentration is 10–100 times higher than that used to eradicate planktonic bacteria. These observations indicate that the concentration of silver in currently available wound dressings is too low for treatment of chronic biofilm wound infections (i.e., mature biofilms) (Bjarnsholt et al. 2007) (See Fig. 2). In another study, it was found that cells in some regions of a 24-h-old biofilm survived 7 days of silver treatment, but that the surviving cells were highly susceptible to tobramycin and ciprofloxacin. The antimicrobial efficacy of the dressings was correlated to the type of base material of the dressing and the silver species loaded (Kostenko et al. 2010).

As stated above, silver containing products are used in wound treatment to combat a broad spectrum of pathogens. However, evidence of their effectiveness in preventing wound infection or promoting healing is lacking. Furthermore, standardized tests are also lacking, so direct comparisons are not possible. In spite of this, a large survey of 26 randomized trials investigated the effects of silver-containing wound dressings and topical agents in preventing wound infection

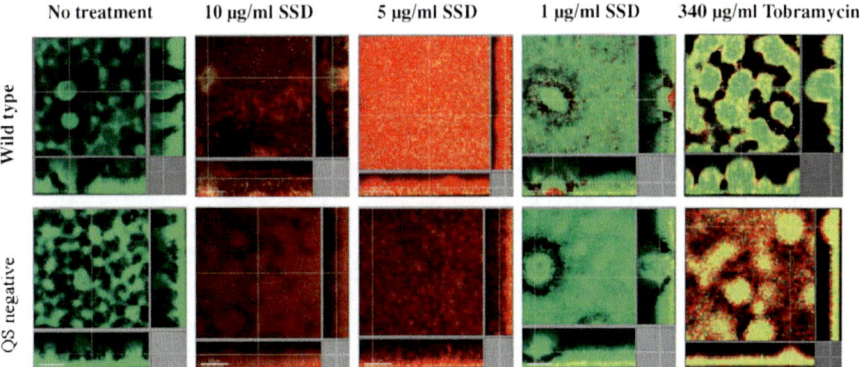

Fig. 2 Silver sulfadiazine and tobramycin treatment of *P. aeruginosa* biofilms. GFP-tagged wild-type *P. aeruginosa* and QS mutant biofilms were grown for 4 days in flow chambers. On the fourth day, silver sulfadiazine (10, 5, or 1 μg/mL) or tobramycin (340 μg/mL) was added to the medium. To assess the extent of the bacterial killing, propidium iodide was added to the media on day 5. The representative pictures show biofilms after 24 h of silver sulfadiazine (SSD) or tobramycin treatment (Bjarnsholt et al. 2007)

and healing of wounds (Storm-Versloot et al. 2010). Among burn wounds only one trial showed fewer infections with silver nitrate when compared with a non-silver dressing. Interestingly, three trials showed significantly more infections with SSD than with the non-silver dressings. In non-burn wounds, most comparisons found no significant differences in infection rates comparing SSD/silver-containing dressings with non-silver dressings. And of most interest, only one comparison showed a significant reduction in healing time using a silver-containing hydrofiber dressing (Storm-Versloot et al. 2010).

Thus, it seems that the clinical competence of silver dressings are limited, which could be due to the widespread presence of biofilms that requires unreachable concentrations of silver. The consequence of an infected wound is a stalled healing process and hence a chronic wound. This is in particular true in wounds infected with *P. aeruginosa* (Bjarnsholt et al. 2008).

7 Antibodies

Antibodies against bacterial components facilitating adhesion and accumulation on surfaces, such as the polysaccharide intercellular adhesin (PIA) and the accumulation-associated protein (Aap) in Staphylococcus aureus (Maira-Litran et al. 2004; Sun et al. 2005) and Opr86 in *P. aeruginosa* (Tashiro et al. 2008), have shown promising results in preventing biofilm formation in vitro.

A successful anti-pseudomonas strategy in CF is polyclonal IgY antibodies from egg yolk (Kollberg et al. 2003; Nilsson et al. 2008). Yolk antibodies, which are essentially an extract of egg yolk in water, should not induce inflammatory reactions because they do not bind to human Fc receptors or complement system (Larsson et al. 1993), and should thus be safe. The IgY egg yolk antibodies were found by Nilsson et al. to bind flagellin as the major antigen (Nilsson et al. 2007). Flagellin is the main protein of the flagella and is crucial for establishing infections in hosts as well as being involved in chemotaxis, motility, and adhesion. As a consequence, anti-Pseudomonas IgY has been shown to prevent adhesion of *Pseudomonas* to dermal epithelial cells in vitro. A secondary positive effect of binding flagellin is the potential to dampen local inflammation since the accessibility of the bacteria to TLR5 is attenuated (Smith et al. 2003; Shanks et al. 2010). Flagella are very abundant in bacteria and hence egg yolk antibodies have been shown to be immunoreactive against several strains of *P. aeruginosa* (Nilsson et al. 2007).

In a long-term study (12 years) of oral treatment with anti-pseudomonas egg yolk antibodies, a significantly lower number of positive *P. aeruginosa* cultures were found in the treated group compared to the control group (2.3 vs. 7 per 100 treatment months). In addition, a lower incidence of chronic *P. aeruginosa* infection was found in the treatment group. Although the data were collected from a small number of patients (17 vs. 23), the data strongly suggest that prophylaxis with a combination of anti-pseudomonas antibodies and antibiotics has great potential (Nilsson et al. 2007, 2008).

Fully human IgG1 monoclonal antibodies, targeting flagellin type b, have recently been found to markedly decrease *P. aeruginosa* motility and to improve survival of mice in a lethal pulmonary mouse model using a multidrug-resistant *P. aeruginosa* strain (Adawi et al. 2012). The authors found that a double dose paradigm administered postinfection, kept 75 % of the mice alive until day 7 compared to 20 % in the formulation- and isotype control (Adawi et al. 2012). Several similar studies have found that anti-flagellin antibodies can reduce mortality and morbidity in murine *P. aeruginosa*-infected burn models (Pollack et al. 1984; Barnea et al. 2006, 2009).

From the information above, it is clear that prophylactic treatment with antibodies has overcome initial disappointing clinical studies and now seems to be a persuasive and promising treatment regime (Bone 1991, 1996).

8 Perspectives for Future Treatment

Infections with aggregating bacteria have proven to be hard to prevent and treat. Recent findings of biofilm and aggregate heterogeneity have opened a window of novel treatment strategies. Research has shown that distinct subpopulations have different susceptibility to antimicrobials and therefore the biofilm should preferably be eradicated with more than one regimen. Combinations of already approved antimicrobials have shown good results in vivo, but new combinations including novel compounds such as QSIs, DNase, silver, or antibodies could ultimately be the end of chronic infections. The synergistic use of these novel drugs in combination with conventional antibiotics could improve the efficacy of treatment by attenuating the renitent biofilm and target several subpopulations within it.

> **Text Box: How to Test Antibiofilm Drugs.**
>
> We have noted that most novel antibacterial compounds and dressings are only tested on either planktonic cultures (including plates) or on very young biofilms (younger than 48 h). Matures biofilms are involved in most chronic infections and we know that the MIC values of antimicrobials increase dramatically in the first days of a biofilm life. Hence, we recommend that, in addition to traditional tests on planktonic cultures, efficacy tests on mature biofilms should be included when testing a novel compound for its antibacterial efficacy.

Some of the most active antibiotics have recently been shown to have dual activities. In addition to working as conventional antibiotics they also function as QSIs (Mizukane et al. 1994; Nalca et al. 2006; Hoffmann et al. 2007; Skindersø et al. 2008). Three out of twelve tested antibiotics (azithromycin, ceftazidime, and ciprofloxacin) could effectively inhibit the production of several QS virulence factors including rhamnolipid and elastase (Skindersø et al. 2008).

These antimicrobials set the bar for future drugs targeting chronic infections by combining several targets in one drug. In addition to the discovery of new drug scaffolds and designing effective combinations of existing compounds, another important task is to identify drug targets that are ubiquitous in many bacteria. Consequently, a drug that would hit a ubiquitous target could then be used to eradicate many different pathogenic organisms and hopefully attract the interest of the pharmaceutical industry. The increase in multiresistant bacteria, the scarcity of newly approved antibiotics, and the desperate lack of leads in the pipeline has raised the need for new strategies against infections. From the data presented, it seems that combinations of treatments seem to be the right way of fighting biofilm infections with *P. aeruginosa* as well as other pathogens.

References

Aaron SD et al (2002) Single and combination antibiotic susceptibilities of planktonic, adherent, and biofilm-grown *Pseudomonas aeruginosa* isolates cultured from sputa of adults with cystic fibrosis. J Clin Microbiol 40(11):4172–4179

Adawi A et al (2012) In vitro and in vivo properties of a fully human IgG1 monoclonal antibody that combats multidrug resistant *Pseudomonas aeruginosa*. Int J Mol Med 30(3):455–464

Afessa B, Green B (2000) Bacterial pneumonia in hospitalized patients with HIV infection: the pulmonary complications, ICU support, and prognostic factors of hospitalized patients with HIV (PIP) study. Chest 117(4):1017–1022

Alhede M et al (2009) *Pseudomonas aeruginosa* recognizes and responds aggressively to the presence of polymorphonuclear leukocytes. Microbiology 155(Pt 11):3500–3508

Alhede M et al (2011) Phenotypes of non-attached *Pseudomonas aeruginosa* aggregates resemble surface attached biofilm. PLoS One 6(11):e27943

Alipour M et al (2009) Importance of DNase and alginate lyase for enhancing free and liposome encapsulated aminoglycoside activity against *Pseudomonas aeruginosa*. J Antimicrob Chemother 64(2):317–325

Allen L et al (2005) Pyocyanin production by *Pseudomonas aeruginosa* induces neutrophil apoptosis and impairs neutrophil-mediated host defenses in vivo. J Immunol 174(6):3643–3649

Allesen-Holm M et al (2006) A characterization of DNA release in *Pseudomonas aeruginosa* cultures and biofilms. Mol Microbiol 59(4):1114–1128

Barbosa FC et al (2001) Subgingival occurrence and antimicrobial susceptibility of enteric rods and pseudomonads from Brazilian periodontitis patients. Oral Microbiol Immunol 16(5):306–310

Barnea Y et al (2006) Efficacy of antibodies against the N-terminal of *Pseudomonas aeruginosa* flagellin for treating infections in a murine burn wound model. Plast Reconstr Surg 117 (7):2284–2291

Barnea Y et al (2009) Therapy with anti-flagellin A monoclonal antibody limits *Pseudomonas aeruginosa* invasiveness in a mouse burn wound sepsis model. Burns 35(3):390–396

Bjarnsholt T, Givskov M (2007) Quorum-sensing blockade as a strategy for enhancing host defences against bacterial pathogens. Philos Trans R Soc Lond B Biol Sci 362(1483):1213–1222

Bjarnsholt T, Givskov M (2008) Quorum sensing inhibitory drugs as next generation antimicrobials: worth the effort? Curr Infect Dis Rep 10(1):22–28

Bjarnsholt T et al (2005a) Pseudomonas aeruginosa tolerance to tobramycin, hydrogen peroxide and polymorphonuclear leukocytes is quorum-sensing dependent. Microbiology 151 (Pt 2):373–383

Bjarnsholt T et al (2005b) Garlic blocks quorum sensing and promotes rapid clearing of pulmonary *Pseudomonas aeruginosa* infections. Microbiology 151(Pt 12):3873–3880

Bjarnsholt T et al (2007) Silver against *Pseudomonas aeruginosa* biofilms. APMIS 115(8):921–928

Bjarnsholt T et al (2008) Why chronic wounds will not heal: a novel hypothesis. Wound Repair Regen 16(1):2–10

Bjarnsholt T et al (2009) *Pseudomonas aeruginosa* biofilms in the respiratory tract of cystic fibrosis patients. Pediatr Pulmonol 44(6):547–558

Bone RC (1991) Monoclonal antibodies to endotoxin. New allies against sepsis? JAMA 266 (8):1125–1126

Bone RC (1996) Why sepsis trials fail. JAMA 276(7):565–566

Bortolussi R et al (1987) Relationship of bacterial growth phase to killing of Listeria monocytogenes by oxidative agents generated by neutrophils and enzyme systems. Infect Immun 55(12):3197–3203

Boucher HW et al (2009) Bad bugs, no drugs: no ESKAPE! An update from the Infectious Diseases Society of America. Clin Infect Dis 48(1):1–12

Bowler PG et al (2012) Multidrug-resistant organisms, wounds and topical antimicrobial protection. Int Wound J 9(4):387–396

Brouqui P et al (1995) Treatment of *Pseudomonas aeruginosa*-infected orthopedic prostheses with ceftazidime-ciprofloxacin antibiotic combination. Antimicrob Agents Chemother 39 (11):2423–2425

Burmølle M et al (2010) Biofilms in chronic infections—a matter of opportunity—monospecies biofilms in multispecies infections. FEMS Immunol Med Microbiol 59(3):324–336

Castellano JJ et al (2007) Comparative evaluation of silver-containing antimicrobial dressings and drugs. Int Wound J 4(2):114–122

Chaw KC et al (2005) Role of silver ions in destabilization of intermolecular adhesion forces measured by atomic force microscopy in *Staphylococcus epidermidis* biofilms. Antimicrob Agents Chemother 49(12):4853–4859

Chiang WC et al (2013) Extracellular DNA shields against aminoglycosides in *Pseudomonas aeruginosa* biofilms. Antimicrob Agents Chemother 57(5):2352–2361

Chopra I (2007) The increasing use of silver-based products as antimicrobial agents: a useful development or a cause for concern? J Antimicrob Chemother 59(4):587–590

Christensen LD et al (2012) Synergistic antibacterial efficacy of early combination treatment with tobramycin and quorum-sensing inhibitors against *Pseudomonas aeruginosa* in an intraperitoneal foreign-body infection mouse model. J Antimicrob Chemother 67(5):1198–1206

Ciofu O (2003) *Pseudomonas aeruginosa* chromosomal beta-lactamase in patients with cystic fibrosis and chronic lung infection. Mechanism of antibiotic resistance and target of the humoral immune response. APMIS Suppl(116): 1–47

Clatworthy AE et al (2007) Targeting virulence: a new paradigm for antimicrobial therapy. Nat Chem Biol 3(9):541–548

Davies DG et al (1998) The involvement of cell-to-cell signals in the development of a bacterial biofilm. Science 280(5361):295–298

Defoirdt T et al (2010) Can bacteria evolve resistance to quorum sensing disruption? PLoS Pathog 6(7):e1000989

Doring G et al (1986) Elastase from polymorphonuclear leucocytes: a regulatory enzyme in immune complex disease. Clin Exp Immunol 64(3):597–605

El-Solh AA et al (2004) Colonization of dental plaques: a reservoir of respiratory pathogens for hospital-acquired pneumonia in institutionalized elders. Chest 126(5):1575–1582

Emerson J et al (2002) Pseudomonas aeruginosa and other predictors of mortality and morbidity in young children with cystic fibrosis. Pediatr Pulmonol 34(2):91–100

Fazli M et al (2009) Nonrandom distribution of *Pseudomonas aeruginosa* and *Staphylococcus aureus* in chronic wounds. J Clin Microbiol 47(12):4084–4089

FitzSimmons SC (1993) The changing epidemiology of cystic fibrosis. J Pediatr 122(1):1–9

Fleiszig SM, Evans DJ (2002) The pathogenesis of bacterial keratitis: studies with *Pseudomonas aeruginosa*. Clin Exp Optom 85(5):271–278

Frederiksen B et al (2006) Effect of aerosolized rhDNase (Pulmozyme) on pulmonary colonization in patients with cystic fibrosis. Acta Paediatr 95(9):1070–1074

Fuqua WC et al (1994) Quorum sensing in bacteria: the LuxR-LuxI family of cell density-responsive transcriptional regulators. J Bacteriol 176(2):269–275

Haagensen JAJ et al (2007) Differentiation and distribution of colistin- and sodium dodecyl sulfate-tolerant cells in *Pseudomonas aeruginosa* biofilms. J Bacteriol 189(1):28–37

Hansen CR et al (2008) Early aggressive eradication therapy for intermittent *Pseudomonas aeruginosa* airway colonization in cystic fibrosis patients: 15 years experience. J Cyst Fibros 7(6):523–530

Hardalo C, Edberg SC (1997) *Pseudomonas aeruginosa*: assessment of risk from drinking water. Crit Rev Microbiol 23(1):47–75

Hawkey PM (2008) The growing burden of antimicrobial resistance. J Antimicrob Chemother 62 (Suppl 1):i1–i9

Hentzer M et al (2003a) Quorum sensing : a novel target for the treatment of biofilm infections. BioDrugs 17(4):241–250

Hentzer M et al (2003b) Attenuation of *Pseudomonas aeruginosa* virulence by quorum sensing inhibitors. EMBO J 22(15):3803–3815

Hoffmann N et al (2007) Azithromycin blocks quorum sensing and alginate polymer formation and increases the sensitivity to serum and stationary-growth-phase killing of *Pseudomonas aeruginosa* and attenuates chronic *P. aeruginosa* lung infection in Cftr(-/-) mice. Antimicrob Agents Chemother 51(10):3677–3687

Høiby N et al (2010) Antibiotic resistance of bacterial biofilms. Int J Antimicrob Agents 35 (4):322–332

Jakobsen TH, Bragason SK et al (2012a) Food as a source for quorum sensing inhibitors: iberin from horseradish revealed as a quorum sensing inhibitor of *Pseudomonas aeruginosa*. Appl Environ Microbiol 78(7):2410–2421

Jakobsen TH, van Gennip M et al (2012b) Ajoene, a sulfur-rich molecule from garlic, inhibits genes controlled by quorum sensing. Antimicrob Agents Chemother 56(5):2314–2325

Jensen PØ et al (2007) Rapid necrotic killing of polymorphonuclear leukocytes is caused by quorum-sensing-controlled production of rhamnolipid by *Pseudomonas aeruginosa*. Microbiology 153(Pt 5):1329–1338

Jesaitis AJ et al (2003) Compromised host defense on *Pseudomonas aeruginosa* biofilms: characterization of neutrophil and biofilm interactions. J Immunol 171(8):4329–4339

Kaplan JB (2009) Therapeutic potential of biofilm-dispersing enzymes. Int J Artif Organs 32 (9):545–554

Kharazmi A (1991) Mechanisms involved in the evasion of the host defence by *Pseudomonas aeruginosa*. Immunol Lett 30(2):201–205

Kharazmi A et al (1984a) Interaction of *Pseudomonas aeruginosa* alkaline protease and elastase with human polymorphonuclear leukocytes *in vitro*. Infect Immun 43(1):161–165

Kharazmi A et al (1984b) *Pseudomonas aeruginosa* exoproteases inhibit human neutrophil chemiluminescence. Infect Immun 44(3):587–591

Kharazmi A et al (1986) Effect of *Pseudomonas aeruginosa* proteases on human leukocyte phagocytosis and bactericidal activity. Acta Pathol Microbiol Immunol Scand C 94(5):175–179

Kharazmi A et al (1989) Effect of *Pseudomonas aeruginosa* rhamnolipid on human neutrophil and monocyte function. APMIS 97(12):1068–1072

Klueh U et al (2000) Efficacy of silver-coated fabric to prevent bacterial colonization and subsequent device-based biofilm formation. J Biomed Mater Res 53(6):621–631

Kohler T et al (2010) Quorum sensing-dependent virulence during *Pseudomonas aeruginosa* colonisation and pneumonia in mechanically ventilated patients. Thorax 65(8):703–710

Kollberg H et al (2003) Oral administration of specific yolk antibodies (IgY) may prevent *Pseudomonas aeruginosa* infections in patients with cystic fibrosis: a phase I feasibility study. Pediatr Pulmonol 35(6):433–440

Kostenko V et al (2010) Impact of silver-containing wound dressings on bacterial biofilm viability and susceptibility to antibiotics during prolonged treatment. Antimicrob Agents Chemother 54 (12):5120–5131

Lansdown AB (2002) Silver. I: its antibacterial properties and mechanism of action. J Wound Care 11(4):125–130

Larsson A et al (1993) Chicken antibodies: taking advantage of evolution–a review. Poult Sci 72 (10):1807–1812

Maira-Litran T et al (2004) Biologic properties and vaccine potential of the staphylococcal poly-N-acetyl glucosamine surface polysaccharide. Vaccine 22(7):872–879

Mizukane R et al (1994) Comparative in vitro exoenzyme-suppressing activities of azithromycin and other macrolide antibiotics against *Pseudomonas aeruginosa*. Antimicrob Agents Chemother 38(3):528–533

Montanaro L et al (2011) Extracellular DNA in biofilms. Int J Artif Organs 34(9):824–831

Mulcahy H et al (2008) Extracellular DNA chelates cations and induces antibiotic resistance in *Pseudomonas aeruginosa* biofilms. PLoS Pathog 4(11):e1000213

Nalca Y et al (2006) Quorum-sensing antagonistic activities of azithromycin in *Pseudomonas aeruginosa* PAO1: a global approach. Antimicrob Agents Chemother 50(5):1680–1688

National Nosocomial Infections Surveillance System (2004) National Nosocomial Infections Surveillance (NNIS) System Report, data summary from January 1992 through June 2004, issued October 2004. Am J Infect Control 32(8):470–485

Nilsson E et al (2007) Pseudomonas aeruginosa infections are prevented in cystic fibrosis patients by avian antibodies binding Pseudomonas aeruginosa flagellin. J Chromatogr B Analyt Technol Biomed Life Sci 856(1–2):75–80

Nilsson E et al (2008) Good effect of IgY against *Pseudomonas aeruginosa* infections in cystic fibrosis patients. Pediatr Pulmonol 43(9):892–899

Oncel S et al (2010) Evaluation of bacterial biofilms in chronic rhinosinusitis. J Otolaryngol Head Neck Surg 39(1):52–55

Pamp SJ et al (2008) Tolerance to the antimicrobial peptide colistin in *Pseudomonas aeruginosa* biofilms is linked to metabolically active cells, and depends on the *pmr* and *mexAB-oprM* genes. Mol Microbiol 68(1):223–240

Parks QM et al (2009) Neutrophil enhancement of *Pseudomonas aeruginosa* biofilm development: human F-actin and DNA as targets for therapy. J Med Microbiol 58(Pt 4):492–502

Parsons D et al (2005) Silver antimicrobial dressings in wound management: a comparison of antibacterial, physical, and chemical characteristics. Wounds 17(8):222–232

Pollack M et al (1984) Immunization with *Pseudomonas aeruginosa* high-molecular-weight polysaccharides prevents death from Pseudomonas burn infections in mice. Infect Immun 43 (2):759–760

Post JC (2001) Direct evidence of bacterial biofilms in otitis media. Laryngoscope 111(12):2083–2094

Rasmussen TB et al (2005) Identity and effects of quorum-sensing inhibitors produced by *Penicillium* species. Microbiology 151(Pt 5):1325–1340

Reyes MP, Lerner AM (1983) Current problems in the treatment of infective endocarditis due to *Pseudomonas aeruginosa*. Rev Infect Dis 5(2):314–321

Robinson PJ (2002) Dornase alfa in early cystic fibrosis lung disease. Pediatr Pulmonol 34(3):237–241

Sapico FL (1996) Microbiology and antimicrobial therapy of spinal infections. Orthop Clin North Am 27(1):9–13

Schaedel C et al (2002) Predictors of deterioration of lung function in cystic fibrosis. Pediatr Pulmonol 33(6):483–491

Shanks KK et al (2010) Interleukin-8 production by human airway epithelial cells in response to *Pseudomonas aeruginosa* clinical isolates expressing type a or type b flagellins. Clin Vaccine Immunol 17(8):1196–1202

Shapiro JA (1998) Thinking about bacterial populations as multicellular organisms. Annu Rev Microbiol 52(1):81–104

Skindersø ME et al (2008) Effects of antibiotics on quorum sensing in *Pseudomonas aeruginosa*. Antimicrob Agents Chemother 52(10):3648–3663

Smith RS, Iglewski BH (2003) *P. aeruginosa* quorum-sensing systems and virulence. Curr Opin Microbiol 6(1):56–60

Smith KD et al (2003) Toll-like receptor 5 recognizes a conserved site on flagellin required for protofilament formation and bacterial motility. Nat Immunol 4(12):1247–1253

Spellberg B et al (2008) The epidemic of antibiotic-resistant infections: a call to action for the medical community from the Infectious Diseases Society of America. Clin Infect Dis 46 (2):155–164

Stewart PS, Costerton JW (2001) Antibiotic resistance of bacteria in biofilms. Lancet 358 (9276):135–138

Storm-Versloot MN et al (2010) Topical silver for preventing wound infection. Cochrane Database Syst Rev(3):CD006478

Stover CK et al (2000) Complete genome sequence of *Pseudomonas aeruginosa* PAO1, an opportunistic pathogen. Nature 406(6799):959–964

Sun D et al (2005) Inhibition of biofilm formation by monoclonal antibodies against *Staphylococcus epidermidis* RP62A accumulation-associated protein. Clin Diagn Lab Immunol 12 (1):93–100

Tacconelli E et al (2009) Epidemiology, medical outcomes and costs of catheter-related bloodstream infections in intensive care units of four European countries: literature- and registry-based estimates. J Hosp Infect 72(2):97–103

Tang JX et al (2005) Anionic poly(amino acid)s dissolve F-actin and DNA bundles, enhance DNase activity, and reduce the viscosity of cystic fibrosis sputum. Am J Physiol Lung Cell Mol Physiol 289(4):L599–L605

Tashiro Y et al (2008) *Opr86* is essential for viability and is a potential candidate for a protective antigen against biofilm formation by *Pseudomonas aeruginosa*. J Bacteriol 190(11):3969–3978

Tetz GV et al (2009) Effect of DNase and antibiotics on biofilm characteristics. Antimicrob Agents Chemother 53(3):1204–1209

Tolker-Nielsen T, Hoiby N (2009) Extracellular DNA and F-actin as targets in antibiofilm cystic fibrosis therapy. Future Microbiol 4(6):645–647

Toy LW, Macera L (2011) Evidence-based review of silver dressing use on chronic wounds. J Am Acad Nurse Pract 23(4):183–192

van Gennip M et al (2009) Inactivation of the rhlA gene in *Pseudomonas aeruginosa* prevents rhamnolipid production, disabling the protection against polymorphonuclear leukocytes. APMIS 117(7):537–546

Walker TS et al (2005) Enhanced *Pseudomonas aeruginosa* biofilm development mediated by human neutrophils. Infect Immun 73(6):3693–3701

Waters CM, Bassler BL (2005) Quorum sensing: cell-to-cell communication in bacteria. Annu Rev Cell Dev Biol 21:319–346

Whitchurch CB et al (2002) Extracellular DNA required for bacterial biofilm formation. Science 295(5559):1487

Withers H et al (2001) Quorum sensing as an integral component of gene regulatory networks in Gram-negative bacteria. Curr Opin Microbiol 4(2):186–193

Wu H et al (2004) Synthetic furanones inhibit quorum-sensing and enhance bacterial clearance in *Pseudomonas aeruginosa* lung infection in mice. J Antimicrob Chemother 53(6):1054–1061

Inhibition of Fungal Biofilms

Christopher G. Pierce, Anand Srinivasan, Priya Uppuluri,
Anand K. Ramasubramanian, and José L. López-Ribot

Abstract Fungal infections constitute a major threat for an expanding population of immunosuppressed patients, as these infections carry unacceptably high morbidity and mortality rates due to, among other reasons, the limited arsenal of antifungal agents. One of the main factors complicating antifungal therapy is the formation of fungal biofilms, resulting in frank resistance to most antifungal drugs, which is multifactorial in nature. Although *Candida albicans* remains the most frequent etiologic agent of fungal biofilm infections, there is an increased recognition that infections caused by other yeasts and filamentous fungi are also associated with the formation of biofilms, both on biomedical devices and host tissues. During the last decade an increasing number of studies have begun to uncover the driving forces behind the formation of fungal biofilms and the molecular basis of biofilm resistance; together with new powerful technologies, they may pave the road for the development of newer therapeutics for the prevention and treatment of these recalcitrant infections.

C.G. Pierce • P. Uppuluri • J.L. López-Ribot (✉)
Department of Biology, The University of Texas at San Antonio, One UTSA Circle, San Antonio, TX 78249, USA

South Texas Center for Emerging Infectious Diseases, The University of Texas at San Antonio, One UTSA Circle, San Antonio, TX 78249, USA
e-mail: jose.lopezribot@utsa.edu

A. Srinivasan • A.K. Ramasubramanian
South Texas Center for Emerging Infectious Diseases, The University of Texas at San Antonio, One UTSA Circle, San Antonio, TX 78249, USA

Department of Biomedical Engineering, The University of Texas at San Antonio, One UTSA Circle, San Antonio, TX 78249, USA

1 Fungal Infections, Fungal Biofilms, and Their Clinical Significance

Fungal infections represent a significant clinical problem as advances in modern medicine prolong the lives of an expanding population of severely compromised patients (Brown et al. 2012). The opportunistic pathogenic fungi *Candida* spp., *Aspergillus* spp., and *Cryptococcus neoformans* are among the most common etiologic agents of mycoses, but infections caused by other yeasts and moulds are on the rise (Brown et al. 2012). Unfortunately, these infections are frequently not recognized (diagnosed) and are treated inadequately, leading to unacceptably high morbidity and mortality rates (Brown et al. 2012). Moreover, these devastating infections place an additional financial burden to our healthcare system (Wilson et al. 2002). Over the last two decades there has been an increasing recognition of the role that biofilms play during fungal infections. Biofilm formation has important clinical implications since sessile cells within these microbial consortia display characteristics that are drastically different from planktonic populations, most notably increased resistance to antifungal drugs. In addition, biofilms provide a safe haven for fungal cells, can become a persistent source of infections, and can adversely affect the function of implanted devices, which further complicates the clinical management of these patients. Overall fungal biofilm formation adversely impacts the health of these patients at an alarmingly increasing rate and with soaring economic consequences.

Candida spp., and in particular *C. albicans,* are the fungal species most frequently associated with formation of biofilms (Kojic and Darouiche 2004; Ramage et al. 2005, 2006, 2009). It is now well established that different forms of candidiasis, now the third to fourth most common nosocomial infection in US hospitals and abroad (Banerjee et al. 1991; Beck-Sague and Jarvis 1993), such as catheter-related candidemia, candiduria, and endocarditis, involve biofilm formation (Kojic and Darouiche 2004; Ramage et al. 2006). Other manifestations such as denture stomatitis and oral and vaginal candidiasis are also associated with biofilm formation (Harriott et al. 2010; Ramage et al. 2004). Somewhat ironically, the increase in *Candida* infections in the last few decades has almost directly paralleled the increase and widespread use of a broad range of medical implant devices, such as stents, shunts, prostheses, implants, endotracheal tubes, pacemakers, and various types of catheters, mainly in populations with impaired host defenses (Ramage et al. 2006). Most notably, *Candida* spp. are the third leading cause of central catheter-related infections, with the overall highest crude mortality (Crump and Collignon 2000). Studies of catheter-related candidiasis have unequivocally shown that retention of vascular catheters is associated with prolonged fungemia, high antifungal therapy failure rates, increased risk of metastatic complications, and death (Pappas et al. 2009; Pfaller and Diekema 2007; Viudes et al. 2002).

Besides *Candida,* other yeasts and filamentous fungi whose biofilm-forming ability are associated with clinical settings have been described and include *Cryptococcus, Aspergillus, Coccidioides, Pneumocystis, Malassezia, Penicillium,*

Histoplasma, Saccharomyces, Trichosporon, Blastoschizomyces, Pneumocystis, and some zygomycetes (reviewed in Fanning and Mitchell 2012; Pitangui et al. 2012; Ramage et al. 2009). For example, the opportunistic yeast *Cryptococcus neoformans* causes life-threatening meningitis in immune deficient individuals, particularly HIV-infected patients. This encapsulated yeast can colonize and subsequently form biofilms on ventricular shunts, peritoneal dialysis fistulas, and cardiac valves (Martinez and Casadevall 2007; Martinez et al. 2006; Ravi et al. 2009). Different *Trichosporon* species (also opportunistic yeasts) have been associated with biofilm formation on catheters, breast implants, and cardiac grafts (Krzossok et al. 2004; Pini et al. 2005; Reddy et al. 2002). Cushion et al. postulated that the attachment and growth of *Pneumocystis* spp. within the lung alveoli resembles a biofilm and developed in vitro methods for the formation and susceptibility testing of *Pneumocystis* biofilms (Cushion and Collins 2011; Cushion et al. 2009). Invasive aspergillosis caused by *Aspergillus* spp. is now a major problem at cancer treatment centers and solid organ transplantation units (Patterson et al. 2000), with recent studies pointing to a role for biofilms in different manifestations of these infections (i.e., prosthetic valve endocarditis) and in the overall pathogenesis of aspergillosis (Loussert et al. 2010; Mowat et al. 2009; Muszkieta et al. 2013). A case report described a patient with recurrent meningitis associated with a *Coccidioides immitis* biofilm at the tip of a ventriculo-peritoneal shunt (Davis et al. 2002). A recent report described *Histoplasma capsulatum* attachment to pneumocyte cells and subsequent biofilm formation (Pitangui et al. 2012). In contrast to *Candida*, which as a strict commensal is only found inside the host, some of these fungi are ubiquitous in nature and it is entirely possible that formation of these attached microbial communities can also contribute to their survival in the environment (Martinez and Casadevall 2007; Ravi et al. 2009; Pini et al. 2005).

2 Fungal Biofilm Formation and Structural Characteristics of Fungal Biofilms

Traditionally, techniques used for fungal biofilm formation were cumbersome and allowed for the formation of only a few biofilms at a time (Ramage et al. 2005). However, more recent research on fungal biofilms has been greatly simplified and expedited by the development of relatively simple methodologies, most notably the 96-well microtiter plate model for the formation of fungal biofilms (Pierce et al. 2008; Ramage et al. 2001a). This technique involves the formation of multiple equivalent fungal biofilms on the bottom of wells of microtiter plates, combined with a colorimetric method that measures the metabolic activities of cells within the biofilm. Although originally developed for *Candida*, it was subsequently adapted for *Cryptococcus, Aspergillus,* and other fungi (Martinez and Casadevall 2007; Mowat et al. 2009; Pierce et al. 2008; Ravi et al. 2009). Overall this microtiter plate-based model of biofilm formation offers an easy, economical, flexible, and robust alternative for biofilm formation that is compatible with the broadly accessible 96-well microplate platform. One additional advantage of this method is that it can also be easily adapted for antifungal susceptibility testing (Pierce et al. 2008; Ramage et al. 2001a).

The architecture of fungal biofilms may differ depending on environmental conditions, the fungal species, and the substrate on which biofilms are formed. From an architectural point of view, mature fungal biofilms exhibit a complex three-dimensional structure and extensive spatial heterogeneity, with atypical microcolony/water channel architecture, and are encased within a matrix of self-produced exopolymeric material (Blankenship and Mitchell 2006; Chandra et al. 2001; Nett and Andes 2006; Ramage et al. 2001b, 2005). Figure 1 shows a *C. albicans* biofilm. In general, the structural features of fungal biofilms seem to represent an ideal spatial arrangement for the uptake of essential nutrients, excretion of metabolic and toxic products, as well as communication between cells and the environment. Fungal biofilm development occurs though different phases, including initial attachment and colonization, proliferation, maturation, and ultimately dispersion so that the "biofilm life cycle" can be repeated all over again (Chandra et al. 2001; Ramage et al. 2009; Uppuluri et al. 2010a) (Fig. 2). In the case of *C. albicans*, the most studied organism in regard to its biofilm-forming ability, biofilm formation is inextricably linked to filamentation, and this may also hold true for other filamentous fungi (Lopez-Ribot 2005; Nobile et al. 2006a; Nobile and Mitchell 2006; Ramage et al. 2002d; Uppuluri et al. 2010b). Also, adhesive interactions, both cell-to-surface and cell-to-cell (Nobile et al. 2006a, b, 2008; Nobile and Mitchell 2005, 2006), as well as quorum sensing mechanisms (Ramage et al. 2002b) play a preponderant role during biofilm development. Although most of the structural information about fungal biofilms comes from in vitro models, even with their intrinsic limitations, the fact that different groups of investigators have reported similar architectural characteristics for biofilms formed in vivo (both in animal models and retrieved directly from patients) is reassuring (Andes et al. 2004; Lazzell et al. 2009; Ricicova et al. 2010; Schinabeck et al. 2004; Shuford et al. 2006).

3 Biofilm Antifungal Drug Resistance

Fungi are eukaryotic and there is a paucity of selective pathogen-specific targets for drug development. Thus, in stark contrast with antibacterial antibiotics, the current list of antifungal drugs is exceedingly short, mostly limited from the clinical point of view (and in particular for the treatment of invasive fungal infections) to three classes of antifungal agents: polyenes, azoles, and echinocandins (Odds et al. 2003; Ostrosky-Zeichner et al. 2010). For example, amphotericin B, a broad-spectrum polyene that binds to ergosterol and compromises membrane integrity, remained the "gold standard" of antifungal therapy during decades after its introduction in the 1950s; but its efficacy is severely limited by its inherent toxicity (Odds et al. 2003).

Fig. 1 A scanning electron microphotograph showing a *C. albicans* biofilm

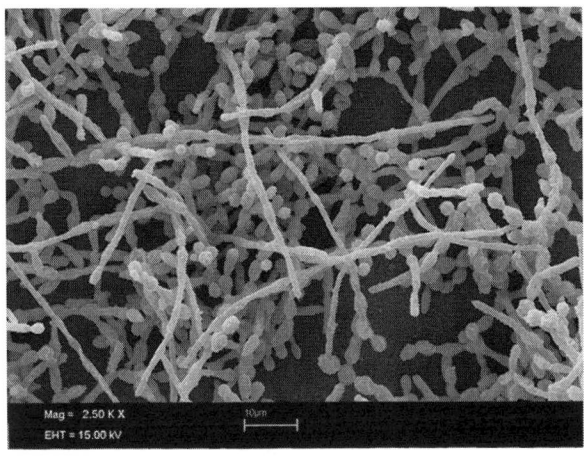

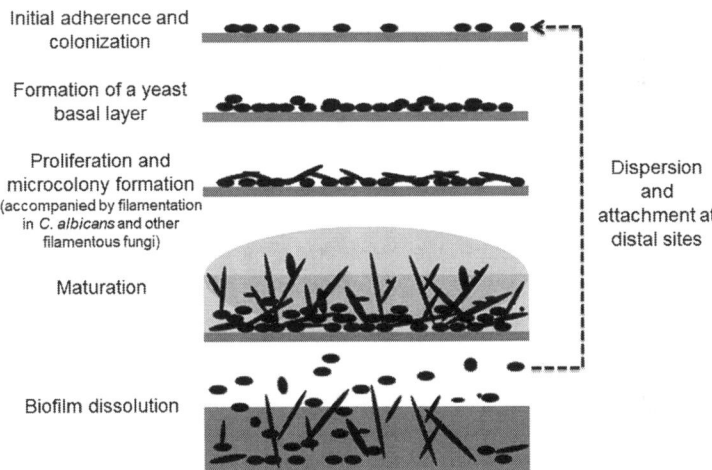

Fig. 2 The different stages of fungal biofilm formation

Azoles (i.e., fluconazole, voriconazole) inhibit ergosterol biosynthesis and were originally developed in the 1980s and 1990s; however, a major problem with these fungistatic agents has been the widespread emergence of resistance (including cross-resistance against multiple azole derivatives) (Odds et al. 2003; Sanglard et al. 2009). The echinocandins (i.e., caspofungin), first introduced in the 2000s, are semisynthetic lipopeptide antibiotics that inhibit the synthesis of 1,3-β-D-glucan, a key structural component of the fungal cell wall. They are generally considered fungicidal against yeast but fungistatic against moulds (Odds et al. 2003; Perlin 2011). Despite their recent introduction, resistance is emerging (Hernandez et al. 2004; Perlin 2011; Wiederhold et al. 2008). From the very early reports on fungal biofilms, it was already demonstrated that sessile cells within these communities show intrinsic resistance to azole derivatives, as well as displaying high levels of resistance against polyenes (Chandra et al. 2001; Ramage et al. 2001a, b). On the other hand, echinocandins seemed to be active against biofilms (Bachmann et al. 2002; Kuhn et al. 2002; Ramage et al. 2002c).

From the clinical perspective, the most salient feature of fungal biofilms is their high levels of resistance to most conventional antifungal agents (Ramage et al. 2012). The current consensus is that fungal biofilm resistance is a complex phenomenon that cannot be explained by one mechanism alone; instead, it is multifactorial in nature and may involve distinct molecular mechanisms of resistance as compared to those displayed by planktonic cells. Potential contributory mechanisms to fungal biofilm resistance include: (1) the increased cellular density of fungal cells within the biofilm, basically a simple concept of "safety in numbers," as elegantly demonstrated by the Chaffin group (Perumal et al. 2007); (2) the existence of subpopulations of "persister" cells, dormant variants of regular cells that form stochastically in microbial populations and are highly tolerant to antifungal treatment and associated with tolerance and chronic infection (LaFleur et al. 2006; Lafleur et al. 2010); (3) differences in the metabolic and physiological status of cells (Baillie and Douglas 1998a, b); (4) the distinct sterol composition of the cell wall membrane of sessile as compared to planktonic cells, in particular decreased ergosterol levels, which may impact amphotericin B activity; (5) the protective effect of the biofilm matrix, most notably glucans in the *Candida* biofilm matrix with the ability to bind azole derivative and also polyenes (Al-Fattani and Douglas 2006; Nett et al. 2007; Nobile et al. 2009); (6) the upregulation of efflux pumps, which may occur physiologically during the biofilm mode of growth as a means to facilitate the removal of toxic products, but may concomitantly result in increased efflux of antifungal molecules (Mukherjee et al. 2003; Ramage et al. 2002a); (7) a highly regulated network orchestrated by the hsp90 molecular chaperone, also with connections to the calcineurin pathway (Robbins et al. 2011). For an excellent, contemporary, and comprehensive review on the topic of biofilm antifungal drug resistance please refer to Ramage et al. (2012).

4 Inhibition of Fungal Biofilms

The high morbidity and mortality rates associated with fungal biofilm infections clearly indicate an urgent and unmet need to develop novel strategies, both preventative (i.e., inhibition of biofilm formation) and therapeutic (i.e., against preformed biofilms), to control fungal biofilms in clinical settings, and this is certainly an area of active research. Much of this work has been facilitated not only by our increasing understanding of mechanisms involved in fungal biofilm development at the cellular and molecular level but also by lessons learned during the clinical management of patients (Nobile et al. 2012; Nobile and Mitchell 2006; Ramage et al. 2009, 2012).

4.1 Conventional Antifungal Agents

As mentioned before, fungal cells within biofilms display dramatically reduced susceptibility to most conventional antifungals, perhaps with the exception of echinocandins. They are up to 1,000 times more resistant to azole derivatives, and they also showed decreased susceptibility to polyenes as compared to their planktonic counterparts (Chanda and Caldwell 2003; Ramage et al. 2001a). In truth, polyenes such as amphotericin B and nystatin exhibit biofilm activity; but unfortunately they do so at concentrations which are exceedingly high and generally considered unsafe due to their intrinsic toxicity (Ramage et al. 2002c). However, liposomal formulations of amphotericin B display unique efficacy against *Candida* biofilms (Kuhn et al. 2002), most likely due to their improved safety profile and perhaps enhanced penetration due to encapsulation. Very importantly, echinocandins display excellent anti-biofilm activity at therapeutic concentrations (Bachmann et al. 2002; Kuhn et al. 2002; Ramage et al. 2002c). Initial experiments with caspofungin (the first echinocandin to reach the market) demonstrated its excellent in vitro activity against *C. albicans* biofilms, and soon after was corroborated for micafungin, anidulafungin, and non-*albicans Candida* spp. (Kucharikova et al. 2011). These observations were further extended to in vivo animal models and to clinical experience with patients (Kucharikova et al. 2010, 2013; Tumbarello et al. 2012). Thus, the excellent anti-biofilm activity of echinocandins has been one of the main factors for them to become first line therapy against candidiasis, particularly when a biofilm etiology (i.e., catheter-related candidemia) is suspected (Pappas et al. 2009; Tumbarello et al. 2012). Although these three echinocandins are considered equally effective, a recent report described drug- and species-specific differences in susceptibility among biofilms of different *Candida* spp., with *C. lusitaniae* and *C. guilliermondii* biofilms generally exhibiting reduced susceptibility (Simitsopoulou et al. 2013). Also, it is important to note that the anti-biofilm activity of echinocandins is not universal, as other fungal biofilms, such as those formed by *A. fumigatus*, are relatively resistant to treatment with this new class of antifungal agents (Fanning and Mitchell 2012; Ramage et al. 2012).

4.2 Combination Therapy

The existence of different classes of antifungal drugs with different molecular targets opened new possibilities for combination therapy against fungal biofilms. An in vitro study using fluconazole, amphotericin B and caspofungin against *Candida* biofilms pointed towards indifference for all antifungal combinations tested (Bachmann et al. 2003). However, the combination between amphotericin B and caspofungin may benefit from a rapid initial killing by the polyene followed by a more sustained effect by the echinocandin (Bachmann et al. 2003; Ramage et al. 2002c). In contrast, there was a trend towards antagonism with the fluconazole/caspofungin combination, which was particularly evident at higher concentrations of the azole (Bachmann et al. 2003). A possibility also is to combine a conventional antifungal with another agent that potentiates its anti-biofilm activity. For example, LaFleur and colleagues performed primary screens for potentiators of the anti-biofilm activity of clotrimazole (an azole antifungal which per se is not effective against *C. albicans* biofilms) (LaFleur et al. 2011). Both calcineurin inhibitors (i.e., cyclosporine, FK506) and hsp90 inhibitors (i.e., geldanamycin), when used in combination with antifungals, were able to overcome biofilm drug resistance (Robbins et al. 2011; Uppuluri et al. 2008). Another possibility is to target the biofilm matrix; for example, Martins and colleagues (Martins et al. 2010, 2012a) demonstrated that addition of DNase (as extracellular DNA is a component of the fungal biofilm matrix) improves the anti-biofilm activity of some antifungal drugs. Martinez et al. tested combinations of conventional antifungals and specific monoclonal antibodies against *C. neoformans* biofilms, but unfortunately the antibodies antagonized the effects of the antifungal agents (Martinez et al. 2006).

4.3 Antifungal Lock Therapy

As mentioned before, yeasts, particularly *Candida* spp., are one of the main causes of catheter-related bloodstream infections (Crump and Collignon 2000; Kojic and Darouiche 2004; Raad 1998; Ramage et al. 2006). The use of antifungal lock solutions (supra-pharmacological concentrations of antifungals locally inside the catheter) is receiving increasing interest as a strategy to prevent and treat these infections (Cateau et al. 2008; Walraven and Lee 2013). Some of the most promising antifungal lock therapy strategies include not only the use of antifungal agents such as amphotericin and echinocandins and antibacterial antibiotics with antifungal effects such as minocycline and rifampin but also the use of antiseptics such as ethanol, anti-microbial peptides, and anti-occlusive agents such as EDTA, heparin, and citrate (Sherertz et al. 2006; Sousa et al. 2011; Walraven and Lee 2013). However, to date, there is little comparative or clinical data between the different potential antifungal lock therapy strategies to permit specific recommendations for its use (Walraven and Lee 2013).

4.4 Search for New Molecules Active Against Fungal Biofilms

Of course, the development of novel anti-biofilm agents may also involve the search for new molecules active against cells within biofilms. Indeed this is currently an area of very active research, which has been greatly facilitated by the development of simple and inexpensive microtiter plate-based models for the formation and antifungal susceptibility testing of fungal biofilms, as mentioned before (Pierce et al. 2008; Ramage et al. 2001a). In particular, there is an increasing interest in the examination of anti-biofilm activity of a variety of natural products (reviewed in Sardi et al. 2013), either by themselves or as potentiators of conventional antifungals (You et al. 2013). There is also considerable interest in screening synthetic small molecule compounds present in different commercially available chemical libraries (Pierce et al. 2011).

The development of a completely new drug is an arduous and extremely costly proposition, as any new medicine has to endure a demanding approval process by the Food and Drug Administration (FDA) in order to make sure that the drug is safe for consumption. Thus, repurposing already FDA-approved drugs as antifungal agents may decrease the time and effort in bringing drugs with novel antifungal activity from the bench to the bedside (Tobinick 2009). In this regard, most recently our group carried out a comprehensive screen of a small molecule library consisting of 1,200 FDA-approved off-patent drugs to identify inhibitors of *C. albicans* biofilm formation (Siles et al. 2013). We found not only several bioactive compounds including well-known antifungals and antiseptics but also several miscellaneous drugs with no previously reported antifungal activity and which may be further developed as anti-biofilms agents.

4.5 Other Strategies for Fungal Biofilm Inhibition

Other strategies for inhibition of fungal biofilms may be as diverse as the development of biomaterials which do not support fungal adherence and colonization or coating of biomaterials to prevent adhesion of fungal cells and subsequent biofilm formation. For example, impregnation of plastics with caspofungin or voriconazole inhibited the development of *Candida* biofilms (Bachmann et al. 2002; Valentin et al. 2012). A novel thin-film coating incorporating different antifungals effectively inhibited *C. albicans* biofilm formation as a potential preventive therapy for denture stomatitis (Redding et al. 2009). *C. albicans* biofilm cells grown on denture acrylic were sensitive to killing by histatin 5, a naturally occurring antimicrobial peptide in saliva, and coating the surface of acrylic with chlorhexidine or histatin 5 prevented biofilm growth (Pusateri et al. 2009). Chitosan, a polymer of chitin isolated from crustacean exoskeletons, damaged *C. albicans* and *C. neoformans* biofilms, both in vitro and in vivo (Martinez et al. 2010a, b).

Another possibility is to use modulators of quorum sensing mechanisms, certainly an area of increasing interest in the prevention and treatment of bacterial biofilms. Farnesol is the main quorum sensing molecule produced by *C. albicans* and it inhibits biofilm formation in vitro, but somewhat surprisingly, exogenous administration of farnesol increased *C. albicans* virulence in vivo (Navarathna et al. 2007). In contrast, a cocktail solution, mimicking the composition of alcohols present in a *C. albicans* culture supernatant (which also includes farnesol), exerted a protective effect against candidiasis in vivo (Martins et al. 2007, 2012b). Some bacteria secrete quorum sensing and other antifungal molecules (e.g., homoserine lactones and phenazines secreted by *Pseudomonas aeruginosa*, mutanobactins secreted by *Streptococcus mutans*) that inhibit or kill *C. albicans* biofilms (Gibson et al. 2009; Hogan et al. 2004; Joyner et al. 2010; Morales et al. 2010; Wang et al. 2012).

Photodynamic therapy (PDT), mediated by the action of reactive oxygen species generated by the photoactivation of a photosensitizer by a light source, represents a relatively new therapeutic technique with potential applications for the treatment of fungal biofilm infections, in particular superficial mycoses such as denture stomatitis and oropharyngeal candidiasis (Costa et al. 2013; Junqueira et al. 2012; Pereira et al. 2011). Another potential strategy, which may be particularly relevant in the case of catheter-associated biofilm infections, is to target dispersion, as fungal cells dispersed from the biofilms are responsible for fungemia, dissemination, extravasation, and ultimately the establishment of foci of invasive mycoses at distal organs, which are the forms associated with the highest mortality rates (Uppuluri et al. 2010a).

5 A Nano-Biofilm Chip for High-Throughput Antifungal Drug Discovery

One of the main impediments for the development of newer antibiotics, including antifungals, has been the fact that conventional microbiological culture techniques are mostly incompatible with modern methodologies for drug discovery that are dominated by high-throughput screening (HTS) and its "hunger for speed." To overcome this major bottleneck, Srinivasan and colleagues recently described the nanoscale culture of *C. albicans*, on a microarray platform (Srinivasan et al. 2011, 2013). The microarray, designated *Ca*BChip (for *Candida albicans* Biofilm Chip), consisted of a standard microscope slide containing 1,200 individual *C. albicans* biofilms ("nano-biofilms"), each with a volume of approximately 30 nL, encapsulated in an inert alginate matrix (see Fig. 3). The authors demonstrated that these nano-biofilms, despite a 3,000-fold miniaturization over conventional biofilms (formed on microtiter plates), are similar in their morphological, architectural, growth, and phenotypic characteristics, including antifungal drug resistance (Srinivasan et al. 2011, 2013). The nanobiofilm chip is amenable to automation and is fully compatible with standard microarray technology and equipment. It allows for rapid and easy handling, minimizing manual labor, and drastically reducing assay costs (Srinivasan et al. 2011, 2013). It enables true high-throughput screening in the search for new anti-biofilm drugs. The techniques should be adaptable to other biofilm-forming species, including polymicrobial biofilms, and should accelerate the antifungal drug discovery process by permitting fast, efficient, and economical screening of thousands of compounds.

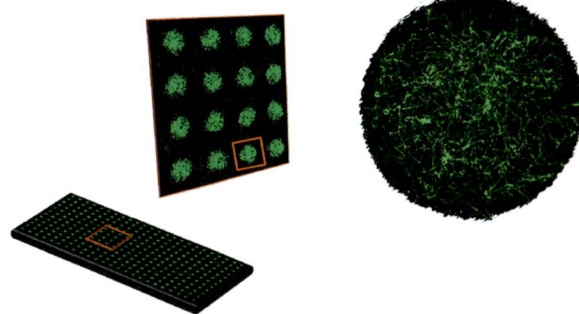

Fig. 3 The *C. albicans* nano-biofilm chip for high-throughput applications

6 Summary

There is an increasing recognition of the role that biofilms play during fungal infections, further complicating antifungal treatment for these already difficult to treat infections. The area of fungal biofilm research has witnessed strong growth in the last two decades, with significant technological advances, the application of powerful molecular technologies, and a vigorous translational emphasis. Hopefully in the not so distant future, we should be able to harness this knowledge for the development of new therapeutic strategies to conquer this formidable challenge.

Acknowledgments Biofilm work in the laboratory is funded by Grant numbered 1R01DE023510 from the National Institute of Dental & Craniofacial Research (to JLL-R) and by the Army Research Office of the Department of Defense under Contract No. W911NF-11-1-0136. Infectious diseases-related work in the A.K.R. laboratory is funded by NIH (SC1HL112629). CGP and AS acknowledge the receipt of predoctoral fellowships from American Heart Association, numbered 51PRE30004 and 13PRE17110093. The funders had no role in study design, data collection and analysis, decision to publish, or preparation of the manuscript, and the content is solely the responsibility of the authors.

References

Al-Fattani MA, Douglas LJ (2006) Biofilm matrix of Candida albicans and Candida tropicalis: chemical composition and role in drug resistance. J Med Microbiol 55(Pt 8):999–1008

Andes D, Nett J, Oschel P, Albrecht R, Marchillo K, Pitula A (2004) Development and characterization of an in vivo central venous catheter Candida albicans biofilm model. Infect Immun 72(10):6023–6031

Bachmann SP, VandeWalle K, Ramage G, Patterson TF, Wickes BL, Graybill JR, Lopez-Ribot JL (2002) In vitro activity of caspofungin against Candida albicans biofilms. Antimicrob Agents Chemother 46(11):3591–3596

Bachmann SP, Ramage G, VandeWalle K, Patterson TF, Wickes BL, Lopez-Ribot JL (2003) Antifungal combinations against Candida albicans biofilms in vitro. Antimicrob Agents Chemother 47(11):3657–3659

Baillie GS, Douglas LJ (1998a) Effect of growth rate on resistance of Candida albicans biofilms to antifungal agents. Antimicrob Agents Chemother 42(8):1900–1905

Baillie GS, Douglas LJ (1998b) Iron-limited biofilms of Candida albicans and their susceptibility to amphotericin B. Antimicrob Agents Chemother 42(8):2146–2149

Banerjee SN, Emori TG, Culver DH, Gaynes RP, Jarvis WR, Horan T, Edwards JR, Tolson J, Henderson T, Martone WJ (1991) Secular trends in nosocomial primary bloodstream infections in the United States, 1980–1989. National Nosocomial Infections Surveillance System. Am J Med 91(3B):86S–89S

Beck-Sague C, Jarvis WR (1993) Secular trends in the epidemiology of nosocomial fungal infections in the United States, 1980–1990. National Nosocomial Infections Surveillance System. J Infect Dis 167(5):1247–1251

Blankenship JR, Mitchell AP (2006) How to build a biofilm: a fungal perspective. Curr Opin Microbiol 9(6):588–594

Brown GD, Denning DW, Gow NA, Levitz SM, Netea MG, White TC (2012) Hidden killers: human fungal infections. Sci Transl Med 4(165):165rv113

Cateau E, Rodier MH, Imbert C (2008) In vitro efficacies of caspofungin or micafungin catheter lock solutions on Candida albicans biofilm growth. J Antimicrob Chemother 62(1):153–155

Chanda SK, Caldwell JS (2003) Fulfilling the promise: drug discovery in the post-genomic era. Drug Discov Today 8(4):168–174

Chandra J, Kuhn DM, Mukherjee PK, Hoyer LL, McCormick T, Ghannoum MA (2001) Biofilm formation by the fungal pathogen Candida albicans: development, architecture, and drug resistance. J Bacteriol 183(18):5385–5394

Costa AC, Pereira CA, Freire F, Junqueira JC, Jorge AO (2013) Methods for obtaining reliable and reproducible results in studies of Candida biofilms formed in vitro. Mycoses 56:614–622

Crump JA, Collignon PJ (2000) Intravascular catheter-associated infections. Eur J Clin Microbiol Infect Dis 19(1):1–8

Cushion MT, Collins MS (2011) Susceptibility of Pneumocystis to echinocandins in suspension and biofilm cultures. Antimicrob Agents Chemother 55(10):4513–4518

Cushion MT, Collins MS, Linke MJ (2009) Biofilm formation by Pneumocystis spp. Eukaryot Cell 8(2):197–206

Davis LE, Cook G, Costerton JW (2002) Biofilm on ventriculo-peritoneal shunt tubing as a cause of treatment failure in coccidioidal meningitis. Emerg Infect Dis 8(4):376–379

Fanning S, Mitchell AP (2012) Fungal biofilms. PLoS Pathog 8(4):e1002585

Gibson J, Sood A, Hogan DA (2009) Pseudomonas aeruginosa-Candida albicans interactions: localization and fungal toxicity of a phenazine derivative. Appl Environ Microbiol 75(2):504–513

Harriott MM, Lilly EA, Rodriguez TE, Fidel PL Jr, Noverr MC (2010) Candida albicans forms biofilms on the vaginal mucosa. Microbiology 156(Pt 12):3635–3644

Hernandez S, Lopez-Ribot JL, Najvar LK, McCarthy DI, Bocanegra R, Graybill JR (2004) Caspofungin resistance in Candida albicans: correlating clinical outcome with laboratory susceptibility testing of three isogenic isolates serially obtained from a patient with progressive Candida esophagitis. Antimicrob Agents Chemother 48(4):1382–1383

Hogan DA, Vik A, Kolter R (2004) A Pseudomonas aeruginosa quorum-sensing molecule influences Candida albicans morphology. Mol Microbiol 54(5):1212–1223

Joyner PM, Liu J, Zhang Z, Merritt J, Qi F, Cichewicz RH (2010) Mutanobactin A from the human oral pathogen Streptococcus mutans is a cross-kingdom regulator of the yeast-mycelium transition. Org Biomol Chem 8(24):5486–5489

Junqueira JC, Jorge AO, Barbosa JO, Rossoni RD, Vilela SF, Costa AC, Primo FL, Goncalves JM, Tedesco AC, Suleiman JM (2012) Photodynamic inactivation of biofilms formed by Candida spp., Trichosporon mucoides, and Kodamaea ohmeri by cationic nanoemulsion of zinc 2,9,16,23-tetrakis(phenylthio)-29H, 31H-phthalocyanine (ZnPc). Lasers Med Sci 27(6):1205–1212

Kojic EM, Darouiche RO (2004) Candida infections of medical devices. Clin Microbiol Rev 17(2):255–267

Krzossok S, Birck R, Henke S, Hof H, van der Woude FJ, Braun C (2004) Trichosporon asahii infection of a dialysis PTFE arteriovenous graft. Clin Nephrol 62(1):66–68

Kucharikova S, Tournu H, Holtappels M, Van Dijck P, Lagrou K (2010) In vivo efficacy of anidulafungin against mature Candida albicans biofilms in a novel rat model of catheter-associated Candidiasis. Antimicrob Agents Chemother 54(10):4474–4475

Kucharikova S, Tournu H, Lagrou K, Van Dijck P, Bujdakova H (2011) Detailed comparison of Candida albicans and Candida glabrata biofilms under different conditions and their susceptibility to caspofungin and anidulafungin. J Med Microbiol 60(Pt 9):1261–1269

Kucharikova S, Sharma N, Spriet I, Maertens J, Van Dijck P, Lagrou K (2013) Activities of systemically administered echinocandins against in vivo mature Candida albicans biofilms developed in a rat subcutaneous model. Antimicrob Agents Chemother 57(5):2365–2368

Kuhn DM, George T, Chandra J, Mukherjee PK, Ghannoum MA (2002) Antifungal susceptibility of Candida biofilms: unique efficacy of amphotericin B lipid formulations and echinocandins. Antimicrob Agents Chemother 46(6):1773–1780

LaFleur MD, Kumamoto CA, Lewis K (2006) Candida albicans biofilms produce antifungal-tolerant persister cells. Antimicrob Agents Chemother 50(11):3839–3846

Lafleur MD, Qi Q, Lewis K (2010) Patients with long-term oral carriage harbor high-persister mutants of Candida albicans. Antimicrob Agents Chemother 54(1):39–44

LaFleur MD, Lucumi E, Napper AD, Diamond SL, Lewis K (2011) Novel high-throughput screen against Candida albicans identifies antifungal potentiators and agents effective against biofilms. J Antimicrob Chemother 66(4):820–826

Lazzell AL, Chaturvedi AK, Pierce CG, Prasad D, Uppuluri P, Lopez-Ribot JL (2009) Treatment and prevention of Candida albicans biofilms with caspofungin in a novel central venous catheter murine model of candidiasis. J Antimicrob Chemother 64(3):567–570

Lopez-Ribot JL (2005) Candida albicans biofilms: more than filamentation. Curr Biol 15(12): R453–R455

Loussert C, Schmitt C, Prevost MC, Balloy V, Fadel E, Philippe B, Kauffmann-Lacroix C, Latge JP, Beauvais A (2010) In vivo biofilm composition of Aspergillus fumigatus. Cell Microbiol 12(3):405–410

Martinez LR, Casadevall A (2007) Cryptococcus neoformans biofilm formation depends on surface support and carbon source and reduces fungal cell susceptibility to heat, cold, and UV light. Appl Environ Microbiol 73(14):4592–4601

Martinez LR, Christaki E, Casadevall A (2006) Specific antibody to Cryptococcus neoformans glucurunoxylomannan antagonizes antifungal drug action against cryptococcal biofilms in vitro. J Infect Dis 194(2):261–266

Martinez LR, Mihu MR, Han G, Frases S, Cordero RJ, Casadevall A, Friedman AJ, Friedman JM, Nosanchuk JD (2010a) The use of chitosan to damage Cryptococcus neoformans biofilms. Biomaterials 31(4):669–679

Martinez LR, Mihu MR, Tar M, Cordero RJ, Han G, Friedman AJ, Friedman JM, Nosanchuk JD (2010b) Demonstration of antibiofilm and antifungal efficacy of chitosan against candidal biofilms, using an in vivo central venous catheter model. J Infect Dis 201(9):1436–1440

Martins M, Henriques M, Azeredo J, Rocha SM, Coimbra MA, Oliveira R (2007) Morphogenesis control in Candida albicans and Candida dubliniensis through signaling molecules produced by planktonic and biofilm cells. Eukaryot Cell 6(12):2429–2436

Martins M, Uppuluri P, Thomas DP, Cleary IA, Henriques M, Lopez-Ribot JL, Oliveira R (2010) Presence of extracellular DNA in the Candida albicans biofilm matrix and its contribution to biofilms. Mycopathologia 169(5):323–331

Martins M, Henriques M, Lopez-Ribot JL, Oliveira R (2012a) Addition of DNase improves the in vitro activity of antifungal drugs against Candida albicans biofilms. Mycoses 55(1):80–85

Martins M, Lazzell AL, Lopez-Ribot JL, Henriques M, Oliveira R (2012b) Effect of exogenous administration of Candida albicans autoregulatory alcohols in a murine model of hematogenously disseminated candidiasis. J Basic Microbiol 52(4):487–491

Morales DK, Jacobs NJ, Rajamani S, Krishnamurthy M, Cubillos-Ruiz JR, Hogan DA (2010) Antifungal mechanisms by which a novel Pseudomonas aeruginosa phenazine toxin kills Candida albicans in biofilms. Mol Microbiol 78(6):1379–1392

Mowat E, Williams C, Jones B, McChlery S, Ramage G (2009) The characteristics of Aspergillus fumigatus mycetoma development: is this a biofilm? Med Mycol 47(Suppl 1):S120–S126

Mukherjee PK, Chandra J, Kuhn DM, Ghannoum MA (2003) Mechanism of fluconazole resistance in Candida albicans biofilms: phase-specific role of efflux pumps and membrane sterols. Infect Immun 71(8):4333–4340

Muszkieta L, Beauvais A, Pahtz V, Gibbons JG, Anton Leberre V, Beau R, Shibuya K, Rokas A, Francois JM, Kniemeyer O, Brakhage AA, Latge JP (2013) Investigation of Aspergillus fumigatus biofilm formation by various "omics" approaches. Front Microbiol 4:13

Navarathna DH, Hornby JM, Krishnan N, Parkhurst A, Duhamel GE, Nickerson KW (2007) Effect of farnesol on a mouse model of systemic candidiasis, determined by use of a DPP3 knockout mutant of Candida albicans. Infect Immun 75(4):1609–1618

Nett J, Andes D (2006) Candida albicans biofilm development, modeling a host-pathogen interaction. Curr Opin Microbiol 9(4):340–345

Nett J, Lincoln L, Marchillo K, Massey R, Holoyda K, Hoff B, VanHandel M, Andes D (2007) Putative role of beta-1,3 glucans in Candida albicans biofilm resistance. Antimicrob Agents Chemother 51(2):510–520

Nobile CJ, Mitchell AP (2005) Regulation of cell-surface genes and biofilm formation by the C. albicans transcription factor Bcr1p. Curr Biol 15(12):1150–1155

Nobile CJ, Mitchell AP (2006) Genetics and genomics of Candida albicans biofilm formation. Cell Microbiol 8(9):1382–1391

Nobile CJ, Andes DR, Nett JE, Smith FJ, Yue F, Phan QT, Edwards JE, Filler SG, Mitchell AP (2006a) Critical role of Bcr1-dependent adhesins in C. albicans biofilm formation in vitro and in vivo. PLoS Pathog 2(7):e63

Nobile CJ, Nett JE, Andes DR, Mitchell AP (2006b) Function of Candida albicans adhesin Hwp1 in biofilm formation. Eukaryot Cell 5(10):1604–1610

Nobile CJ, Schneider HA, Nett JE, Sheppard DC, Filler SG, Andes DR, Mitchell AP (2008) Complementary adhesin function in C. albicans biofilm formation. Curr Biol 18(14):1017–1024

Nobile CJ, Nett JE, Hernday AD, Homann OR, Deneault JS, Nantel A, Andes DR, Johnson AD, Mitchell AP (2009) Biofilm matrix regulation by Candida albicans Zap1. PLoS Biol 7(6):e1000133

Nobile CJ, Fox EP, Nett JE, Sorrells TR, Mitrovich QM, Hernday AD, Tuch BB, Andes DR, Johnson AD (2012) A recently evolved transcriptional network controls biofilm development in Candida albicans. Cell 148(1–2):126–138

Odds FC, Brown AJ, Gow NA (2003) Antifungal agents: mechanisms of action. Trends Microbiol 11(6):272–279

Ostrosky-Zeichner L, Casadevall A, Galgiani JN, Odds FC, Rex JH (2010) An insight into the antifungal pipeline: selected new molecules and beyond. Nat Rev Drug Discov 9(9):719–727

Pappas PG, Kauffman CA, Andes D, Benjamin DK Jr, Calandra TF, Edwards JE Jr, Filler SG, Fisher JF, Kullberg BJ, Ostrosky-Zeichner L, Reboli AC, Rex JH, Walsh TJ, Sobel JD (2009) Clinical practice guidelines for the management of candidiasis: 2009 update by the Infectious Diseases Society of America. Clin Infect Dis 48(5):503–535

Patterson TF, Kirkpatrick WR, White M, Hiemenz JW, Wingard JR, Dupont B, Rinaldi MG, Stevens DA, Graybill JR (2000) Invasive aspergillosis. Disease spectrum, treatment practices, and outcomes. I3 Aspergillus Study Group. Medicine (Baltimore) 79(4):250–260

Pereira CA, Romeiro RL, Costa AC, Machado AK, Junqueira JC, Jorge AO (2011) Susceptibility of Candida albicans, Staphylococcus aureus, and Streptococcus mutans biofilms to photodynamic inactivation: an in vitro study. Lasers Med Sci 26(3):341–348

Perlin DS (2011) Current perspectives on echinocandin class drugs. Future Microbiol 6(4):441–457

Perumal P, Mekala S, Chaffin WL (2007) Role for cell density in antifungal drug resistance in Candida albicans biofilms. Antimicrob Agents Chemother 51(7):2454–2463

Pfaller MA, Diekema DJ (2007) Epidemiology of invasive candidiasis: a persistent public health problem. Clin Microbiol Rev 20(1):133–163

Pierce CG, Uppuluri P, Tristan AR, Wormley FL Jr, Mowat E, Ramage G, Lopez-Ribot JL (2008) A simple and reproducible 96-well plate-based method for the formation of fungal biofilms and its application to antifungal susceptibility testing. Nat Protoc 3(9):1494–1500

Pierce CG, Saville SP, Lopez-Ribot JL (2011) Characterization of small molecule inhibitors of Candida albicans biofilm formation using phenotype-based high content screening: implications for antifungal drug development. Paper presented at FEBS Advanced Lecture Course on Human Fungal Pathogens, Le Colle-sur-Loup, France

Pini G, Faggi E, Donato R, Fanci R (2005) Isolation of Trichosporon in a hematology ward. Mycoses 48(1):45–49

Pitangui NS, Sardi JC, Silva JF, Benaducci T, Moraes da Silva RA, Rodriguez-Arellanes G, Taylor ML, Mendes-Giannini MJ, Fusco-Almeida AM (2012) Adhesion of Histoplasma capsulatum to pneumocytes and biofilm formation on an abiotic surface. Biofouling 28(7): 711–718

Pusateri CR, Monaco EA, Edgerton M (2009) Sensitivity of Candida albicans biofilm cells grown on denture acrylic to antifungal proteins and chlorhexidine. Arch Oral Biol 54(6):588–594

Raad I (1998) Intravascular-catheter-related infections. Lancet 351(9106):893–898

Ramage G, Vande Walle K, Wickes BL, Lopez-Ribot JL (2001a) Standardized method for in vitro antifungal susceptibility testing of Candida albicans biofilms. Antimicrob Agents Chemother 45(9):2475–2479

Ramage G, Vandewalle K, Wickes BL, Lopez-Ribot JL (2001b) Characteristics of biofilm formation by Candida albicans. Rev Iberoam Micol 18(4):163–170

Ramage G, Bachmann S, Patterson TF, Wickes BL, Lopez-Ribot JL (2002a) Investigation of multidrug efflux pumps in relation to fluconazole resistance in Candida albicans biofilms. J Antimicrob Chemother 49(6):973–980

Ramage G, Saville SP, Wickes BL, Lopez-Ribot JL (2002b) Inhibition of Candida albicans biofilm formation by farnesol, a quorum-sensing molecule. Appl Environ Microbiol 68(11): 5459–5463

Ramage G, VandeWalle K, Bachmann SP, Wickes BL, Lopez-Ribot JL (2002c) In vitro pharmacodynamic properties of three antifungal agents against preformed Candida albicans biofilms determined by time-kill studies. Antimicrob Agents Chemother 46(11):3634–3636

Ramage G, VandeWalle K, Lopez-Ribot JL, Wickes BL (2002d) The filamentation pathway controlled by the Efg1 regulator protein is required for normal biofilm formation and development in Candida albicans. FEMS Microbiol Lett 214(1):95–100

Ramage G, Tomsett K, Wickes BL, Lopez-Ribot JL, Redding SW (2004) Denture stomatitis: a role for Candida biofilms. Oral Surg Oral Med Oral Pathol Oral Radiol Endod 98(1):53–59

Ramage G, Saville SP, Thomas DP, Lopez-Ribot JL (2005) Candida biofilms: an update. Eukaryot Cell 4(4):633–638

Ramage G, Martinez JP, Lopez-Ribot JL (2006) Candida biofilms on implanted biomaterials: a clinically significant problem. FEMS Yeast Res 6(7):979–986

Ramage G, Mowat E, Jones B, Williams C, Lopez-Ribot J (2009) Our current understanding of fungal biofilms. Crit Rev Microbiol 35(4):340–355

Ramage G, Rajendran R, Sherry L, Williams C (2012) Fungal biofilm resistance. Int J Microbiol 2012:528521

Ravi S, Pierce C, Witt C, Wormley FL Jr (2009) Biofilm formation by Cryptococcus neoformans under distinct environmental conditions. Mycopathologia 167(6):307–314

Redding S, Bhatt B, Rawls HR, Siegel G, Scott K, Lopez-Ribot J (2009) Inhibition of Candida albicans biofilm formation on denture material. Oral Surg Oral Med Oral Pathol Oral Radiol Endod 107(5):669–672

Reddy BT, Torres HA, Kontoyiannis DP (2002) Breast implant infection caused by Trichosporon beigelii. Scand J Infect Dis 34(2):143–144

Ricicova M, Kucharikova S, Tournu H, Hendrix J, Bujdakova H, Van Eldere J, Lagrou K, Van Dijck P (2010) Candida albicans biofilm formation in a new in vivo rat model. Microbiology 156(Pt 3):909–919

Robbins N, Uppuluri P, Nett J, Rajendran R, Ramage G, Lopez-Ribot JL, Andes D, Cowen LE (2011) Hsp90 governs dispersion and drug resistance of fungal biofilms. PLoS Pathog 7(9): e1002257

Sanglard D, Coste A, Ferrari S (2009) Antifungal drug resistance mechanisms in fungal pathogens from the perspective of transcriptional gene regulation. FEMS Yeast Res 9(7):1029–1050

Sardi JC, Scorzoni L, Bernardi T, Fusco-Almeida AM, Mendes Giannini MJ (2013) Candida species: current epidemiology, pathogenicity, biofilm formation, natural antifungal products and new therapeutic options. J Med Microbiol 62(Pt 1):10–24

Schinabeck MK, Long LA, Hossain MA, Chandra J, Mukherjee PK, Mohamed S, Ghannoum MA (2004) Rabbit model of Candida albicans biofilm infection: liposomal amphotericin B antifungal lock therapy. Antimicrob Agents Chemother 48(5):1727–1732

Sherertz RJ, Boger MS, Collins CA, Mason L, Raad II (2006) Comparative in vitro efficacies of various catheter lock solutions. Antimicrob Agents Chemother 50(5):1865–1868

Shuford JA, Rouse MS, Piper KE, Steckelberg JM, Patel R (2006) Evaluation of caspofungin and amphotericin B deoxycholate against Candida albicans biofilms in an experimental intravascular catheter infection model. J Infect Dis 194(5):710–713

Siles SA, Srinivasan A, Pierce CG, Lopez-Ribot JL, Ramasubramanian AK (2013) High-throughput screening of a collection of known pharmacologically active small compounds for the identification of Candida albicans biofilm inhibitors. Antimicrob Agents Chemother 57:3681–3687

Simitsopoulou M, Peshkova P, Tasina E, Katragkou A, Kyrpitzi D, Velegraki A, Walsh TJ, Roilides E (2013) Species-specific and drug-specific differences in susceptibility of Candida biofilms to echinocandins: characterization of less common bloodstream isolates. Antimicrob Agents Chemother 57(6):2562–2570

Sousa C, Henriques M, Oliveira R (2011) Mini-review: antimicrobial central venous catheters–recent advances and strategies. Biofouling 27(6):609–620

Srinivasan A, Uppuluri P, Lopez-Ribot J, Ramasubramanian AK (2011) Development of a high-throughput Candida albicans biofilm chip. PLoS One 6(4):e19036

Srinivasan A, Leung KP, Lopez-Ribot JL, Ramasubramanian AK (2013) High-throughput nano-biofilm microarray for antifungal drug discovery. mBio 4:e00331-13

Tobinick EL (2009) The value of drug repositioning in the current pharmaceutical market. Drug News Perspect 22(2):119–125

Tumbarello M, Fiori B, Trecarichi EM, Posteraro P, Losito AR, De Luca A, Sanguinetti M, Fadda G, Cauda R, Posteraro B (2012) Risk factors and outcomes of candidemia caused by biofilm-forming isolates in a tertiary care hospital. PLoS ONE 7(3):e33705

Uppuluri P, Nett J, Heitman J, Andes D (2008) Synergistic effect of calcineurin inhibitors and fluconazole against Candida albicans biofilms. Antimicrob Agents Chemother 52(3):1127–1132

Uppuluri P, Chaturvedi AK, Srinivasan A, Banerjee M, Ramasubramaniam AK, Köhler JR, Kadosh D, Lopez Ribot JL (2010a) Dispersion as an important step in the Candida albicans biofilm developmental cycle. PLoS Pathog 6:e1000828

Uppuluri P, Pierce CG, Thomas DP, Bubeck SS, Saville SP, Lopez-Ribot JL (2010b) The transcriptional regulator Nrg1p controls Candida albicans biofilm formation and dispersion. Eukaryot Cell 9(10):1531–1537

Valentin A, Canton E, Peman J, Martinez JP (2012) Voriconazole inhibits biofilm formation in different species of the genus Candida. J Antimicrob Chemother 67(10):2418–2423

Viudes A, Peman J, Canton E, Ubeda P, Lopez-Ribot JL, Gobernado M (2002) Candidemia at a tertiary-care hospital: epidemiology, treatment, clinical outcome and risk factors for death. Eur J Clin Microbiol Infect Dis 21(11):767–774

Walraven CJ, Lee SA (2013) Antifungal lock therapy. Antimicrob Agents Chemother 57(1):1–8

Wang X, Du L, You J, King JB, Cichewicz RH (2012) Fungal biofilm inhibitors from a human oral microbiome-derived bacterium. Org Biomol Chem 10(10):2044–2050

Wiederhold NP, Grabinski JL, Garcia-Effron G, Perlin DS, Lee SA (2008) Pyrosequencing to detect mutations in FKS1 that confer reduced echinocandin susceptibility in Candida albicans. Antimicrob Agents Chemother 52(11):4145–4148

Wilson LS, Reyes CM, Stolpman M, Speckman J, Allen K, Beney J (2002) The direct cost and incidence of systemic fungal infections. Value Health 5(1):26–34

You J, Du L, King JB, Hall BE, Cichewicz RH (2013) Small-molecule suppressors of Candida albicans biofilm formation synergistically enhance the antifungal activity of amphotericin B against clinical Candida isolates. ACS Chem Biol 8(4):840–848

Biofilm Control Strategies in Dental Health

Jorge Frias-Lopez

Abstract The human oral microbiota is one of the most diverse biofilms colonizing the human body consisting of over 700 individual taxa. Two of the most common human diseases, caries and periodontal disease, are the result of a healthy microbial biofilm becoming a dysbiotic one due to mechanisms not completely understood. These are special cases of infectious diseases in which the origin of the infection is not an exogenous organism but a commensal organism that somehow overgrows and modifies the features of the microbial biofilm to its advantage. Dental Unit Water Systems (DUWS) deserve special consideration when studying biofilm control methods in oral health. In dentistry, dental chair units (DCU) are equipped with complex networks of plastic pipes that supply water to the DCU instruments and constitute an ideal environment for the growth of biofilms, especially bacterial biofilms. In the present chapter we present a brief overview of different methods that have been used in dentistry to control biofilm growth both in the oral cavity as well as in DUWS, with special focus on new strategies whose goal it is to modulate the composition and growth of the biofilm rather than the complete removal of the microbial community.

1 Introduction

The origins of oral microbiology are intertwined with the beginnings of microbiology as a science. Antonie van Leeuwenhoek (1632–1723) was the first scientist to describe microbes sampled from the oral biofilm (commonly called dental plaque). Using his primitive microscope, he was able to describe with high precision a large number of microorganisms of different shapes that were centuries later identified as common inhabitants of the oral biofilm growing on teeth, also known as dental

---

J. Frias-Lopez (✉)
The Forsyth Institute, 245 First Street, Cambridge, MA 02142, USA
e-mail: jfrias@forsyth.org

plaque. In September 1683, Leeuwenhoek wrote to the Royal Society about his observations on the plaque between his own teeth, "didn't clean my teeth for three days and then took the material that has lodged in small amounts on the gums above my front teeth... I found a few living animalcules" (He and Shi 2009). With those simple descriptions Antonie van Leeuwenhoek initiated the study of oral microbial ecology and gave us a first indication of the complexity of the bacterial community colonizing our oral cavity (Bardell 1983). In recent years, dental plaque has been evaluated and discussed as a biofilm that may contain bacteria, archaea, viruses, and yeast (Marsh 2004, 2006). Nonetheless, the vast majority of the oral biofilm biomass is of bacterial origin.

In 2002, Donlan and Costerton defined a biofilm as "a microbially derived sessile community characterized by cells that are irreversibly attached to a substratum or interface or to each other, are embedded in a matrix of extracellular polymeric substances that they have produced, and exhibit an altered phenotype with respect to growth rate and gene transcription." (Donlan and Costerton 2002). In fact, a biofilm is an accumulation of microbial cells within a matrix, optimizing the use of the available nutritional resources. The physiology of bacteria growing in biofilms is completely different than the physiology of bacteria growing in planktonic conditions. The susceptibility of microorganisms to antimicrobials is considerably reduced in biofilms (Larsen and Fiehn 1996; Roberts and Mullany 2010). It has also been shown, in established biofilms, that the effect of antimicrobials on bacterial vitality is limited to the most superficial layers (Zaura-Arite et al. 2001). Thus, the efficacy of antimicrobials against existing biofilms appears to be limited.

Dental plaque incorporates all of the features of biofilm architecture and microbial community interaction and its establishment follows an ordered sequence of events that results in a well organized and distinct architecture (Kolenbrander 2000; Kolenbrander et al. 2002, 2010) (See Fig. 1). The dental biofilm is different in that it is extremely complex with more than 600 contributing oral bacterial taxa in the oral cavity (Paster et al. 2006). Many of them are not cultivable species, and the only information we possess about them derives from their 16S rRNA phylogenetic affiliation (Dewhirst et al. 2010; Paster et al. 2001, 2006). Moreover, the possibility exists that novel microorganisms may yet be found and that they may be important in oral diseases.

Only 20–25 % of the oral environment is tooth surface, the rest are mucosal surfaces that may be important contributors to periodontal microbial biofilms. For tooth surfaces, pellicle formation is the preconditioning stage that defines the reversible–irreversible attachment of the colonizing bacteria. Attachment is defined as a slime layer forming around the colonizing pioneer bacteria, which consists mainly of Gram-positive cocci and rods that divide and form microcolonies. If this early supragingival plaque is unregulated, owing to the absence of effective oral hygiene, the bacterial composition can mature into a more complex flora in a three-stage scenario. The first stage is predominantly Gram-positive cocci and is represented by the streptococcal species, the second stage is cross-linking via fusobacterium species, and the third stage is predominantly Gram-negative organisms.

Biofilm Control Strategies in Dental Health

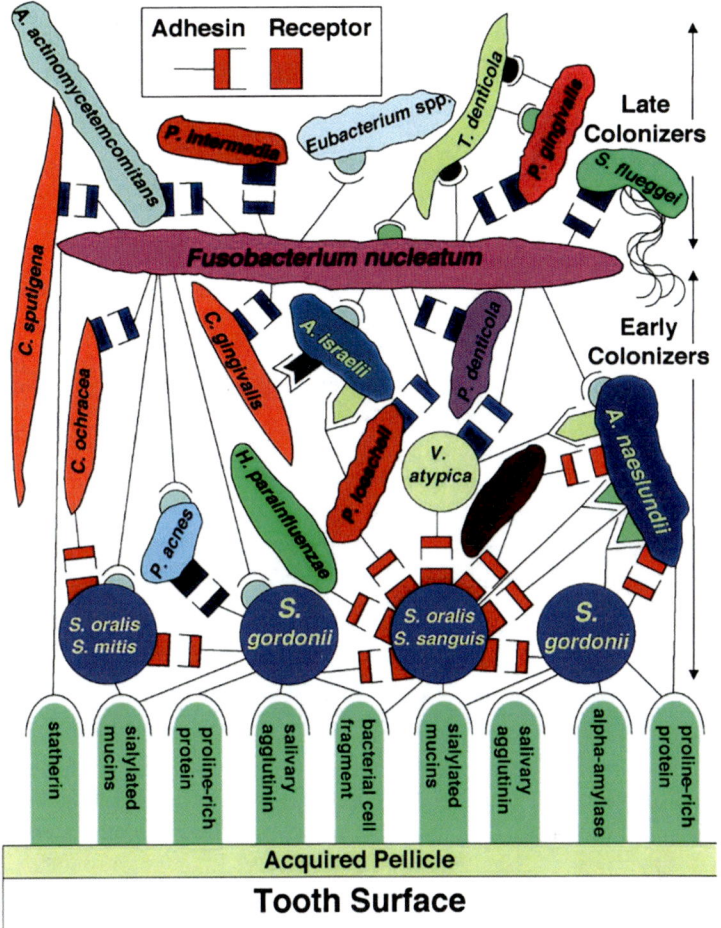

Fig. 1 Dental biofilm formation follows a defined sequence of events, and the biofilm has a well-defined architecture. Model of oral biofilm showing the dynamics of formation from the bottom with oral streptococci to the top with a more mature biofilm and the presence of gram-negative bacteria. Species–species interactions are mediated by specific molecules, and the final result is a biofilm with well-defined architecture (From Kolenbrander et al. 2002)

Mature oral biofilms are robust and resilient, acting as reservoirs of antibiotic resistance and virulence in deep periodontal pockets. Their uncontrolled growth eventually may lead to disease. A defining characteristic of the multispecies dental plaque biofilm, as well as other microbial biofilms, is communication either from cell to cell or from microcommunity to macrocommunity. This dynamic communication, called "quorum sensing," provides a mechanism for bacteria to monitor each other's presence and to modulate gene expression in response to changes in population density. With the advent of Next Generation Sequencing technologies, there has been a renewed interest on studying the composition of the oral microbiome (Peterson et al. 2013; Wade 2013; Zaura et al. 2009). These studies have confirmed the high diversity of the oral microbiome as well as the main traits in composition already described in previous studies where the predominant taxa in healthy samples belonged to Firmicutes (genus *Streptococcus*, family *Veillonellaceae*, genus *Granulicatella*), Proteobacteria (genus *Neisseria*, *Haemophilus*), Actinobacteria (genus *Corynebacterium*, *Rothia*, *Actinomyces*), while Bacteroidetes (genus *Prevotella*, *Capnocytophaga*, *Porphyromonas*) and Fusobacteria (genus *Fusobacterium*) appear to be more associated with disease.

Important factors for shaping the composition and structure of oral biofilms are the highly variable conditions in its environment. Access to nutrients follows the daily intake by the host, with long periods of low concentrations of nutrients followed by short periods of high concentrations of a large variety of different nutrients. These fluctuations have a major impact in selecting the species of bacteria that can survive and colonize the oral cavity.

In this brief review we present the main strategies that have been used to control the growth and composition of oral biofilms and other biofilms important in dental health and a final summary of future directions that could lead to better control of the communities present in the oral cavity.

2 Biofilm-Associated Oral Diseases

In healthy individuals the oral microbial biofilm is in homeostasis with its environment and the host. However, changes in response to variations in host physiology and environmental factors may lead to a shift from a health-associated biofilm to a disease-associated biofilm. The nature of these changes is not completely understood and is the focus of an important part of research in oral microbiology studies.

Two of the most common human diseases, caries and periodontal disease, are the result of a healthy microbial biofilm becoming a dysbiotic one due to mechanisms not completely understood. These are special cases of infectious diseases in that the origin of the infection is not an exogenous organism but a commensal organism that somehow overgrows and modifies the features of the microbial biofilm to its advantage.

Dental caries is caused by an overgrowth of acidogenic organisms usually linked to a diet rich in fermentable carbohydrates that can be used as nutrients by the microorganisms growing in the oral biofilm, resulting in the release of lactic acid to the surface of the tooth and leading to the formation of cavities characteristic of dental caries. Most cases of caries can be associated with the presence of high numbers of *Streptococcus mutans* or *Lactobacillus* as the main organisms responsible for the production of acid on the teeth (Loesche et al. 1975; Takahashi and Nyvad 2011). Nonetheless, the colonization of *S. mutans* is the result of ecological changes in the environment and shifts in the structure of the biofilm. Among important virulence factors of this pathogen, its ability to form and sustain a biofilm is vital not only to its survival and persistence in the oral cavity but also for its pathogenicity as well. Under low sugar concentrations, acidification is rare and short-lived but if high sugar concentrations persist the acidogenicity of the environment increases, selecting for aciduric strains. *S. mutans* and *Streptococcus gordonii*, a non-cariogenic streptococci, are found in inverse proportion in the oral biofilm (Loesche et al. 1975). *S. gordonii* can antagonize the colonization of *S. mutans* by inactivating competence stimulating peptide (CSP) and thus maintaining a healthy biofilm. Under prolonged acidogenic conditions, mutans streptococci, lactobacilli, as well as aciduric strains of non-mutans streptococci, *Actinomyces*, bifidobacteria, and yeast become dominant and lead to the establishment of a cariogenic biofilm that can cause disease (Takahashi and Nyvad 2011).

The other most common oral disease, periodontitis, is responsible for half of all tooth loss in adults and is a widespread and serious health condition that occurs in moderate form in 39 % of American adults and in severe form in 9 % of adults, with prevalence of 70 % in adults older than 65 (Eke et al. 2012). Periodontal disease is a polymicrobial bacterial biofilm-mediated pathology that leads to a progressive loss of the bone which, if left untreated, results in loosening and eventual tooth loss (Albandar et al. 1999; Oliver et al. 1998). Löe and collaborators demonstrated the role that accumulation of dental plaque has as the etiological agent of gingivitis, the first stage in developing periodontitis (Lang et al. 1973). Thus, controlling the growth of dental plaque is a key element in maintaining dental health (Listgarten 1988).

Periodontitis is triggered mainly by the presence of a complex microbial biofilm that colonizes the sulcular space between the tooth surface and the gingival margin. This biofilm undergoes a change in composition from health to the most severe forms of periodontitis. As briefly described above, during the succession of colonization of the teeth, the early colonizers are predominantly Gram-positive and later shift to a more Gram-negative community. Socransky, Haffajee, and colleagues defined the organisms within the subgingival microbiota, placing them in five "complexes." This concept emphasized that microorganisms create their own habitat, interact with each other, and are implicated in disease severity (Socransky et al. 1998; Socransky and Haffajee 2005; Haffajee et al. 2008) (Fig. 2). The organisms in the plaque reflected the environmental conditions. The most virulent combinations were strict anaerobes, and the less virulent microorganisms thrived in a relatively low-oxygen (microaerophilic) environment. In a detailed analysis using a checkerboard DNA–DNA hybridization approach of more than 13,000 subgingival samples from nearly 200 adults, Socransky and colleagues demonstrated that certain bacterial complexes were associated with either health or disease (Socransky et al. 1998). The presence of certain complexes such as the "red complex" (*Porphyromonas gingivalis*, *Tannerella forsythia*, and *Treponema denticola*) were associated more commonly with clinical indicators of periodontal diseases and were detected rarely in the absence of bacteria from other complexes.

Finally, though it is not a disease, halitosis or bad breath is a common medical and social problem caused in most cases by the resident biofilm in the dorsal surface of the tongue (Zalewska et al. 2012). These bacteria produce volatile sulfur compounds (VSC) that are the cause of halitosis.

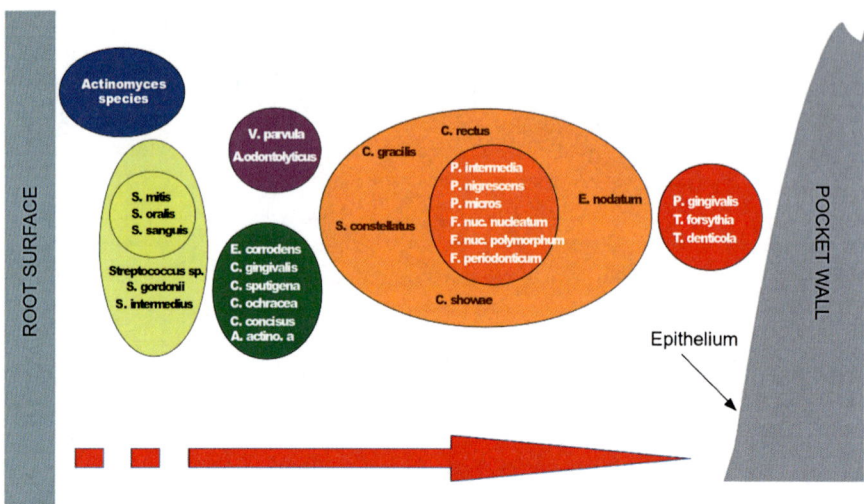

Fig. 2 Different bacterial species from subgingival plaque samples appear together forming different well-defined clusters. Figure shows the different clusters or "complexes" of bacteria that appear in association in subgingival plaque as defined by Socransky et al. (1998). The red cluster always appears associated with severe periodontitis and not in healthy samples. The *red arrow* shows the succession of complexes from health to disease as well as their location in the periodontal pocket (Adapted from Socransky et al. 1998 with permissions)

The objective in the prevention of caries and periodontal diseases is therefore to maintain the equilibrium between the host and the resident microflora. Strategies may aim at reducing the total biofilm mass or the levels of specific groups of pathogens, but not in totally eradicating the biofilm.

3 Microbial Biofilm Formation and Contamination in Dental-Unit Water Systems

Dental Unit Water Systems (DUWS) deserve a special consideration when studying biofilm control methods in oral health. In dentistry, dental chair units (DCU) are equipped with complex networks of plastic pipes that supply water to the DCU instruments and constitute an ideal environment for the growth of biofilms, especially of bacterial biofilms. The first reports on DUWS contamination were published in the early 1960s. Since then a large number of publications reporting DUWS contamination have populated the specialized literature (Coleman et al. 2009; O'Donnell et al. 2011).

Although the number of bacteria in water supplies is relatively low, it is common for them to be able to attach to the surface of the plastic tubing system of the DUWS. Once they attach they start secreting exopolysaccharides that protect them against the action of antimicrobial agents, facilitating the establishment of a mature biofilm. The fact that water is not constantly running and could be stagnant for long periods of time further increases the odds of a biofilm growing in the DUWS.

The initial colonizers, mostly environmental Gram-negative heterotrophic bacteria, are in general not pathogenic organisms and they are not considered a public health threat. The real problem resides in the fact that important pathogens can attach to this nonpathogenic biofilm and colonize it, creating a focus of infection due to the resilience of biofilms. Known human bacterial pathogens recovered from DUWS include *Pseudomonas* species, particularly *P. aeruginosa, Legionella* species, particularly *L. pneumophila*, and nontuberculosis mycobacterial species (Coleman et al. 2009; O'Donnell et al. 2011). Moreover, DUWS output water can also be a major source of bacterial endotoxins released from the cell walls of Gram-negative bacteria, which can create serious health problems in certain groups of patients (e.g., asthmatic patients) and stimulate the release of pro-inflammatory cytokines in gingival tissue during oral surgery.

Nonchemical methods to control bacterial growth in DUWS, such as flushing or adding filters to the tubing, can be useful to reduce the bacterial content on the output water but have no effect on the bacterial community growing in the biofilm. If a pathogen has colonized the pipe system of a DUWS, these methods do not have any effect in controlling its presence, and the DUWS should be treated with specific chemicals that also target bacteria growing in biofilms.

4 Biofilm Control Strategies in Oral Health

In the following sections we present a brief overview of different strategies for controlling biofilms important to oral health in a broad sense. This discussion will not just consider methods that completely remove the biofilm but rather, in the case of the oral biofilm, methods that control the biofilm to restore homeostasis, which should be the ultimate goal of all these treatments. The breath of approaches is summarized in Table 1. We divided the different approaches into three main categories: physical, chemical, and biological control of oral biofilms. However, we have not included other strategies that may restore biofilm homeostasis but do not use a direct targeting of the biofilm, such as changing diet habits to a low-sugar intake, which may be the best prevention of caries but does not specifically target the oral biofilm to prevent disease. In some cases the distinction between physical, chemical, or biological treatments is difficult to delimit. For instance, tooth brushing represents a physical control of the oral biofilm but is almost always done in the presence of some antimicrobial agent contained in the toothpaste. In those cases where delimiting the nature of the approach is blurry, we discuss their properties together.

Table 1 Summary of methods to control biofilm growth in oral health

Physical methods	Comments	References
Mechanical plaque control: tooth brushing and flossing	In combination with toothpaste is the most common way of controlling dental plaque	Andlaw, (1978), Barbier et al. (2010), Iacono et al. (1998), Mandel (1994)
Scaling and root planing	Removal of plaque and calculus to treat periodontitis	
Ultrasonic debridement	Professional removal of plaque and calculus	
Photodynamic therapy	Use of a chemical compound called "photosensitizer" which absorbs light. It is preferentially taken up by bacteria and subsequently activated by light of the appropriate wavelength	Soukos and Goodson (2011), Vlacic et al. (2007), Williams et al. (2004)
Nanoparticles	Used in photodynamic therapy to encapsulate the "photosensitizer" or to deliver antimicrobial compounds	Allaker (2010), Hannig and Hannig (2010), Hetrick et al. (2009)
Chemical methods		
Antiseptics	See Table 3	
Antibiotics	Penicillin (Amoxicillin Metronidazole Tetracycline Azithromycin Ciprofloxacine	Leszczyńska et al. (2011), Payne and Golub (2011), Rams et al. (2013), Sgolastra et al. (2011), Slots (2012)
Antimicrobial peptides	A special class of broad spectrum antibiotics that have attracted a great interest to control oral biofilms	Eckert et al. (2006), Folkesson et al. (2008), Helmerhorst et al. (1999), Lucchese et al. (2012)
Natural products	Extracts containing plant polyphenols	Ferrazzano et al. (2011), Hamilton-Miller (2001), Hannig et al. (2008), Löhr et al. (2011), Tagashira et al. (1997), Xu et al. (2011), Yano et al. (2012), Yoo et al. (2011)
Quorum sensing inhibitors	Quorum sensing is a complex regulatory process dependent on bacterial cell density that controls biofilm formation. Inhibition of the process will disturb the oral biofilm	Asahi et al. (2010), He et al. (2012), Jang et al. (2013), LoVetri and Madhyastha (2010), Romero et al. (2012)

Biological methods		
Vaccination		Culshaw et al. (2007), Liu et al. (2010), Niu et al. (2009), O-Brien-Simpson et al. (2003), Polak et al. (2010), Rajapakse et al. (2002), Russell et al. (2004), Shi et al. (2012), Sun et al. (2012), Takamatsu-Matsushita et al. (1996), Yu et al. (2011), Zhang et al. (2009)
Probiotics	Live microorganisms administered in adequate amounts conferring beneficial health effects on the host. Usually are *Lactobacillus* or *Bifidobacterium* species	Hillman (2002), Iwamoto et al. (2010), Jindal et al. (2011), Juneja and Kakade (2012), Mayanagi et al. (2009), Näse et al. (2001), Vivekananda et al. (2010)

5 Physical Methods to Control Oral Biofilms

5.1 Mechanical Plaque Control

One of the most commonly used, simpler and more efficient strategies for oral biofilm control is the physical removal of dental plaque by mechanical methods. Mechanical plaque control is highly effective for the prevention and control of periodontal disease, but it requires a well motivated patient who uses the devices in a proper fashion for a sufficient duration of time and with adequate frequency, which is seldom the case. The toothbrush is the most effective device for the removal of dental plaque without requiring professional cleaning. Flossing and interdental cleaning with brushes can remove plaque from proximal tooth surfaces, and there is evidence that it can reduce caries (Andlaw 1978) and gingivitis incidence (Iacono et al. 1998). Tooth brushing is usually accompanied by the use of a toothpaste that contains an antimicrobial agent. Common antimicrobial agents added to toothpastes are fluoride salts and triclosan. The efficacy of fluoride toothpaste in reducing dental caries is well established (Andlaw 1978; Barbier et al. 2010), while triclosan has been proven effective in controlling plaque growth (Mandel 1994; Phan and Marquis 2006; Teles and Teles 2009). Triclosan is a polychloro phenoxy phenol that inhibits fatty acid biosynthesis in bacteria and is used as a disinfectant in a large variety of products (Wright and Reynolds 2007).

Even with good oral hygiene, accumulation and mineralization of plaque can occur. In these cases professional cleaning is necessary and is usually performed using manual, sonic, or ultrasonic scalers to remove calculus and plaque to maintain oral health.

Once a pathogenic biofilm has colonized the teeth, causing either caries or periodontitis, dental plaque removal has to be performed by specialized professionals to guarantee the complete removal of the pathogenic community. Common procedures used in treating periodontitis are scaling and root planing. This consists of removal of plaque and calculus inside the periodontal pocket, between the gums and the teeth, eliminating as much as possible of the oral community, thus restarting the process of colonization by health-associated organisms.

Although these mechanical methods are highly efficient in controlling oral biofilms, they are hampered by the fact that few people follow a thorough toothbrush protocol that guarantees removal of most of the accumulated dental plaque, especially in interdental sites, the elimination of the oral biofilm is extremely difficult. The development of alternative antibacterial therapeutic strategies, which complement mechanical cleaning with other methods that control microbial growth in the oral cavity, therefore becomes important.

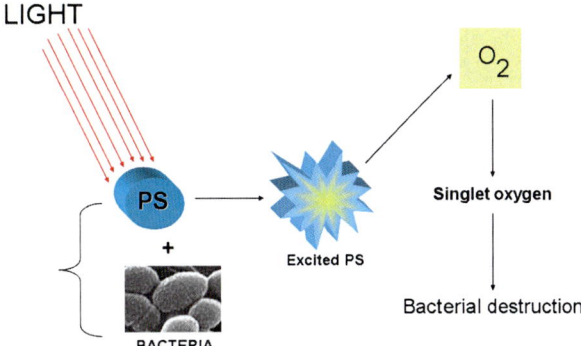

Fig. 3 Mechanism of photodynamic therapy. A photosensitizer is taken up by bacteria and gets activated by the appropriate wavelength. The photosensitizer transfers the energy to molecular oxygen and generates free radicals that will kill bacteria that have incorporated the photosensitizer (Adapted from Soukos and Goodson (2011) with permissions)

5.2 Photodynamic Therapy

Photodynamic therapy refers to the use of a chemical compound called a photosensitizer, which absorbs light and is preferentially taken up by bacteria and subsequently activated by light of the appropriate wavelength in the presence of oxygen to generate oxygen free radicals that are toxic to microorganisms (Soukos and Goodson 2011) (Fig. 3). One of the advantages of using photodynamic therapy is that, because of the molecular nature of the oxygen free radicals, it is unlikely that microorganisms will develop resistance to the cytotoxic action of these compounds. Photodynamic therapy has emerged as an alternative to antimicrobial regimes designed to complement mechanical methods in eliminating pathogenic components of the biofilm. Unlike mechanical methods that do not discriminate which organisms are removed, photodynamic therapy targets the organisms that take up the phothosensitizer but not the rest of the biofilm, thus allowing for a more targeted control of the oral biofilm. For a thorough review of photosensitizers and their clinical use in oral health, I would address the reader to the review by Soukos and Goodson on the use of photodynamic therapy to control oral biofilms in disease (Soukos and Goodson 2011).

Several laboratories have demonstrated the susceptibility of cariogenic bacteria, both in suspension and biofilms, to photodynamic therapy using toluidine blue O or disulfonated aluminum phthalocyanine (AlPcS2) as photosensitizers. Toluidine blue O-induced photodynamic therapy was able to achieve a 10-fold reduction of *S. mutans* when the organism was embedded in a collagen matrix mimicking carious dentin or present in decayed teeth (Williams et al. 2004). Moreover, the combined application of photodynamic therapy and casein phosphopeptide-amorphous calcium phosphate, a compound with established remineralization capabilities (Reynolds 1997), proved to be a successful treatment approach in removing the cariogenic bacteria and arresting root surface caries in vivo (Vlacic et al. 2007).

Organisms associated with periodontal disease have also been targeted by photodynamic therapy strategies, especially to reduce their number after scaling and root planing. Methylene blue has been extensively used as a photosensitizer, being applied directly to the dental pockets and exposed to red light via a fiber optic. Other photosensitizers have also been used in experimental settings on planktonic organisms, biofilms, and model animals. Although, in general there is a decrease in the number of bacteria, the elimination of periodontal pathogens is not complete and the reduction of total biofilm load is limited.

Finally, an exciting line of research in phototherapy is based on the observation that some organisms in the mouth already have "natural" photosensitizers in their cells; hence, phototherapy can be applied without the addition of any chemicals. Thus, oral black-pigmented bacteria (species of *Porphyromonas* and *Prevotella*), which are important organisms associated with periodontal disease, have endogenous porphyrins that when exposed to blue light release oxygen radicals that kill them. These kinds of approaches are advantageous because they are less aggressive and specifically target the group of bacteria associated with disease, while leaving the rest of the biofilm intact and thus shifting the composition from a disease-associated biofilm to a health-associated one. Moreover, the wavelengths used in phototherapy are not harmful to host cells (Soukos and Goodson 2011).

5.3 Nanoparticles and the Control of Oral Biofilms

In recent years bio-nanotechnology has increasingly been in the spotlight as a new way of treating medical conditions, either as a more specific delivery system for drugs or as a technology that allows for the creation of better biomaterials to be used in medicine. Nanotechnology refers broadly to a field of applied science and technology whose unifying theme is the control of matter on the atomic and molecular scale, between approximately 1 and 100 nm. One key element in designing nanoparticles is that the materials used for their synthesis must be safe to use in the human body, which limits the number of compounds available.

In photodynamic therapy, nanoparticles made of the biodegradable polymer poly(D,L-lactide-co-glycolide) (PLGA) have been used to encapsulate different photosensitizers and deliver them to the pocket before shining it with light (Allaker 2010). The use of nanoparticles solves one of the major problems of using photosensitizers in a clinical setting, which is their poor penetration in to the biofilm, thus reducing the effectiveness of the treatment. Encapsulating these photosensitizers in nanoparticles facilitates their penetration and increases the phototoxicity of the treatment and the reduction of oral biofilm biomass.

Another use of nanoparticles is in dental materials with antimicrobial activity such as filling materials, cements, sealants, materials for temporary restorations, coating materials, and adhesives. These dental materials contain a compound linked to the inert matrix that has antimicrobial activity and will kill bacteria that try to colonize. Among the materials used for this purpose, silver and copper have received the most attention. For centuries it has been known that certain metals have antibacterial activity, although their mechanisms of action are not completely understood. Nonetheless, studies have shown that the positive charge on the metal ion is critical for antimicrobial activity, allowing for the electrostatic attraction between the negative charge of the bacterial cell membrane and positively charged nanoparticles. In terms of the molecular mechanisms of inhibitory action of silver ions on microorganisms, it is known that DNA loses its ability to replicate in the presence of these ions (Feng et al. 2000), and the expression of ribosomal subunit proteins and other cellular proteins and enzymes necessary for ATP production becomes inactivated (Yamanaka et al. 2005). Most of the experimental results involving metal nanoparticles have been obtained in vitro, and biosafety concerns should be addressed before they can be used in vivo.

Other materials with better biocompatibility have been used to produce nanoparticles that control oral biofilms. Yudovin-Farber et al. have recently developed quaternary ammonium poly(ethylene imine) (QA-PEI) nanoparticles as an antimicrobial to incorporate into restorative composite resins (Yudovin-Farber et al. 2008). This may have distinct advantages over the composite resins currently used to restore hard tissues, which are known for allowing the buildup of biofilms on both teeth and the restorative material. Particles based upon the element silicon, for the rapid delivery of antimicrobial and anti-adhesive capabilities to the desired site within the oral cavity, have received much attention for their physical properties. The use of silica nanoparticles to polish the tooth surface may help protect against damage by cariogenic bacteria, presumably because the bacteria can be removed more easily. This effect is due to the forces of interaction, which are weaker between an organic substance and flat surfaces in comparison with rough corrugated surfaces. Other novel systems based upon silica have been investigated with respect to the control of oral biofilms. The use of nitric oxide (NO)-releasing silica nanoparticles to kill biofilm-based microbial cells has recently been reported (Hetrick et al. 2009). Finally, the application of hydroxyapatite nanoparticles has been shown to affect oral biofilm formation and provide a remineralization capability (Hannig and Hannig 2010).

6 Chemical Methods to Control Oral Biofilms

Chemical control is the most common approach to control dental biofilms after mechanical plaque removal. The compounds used in biofilm control are designed to prevent the formation of the biofilm and/or to remove an established biofilm. As mentioned previously, mechanical methods are not enough to control the growth of oral health related biofilms, hence the need for complementing mechanical removal with other approaches. In the case of DUWS, physical methods are not effective against the established biofilm and only by using chemical compounds can the biofilm be controlled. For treatment of DUWS, biosafety issues are not an important concern and toxic compounds have been used to remove biofilms. However, in the treatment of human patients, biosafety of the products used is one main consideration in devising a strategy for controlling biofilms, limiting the number of options to chemical compounds with low or no toxicity.

6.1 Chemical Control of Biofilms in DUWS

Biofilm control of DUWS deserves a separate section given that the list of chemical agents used includes compounds that cannot be used as antiplaque agents due to their toxicity. Table 2 is based on O'Donnell et al. (2011) and shows a summary of compounds that have been used for controlling biofilms in DUWS. In all cases these compounds minimize the contamination of output water but not all of them have an effect on biofilm removal. Of all these compounds, chlorhexidine gluconate, glutaraldehyde, sodium hypochlorite, hydrogen peroxide, hydrogen peroxide and silver, alkaline peroxide, sodium fluoride and EDTA showed efficacy in controlling biofilms with different degrees of success (O'Donnell et al. 2011). However, once the biofilm is established it is almost impossible to completely remove it from the water system. A more efficient approach would be to devise materials for manufacturing pipes, similar to what we described in the previous section on nanoparticles, which contain compounds that prevent or reduce the formation of the bacterial biofilm.

6.2 Use of Antiseptics in Plaque Control

Many chemicals with antimicrobial properties have been used historically to prevent and treat oral diseases. However, it was not until the 1960s, with progress in understanding the bacterial etiology of caries and periodontal diseases, that the interest in antimicrobials regained momentum. In a series of seminal studies, Löe et al. demonstrated that chlorhexidine mouth rinses could inhibit the development of plaque and gingivitis even in the absence of oral hygiene (Löe and Schiott 1970).

Table 2 Chemical agents used to treat contamination of Dental Unit Water Systems (DUWS)

Compound	Biofilm removal	Output water contamination	Treatment
Chlorexidine gluconate, chlorhexidine gluconate, and alcohol	Variable	Effective	Intermittent
Activated chlorine dioxide	Not effective	Effective	Intermittent
Chlorine dioxide and sodium phosphate mouth rinse	Not effective	Effective	Residual or continuous
Glutaraldehyde, glutaraldehyde, and quaternary ammonium salts (very toxic)	Variable	Effective	Intermittent
Sodium hypochlorite and citric acid	Not effective	Effective	Intermittent
Hydrogen peroxide	Effective	Effective	Intermittent
Hydrogen peroxide and silver			Residual or continuous
Alkaline peroxide			
Electrochemically activated solutions	Very effective	Very effective	Residual or continuous
Paracetic acid	Not effective	Not effective	Intermittent
Povidone-iodine	Not effective	Effective	Intermittent
Sodium fluoride	Partial elimination	Effective	Intermittent
Sodium perborate	Not effective	Variable	Intermittent
EDTA	Effective	Effective	Intermittent
Citric acid and sodium-p-toluol-sulfonechloramide and sodium EDTA	Not effective	Effective	Residual or continuous
Sodium-p-toluol-sulfonechloramide and sodium EDTA	Not effective	Effective	Residual or continuous
P-hydroxybenzoeicacidester, polyaminoprophylbiguanid, 1,2-prophyenglycol	Not effective	Effective	

Adapted from O'Donnell et al. (2011). *"Management of dental unit waterline biofilms in the 21st century"*

The list of antiplaque agents used in toothpastes and mouth rinses is long. Thus, for the purpose of this review, we will comment on the most commonly used antimicrobial compounds used in oral health products (Table 3).

As of today, the most effective antimicrobial agent for plaque control has been chlorhexidine (Baehni and Takeuchi 2003). Chlorhexidine is among the most tested compound, and its antiplaque properties are well known. One main feature of chlorhexidine that makes it so successful in controlling plaque is that it is adsorbed on to the enamel surface or the salivary pellicle on the teeth so that bacterial adhesion is inhibited (Pratten et al. 1998; Rölla and Melsen 1975). Moreover, once chlorhexidine attaches to the surface of the teeth it stays there for long periods of time before it is washed out by saliva flow, thus maintaining its antiplaque activity even long after the mouthwash has been used.

At low concentrations chlorhexidine is bacteriostatic against most oral bacteria and can interfere with the metabolism of oral bacteria by inhibiting sugar transport and acid production in cariogenic streptococci, various membrane functions in streptococci, and a major protease (gingipain) in the periodontal pathogen

Table 3 Antiplaque compounds used commonly in mouthwashes and toothpastes

	Mechanims of action	References
Chlorhexidine	Inhibits bacterial adhesion. Bacteriostatic at high concentrations. At sublethal concentrations inhibits: (a) sugar transport and acid production in cariogenic streptococci, (b) various membrane functions in streptococci, including inhibiting enzymes responsible for maintaining an appropriate intracellular pH, and (c) a major protease (gingipain) in the periodontal pathogen, *Porphyromonas gingivalis*	Baehni and Takeuchi (2003), Hope and Wilson (2004), Houari and Di Martino (2007), Löe and Schiott (1970), Modesto and Drake (2006), Pratten et al. (1998), Rölla and Melsen (1975)
Triclosan	Inhibits fatty acid biosynthesis and FabI-related enoyl-ACP reductase enzymes	Binney et al. (1997), Heath et al. (2001), Owens et al. (1997)
"Essential oils" (Listerine)	Bacterial cell wall destruction, bacterial enzymatic inhibition, and extraction of bacterial lipopolysaccharides	Leszczyńska et al. (2011), Mandel (1994)
Hexetidine	Competitive action with thiamine	Afennich et al. (2011)
Cetylpyridinium chloride (CPC)	Membrane destabilization	Leszczyńska et al. (2011)
Amine fluoride/ stannous fluoride	Stannous ions bind to lipotechoic acid on the surface of Gram-positive bacteria and reverse the charge on the surface of the cell or tin ions displace calcium ions, altering enzyme functions in the cell	Bansal et al. (1990), Bullock et al. (1989), Kay and Wilson (1988), Mayhew and Brown (1981)

Porphyromonas gingivalis. In several biofilm models, chlorhexidine was shown to inhibit bacterial growth and biofilm formation (Houari and Di Martino 2007; Modesto and Drake 2006). At high concentrations, the agent is bactericidal and acts as a detergent by damaging the bacterial cell membrane and interfering with biofilm formation (Hope and Wilson 2004). However, because of its attachment to the surface of teeth, chlorhexidine has one negative effect that prevents its daily use, namely a brown-yellow staining on the teeth if it is used for long periods of time.

Triclosan is another widely used antimicrobial in oral health products to prevent plaque formation. Several large clinical trials have shown that toothpastes containing triclosan and zinc citrate significantly reduced plaque and gingival scores (Binney et al. 1997; Owens et al. 1997). The mechanism of action of triclosan is by inhibiting FabI-related enoyl-ACP reductases, key enzymes in fatty acid biosynthesis essential for membrane formation (Heath et al. 2001; Moir 2005).

"Listerine" is the commercial name of a commonly used mouth rinse that contains "essential oils" as active antimicrobial compounds. In its formulation we found thymol (0.06 %), eucalyptol (0.09 %), methyl salicylate (0.06 %) and menthol (0.04 %). Listerine acts on the biofilm by destabilizing the bacterial cell wall (Leszczyńska et al. 2011).

Hexetidine is also a common antiplaque agent added to mouthwashes. Hexetidine is a very safe oral antiseptic with broad antibacterial and antifungal activity in vivo and in vitro. Hexetidine has lower antiplaque activity than chlorhexidine but without the negative staining effect (Afennich et al. 2011). Given hexetidine is a pyrimidine derivative, the most likely mechanism of action is by exerting a competitive action with thiamine.

Cetylpyridinium chloride (CPC) is a quaternary ammonium compound with broad spectrum antibacterial activity that adsorbs readily to oral surfaces. This molecule has both hydrophilic and hydrophobic groups, providing the possibility for hydrophilic and hydrophobic interactions. The positively charged hydrophilic region of the CPC molecule has high binding affinity for bacterial cells whose surface has net negative charge. The strong positive charge and hydrophobic region of CPC enables the compound to interact with the microbial cell surface and integrate into the cytoplasmic membrane. As a result of this interaction, there is disruption of membrane integrity resulting in cell death (Leszczyńska et al. 2011; Tattawasart et al. 2000).

Finally, both amine fluoride (Kay and Wilson 1988) and stannous fluoride (Mayhew and Brown 1981) possess bactericidal activity against oral bacteria. In addition, amine fluoride has been shown to inhibit the growth of mixed bacterial populations found in subgingival plaque (Bansal et al. 1990; Bullock et al. 1989). Additionally, stannous and amine fluorides can also inhibit the adhesion of *Streptococcus sanguis* to glass conditioned with either saliva or bovine serum albumin.

6.3 Antibiotics and the Oral Biofilm

The use of antibiotics in oral medicine has a long history, especially as prophylactic agents for preventive management. In addition, antibiotics are used for therapeutic reasons in cases where infections of oral hard and soft tissues, such as teeth and gingiva, cannot be controlled by local debridement and can spread to distant organs and therefore require supplemental therapy. Nonetheless, the use of antibiotics in dental practice has been restricted by clinicians to limit spreading antibiotic resistance among oral isolates. Moreover, bacteria growing in biofilms are more resistant to antibiotics than their planktonic counterparts. Antibiotic concentrations necessary to inhibit biofilms can be 10–1,000 times higher than those needed to inhibit bacteria growing planktonically (Simões 2011). The mechanism by which bacteria become more resistant in the biofilm is not completely understood and is likely multifactorial. Lewis has proposed that a fraction of the biofilm are persister or dormant variants of regular cells that have a reduced metabolism by a toxin–antitoxin mechanism and thus are less affected by the presence of antibiotics (Lewis 2010).

A wide variety of antibiotics have been used in oral health. The list comprises among others: amoxicillin, tetracycline, minocycline, doxycycline, azithromycin, clindamycin, and metronidazole (Leszczyńska et al. 2011). Amoxicillin is one of the most commonly used antibiotics. It is a semisynthetic penicillin with broad antimicrobial spectrum and is used in periodontology to fight some subgingival bacterial species. However, many bacteria isolated in subgingival plaque samples are resistant to amoxicillin (Handal et al. 2004; Rams et al. 2013; van Winkelhoff et al. 1997). Nonetheless, the combination of amoxicillin, metronidazole with scaling and root planing leads to a beneficial change in the composition of the subgingival microbiota by reducing the concentration of anaerobic pathogens such as *P. gingivalis* and *P. intermedia* and allowing the growth of host-compatible species (Mestnik et al. 2012; Socransky et al. 2013).

The combination ciprofloxacin–metronidazole has proven to be effective in elderly patients and patients that have enteric rods in their subgingival plaque (Slots 2012). Azithromycin is effective against Gram-negative aerobic and anaerobic bacteria, has a long half-life in periodontal tissues, and has been successfully used as an adjuvant in the treatment of chronic periodontitis (Muniz et al. 2013). Clindamycin is a pyranoside antibiotic that has been tested in several clinical studies. Nevertheless, resistance of oral microorganisms to clindamycin seems to be widespread (Skucaite et al. 2010), and thus it is only recommended in cases where there is intolerance to other antibiotics. Tetracycline and its homologues are commonly used in dental practice as a prophylactic agent and for treatment of oral infections (Payne and Golub 2011; Sgolastra et al. 2011). These compounds have antimicrobial and antiinflammatory activities. However, tetracyclines have a large number of adverse effects such as nausea, vomiting and have negative interaction with penicillins.

A more controlled delivery system for antibiotics, and by that matter antiseptics, is by using a local drug delivery system, where the active agent is placed directly into the diseased pocket to treat periodontitis (i.e., analogous to topical application). Using this system allows for a reduction in the total amount of antibiotic used, lowering the potential side effects, and also preventing the negative consequences of systemic antibiotics, e.g., disrupting the gut flora. These drug delivery systems include fibers, strips, gels, etc., with a polymer matrix that contains the compound to be released. Currently, there are commercial systems that contain tetracycline, doxycycline, metronidazole, minocycline, and azithromycin (Leszczyńska et al. 2011).

The main objection to the extensive use of antibiotics to control oral biofilms is the potential spread of antibiotic resistance and the public health implications of their uncontrolled use. The oral biofilm is a perfect setting for the transmission of antibiotic resistance genes due to its complexity in composition and the large number of bacteria that colonize it, especially in the case of the subgingival plaque (Mullany et al. 2012; Roberts and Mullany 2010).

Bacterial resistance to antibiotics is acquired by different mechanisms. Resistance to penicillin is primarily acquired by alteration of the penicillin-binding proteins and/or production of β-lactamases. However, a very small proportion of the subgingival microbiota are resistant to penicillins. Bacteria become resistant to tetracyclines or macrolides by limiting their access to the cell, by altering the ribosome in order to prevent effective binding of the drug, or by producing tetracycline/macrolide-inactivating enzymes (Soares et al. 2012) Periodontal pathogens are frequently resistant to these drugs. Moreover in the case of tetracyclines, their use could actually initiate the mobilization of elements carrying resistance genes (Salyers and Shoemaker 1997).

6.3.1 Antimicrobial Peptides

Antimicrobial peptides (AMPs) are a special class of broad spectrum antibiotics that has attracted a great deal of interest in controlling oral biofilms and have great potential as new therapeutic compounds against bacterial pathogens. They are naturally synthesized molecules produced by a wide range of organisms and also an important element of the innate immune system of eukaryotes. AMPs mount a rapid, nonspecific response against colonization by a pathogen, especially in the early stages of invasion. Their mechanism of action is based on the fact that mammalian cells have no net charge on their membranes, while bacterial cells have a negative net charge. This interaction causes the antimicrobial peptides to produce pores in the cell wall, thus causing bacterial cell death. There are also antimicrobial peptides whose targets are not bacteria. Antimicrobial peptides against virus, fungi, protists, parasites, and even against insects and tumor cells have been identified (Lucchese et al. 2012).

Gingival epithelial cells shape the oral microbiome by secreting antimicrobial peptides directly into the gingival crevice, which inhibits the growth of pathogens (da Silva et al. 2012). The oral epithelium produces two important AMPs: β-defensins and LL-37. Both are produced in the subgingival epithelium and may play an important role in the homeostasis of dental plaque. However, the activity of these peptides against bacteria growing in structured biofilms is limited (Folkesson et al. 2008), in part compromised by the production of exopolysaccharides by the biofilm and also by the emergence of resistance against AMPs.

To increase the effectiveness of peptides against oral biofilms, one strategy has been to design synthetic peptides, based on the knowledge gained using natural peptides, to improve their efficacy in controlling biofilms (Helmerhorst et al. 1999; Younson and Kelly 2004). Some of these synthetic peptides inhibit colonization of the biofilm by the pathogenic organism while others have direct antimicrobial activities killing the target of interest.

A critical early step in any bacterial infection is adherence of the pathogen to the host. Blocking this first step of colonization can prevent invasion of the plaque by the pathogen and thus maintain a healthy biofilm. One way to prevent this adhesion is to use adhesion epitopes to block receptor sites that pathogens utilize to adhere to

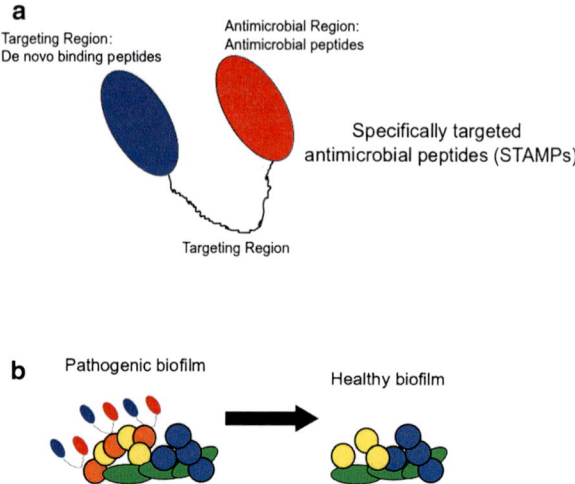

Fig. 4 Illustration of the basic structure and mechanism of action of the specifically targeted antimicrobial peptides (STAMPs). Adapted from Eckert et al. (2012). (**a**) Basic structure of STAMPs showing the targeting region specific for the targeted organisms and the antimicrobial region consisting of a peptide with killing activity. (**b**) STAMPs specifically target the pathogen of interest while leaving the healthy biofilm intact (Eckert et al. 2012)

other organisms in the biofilm. Most of these efforts have focused on preventing the attachment of *S. mutans* to control the cariogenic biofilm. In the case of *S. mutans* initial adherence to the tooth surface is mediated by an adhesin that is expressed on the surface of the bacterium, the streptococcal antigen I/II (SA I/II). A synthetic peptide (p1025) corresponding to residues 1025–1044 of the adhesin has been used to block adhesion of *S. mutans* to teeth but had no effect on adhesion and colonization of other health-associated organisms such as *Actinomyces* (Kelly et al. 1999; Younson and Kelly 2004).

Synthetic peptides can also be used to kill the pathogen of interest. Histatin-derived peptides have antibacterial activity in in vitro oral biofilm models as well as ex vivo treatment of plaque bacteria (Helmerhorst et al. 1999).

A more targeted approach with great potential has been developed by Eckert et al. (2006). This technique called Specifically Targeted Antimicrobial Peptides (STAMPs) is based on the fusion of a species-specific targeting peptide domain with a wide-spectrum antimicrobial peptide domain. The targeting domain provides specific binding to a selected pathogen and facilitates the targeted delivery of an attached antimicrobial peptide (Fig. 4).

6.4 Natural Products

The use of natural products to prevent or treat diseases dates back thousands of years in almost every civilization around the world. The term "natural product" refers to a substance produced by living organisms that has distinct pharmacological effects. Due to this broad definition natural products provide a diverse array of chemical structures and possess a plethora of biological activities. A number of natural products that possess the ability to inhibit or disperse bacterial biofilms have been used as the starting points for medicinal chemistry programs in which synthetic manipulation of the natural product scaffold has allowed for the design of more active compounds. Much of the natural product inspiration for these programs has come from compounds isolated from plants and marine organisms that use them as chemical warfare against other organisms. One of the main advantages of using natural products for biofilm control in oral formulations is their low or nonexistent toxicity.

In most cases, the active component in natural products used in oral health is polyphenol. Polyphenols constitute one of the most common groups of substances in plants and are considered secondary metabolites involved in chemical defense. The number of known plant polyphenols is enormous, including a wide variety of molecules that contain at least one aromatic ring with one or more hydroxyl groups in addition to other substitutions. A variety of potential mechanisms of action by which polyphenols exert antimicrobial effects have been reported (Ferrazzano et al. 2011; Furiga et al. 2008; Grenier and La 2011; Hannig et al. 2008). They have the ability to inactivate bacterial toxins, which has generated increasing interest because plant polyphenols could represent a new source of agents to fight antibiotic-resistant human pathogens. For instance, apple peel polyphenol-rich extract has been shown to inhibit vacuolation by the vacuolating bacterial toxin (VacA) of *Helicobacter pilori* (Pastene et al. 2010).

A large variety of polyphenols, from as diverse origins as tea, grape juice, cocoa, coffee, and red wine, have been shown to inhibit initial adherence to the tooth surface, especially in the case of the cariogenic bacterium *S. mutans* (Ferrazzano et al. 2011; Hannig et al. 2008). For example, *S. mutans* produces glucan that facilitates its attachment to teeth. Glycosyltransferase (GTF) is an essential enzyme in the production of glucan, and it is inhibited by polyphenols from tea, hampering the initial stages of biofilm colonization by *S. mutans* (Hamilton-Miller 2001). Polyphenols from apples also inhibit *S. mutans* GTFs. A high molecular weight hop bract polyphenol (HBP) also inhibited adherence of *S. mutans* via the action of GTFs involved in water-insoluble glucan synthesis, but did not suppress the growth or acid production of the bacteria (Tagashira et al. 1997). Other polyphenols with similar activities have been isolated form cranberries (Yoo et al. 2011), grapes (Yano et al. 2012), tea (Xu et al. 2011; Yano et al. 2012), cacao (Ito et al. 2003), and red wine (Furiga et al. 2008).

There are also reports that polyphenols inhibit adhesion of periodontal pathogens. Polyphenol-enriched extract from the plant *Myrothamnus flabellifolia* inhibited *P. gingivalis* adhesion and cell invasion by interaction with outer membrane proteins of the bacterial cell (Löhr et al. 2011). (−)-Epigallocatechin gallate (EGCg), the dominant component of tea polyphenols, inhibited the growth and adherence of *P. gingivalis* onto buccal epithelial cells (Sakanaka et al. 1996). Finally, polyphenols from cranberries inhibited *P. gingivalis* biofilm formation by an unknown mechanism (Yamanaka et al. 2007).

Natural compounds also have antibacterial activities. *Camellia sinensis* (used to make green tea and oolong tea) has been widely studied for its antibacterial activities against caries-related bacteria, using both in vitro and in vivo models (Ferrazzano et al. 2009; Hassani et al. 2008). Although the composition of green tea is complex, specific catechins from green tea have been associated with antibacterial activity against *S. mutans* and *Streptococcus sobrinus* (Nakahara et al. 1993). This inhibitory effect appears to be related to the presence of three hydroxy moieties 3', 4', and 5' on the B ring of the catechin and epicatechin molecular structure (Miyake et al. 2011). However, the exact mechanisms of action and their putative target(s) remain to be elucidated. Among the polyphenols from green tea with potential use in periodontal disease treatment, Epigallocatechin 3 gallate (EGCG) and Epicatechin 3 Gallate (ECG) are the most predominant catechins. These catechins have antioxidant, antimicrobial, anticollagenase, antimutagenic, and c hemopreventive properties (Venkateswara et al. 2011).

Several groups have identified natural plant-derived inhibitors effective on *P. gingivalis* proteases. More specifically, polyphenols isolated from cranberry and green tea were found to inhibit Arg-gingipain and Lys-gingipain produced by *P. gingivalis* (Yamanaka et al. 2007). These enzymes have been suggested to play multiple roles in the pathogenic process of periodontitis. Additionally, green tea polyphenols inhibit the production of toxic end metabolites by *P. gingivalis* such as n-butyric, phenylacetic, and propionic acid (Sakanaka and Okada 2004).

Finally, one interesting natural product with antimicrobial activity and a very different origin, which has been used experimentally to inhibit oral bacteria, is honey. Although the number of studies is small, honey has been shown to have antimicrobial activity against *S. mutans* (Ahmadi-Motamayel et al. 2013) and potential use in periodontal disease (Molan 2001).

6.5 Quorum Sensing

Quorum sensing (QS) is a complex regulatory process dependent on bacterial cell density (Miller and Bassler 2001) and is typically involved in the regulation of genes involved in biofilm maturation and maintenance (Hammer and Bassler 2003; Joo and Otto 2012). QS controls the production of virulence factors in both Gram-positive and Gram-negative pathogenic bacteria (Rutherford and Bassler 2012). Thus, inhibitors of QS, in addition to possessing anti-biofilm activity, could also counteract bacterial pathogenicity.

The majority of bacteria present in the oral biofilm can sense their microenvironment using QS mechanisms. In Gram-negative bacteria, autoinducers, the molecules used in QS, belong to the chemical class of the acyl-homoserine lactones (AHLs) (Fuqua and Greenberg 1998), in Gram-positive bacteria autoinducing peptides (AIPs) control QS mechanisms (Sturme et al. 2002), while a third class of autoinducers, autoinducer-2 (AI-2), is conserved in both Gram-negative and Gram-positive bacteria (Rickard et al. 2006). This latter autoinducer, AI-2, is especially important in multispecies biofilms such as dental plaque where Gram-negative and Gram-positive bacteria coexist. AI-2 promotes biofilm formation and maturation in multispecies communities. The *lux*S gene, responsible for the expression of AI-2, is conserved among many species of bacteria, including *S. mutans*, *S. gordonii*, *Streptococcus oralis*, *P. gingivalis*, *A. actinomycetemcomitans*, and other oral microorganisms (Frias et al. 2001; Jakubovics 2010).

The species-specific QS autoinducers *N*-acyl homoserine lactones (autoinducer-1) have not been identified in oral bacteria (Jakubovics and Kolenbrander 2010). However, many oral microorganisms produce and/or respond to the interspecies signal AI-2. AI-2 is the collective term given to a number of molecules that spontaneously form an equilibrium when 4,5-dihydroxy-2,3-pentanedione (DPD) is dissolved in water. Bacteria produce AI-2 during amino acid metabolism as a product of the enzyme encoded by the luxS gene. AI-2 plays an important role in regulating the essential activities of oral pathogens. AI-2 QS regulates iron acquisition in the periodontal pathogens *P. gingivalis* and *A. actinomycetemcomitans* (Shao and Demuth 2010) and modulates protease and haemagglutinin activities in *P. gingivalis* (Burgess et al. 2002). Additionally, AI-2 also regulates biofilm formation in oral pathogens. *A. actinomycetemcomitans lux*S mutants are capable of forming a mature biofilm, but they exhibit significantly lower total biomass and biofilm depth when compared with the wild-type strain (Shao and Demuth 2010). Similarly, *lux*S mutants of *S. mutans* form a defective biofilm compared to wild type due to a decrease in glycosyltransferase activity, suggesting that the activity of this enzyme is controlled by AI-2 (Huang et al. 2009; Yoshida et al. 2005).

QS inhibitors have the potential to control biofilm growth and maturation with the advantage of reducing the generation of mutants resistant to the treatment. The increasing interest in interfering with these signaling systems as a way of controlling biofilms (Quorum Quenching) is reflected by the increasing number of patents filed using this approach (more than 45 since 2009) (Romero et al. 2012).

Recently, it was shown that two QS inhibitors (5Z)-4-bromo-5-(bromomethylene)-2(5H)-furanone (furanone compound) and D-ribose inhibited dual biofilm formation between *Fusobacterium nucleatum* and members of the "red complex" (*Porphyromonas gingivalis*, *Treponema denticola*, and *Tannerella forsythia*) (Jang et al. 2013). *Fusobacterium nucleatum* is the major coaggregation bridge organism that links early colonizing commensals and late pathogenic colonizers in dental biofilms via the accretion of periodontopathogens from the "red complex." Disturbance of coaggregation and biofilm formation between these organisms may have an impact in controlling pathogenic biofilm. He and collaborators have shown that by using the synthetic QS inhibitor furanone C-30, biofilm formation by *Streptococcus mutans* was significantly reduced (He et al. 2012). Although there are no reports that suggest production of AHLs by *P. gingivalis*, it has been shown that synthetic *N*-acyl HSL analogues can inhibit biofilm formation by this organism (Asahi et al. 2010). The mechanisms by which these analogues interfere with biofilm formation in *P. gingivalis* are still unknown.

Although *S. mutans* possesses the AI-2 system (Merritt et al. 2003), its primary QS system is comprised of the Competence Stimulating Peptide (CSP) and the ComD/ComE two-component signal transduction system. In addition to biofilm formation, the CSP-mediated QS system in *S. mutans* also affects its acidogenicity, aciduricity, genetic transformation, and bacteriocin production (Senadheera and Cvitkovitch 2008). CSP is synthesized in the cell and released into the extracellular medium. When the bacterial density increases, CSP molecules in the external environment reach a threshold concentration and, as with other two component systems, the expression profiles of bacteria producing the CSP molecules is modulated. When used at high concentrations, CSP can actually contribute to cell death in *S. mutans*. Analogues of this QS peptide decrease biofilm formation of various *Streptococcus* species and can potentially be used to control cariogenic biofilms (LoVetri and Madhyastha 2010).

7 Biological Methods to Control Oral Biofilms

7.1 Vaccination

As antibacterial agents can be rendered ineffective by the development of resistance in target organisms, it can be difficult to maintain a therapeutic concentration in the oral cavity and can be toxic to the host. Thus, there is a need to develop alternative approaches for treatment. Vaccines are proposed to be an effective therapy to control oral biofilm and target oral pathogens. Although viruses have been implicated in some cases of periodontal disease (Grinde and Olsen 2010; Slots 2010) we will only focus on the use of vaccines against the bacterial fraction of the oral biofilm.

An example of a recent vaccine whose strategy is to control oral biofilm rather than target specific pathogens is a vaccine against the oral bacterium *Fusobacterium nucleatum*. As previously mentioned, this organism plays a central bridging role in the structure of pathogenic periodontal biofilms (Kolenbrander 2000; Kolenbrander et al. 2002). An antibody generated against the FomA outer membrane protein of *F. nucleatum* significantly abrogated bacterial co-aggregation, biofilm formation, and the production of volatile sulfur compounds that cause halitosis (Liu et al. 2010).

Numerous studies have documented effective vaccination against oral pathogens. Most of these vaccines were developed based on the identification of virulence factors that stimulate the induction of salivary immunoglobulin A antibody responses. Primary targets have been cell-surface fibrillar proteins, which mediate adherence to the salivary pellicle, and GFT enzymes, which synthesize adhesive glucans and allow for microbial accumulation. Immunization when infants are about 1 year old can establish effective immunity against ensuing colonization attempts by mutans streptococci. Intranasal vaccines, targeting the flagellin of *S. mutans*, have shown reduction of caries in animals (Shi et al. 2012; Sun et al. 2012). Streptococcal GTF have also been demonstrated to be effective components of dental caries vaccines, and different peptides from these proteins have been used to design them (Culshaw et al. 2007; Russell et al. 2004).

Another approach in designing vaccines against caries has been the use of DNA vaccines. These vaccines are composed of bacterial plasmids that are delivered to the host, after which some cells uptake the plasmids and begin producing the antigen of interest. Expression plasmid DNA contains antigen-encoding sequences cloned under heterologous promoter control that are delivered by different techniques and lead to antigen expression in transfected cells in vivo (Williams et al. 2009). Several DNA vaccines have been designed for use in caries therapy. A fusion anti-caries DNA vaccine (pGJA-P/VAX) that encodes two important antigenic domains (PAc and GLU) of *S. mutans* was successful in reducing the levels of dental caries caused by *S. mutans* in gnotobiotic animals (Niu et al. 2009). Two other vaccines, pGJGAC/VAX and pGJGA-5C/VAX, constructed by cloning different sections of the catalytic regions of GTFs, protected against cariogenic bacteria, and specifically against *S. sobrinus* (Sun et al. 2009).

Most immunization approaches, both active and passive, against periodontitis have been focused on *P. gingivalis* and *A. actinomycetemcomitans*. As noted above, *P. gingivalis* has been implicated as a major periodontopathogen in human periodontitis. In this context, it has developed a variety of survival strategies enabling it to evade host defense mechanisms. Virulence factors of *P. gingivalis* include cysteine proteases, fimbriae, capsular polysaccharide (CPS), lipopolysaccharide, and outer membrane vesicles (Holt et al. 1999).

There have also been attempts to use inactivated whole cells of *P. gingivalis* as antigens in vaccine development. In a recent study, a mixed vaccine of whole *P. gingivalis* and *F. nucleatum* cells suppressed inflammation but failed to prevent disease progression in an animal model system (Polak et al. 2010).

RgpA and Kgp are polyprotein proteinases with C-terminal adhesin domains that are proteolytically processed. An RgpA-Kgp complex vaccine produced a high antibody titer in animals which protected them from a *P. gingivalis* challenge (O-Brien-Simpson et al. 2003; Rajapakse et al. 2002). FimA and the 40-kDa outer membrane protein of *P. gingivalis* have also been used in designing vaccines against periodontitis (Lucchese et al. 2013; Namikoshi et al. 2003).

As in the case of caries vaccines, a new generation of DNA vaccines have also been devised against FimA (Yu et al. 2011) and the 40-kDa outer membrane (Zhang et al. 2009) of *P. gingivalis,* eliciting a protective immune response.

A. actinomycetemcomitans is considered another important pathogen in human periodontal disease, especially in the localized form of juvenile aggressive periodontitis. Honma et al. demonstrated that high salivary IgA response could be induced against a fimbrial synthetic peptide by intranasal mucosal immunization (Honma et al. 1999). Harano et al. using olegopeptides from frimbriae, prepared an antiserum that blocked the adhesion of the organism to saliva-coated hydroxyapatite beads, to buccal epithelial cells, and to a fibroblast cell line (Harano et al. 1995). Also, subcutaneous and intranasal immunization of mice with a capsular serotype b-specific polysaccharide antigen of *A. actinomycetemcomitans* resulted in specific antibodies that efficiently opsonized the organism (Takamatsu-Matsushita et al. 1996). Antibodies elicited against fimbriae composed of a 54 kDa protein derived from *A. actinomycetemcomitans* protected against continued infection by this microorganism. IgG responses to fimbriae antigen elicited by the initial contact with *A. actinomycetemcomitans* may play an important role for eliminating organisms from the periodontal pockets of patients harboring high IgG antibody against these antigens (Ishikawa et al. 1997). However, relatively few studies have been conducted on developing vaccines against *A. actinomycetemcomitans*.

7.2 Probiotics

Total elimination of the oral biofilm is neither desirable nor possible, and therefore replacement strategies with "probiotics" have been the subject of extensive research. Recently, probiotics have been gaining interest for alleviating oral and other health disorders. The Food and Agriculture Organization of the United Nations has defined probiotics as "live microorganisms administered in adequate amounts conferring beneficial health effect on the host". Probiotics are naturally found in food products such as yogurt and milk and have yet to cause the serious side effects that are associated with currently available antimicrobials.

Lactobacillus and *Bifidobacterium* are the most commonly used genus of bacteria in probiotic formulations and they have been used with positive results for a large number of different health disorders (Girardin and Seidman 2011; Kruis 2012; Travers et al. 2011; Twetman and Stecksén-Blicks 2008; Uccello et al. 2012).

The balance between beneficial and pathogenic bacteria is essential in order to maintain oral health. Therefore, the oral cavity has recently been suggested as a relevant target for probiotic applications. Although it is a promising concept, there is still not conclusive evidence that current probiotics have any beneficial effect on oral diseases and further studies are needed to assess its value as a therapy.

One can conclude that there are three main goals that probiotics should achieve to prove successful as therapeutics for gingivitis and periodontitis. The first goal involves the modulation of the host's inflammatory processes. The second goal involves the reduction of plaque formation. Finally, the third goal is to reduce the presence and numbers of disease promoting microorganisms. Given our focus on biofilm control we are going to skip modulation of the host's inflammatory response and focus on the two latter goals.

One of the problems of probiotic therapy is colonization of the biofilm by the probiotic bacterium. Long-term establishment of probiotics in the oral biofilm is difficult, and detectable levels are commonly only found at the beginning of the treatment (Twetman and Stecksén-Blicks 2008). Probably, the most successful example of the use of probiotics in treating a disease is using "fecal transplantation" for treating recurrent *Clostridium difficile* infections (Aroniadis and Brandt 2013). Fecal microbiota transplantation (FMT) has been used as a treatment to reconstitute the normal microbial homeostasis after the gastrointestinal microbiota has been eliminated by antibiotic treatments, allowing for the over growth of *C. difficile*. It may be possible that for the effective action of probiotics, the normal microbiota has first to be reduced allowing for the establishment of the desired bacteria.

Although still in its infancy there are a few examples of the use of probiotics for controlling oral biofilm. In a randomized clinical trial, Vivekannanda and collaborators demonstrated that *L. reuteri* inhibited plaque accumulation and reduced inflammation in patients with chronic periodontitis (Vivekananda et al. 2010). A probiotic therapeutic formulation for oral diseases should also be capable of reducing the prevalence of pathogenic microorganisms. Mayanagi et al. used *L. salivarius* as a probiotic to demonstrate the inhibition of a series of oral pathogens, specifically: *A. actinomycetemcomitans*, *Prevotella intermedia*, *P. gingivalis*, *Treponema denticola*, and *Tannerella forsythia* (Mayanagi et al. 2009). At the end of the study, there was a significant decrease in the numbers of all of the periodontal pathogens in the test group compared with the control group. In another study, Iwamoto et al. found that *L. salivarius* also had beneficial effects on halitosis and periodontal pocket bleeding upon probing (Iwamoto et al. 2010).

As aforementioned, the goals for the successful prevention of dental caries by a therapeutic are to inhibit the proliferation of *S. mutans* and to inhibit its adherence to the oral surface. Several probiotics have been assessed in clinical trials aimed to prevent caries. *Bacillus coagulans* have been shown to have an inhibitory effect on *S. mutans* salivary counts in children (Jindal et al. 2011). In another study, Juneja and Kakade showed a statistically significant reduction in salivary mutans streptococci counts in children immediately after consumption of probiotic *Lactobacillus rhamnosus* containing milk (Juneja and Kakade 2012), suggesting that adding probiotics to milk may be a safe and easy way to prevent caries. Nase et al. showed that long-term consumption of *L. rhamnosus* in milk had beneficial effects in preventing dental caries especially in 3- to 4-year-old children (Näse et al. 2001).

In addition to the classic probiotic strains, other oral residents or genetically modified strains have also been tested for their ability to inhibit cariogenic microbes. Hillman and colleagues engineered an *S. mutans* strain deficient in acid lactic production that produced a bacteriocin active against other *S. mutans* strains and could be introduced into the oral cavity to replace the naturally occurring pathogenic strains (Hillman 2002; Hillman et al. 2000).

8 Future Directions

Total elimination of the oral biofilm is neither desirable nor possible. In the future, strategies to control oral biofilms should aim for preserving a healthy community and eliminating organisms with pathogenic potential or those that could shift the homeostasis of the health-associated biofilm towards a dysbiotic one. The motivation behind devising new strategies for controlling oral biofilms is twofold. On one hand widespread antibiotic resistance is a major incentive for the investigation of novel ways to treat or prevent infections. On the other hand, as mentioned above, mechanical cleaning, although effective, is not performed at the level needed to maintain a healthy oral biofilm without additional treatments.

Photodynamic therapy and the use of nanoparticles are promising areas of research in the control of oral biofilms, especially when targeting oral pathogens using the strategies mentioned above. The search for new inhibitory biofilm compounds should continue but focus on shifting the biofilm to health rather than trying to completely eliminate the oral biofilm from our mouths. Preventing the attachment of specific pathogens is a promising strategy that fulfills all the requirements for a good treatment. We now have tools available to perform *in silico* screening for new products based on our previous knowledge. Chemoinformatics-assisted technologies have been used to develop new anti-biofilm compounds against *Staphylococcus aureus* (Dürig et al. 2010).

In the field of vaccination against oral pathogens, stimulation of antigen-specific T-cells polarized toward helper T-cells with a regulatory phenotype is a new and promising field of research. Targeting not only a single pathogen but also polymicrobial organisms, and targeting not only periodontal disease but also periodontal disease-triggered systemic disease, could be a feasible goal (Choi and Seymour 2010).

The use of probiotics is another promising area of research in the field of oral biofilm control. However, many of the clinical studies are pilot in nature, therefore, larger clinical trials, using probiotic strains with proven periodontal probiotic effects in vitro, are needed. Furthermore, administration of probiotics is another area that could be optimized. As mentioned above, "fecal transplant" has been successful in part because of the lack of an established, stable community already colonizing the intestinal epithelium. Thus, removing the microbiota before using probiotics in the oral cavity may help to establish the desired, healthy microbiota and inhibit colonization by oral pathogens.

Finally, a topic we did not describe, but that could have potential in controlling bacterial biofilms, is the use of bacteriophage therapy. Recently, Castillo-Ruiz showed that a bacteriophage against *A. actinomycetemcomitans* killed up to 99 % of cells in a biofilm, opening a window to the possibility of using phage to modulate the oral biofilm (Castillo-Ruiz et al. 2011).

Bibliography

Afennich F, Slot DE, Hossainian N, Van der Weijden GA (2011) The effect of hexetidine mouthwash on the prevention of plaque and gingival inflammation: a systematic review. Int J Dent Hyg 9:182–190

Ahmadi-Motamayel F, Hendi SS, Alikhani MY, Khamverdi Z (2013) Antibacterial activity of honey on cariogenic bacteria. J Dent (Tehran) 10:10–15

Albandar JM, Brunelle JA, Kingman A (1999) Destructive periodontal disease in adults 30 years of age and older in the United States, 1988–1994. J Periodontol 70:13–29

Allaker RP (2010) The use of nanoparticles to control oral biofilm formation. J Dent Res 89:1175–1186

Andlaw RJ (1978) Oral hygiene and dental caries–a review. Int Dent J 28:1–6

Aroniadis OC, Brandt LJ (2013) Fecal microbiota transplantation: past, present and future. Curr Opin Gastroenterol 29:79–84

Asahi Y, Noiri Y, Igarashi J, Asai H, Suga H, Ebisu S (2010) Effects of N-acyl homoserine lactone analogues on Porphyromonas gingivalis biofilm formation. J Periodont Res 45:255–261

Baehni PC, Takeuchi Y (2003) Anti-plaque agents in the prevention of biofilm-associated oral diseases. Oral Dis 9(Suppl 1):23–29

Bansal GS, Newman HN, Wilson M (1990) The survival of subgingival plaque bacteria in an amine fluoride-containing gel. J Clin Periodontol 17:414–418

Barbier O, Arreola-Mendoza L, Del Razo LM (2010) Molecular mechanisms of fluoride toxicity. Chem Biol Interact 188:319–333

Bardell D (1983) The roles of the sense of taste and clean teeth in the discovery of bacteria by Antoni van Leeuwenhoek. Microbiol Rev 47:121–126

Binney A, Addy M, Owens J, Faulkner J (1997) A comparison of triclosan and stannous fluoride toothpastes for inhibition of plaque regrowth. A crossover study designed to assess carry over. J Clin Periodontol 24:166–170

Bullock S, Newman HN, Wilson M (1989) The in-vitro effect of an amine fluoride gel on subgingival plaque bacteria. J Antimicrob Chemother 23:59–67

Burgess NA, Kirke DF, Williams P, Winzer K, Hardie KR, Meyers NL, Aduse-Opoku J, Curtis MA, Camara M (2002) LuxS-dependent quorum sensing in Porphyromonas gingivalis modulates protease and haemagglutinin activities but is not essential for virulence. Microbiology 148:763–772

Castillo-Ruiz M, Vinés ED, Montt C, Fernández J, Delgado JM, Hormazábal JC, Bittner M (2011) Isolation of a novel Aggregatibacter actinomycetemcomitans serotype b bacteriophage capable of lysing bacteria within a biofilm. Appl Environ Microbiol 77:3157–3159

Choi J-I, Seymour GJ (2010) Vaccines against periodontitis: a forward-looking review. J Periodontal Implant Sci 40:153–163

Coleman DC, O'Donnell MJ, Shore AC, Russell RJ (2009) Biofilm problems in dental unit water systems and its practical control. J Appl Microbiol 106:1424–1437

Culshaw S, Larosa K, Tolani H, Han X, Eastcott JW, Smith DJ, Taubman MA (2007) Immunogenic and protective potential of mutans streptococcal glucosyltransferase peptide constructs selected by major histocompatibility complex class II allele binding. Infect Immun 75:915–923

Da Silva BR, de Freitas VAA, Nascimento-Neto LG, Carneiro VA, Arruda FVS, de Aguiar ASW, Cavada BS, Teixeira EH (2012) Antimicrobial peptide control of pathogenic microorganisms of the oral cavity: a review of the literature. Peptides 36:315–321

Dewhirst FE, Chen T, Izard J, Paster BJ, Tanner ACR, Yu W-H, Lakshmanan A, Wade WG (2010) The human oral microbiome. J Bacteriol 192:5002–5017

Donlan RM, Costerton JW (2002) Biofilms: survival mechanisms of clinically relevant microorganisms. Clin Microbiol Rev 15:167–193

Dürig A, Kouskoumvekaki I, Vejborg RM, Klemm P (2010) Chemoinformatics-assisted development of new anti-biofilm compounds. Appl Microbiol Biotechnol 87:309–317

Eckert R, He J, Yarbrough DK, Qi F, Anderson MH, Shi W (2006) Targeted killing of Streptococcus mutans by a pheromone-guided "smart" antimicrobial peptide. Antimicrob Agents Chemother 50:3651–3657

Eckert R, Sullivan R, Shi W (2012) Targeted antimicrobial treatment to re-establish a healthy microbial flora for long-term protection. Adv Dent Res 24:94–97

Eke PI, Dye BA, Wei L, Thornton-Evans GO, Genco RJ (2012) Prevalence of periodontitis in adults in the United States: 2009 and 2010. J Dent Res 91(10):914–920

Feng QL, Wu J, Chen GQ, Cui FZ, Kim TN, Kim JO (2000) A mechanistic study of the antibacterial effect of silver ions on Escherichia coli and Staphylococcus aureus. J Biomed Mater Res 52:662–668

Ferrazzano GF, Amato I, Ingenito A, De Natale A, Pollio A (2009) Anti-cariogenic effects of polyphenols from plant stimulant beverages (cocoa, coffee, tea). Fitoterapia 80:255–262

Ferrazzano GF, Amato I, Ingenito A, Zarrelli A, Pinto G, Pollio A (2011) Plant polyphenols and their anti-cariogenic properties: a review. Molecules 16:1486–1507

Folkesson A, Haagensen JAJ, Zampaloni C, Sternberg C, Molin S (2008) Biofilm induced tolerance towards antimicrobial peptides. PLoS ONE 3:e1891

Frias J, Olle E, Alsina M (2001) Periodontal pathogens produce quorum sensing signal molecules. Infect Immun 69:3431–3434

Fuqua C, Greenberg EP (1998) Self perception in bacteria: quorum sensing with acylated homoserine lactones. Curr Opin Microbiol 1:183–189

Furiga A, Lonvaud-Funel A, Dorignac G, Badet C (2008) In vitro anti-bacterial and anti-adherence effects of natural polyphenolic compounds on oral bacteria. J Appl Microbiol 105:1470–1476

Girardin M, Seidman EG (2011) Indications for the use of probiotics in gastrointestinal diseases. Dig Dis 29:574–587

Grenier D, La VD (2011) Proteases of Porphyromonas gingivalis as important virulence factors in periodontal disease and potential targets for plant-derived compounds: a review article. Curr Drug Targets 12:322–331

Grinde B, Olsen I (2010) The role of viruses in oral disease. J Oral Microbiol 2

Haffajee AD, Socransky SS, Patel MR, Song X (2008) Microbial complexes in supragingival plaque. Oral Microbiol Immunol 23:196–205

Hamilton-Miller JM (2001) Anti-cariogenic properties of tea (Camellia sinensis). J Med Microbiol 50:299–302

Hammer BK, Bassler BL (2003) Quorum sensing controls biofilm formation in Vibrio cholerae. Mol Microbiol 50:101–104

Handal T, Olsen I, Walker CB, Caugant DA (2004) Beta-lactamase production and antimicrobial susceptibility of subgingival bacteria from refractory periodontitis. Oral Microbiol Immunol 19:303–308

Hannig C, Hannig M (2010) Natural enamel wear–a physiological source of hydroxylapatite nanoparticles for biofilm management and tooth repair? Med Hypotheses 74:670–672

Hannig C, Spitzmüller B, Al-Ahmad A, Hannig M (2008) Effects of Cistus-tea on bacterial colonization and enzyme activities of the in situ pellicle. J Dent 36:540–545

Harano K, Yamanaka A, Okuda K (1995) An antiserum to a synthetic fimbrial peptide of Actinobacillus actinomycetemcomitans blocked adhesion of the microorganism. FEMS Microbiol Lett 130:279–285

Hassani AS, Amirmozafari N, Ordouzadeh N, Hamdi K, Nazari R, Ghaemi A (2008) Volatile components of Camellia sinensis inhibit growth and biofilm formation of oral streptococci in vitro. Pak J Biol Sci 11:1336–1341

He X, Shi W (2009) Oral microbiology: past, present and future. Int J Oral Sci 1:47–58

He Z, Wang Q, Hu Y, Liang J, Jiang Y, Ma R, Tang Z, Huang Z (2012) Use of the quorum sensing inhibitor furanone C-30 to interfere with biofilm formation by Streptococcus mutans and its luxS mutant strain. Int J Antimicrob Agents 40:30–35

Heath RJ, White SW, Rock CO (2001) Lipid biosynthesis as a target for antibacterial agents. Prog Lipid Res 40:467–497

Helmerhorst EJ, Hodgson R, van't Hof W, Veerman EC, Allison C, Nieuw Amerongen AV (1999) The effects of histatin-derived basic antimicrobial peptides on oral biofilms. J Dent Res 78:1245–1250

Hetrick EM, Shin JH, Paul HS, Schoenfisch MH (2009) Anti-biofilm efficacy of nitric oxide-releasing silica nanoparticles. Biomaterials 30:2782–2789

Hillman JD (2002) Genetically modified Streptococcus mutans for the prevention of dental caries. Antonie Van Leeuwenhoek 82:361–366

Hillman JD, Brooks TA, Michalek SM, Harmon CC, Snoep JL, van Der Weijden CC (2000) Construction and characterization of an effector strain of Streptococcus mutans for replacement therapy of dental caries. Infect Immun 68:543–549

Holt SC, Kesavalu L, Walker S, Genco CA (1999) Virulence factors of Porphyromonas gingivalis. Periodontology 2000(20):168–238

Honma K, Kato T, Okuda K (1999) Salivary immunoglobulin A production against a synthetic oligopeptide antigen of Actinobacillus actinomycetemcomitans fimbriae. Oral Microbiol Immunol 14:288–292

Hope CK, Wilson M (2004) Analysis of the effects of chlorhexidine on oral biofilm vitality and structure based on viability profiling and an indicator of membrane integrity. Antimicrob Agents Chemother 48:1461–1468

Houari A, Di Martino P (2007) Effect of chlorhexidine and benzalkonium chloride on bacterial biofilm formation. Lett Appl Microbiol 45:652–656

Huang Z, Meric G, Liu Z, Ma R, Tang Z, Lejeune P (2009) luxS-based quorum-sensing signaling affects biofilm formation in Streptococcus mutans. J Mol Microbiol Biotechnol 17:12–19

Iacono VJ, Aldredge WA, Lucks H, Schwartzstein S (1998) Modern supragingival plaque control. Int Dent J 48:290–297

Ishikawa I, Nakashima K, Koseki T, Nagasawa T, Watanabe H, Arakawa S, Nitta H, Nishihara T (1997) Induction of the immune response to periodontopathic bacteria and its role in the pathogenesis of periodontitis. Periodontol 2000 14:79–111

Ito K, Nakamura Y, Tokunaga T, Iijima D, Fukushima K (2003) Anti-cariogenic properties of a water-soluble extract from cacao. Biosci Biotechnol Biochem 67:2567–2573

Iwamoto T, Suzuki N, Tanabe K, Takeshita T, Hirofuji T (2010) Effects of probiotic Lactobacillus salivarius WB21 on halitosis and oral health: an open-label pilot trial. Oral Surg Oral Med Oral Pathol Oral Radiol Endod 110:201–208

Jakubovics NS (2010) Talk of the town: interspecies communication in oral biofilms. Mol Oral Microbiol 25:4–14

Jakubovics N, Kolenbrander P (2010) The road to ruin: the formation of disease-associated oral biofilms. Oral Dis 16:729–739

Jang Y-J, Choi Y-J, Lee S-H, Jun H-K, Choi B-K (2013) Autoinducer 2 of Fusobacterium nucleatum as a target molecule to inhibit biofilm formation of periodontopathogens. Arch Oral Biol 58:17–27

Jindal G, Pandey RK, Agarwal J, Singh M (2011) A comparative evaluation of probiotics on salivary mutans streptococci counts in Indian children. Eur Arch Paediatr Dent 12:211–215

Joo H-S, Otto M (2012) Molecular basis of in vivo biofilm formation by bacterial pathogens. Chem Biol 19:1503–1513

Juneja A, Kakade A (2012) Evaluating the effect of probiotic containing milk on salivary mutans streptococci levels. J Clin Pediatr Dent 37:9–14

Kay HM, Wilson M (1988) The in vitro effects of amine fluorides on plaque bacteria. J Periodontol 59:266–269

Kelly CG, Younson JS, Hikmat BY, Todryk SM, Czisch M, Haris PI, Flindall IR, Newby C, Mallet AI, Ma JK-C et al (1999) A synthetic peptide adhesion epitope as a novel antimicrobial agent. Nat Biotechnol 17:42–47

Kolenbrander PE (2000) Oral microbial communities: biofilms, interactions, and genetic systems. Annu Rev Microbiol 54:413–437

Kolenbrander PE, Andersen RN, Blehert DS, Egland PG, Foster JS, Palmer RJ Jr (2002) Communication among oral bacteria. Microbiol Mol Biol Rev 66:486–505

Kolenbrander PE, Palmer RJ Jr, Periasamy S, Jakubovics NS (2010) Oral multispecies biofilm development and the key role of cell-cell distance. Nat Rev Microbiol 8:471–480

Kruis W (2012) Specific probiotics or "fecal transplantation. Dig Dis 30(Suppl 3):81–84

Lang NP, Cumming BR, Löe H (1973) Toothbrushing frequency as it relates to plaque development and gingival health. J Periodontol 44:396–405

Larsen T, Fiehn NE (1996) Resistance of Streptococcus sanguis biofilms to antimicrobial agents. APMIS 104:280–284

Leszczyńska A, Buczko P, Buczko W, Pietruska M (2011) Periodontal pharmacotherapy – an updated review. Adv Med Sci 56:123–131

Lewis K (2010) Persister cells. Annu Rev Microbiol 64:357–372

Listgarten MA (1988) The role of dental plaque in gingivitis and periodontitis. J Clin Periodontol 15:485–487

Liu P-F, Shi W, Zhu W, Smith JW, Hsieh S-L, Gallo RL, Huang C-M (2010) Vaccination targeting surface FomA of Fusobacterium nucleatum against bacterial co-aggregation: implication for treatment of periodontal infection and halitosis. Vaccine 28:3496–3505

Löe H, Schiott CR (1970) The effect of mouthrinses and topical application of chlorhexidine on the development of dental plaque and gingivitis in man. J Periodont Res 5:79–83

Loesche WJ, Rowan J, Straffon LH, Loos PJ (1975) Association of Streptococcus mutants with human dental decay. Infect Immun 11:1252–1260

Löhr G, Beikler T, Podbielski A, Standar K, Redanz S, Hensel A (2011) Polyphenols from Myrothamnus flabellifolia Welw. inhibit in vitro adhesion of Porphyromonas gingivalis and exert anti-inflammatory cytoprotective effects in KB cells. J Clin Periodontol 38:457–469

LoVetri K, Madhyastha S (2010) Antimicrobial and antibiofilm activity of quorum sensing peptides and Peptide analogues against oral biofilm bacteria. Methods Mol Biol 618:383–392

Lucchese A, Guida A, Petruzzi M, Capone G, Laino L, Serpico R (2012) Peptides in oral diseases. Curr Pharm Des 18:782–788

Lucchese A, Guida A, Capone G, Petruzzi M, Lauritano D, Serpico R (2013) Designing a peptide-based vaccine against Porphyromonas gingivalis. Front Biosci (Schol Ed) 5:631–637

Mandel ID (1994) Antimicrobial mouthrinses: overview and update. J Am Dent Assoc 125(Suppl 2): 2S–10S

Marsh PD (2004) Dental plaque as a microbial biofilm. Caries Res 38:204–211

Marsh PD (2006) Dental plaque as a biofilm and a microbial community – implications for health and disease. BMC Oral Health 6(Suppl 1):S14

Mayanagi G, Kimura M, Nakaya S, Hirata H, Sakamoto M, Benno Y, Shimauchi H (2009) Probiotic effects of orally administered Lactobacillus salivarius WB21-containing tablets on periodontopathic bacteria: a double-blinded, placebo-controlled, randomized clinical trial. J Clin Periodontol 36:506–513

Mayhew RR, Brown LR (1981) Comparative effect of SnF2, NaF, and SnCl2 on the growth of Streptococcus mutants. J Dent Res 60:1809–1814

Merritt J, Qi F, Goodman SD, Anderson MH, Shi W (2003) Mutation of luxS affects biofilm formation in Streptococcus mutans. Infect Immun 71:1972–1979

Mestnik MJ, Feres M, Figueiredo LC, Soares G, Teles RP, Fermiano D, Duarte PM, Faveri M (2012) The effects of adjunctive metronidazole plus amoxicillin in the treatment of generalized aggressive periodontitis: a 1-year double-blinded, placebo-controlled, randomized clinical trial. J Clin Periodontol 39:955–961

Miller MB, Bassler BL (2001) Quorum sensing in bacteria. Annu Rev Microbiol 55:165–199

Miyake T, Yasukawa K, Inouye K (2011) Analysis of the mechanism of inhibition of human matrix metalloproteinase 7 (MMP-7) activity by green tea catechins. Biosci Biotechnol Biochem 75:1564–1569

Modesto A, Drake DR (2006) Multiple exposures to chlorhexidine and xylitol: adhesion and biofilm formation by Streptococcus mutans. Curr Microbiol 52:418–423

Moir DT (2005) Identification of inhibitors of bacterial enoyl-acyl carrier protein reductase. Curr Drug Targets Infect Disord 5:297–305

Molan PC (2001) The potential of honey to promote oral wellness. Gen Dent 49:584–589

Mullany P, Allan E, Warburton PJ (2012) Tetracycline resistance genes and mobile genetic elements from the oral metagenome. Clin Microbiol Infect 18(Suppl 4):58–61

Muniz FWMG, de Oliveira CC, de Sousa Carvalho R, Moreira MMSM, de Moraes MEA, Martins RS (2013) Azithromycin: a new concept in adjuvant treatment of periodontitis. Eur J Pharmacol 705:135–139

Nakahara K, Kawabata S, Ono H, Ogura K, Tanaka T, Ooshima T, Hamada S (1993) Inhibitory effect of oolong tea polyphenols on glycosyltransferases of mutans Streptococci. Appl Environ Microbiol 59:968–973

Namikoshi J, Otake S, Maeba S, Hayakawa M, Abiko Y, Yamamoto M (2003) Specific antibodies induced by nasally administered 40-kDa outer membrane protein of Porphyromonas gingivalis inhibits coaggregation activity of P. gingivalis. Vaccine 22:250–256

Näse L, Hatakka K, Savilahti E, Saxelin M, Pönkä A, Poussa T, Korpela R, Meurman JH (2001) Effect of long-term consumption of a probiotic bacterium, Lactobacillus rhamnosus GG, in milk on dental caries and caries risk in children. Caries Res 35:412–420

Niu Y, Sun J, Fan M, Xu Q-A, Guo J, Jia R, Li Y (2009) Construction of a new fusion anti-caries DNA vaccine. J Dent Res 88:455–460

O'Donnell MJ, Boyle MA, Russell RJ, Coleman DC (2011) Management of dental unit waterline biofilms in the 21st century. Future Microbiol 6:1209–1226

O-Brien-Simpson NM, Veith PD, Dashper SG, Reynolds EC (2003) Porphyromonas gingivalis gingipains: the molecular teeth of a microbial vampire. Curr Protein Pept Sci 4:409–426

Oliver RC, Brown LJ, Loe H (1998) Periodontal diseases in the United States population. J Periodontol 69:269–278

Owens J, Addy M, Faulkner J (1997) An 18-week home-use study comparing the oral hygiene and gingival health benefits of triclosan and fluoride toothpastes. J Clin Periodontol 24:626–631

Pastene E, Speisky H, García A, Moreno J, Troncoso M, Figueroa G (2010) In vitro and in vivo effects of apple peel polyphenols against Helicobacter pylori. J Agric Food Chem 58:7172–7179

Paster BJ, Boches SK, Galvin JL, Ericson RE, Lau CN, Levanos VA, Sahasrabudhe A, Dewhirst FE (2001) Bacterial diversity in human subgingival plaque. J Bacteriol 183:3770–3783

Paster BJ, Olsen I, Aas JA, Dewhirst FE (2006) The breadth of bacterial diversity in the human periodontal pocket and other oral sites. Periodontol 2000 42:80–87

Payne JB, Golub LM (2011) Using tetracyclines to treat osteoporotic/osteopenic bone loss: from the basic science laboratory to the clinic. Pharmacol Res 63:121–129

Peterson SN, Snesrud E, Liu J, Ong AC, Kilian M, Schork NJ, Bretz W (2013) The dental plaque microbiome in health and disease. PLoS ONE 8:e58487

Phan T-N, Marquis RE (2006) Triclosan inhibition of membrane enzymes and glycolysis of Streptococcus mutans in suspensions and biofilms. Can J Microbiol 52:977–983

Polak D, Wilensky A, Shapira L, Weiss EI, Houri-Haddad Y (2010) Vaccination of mice with Porphyromonas gingivalis or Fusobacterium nucleatum modulates the inflammatory response, but fails to prevent experimental periodontitis. J Clin Periodontol 37:812–817

Pratten J, Wills K, Barnett P, Wilson M (1998) In vitro studies of the effect of antiseptic-containing mouthwashes on the formation and viability of Streptococcus sanguis biofilms. J Appl Microbiol 84:1149–1155

Rajapakse PS, O'Brien-Simpson NM, Slakeski N, Hoffmann B, Reynolds EC (2002) Immunization with the RgpA-Kgp proteinase-adhesin complexes of Porphyromonas gingivalis protects against periodontal bone loss in the rat periodontitis model. Infect Immun 70:2480–2486

Rams TE, Degener JE, van Winkelhoff AJ (2013) Prevalence of β-lactamase-producing bacteria in human periodontitis. J Periodont Res 48(4):493–499

Reynolds EC (1997) Remineralization of enamel subsurface lesions by casein phosphopeptide-stabilized calcium phosphate solutions. J Dent Res 76:1587–1595

Rickard AH, Palmer RJ Jr, Blehert DS, Campagna SR, Semmelhack MF, Egland PG, Bassler BL, Kolenbrander PE (2006) Autoinducer 2: a concentration-dependent signal for mutualistic bacterial biofilm growth. Mol Microbiol 60:1446–1456

Roberts AP, Mullany P (2010) Oral biofilms: a reservoir of transferable, bacterial, antimicrobial resistance. Expert Rev Anti Infect Ther 8:1441–1450

Rölla G, Melsen B (1975) On the mechanism of the plaque inhibition by chlorhexidine. J Dent Res 54 Spec No B:B57–B62

Romero M, Acuña L, Otero A (2012) Patents on quorum quenching: interfering with bacterial communication as a strategy to fight infections. Recent Pat Biotechnol 6:2–12

Russell MW, Childers NK, Michalek SM, Smith DJ, Taubman MA (2004) A caries vaccine? The state of the science of immunization against dental caries. Caries Res 38:230–235

Rutherford ST, Bassler BL (2012) Bacterial quorum sensing: its role in virulence and possibilities for its control. Cold Spring Harb Perspect Med 2(11)

Sakanaka S, Okada Y (2004) Inhibitory effects of green tea polyphenols on the production of a virulence factor of the periodontal-disease-causing anaerobic bacterium Porphyromonas gingivalis. J Agric Food Chem 52:1688–1692

Sakanaka S, Aizawa M, Kim M, Yamamoto T (1996) Inhibitory effects of green tea polyphenols on growth and cellular adherence of an oral bacterium, Porphyromonas gingivalis. Biosci Biotechnol Biochem 60:745–749

Salyers AA, Shoemaker NB (1997) Conjugative transposons. Genet Eng (N Y) 19:89–100

Senadheera D, Cvitkovitch DG (2008) Quorum sensing and biofilm formation by Streptococcus mutans. Adv Exp Med Biol 631:178–188

Sgolastra F, Petrucci A, Gatto R, Giannoni M, Monaco A (2011) Long-term efficacy of subantimicrobial-dose doxycycline as an adjunctive treatment to scaling and root planing: a systematic review and meta-analysis. J Periodontol 82:1570–1581

Shao H, Demuth DR (2010) Quorum sensing regulation of biofilm growth and gene expression by oral bacteria and periodontal pathogens. Periodontol 2000 52:53–67

Shi W, Li YH, Liu F, Yang JY, Zhou DH, Chen YQ, Zhang Y, Yang Y, He BX, Han C et al (2012) Flagellin enhances saliva IgA response and protection of anti-caries DNA vaccine. J Dent Res 91:249–254

Simões M (2011) Antimicrobial strategies effective against infectious bacterial biofilms. Curr Med Chem 18:2129–2145

Skucaite N, Peciuliene V, Vitkauskiene A, Machiulskiene V (2010) Susceptibility of endodontic pathogens to antibiotics in patients with symptomatic apical periodontitis. J Endod 36:1611–1616

Slots J (2010) Human viruses in periodontitis. Periodontol 2000 53:89–110

Slots J (2012) Low-cost periodontal therapy. Periodontol 2000 60:110–137

Soares GMS, Figueiredo LC, Faveri M, Cortelli SC, Duarte PM, Feres M (2012) Mechanisms of action of systemic antibiotics used in periodontal treatment and mechanisms of bacterial resistance to these drugs. J Appl Oral Sci 20:295–309

Socransky SS, Haffajee AD (2005) Periodontal microbial ecology. Periodontol 2000 38:135–187

Socransky SS, Haffajee AD, Cugini MA, Smith C, Kent RL Jr (1998) Microbial complexes in subgingival plaque. J Clin Periodontol 25:134–144

Socransky SS, Haffajee AD, Teles R, Wennstrom JL, Lindhe J, Bogren A, Hasturk H, van Dyke T, Wang X, Goodson JM (2013) Effect of periodontal therapy on the subgingival microbiota over a 2-year monitoring period. I. Overall effect and kinetics of change. J Clin Periodontol 40(8):771–780

Soukos NS, Goodson JM (2011) Photodynamic therapy in the control of oral biofilms. Periodontol 2000 55:143–166

Sturme MHJ, Kleerebezem M, Nakayama J, Akkermans ADL, Vaugha EE, de Vos WM (2002) Cell to cell communication by autoinducing peptides in Gram-positive bacteria. Antonie Van Leeuwenhoek 81:233–243

Sun J, Yang X, Xu Q-A, Bian Z, Chen Z, Fan M (2009) Protective efficacy of two new anti-caries DNA vaccines. Vaccine 27:7459–7466

Sun Y, Shi W, Yang JY, Zhou DH, Chen YQ, Zhang Y, Yang Y, He BX, Zhong MH, Li YM et al (2012) Flagellin-PAc fusion protein is a high-efficacy anti-caries mucosal vaccine. J Dent Res 91:941–947

Tagashira M, Uchiyama K, Yoshimura T, Shirota M, Uemitsu N (1997) Inhibition by hop bract polyphenols of cellular adherence and water-insoluble glucan synthesis of mutans streptococci. Biosci Biotechnol Biochem 61:332–335

Takahashi N, Nyvad B (2011) Critical reviews in oral biology & medicine: the role of bacteria in the caries process: ecological perspectives. J Dent Res 90:294–303

Takamatsu-Matsushita N, Yamaguchi N, Kawasaki M, Yamashita Y, Takehara T, Koga T (1996) Immunogenicity of Actinobacillus actinomycetemcomitans serotype b-specific polysaccharide-protein conjugate. Oral Microbiol Immunol 11:220–225

Tattawasart U, Maillard JY, Furr JR, Russell AD (2000) Outer membrane changes in Pseudomonas stutzeri resistant to chlorhexidine diacetate and cetylpyridinium chloride. Int J Antimicrob Agents 16:233–238

Teles RP, Teles FRF (2009) Antimicrobial agents used in the control of periodontal biofilms: effective adjuncts to mechanical plaque control? Braz Oral Res 23(Suppl 1):39–48

Travers M-A, Florent I, Kohl L, Grellier P (2011) Probiotics for the control of parasites: an overview. J Parasitol Res 2011:610769

Twetman S, Stecksén-Blicks C (2008) Probiotics and oral health effects in children. Int J Paediatr Dent 18:3–10

Uccello M, Malaguarnera G, Basile F, D'agata V, Malaguarnera M, Bertino G, Vacante M, Drago F, Biondi A (2012) Potential role of probiotics on colorectal cancer prevention. BMC Surg 12(Suppl 1):S35

Van Winkelhoff AJ, Winkel EG, Barendregt D, Dellemijn-Kippuw N, Stijne A, van der Velden U (1997) beta-Lactamase producing bacteria in adult periodontitis. J Clin Periodontol 24:538–543

Venkateswara B, Sirisha K, Chava VK (2011) Green tea extract for periodontal health. J Indian Soc Periodontol 15:18–22

Vivekananda MR, Vandana KL, Bhat KG (2010) Effect of the probiotic Lactobacilli reuteri (Prodentis) in the management of periodontal disease: a preliminary randomized clinical trial. J Oral Microbiol 2

Vlacic J, Meyers IA, Walsh LJ (2007) Combined CPP-ACP and photoactivated disinfection (PAD) therapy in arresting root surface caries: a case report. Br Dent J 203:457–459

Wade WG (2013) The oral microbiome in health and disease. Pharmacol Res 69:137–143

Williams JA, Pearson GJ, Colles MJ, Wilson M (2004) The photo-activated antibacterial action of toluidine blue O in a collagen matrix and in carious dentine. Caries Res 38:530–536

Williams JA, Carnes AE, Hodgson CP (2009) Plasmid DNA vaccine vector design: impact on efficacy, safety and upstream production. Biotechnol Adv 27:353–370

Wright HT, Reynolds KA (2007) Antibacterial targets in fatty acid biosynthesis. Curr Opin Microbiol 10:447–453

Xu X, Zhou XD, Wu CD (2011) The tea catechin epigallocatechin gallate suppresses cariogenic virulence factors of Streptococcus mutans. Antimicrob Agents Chemother 55:1229–1236

Yamanaka M, Hara K, Kudo J (2005) Bactericidal actions of a silver ion solution on Escherichia coli, studied by energy-filtering transmission electron microscopy and proteomic analysis. Appl Environ Microbiol 71:7589–7593

Yamanaka A, Kouchi T, Kasai K, Kato T, Ishihara K, Okuda K (2007) Inhibitory effect of cranberry polyphenol on biofilm formation and cysteine proteases of Porphyromonas gingivalis. J Periodont Res 42:589–592

Yano A, Kikuchi S, Takahashi T, Kohama K, Yoshida Y (2012) Inhibitory effects of the phenolic fraction from the pomace of Vitis coignetiae on biofilm formation by Streptococcus mutans. Arch Oral Biol 57:711–719

Yoo S, Murata RM, Duarte S (2011) Antimicrobial traits of tea- and cranberry-derived polyphenols against Streptococcus mutans. Caries Res 45:327–335

Yoshida A, Ansai T, Takehara T, Kuramitsu HK (2005) LuxS-based signaling affects Streptococcus mutans biofilm formation. Appl Environ Microbiol 71:2372–2380

Younson J, Kelly C (2004) The rational design of an anti-caries peptide against Streptococcus mutans. Mol Divers 8:121–126

Yu F, Xu Q-A, Chen W (2011) A targeted fimA DNA vaccine prevents alveolar bone loss in mice after intra-nasal administration. J Clin Periodontol 38:334–340

Yudovin-Farber I, Beyth N, Nyska A, Weiss EI, Golenser J, Domb AJ (2008) Surface characterization and biocompatibility of restorative resin containing nanoparticles. Biomacromolecules 9:3044–3050

Zalewska A, Zatoński M, Jabłonka-Strom A, Paradowska A, Kawala B, Litwin A (2012) Halitosis–a common medical and social problem. A review on pathology, diagnosis and treatment. Acta Gastroenterol Belg 75:300–309

Zaura E, Keijser BJF, Huse SM, Crielaard W (2009) Defining the healthy "core microbiome" of oral microbial communities. BMC Microbiol 9:259

Zaura-Arite E, van Marle J, ten Cate JM (2001) Conofocal microscopy study of undisturbed and chlorhexidine-treated dental biofilm. J Dent Res 80:1436–1440

Zhang T, Hashizume T, Kurita-Ochiai T, Yamamoto M (2009) Sublingual vaccination with outer membrane protein of Porphyromonas gingivalis and Flt3 ligand elicits protective immunity in the oral cavity. Biochem Biophys Res Commun 390:937–941

Control of Polymicrobial Biofilms: Recent Trends

Derek S. Samarian, Kyung Rok Min, Nicholas S. Jakubovics, and Alexander H. Rickard

Abstract Biofilms represent the dominant mode of bacterial existence in natural and man-made environments. Bacteria within biofilms possess collective biofilm-imposed properties that make them distinct from their planktonic counterparts. One key property is an enhanced resistance to antimicrobials. Previous strategies to treat biofilms have focused on either single or combined chemical treatments or physical removal. Considering that many chronic bacterial illnesses are associated with multispecies biofilms, such approaches may not be effective because juxtaposed species can act synergistically to enhance recalcitrance to treatments, resulting in treatment failure. This section will introduce the reader to the processes leading to the development of human-associated polymicrobial biofilms with a particular emphasis on multispecies succession, ecology, and integration by pathogenic bacteria. Then, cognizant of processes and properties, newly developed or promising approaches to control pathogenic biofilm communities will be considered. These approaches will either be preventative or therapeutic and based upon the manipulation of biological processes (e.g., cell–cell signaling, coaggregation, treatment with phage) or based purely upon technological advances (e.g., cold plasma, modified-surface technologies, nanoparticles). The advantages and disadvantages of the different approaches will be discussed and future prospects considered. Recognizing the current issues associated with the spread of antimicrobial resistance and the overall recalcitrance of biofilms, we propose that new technologies to

---

D.S. Samarian • A.H. Rickard (✉)
Department of Epidemiology, University of Michigan School of Public Health, SPH I, Room 5646, 1415 Washington Heights, Ann Arbor, MI 48109-2029, USA
e-mail: alexhr@umich.edu

K.R. Min
Department of Environmental Health Sciences, University of Michigan School of Public Health, Ann Arbor, MI, USA

N.S. Jakubovics
Newcastle University, Oral Biology, School of Dental Sciences, Newcastle upon Tyne, UK

prevent disease or approaches to craft communities to maintain health should be a mainstay of current biofilm research.

1 Introduction

Biofilms are surface-attached or interface-attached aggregated communities of bacteria (Costerton 1995). Most biofilms in medical and natural environments are composed of multiple species of bacteria (Stoodley et al. 2002) and are often referred to as "polymicrobial biofilms" (Peters et al. 2012; Wolcott et al. 2013). Examples of such communities include dental plaque biofilms, which can contain up to 500 species of closely located but taxonomically disparate species, and chronic wound biofilms which can also contain hundreds of species (Kolenbrander et al. 2006; Smith et al. 2010). A major issue associated with the formation of these communities is their antimicrobial recalcitrance. Biofilm bacteria are up to 1,000 times more tolerant to antimicrobials than their planktonic counterparts (Mah and O'Toole 2001; Gilbert et al. 2002) and resist abrasive removal (Gibson et al. 1999; Paranhos et al. 2007; He and Shi 2009). Such a pronounced resistance underlies why 23.7 % of US adults have untreated dental caries and 38.5 % have moderate to severe periodontitis (Eke et al. 2012; National Center for Health Statistics 2012). Similarly, the recalcitrance of chronic wound biofilms accounts for the 6.5 million patients that are treated every year in the USA, and a financial burden of greater than $25 billion is spent annually on treatment of chronic wounds (Sen et al. 2009). It should be noted that from a public health perspective, the burden of oral healthcare and wound treatment/prevention is rapidly growing, in part due to biofilms and the spread of antimicrobial resistance, but also because of the dramatic increases in the cost of healthcare, number of aging population, incidence of diabetes, and increasing numbers suffering from obesity (Alanis 2005; Howell-Jones et al. 2005; Cosgrove 2006; Allukian & Adekugbe 2008; Fisher et al. 2009; Tsai et al. 2011). Thus, before considering strategies to control biofilm communities, an understanding and appreciation of the processes that contribute to biofilm development and persistence are essential.

Since the first descriptions of biofilms, there has been a continual and recently increasing interest in the processes that are important in biofilm development (Costerton et al. 1987; Hall-Stoodley and Stoodley 2002, Kolenbrander et al. 2006). While the excellent and often-cited model for biofilm development is that of *Pseudomonas aeruginosa* described by Davies and colleagues (Stoodley et al. 2002; Davies 2003), this model only accounts for the formation of single-species biofilms. A more realistic model, however, is one that considers multiple species. One such model by Rickard et al. (2003), that is based upon models of oral biofilm formation as well as those of multispecies biofilms from other environments (Callow 1993; Marsh and Bradshaw 1995; Wimpenny 1996; Palmer and White 1997), considers biofilm development as a successional process (Fig. 1). Succession is the ordered and often reproducible process (from environment to environment) by which primary colonizers adhere to *conditioned* surfaces and the growth of these pioneering species, in turn, facilitates the integration of additional species. A conditioned surface is usually one with adsorbed proteins, polysaccharides, and other organic materials. Primary colonizing bacteria can adhere to the conditioning film by nonspecific electrostatic forces, Lifshitz-van der Waals forces, and Lewis acid–base forces. Similar nonspecific forces enable the later colonizers, which are often referred to as secondary colonizers, but also highly specific coaggregation interactions contribute to the sequential integration of species (Rickard et al. 2003; Hojo et al. 2009) (Fig. 1). Throughout the process of biofilm development, cells can also produce extracellular polymeric substances (EPS) that include proteins, polysaccharides, and DNA which, among other roles, stabilize and strengthen the biofilm (Flemming and Wingender 2010; Jakubovics et al. 2013). In this model, secondary colonizers can be species that facilitate the integration of pathogens. While such a process has been described in dental plaque biofilms (Diaz et al. 2006; Kolenbrander et al. 2006; Palmer et al. 2006) as well as in freshwater biofilm communities (Martiny et al. 2003; Lyautey et al. 2005), studies of other biofilms, especially those found in environments where flowing liquid is less pronounced (e.g., skin/wound biofilm communities), are still in their infancy. Nonetheless, the concept of successional biofilm development has expanded into many interdisciplinary research fields over the last 30 years and the cellular biofilm-specific mechanisms that promote such ordered integrations have also been studied, identified, and compared (Kjelleberg and Molin 2002; Rickard et al. 2003; Raes and Bork 2008; Dobretsov et al. 2009; Schillinger et al. 2012). This increased understanding of mechanisms has been fueled by the recognition that alternative strategies, that are likely to be highly effective, can be developed to prevent biofilms from being problematic.

2 Relevance of Polymicrobial Biofilms to Human Health

Before addressing new techniques that manipulate or circumvent biofilm-specific properties, it is important to briefly describe current approaches to control biofilms and how effective these methods are in preventing or treating diseases. It should be added that most approaches to date are developmentally polar: current technologies aim to prevent initial biofilm formation (e.g., anti-biofilm coatings) or maximize inactivation of mature biofilms (e.g., biofilm permeabilizers). Technologies that either interfere with or retard biofilm development or that endeavor to bypass intrinsic biofilm recalcitrance are only now becoming realized.

The concept of preventing biofilm formation is not new. One of the first approaches to intentionally prevent biofilm formation was developed by the ancient Carthaginians and Phoenicians who used pitch and possibly copper-based sheathing material on the hulls of their ships (Anon. 1952). In the medical fields, one of the oldest techniques known is debridement, which is the surgical removal of dead tissue from a wound that can sometimes be combined with chemical treatments to eradicate biofilms that hinder the healing process. Notably, one of the first advocates for combinational debridement was Alexis Carrel, who in the early 1900s

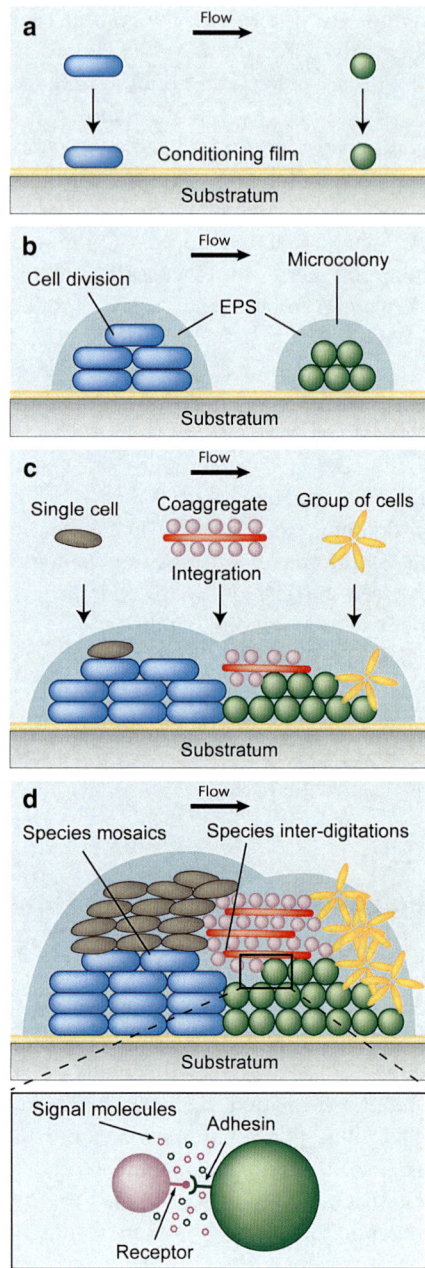

Fig. 1 Diagrammatic example demonstrating the successional development of a multispecies biofilm under flowing conditions. (**a**) Initial attachment of primary colonizing species to a conditioning film. (**b**) Growth of adherent cells and simultaneous expression of EPS to form micro-colonies. (**c**) Further growth of adherent primary colonizing species and recruitment of secondary species. Recruitment can be through specific coaggregation interactions (also called

treated wounds with sodium hypochlorite to enhance mechanical debridement of dead tissue and slow bacterial regrowth to facilitate the healing process (Carrel and Dehelly 1917; Hirsch 2008). Carrel's 227-page monograph described in exquisite detail the decision-making process as to how to treat wounds and also compared this strategy to others, such as those that make use of silver nitrate and hydrogen peroxide. This combinational type of approach was particularly useful during World War I, where wounds were common place and amputations were often considered the only resort for severe wounds. However, it should be noted that there were many who considered the use of antimicrobials in the debridement process as being superfluous, especially in the treatment of war wounds. During this time, Burghard, Leishman, Moynihan, and Wright wrote "the treatment of suppurating wounds by means of antiseptics is illusory, and that belief in its efficacy is founded upon false reasoning." (Burghard et al. 1915; Hirsch 2008). However, such opinion was not shared universally and in time it was recognized that combining mechanical removal of dead material and associated biofilm with chemical treatment was in fact beneficial. Interestingly, however, Burghard and colleagues were actually right in their suggestion that the benefits conferred by antimicrobial treatments can be limited, especially in the absence of other treatments, such as mechanical removal. Why? Because undisturbed biofilm bacteria display properties at the cellular and biofilm levels that individually and collectively contribute to antimicrobial resistance.

Antimicrobials, whether antibiotics or biocides, represent the current mainstay of most chemical treatment strategies to control biofilms. However, as hinted above, most antimicrobials have a significant *Achilles' heel*; their effectiveness against bacteria within undisturbed biofilms is significantly reduced when compared to planktonic bacteria. This tolerance is due to changes in bacterial cellular properties and the gross biofilm state (Fig. 2). When considering the biofilm as a whole, penetration of an antimicrobial can be hindered due to the expression of extracellular polymeric substances that bind to the antimicrobial and/or by the whole cells adsorbing or sequestering/inactivating the antimicrobial (Xu et al. 1996; Anderl et al. 2000; Gilbert et al. 2002). This effect can be further enhanced in multispecies (polymicrobial) biofilms whereby one species preferentially binds or removes a given antimicrobial and thus protects another species (Leriche et al. 2003; Schwering et al. 2013). Conceivably, this would be enhanced further if the two species were intimately associated in close proximity, as opposed to distantly located within the same biofilm (Gilbert et al. 2002). This failure to

---

Fig. 1 (continued) coadhesion) and also through nonspecific interactions. Recruited secondary colonizing cells can be as single cells, coaggregates, or aggregates of genetically identical cells (called autoaggregates). (**d**) Species within the developing polymicrobial biofilm expand and component species interact. Interactions can be positive or negative and mediated, for example in the expanded box, through the production and detection of cell–cell signal molecules. Coaggregation interactions may enhance the close juxtaposition of interacting species through adhesins expressed on one cell surface binding to cognate receptors expressed on a partner cell surface. Multiple cell–cell signal molecules may be involved in the interaction (shown as *green* and *purple signals*). Diagram modified with permission from Rickard et al. (2003)

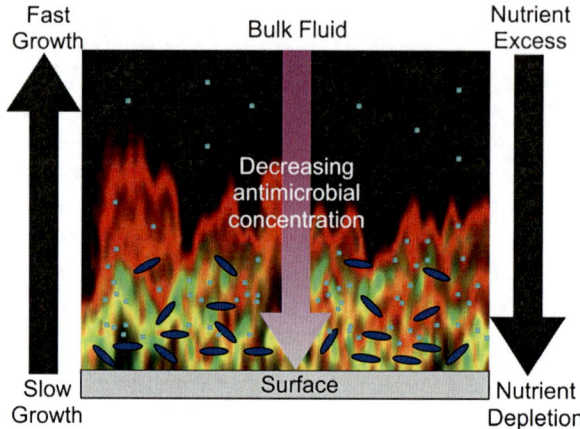

Fig. 2 Diagrammatic representation of the possible factors contributing to antimicrobial resistance by biofilm bacteria, resulting in poor penetration and uneven kill. Biofilm is presented in X–Y field of view. Key factors are growth rate (fastest at top of biofilm), nutrient depletion (closest to the base of biofilm), the production of cell–cell signal molecules (collectively shown as *turquoise dots*) that accumulate in the biofilm and some of which are released into the planktonic phase, and the generation of persister cells that dominate in the depths of the biofilm (*blue cells*). Image modified with permission from McBain et al. (2013)

penetrate, or retardation of penetration, is often referred to as *reaction diffusion limitation* (Xu et al. 1996; Stewart et al. 1998). However, this process alone is not enough to account for the enhanced tolerance to antimicrobials by biofilm bacteria, especially for antimicrobials that are less reactive to biofilm components. Another contributing factor is the altered growth rates, which often play an important role in the susceptibility of a bacterium to an antimicrobial (Brown et al. 1988, Evans et al. 1990a, b). Bacteria in biofilms are heterogeneous with respect to growth rates, and this is due to a number of reasons including changes in the concentrations of dissolved oxygen, nutrient, and metabolites within biofilms. For example, in an aerobic wastewater-like environment, generally the amount of available oxygen at the base of a biofilm will be less than at the top of the biofilm (Yu et al. 2004). Indeed, it is this type of heterogeneity that provides different ecological niches within a biofilm system and the expansion of bacterial populations that will benefit from that niche (Ziegler et al. 2013). As a consequence, stratified communities with different susceptibilities and growth rates develop. An extreme form of growth-rate-induced reduction in antimicrobial susceptibility, at the cellular level, has been described as the *persister* phenotype (Spoering and Lewis 2001; Keren et al. 2004). First observed by Bigger (1944), it has since become evident that in any bacterial population that is susceptible to a given antimicrobial, there is a subpopulation that survives even the harshest of treatments by this antimicrobial. This subpopulation can then regrow in the absence of the antimicrobial, but the regrown population still retains the original wild-type susceptibility (i.e., this phenomenon is not due to the selection of a resistant mutant cell line) (Sufya et al. 2003). Such a phenomenon, which may well be linked to extremely slow growth rates or a form of quiescence

(Gilbert et al. 2002; Roberts and Stewart 2005; Klapper et al. 2007; Lewis 2010), is enriched in biofilms as well as aged (stationary phase) planktonic cultures (Spoering and Lewis 2001; Shapiro et al. 2011; Wang and Wood 2011). Two other biofilm-specific phenotypes that have received attention in association with biofilm tolerance are the expression of efflux pumps and the production of cell–cell signaling molecules. Efflux pumps can expel chemically unrelated antimicrobial agents from the bacterial cell, and their expression may be specific to the biofilm mode of growth or induced by antimicrobials and the biofilm lifestyle (De Kievit et al. 2001; Gillis et al. 2005; Hoiby et al. 2010; Coenye et al. 2011). For example, Gillis and coworkers (2005) demonstrated, using DNA microarrays, that *P. aeruginosa* PAO1 cells in biofilms that were exposed to azithromycin showed upregulation of transcripts encoding for restriction-nodulation-cell division (RND) efflux pumps. Specifically, both the *mexAB-oprM* and *mexCD-oprJ* operons that code for efflux pumps were required for biofilm formation in the presence of azithromycin, but *mexCD-oprJ* was a biofilm-specific mechanism for azithromycin resistance while *mexAB-oprM* was indicated to be important for azithromycin resistance in both planktonic and biofilm communities. Efflux pumps can also be involved in cell–cell signaling (Evans et al. 1998; Pearson et al. 1999; Herzberg et al. 2006; Buroni et al. 2009; Lamarche and Deziel 2011).

Cell–cell signaling, the production and detection of low molecular weight signal molecules, has received increasing attention in the last decade (Dickschat 2010). In particular, this is because biofilms not only concentrate cells but also concentrate the signal molecules that are produced by the cells (Alberghini et al. 2009; Kolenbrander et al. 2010). Thus, the localized increase in density of cell–cell signal molecules within a biofilm can act as a threshold-based queue for the expression of biofilm-specific phenotypes and probably contributes to the formation of species mosaics (Fig. 1) (Gu et al. 2013). As a consequence of changes in cellular properties, and conceivably spatial species patterning, biofilms may also have enhanced tolerance levels and/or alter the expression of virulence factors. For example, numerous research studies have indicated that the intraspecies and interspecies cell–cell signaling molecule autoinducer-2 (AI-2) is responsible for a reduced susceptibility to antimicrobials, and the threshold concentration of AI-2 is likely only reached in biofilms (Ahmed et al. 2007, 2009; Roy et al. 2011, 2013). While few examples exist to date, and considering the multispecies nature of many biofilms, it would be interesting to examine the effects of cell–cell signaling molecules that are produced by one species on another species that does not produce them (e.g., AHLs are not produced by any Gram positive species), to examine the effects of foreign signal molecules on gene expression. A fascinating study based around such a concept was presented by Duan and colleagues (2003), which indicated that non-AI-2 producing *P. aeruginosa* (*P. aeruginosa* does not possess the *luxS* gene responsible for the production of AI-2) can detect AI-2 produced by other bacteria in the human lung of cystic fibrosis patients. Genome-wide transcriptional analysis demonstrated that approximately 4 % of the *P. aeruginosa* genome responded to the presence of other bacterial species and further experiments using exogenously added AI-2 indicated that some of these genes were influenced by AI-2, including a number of virulence factors.

Although only the smallest of a glimpse of the now burgeoning field of polymicrobial research is given in the above paragraphs, it is now clear that polymicrobial biofilms are not random assemblages of microorganisms and the component species are far from solitary units of life (Rickard et al. 2003; Percival et al. 2010; Bandara et al. 2012). Polymicrobial biofilms are interactive communities which are exquisitely structured through spatiotemporal developmental processes (Palmer and White 1997; Teles et al. 2013). Such communities possess individual and collective properties that negate or reduce the effectiveness of antimicrobials, especially those that may only target one or a couple of species in that community, because of interspecies protection (Bridier et al. 2011). With this understanding, it is the aim of the following sections that are separated into *Biological strategies* and *Technological strategies* to alert the reader to some up-and-coming technologies to control polymicrobial biofilm communities. These sections will be far from all-encompassing, but it is hoped the reader will gain insight into the multitude of techniques and technologies that are currently being developed to control polymicrobial biofilms. It is possible that one or several combinations of these techniques can become a mainstay approach to controlling biofilms in the future.

3 Biological Strategies

Microbiology as a field of research has developed rapidly in the last few decades and our appreciation of the social and interactive nature of microorganisms, as well as the biofilm *forum* in which they commune, has allowed us to interrogate and discover possible chinks in the biofilm armor. One fundamental paradigm shift in anti-biofilm approaches has been fueled by the realization that a one-step approach to kill biofilm microorganisms is often not sufficient and sometimes not warranted. Indeed, removal of biofilm (or removal of certain species) may be all that is required if the component cells are washed away such as in the human oral cavity. However, if the dispersed bacteria are likely to add additional problems, such as spread of species to other areas of the human body, then a combined or a two-step approach could be used in conjunction with an antimicrobial (since planktonic bacteria are more susceptible to antimicrobials than their planktonic counterparts). Furthermore, if the biology of polymicrobial biofilms can be manipulated by altering the interactions between component species, by chemical means, or through the introduction of another species, then it is possible that such an altered community will be easier to control.

3.1 Biofilm Dispersal/De-adhesion

The concept of manipulating biofilms using technologies to cause a biofilm to be destabilized and disperse is not new (Kaplan 2010). At least three distinct modes of biofilm dispersal have been identified: *erosion* through de-adhesion events which increases with greater biofilm biomass and increasing fluid shear (Characklis 1990), *sloughing* through the loss of large multicellular aggregates of the biofilm community due to localized physiological and physicochemical stresses (Lappin-Scott and Bass 2001), and *seeding* which represents a triggered loss of cells from the biofilm (Boles and Horswill 2008; Davies and Marques 2009).

While some currently used antimicrobials may have de-adhesion activities, by altering the membrane properties of biofilm bacteria resulting in the loss of cell–cell and/or cell–surfaces adhesion (Neu 1996; Rao et al. 2011), more efficacious approaches are currently being investigated. In particular, unlike antimicrobials that may have secondary dispersal effects, the use of non-antimicrobial dispersal agents will not likely have the associated problems of developing antimicrobial resistance and will potentially be less noxious to the environment/host in which it is deployed. One notable example of a dispersal agent is Dispersin B. Dispersin B, also known as DspB, is a 42 kDa glycoside hydrolase that was identified as being produced by the human oral pathogen *Aggregatibacter* (*Actinobacillus*) *actinomycetemcomitans* (Kaplan et al. 2003). The enzyme catalyzes the hydrolysis of poly-*N*-acetylglucosamine (PNAG), one of a number of polysaccharides present in EPS that is produced by a broad taxonomic range of Gram-positive and Gram-negative biofilm forming species (Itoh et al. 2005; Chaignon et al. 2007). Interestingly, Dispersin B has been shown to not only disperse biofilm species but also enhance penetration of cetylpyridinium chloride (Ganeshnarayan et al. 2009). This latter point is particularly interesting as it raises the possibility of synergy between Dispersin B (or any other EPS degrading enzyme) and other anti-biofilm agents and/or antimicrobials. A fascinating example of such a combinational approach was presented by Lu and Collins (2007), who showed that by genetically engineering bacteriophage T7 to express Dispersin B, treated *E. coli* biofilms were reduced by 4.5 orders of magnitude, which was about two orders of magnitude better than treating with non-engineered phage (Lu and Collins 2007). While phage technology represents an interesting approach to control biofilms, albeit potentially restricted in taxonomic breadth, the specificity of phage for host bacteria is potentially an advantage or disadvantage depending upon applications and species composition of a polymicrobial biofilm. This combinational approach acts in a multifaceted manner that is potentially self-sustaining. Specifically, engineered phage remains in biofilms as long as suitable non-dispersed cells are present. Therefore, this approach bypasses the problems associated with contact time—the requirement for a treatment regimen to be sustained for a period of time in order to be effective.

One limitation of polysaccharide lyases such as Dispersin-B is that they exhibit a high degree of substrate specificity. This is due to the complexity of carbohydrates, which arises from the multiple linkages that are possible between monosaccharide units. For example, there are nine naturally occurring disaccharides formed by the linkage of two glucose residues (Rüdiger and Gabius 2009). By contrast, proteins and nucleic acids are linear chains that can be digested relatively easily by enzymes. There is now strong evidence that extracellular DNA (eDNA) is an important structural component in many different biofilms (Jakubovics et al. 2013). This was elegantly shown by Whitchurch et al. (2002), who observed that eDNA was abundant in the extracellular matrix of model *P. aeruginosa* biofilms. Treating biofilms with bovine DNase I removed bacterial cells from the surface. A number of bacteria produce extracellular DNase that digest eDNA in biofilms, and these have potential for exploitation as anti-biofilm agents. For example, studies have demonstrated the effectiveness of NucB, a DNase from a marine isolate of *Bacillus licheniformis*, at removing microbial cells from single- or mixed-species biofilms (Shakir et al. 2012; Shields et al. 2013). In fact, NucB has been shown to cause significant release of microorganisms from naturally occurring mixed-species biofilms on the surfaces of tracheoesophageal speech valves recovered from patients (Shakir et al. 2012). Such a finding not only demonstrates the potential application of NucB against polymicrobial biofilms but also highlights the importance of eDNA in the EPS of biofilms formed on surfaces held within the human body. The role of eDNA in biofilm development and structural integrity is receiving increasing attention, especially as a target for biofilm destabilization (Das et al. 2010; Jakubovics et al. 2013; Peterson et al. 2013). In fact, the potential for the use of DNases in vivo has been demonstrated by a recent study using a mouse model of diabetes-associated *P. aeruginosa* wound infections (Watters et al. 2013). Here, the addition of DNase significantly increased the susceptibility of wound biofilms to gentamicin. It is noteworthy that a DNase (Streptodornase) has been used for many years in the form of "Varidase" for the treatment of chronic wounds (Smith et al. 2013). However, Varidase is a crude mixture of many different components, and it is not yet clear to what extent the streptodornase contributes to the overall efficacy of Varidase in managing wound infections.

An interesting nonenzyme-based approach to potentially disrupt polymicrobial biofilms has been presented by Davies and Marques (2009). The team demonstrated that *P. aeruginosa* produces cis-2-decenoic acid, which induced the dispersion of established biofilms and also inhibited biofilm development (Davies and Marques 2009). Although the precise mechanism of action has yet to be elucidated, of particular interest is that the team demonstrated that the molecule was active against a myriad of Gram-positive and Gram-negative species and the yeast *Candida albicans*. Considering that the activity is in the nanomolar range, the potential benefits of using cis-2-decenoic acid on its own or in combinational therapies against polymicrobial biofilms are intriguing.

3.2 Cell–Cell Signaling Inhibitors

Quorum sensing, also known as cell–cell signaling, is crucial for the development of biofilms (Dickschat 2010). So much so that the term *Sociomicrobiology*, which describes cell–cell signaling in biofilms, has been proposed to be a unifying term (Parsek and Greenberg 2005). Over the last four decades there has been increasing interest in this avenue of research both from fundamental and applied perspectives (Raina et al. 2009; Shank and Kolter 2009). Similarly, a perusal of published patents reveals considerable interest in cell–cell signaling inhibitor technology. There are now numerous types of cell–cell signaling molecules that have been identified and while this section cannot adequately cover every one, a notable few will be highlighted for which inhibitors have been developed.

Acyl homoserine lactones (AHLs) and autoinducer-2 (AI-2) have been the hot topic of study over the last decade. AHLs are produced solely by Gram-negative species while AI-2 is produced by both Gram-positive and Gram-negative species. While one would automatically consider AI-2 to be more important, this conclusion should not be hastily decided upon. A key reason why AHLs are of tremendous importance, apart from being discovered first, include the demonstration that many Gram-negative pathogens (with the exception of a few, including certain Gram-negative periodontal pathogens) use AHLs for intraspecies communication. Different species often produce one or more different forms of AHLs, and it is unclear the degree of interspecies communication that AHLs confer, although examples do exist (Stickler et al. 1998; Bernier et al. 2008). Fortunately, from the standpoint of creating inhibitors, all AHLs consist of a homoserine lactone ring moiety but differ with respect to the length, degree of saturation, and specific substitutions within an attached acyl side chain. The differences in AI-2 structure are not different between species, as one would expect with a cell–cell signaling system that is proposed to be a "universal signaling system" (Xavier and Bassler 2003), although there is a much more subtle process occurring. Specifically, AI-2 is actually an umbrella term for a collection of inter-convertible forms that are derived from the molecule 4, 5-dihydroxy-2,3-pentanedione (DPD) which are in equilibrium (Chen et al. 2002; Miller et al. 2004; Thiel et al. 2009). Different forms are recognized by different species and forms and equilibrium are modifiable by the environment (Chen et al. 2002; Miller et al. 2004). Because AHLs and AI-2 are responsible for a swathe of biofilm properties that are (1) species dependent, (2) concentration dependent, (3) form/structure dependent, and (4) environment dependent, unifying approaches using inhibitors to control polymicrobial biofilms are not entirely clear. That being said, as a matter of completeness and recognition, excellent studies describing inhibitors for either AI-2 or AHLs are available and these either act on the proteins that produce the signal molecules (Chung et al. 2011), act on the receptors that facilitate recognition (Jiang and Li 2013), are modified to enable smart targeting of one species over another (Guo et al. 2012), or are tethered to antimicrobials to enable enhanced uptake and killing (Eckert et al. 2006). This latter approach, to tether antimicrobials to signal molecules, has received increasing attention in part due to intriguing work by Eckert et al. (2006). This group presented a fascinating study that showed that cell–cell signaling by the cariogenic organism *S. mutans* can be manipulated in a manner similar to that used by the ancient Greeks to enter the city of Troy by using the mythological Trojan horse. Specifically, the research team demonstrated a radically new class of pathogen-selective molecules, which they called *selectively targeted antimicrobial peptides* (STAMPs). This technology was developed by creating a fusion of a species-specific targeting peptide domain (in this case, competence stimulating peptide [CSP] molecules produced by *S. mutans*) with an antimicrobial peptide. CSP-based STAMPs were internalized by *S. mutans* and this resulted in their death. Because CSPs are strain/species specific (Suntharalingam and Cvitkovitch 2005), *S. mutans* was shown to be specifically targeted while other oral streptococci were unaffected. The team (Eckert et al. 2006) also suggested that, in the future, STAMPs could be produced by certain nonpathogenic resident oral species to act as artificially created probiotic organisms.

While not addressed in this section, it should be noted that the understanding of novel approaches to inhibit the complex autoinducer peptide cell–cell signaling system of the staphylococci are coming to fruition. Because of the strain/species specificity of this type of cell–cell signaling system, interesting targeted approaches have the potential to be developed (the reader is directed to the section on *Staphylococcus aureus* biofilm inhibitors in Chap. 11, by Alex Horswill).

3.3 Manipulation of Coaggregation

Coaggregation is the specific recognition and adhesion of different species of bacteria to one another (Kolenbrander 1988). Originally discovered in the early 1970s (Gibbons and Nygaard 1970; Cisar et al. 1979) and considered to be a simple mechanism by which bacteria can integrate into dental plaque (Kolenbrander and London 1993), it has become clear that coaggregation is likely essential for numerous roles in maintaining homeostasis in multispecies biofilms in a variety of environments and between taxonomically disparate species (Bos et al. 1999; Rickard et al. 2003; Kolenbrander et al. 2006; Hojo et al. 2009; Shirtliff et al. 2009). Thus, if coaggregation is integral to polymicrobial biofilm development and the interaction of different species in a biofilm, approaches to prevent, restructure, or destabilize polymicrobial biofilms through manipulating coaggregation interactions would offer attractive alternatives to traditional anti-biofilm/antimicrobial strategies.

A key role for coaggregation is to facilitate cellular juxtaposition. Work by oral microbiologists has shown that coaggregation has roles in bringing species in close proximity to facilitate cell–cell signaling (Fig. 1) and also for the exchange of metabolites (Kolenbrander et al. 2010). Indeed, work by Egland et al. (2004) has shown that in order for biofilm populations of the oral bacterium *Veillonella atypica* PK1910 to grow in biofilms, the species need to be juxtaposed to *S. gordonii* V288. This pair, coincidently, coaggregates strongly, and the authors speculated that coaggregation is required to enhance the growth of *V. atypica* (Egland et al. 2004; Kolenbrander et al. 2010).

The importance of coaggregation in supporting the retention of species is not limited to dental plaque biofilms (Rickard et al. 2003). Recent work by Min and Rickard (2009) demonstrated that coaggregation was also required by the freshwater bacterium *Sphingomonas natatoria* 2.1 in order to compete with *Micrococcus luteus* 2.13. A spontaneous coaggregation-deficient mutant of *S. natatoria* 2.1 was not only unable to form substantial biofilms with *M. luteus* 2.13, but it was also impaired in forming single-species biofilms suggesting a dual role for coaggregation (cell–surface adhesion and cell–cell adhesion/coaggregation). Thus, coaggregation could be targeted to prevent initial and later colonizing events in biofilms. Considering the difficulties in treating freshwater biofilms and the potential for pathogens or problematic organisms that reside in freshwater biofilms in public, industrial, and healthcare settings (Exner et al. 2005; Simoes et al. 2008; Declerck 2010; Vornhagen et al. 2013), such an approach to inhibit pathogen integration that do not require the use of antimicrobials is extremely attractive.

Coaggregation may have potential to be used in a more manipulative fashion, within polymicrobial biofilm communities. It is possible that the use of a coaggregating organism that is antagonistic to pathogens will represent a *guerrilla warfare approach* to preventing the colonization or expansion of pathogenic populations in polymicrobial biofilms. For example, building upon early work by Reid and coworkers (1988), recent work by Younes et al. (2012) has raised the potential utility of coaggregation as a tool to manipulate urogenital biofilm communities. Specifically, coaggregating urogenital lactobacilli have been shown to coaggregate with urogenital pathogens and inactivate their pathogenic potential (Reid et al. 1988; Reid and Burton 2002; Younes et al. 2012). Younes and colleagues stated that coaggregation "creates a hostile microenvironment around a pathogen. With antimicrobial options fading, it therewith becomes increasingly important to identify lactobacilli that bind strongly with pathogens" (Younes et al. 2012).

Unfortunately, approaches to use coaggregation to control polymicrobial biofilms have yet to gain substantial traction. This is likely because coaggregation was originally considered to be a phenomenon unique to the human oral cavity (Handely et al. 2001). Not until the mid-1990s did coaggregation begin to receive attention by researchers in other fields, and this attention has since accelerated (Vandevoorde et al. 1992; Kmet et al. 1995; Drago et al. 1997; Kmet and Lucchini 1997; Egwari et al. 2000; Rickard et al. 2002; Malik and Kakii 2003; Edwards et al. 2006; Hill et al. 2010). This is in part likely driven by increasing interest in polymicrobial biofilm communities and the developmental processes that lead to their development. It has now become increasingly clear that the vast majority of coaggregation interactions between bacteria are mediated by protein lectin-like adhesins and receptor polysaccharides on partner cells (Rickard et al. 2003). In many instances, coaggregation can be inhibited by the addition of simple sugars or chelating agents such as EDTA—which presumably results in competitive inhibition and the alteration of structure, respectively, of receptor polysaccharides. While yet to be evaluated in complex polymicrobial communities, it is possible that the addition of sugars or chelators, that inhibit coaggregation, may retard biofilm development. Further stressing the importance of coaggregation and the possible usefulness of inhibiting the activity of adhesins or receptors, de Toleedo and colleagues (2012) reported that oral *S. oralis* coaggregation receptor polysaccharides induce inflammatory responses in human aortic endothelial cells and may contribute to the development of infective endocarditis and atherosclerosis, both of which can contain significant polymicrobial biofilm communities (Fitzgerald et al. 2006; Kozarov et al. 2006; Rivera et al. 2013). Thus, control of coaggregation in polymicrobial dental plaque biofilms may have health relevance to systemic health, and coaggregation interactions may be a reason why poor oral health is a risk factor for poor systemic health (Rogers 1976; Seymour et al. 2007; Kebschull et al. 2010).

4 Technology-Based Strategies

Technology has always been at the forefront of approaches to control biofilms. From the use of scouring pigs used for cleaning pipelines (Tiratsoo 1999) to the ultrasonic dental hygiene devices that are used on patients in the dentist's chair (Walmsley et al. 1992), there has often been a link between mechanical-based and electrical-based technology and the use of chemical antimicrobials/anti-biofilm agents. Over the last decade, biological-based and technological-based technologies have begun to diverge in some instances. This divergence has both advantages and disadvantages, but in general newer electrical or mechanical technologies have the potential to meet or even exceed the ability to control polymicrobial communities. Obviously, a key concern with such a divergence is that research teams cannot cope with a broad multi-disciplinary scope. Such a possible problem was recognized by the Association of Professors of Medicine (APM) and the National Institutes of Health (NIH) in the early 2000s. Subsequently, the APM made recommendations to the NIH to support multi-disciplinary teams to overcome structural barriers to team science (Crist et al. 2003). Similar recommendations and supporting infrastructure changes were also implemented outside the USA by other countries and this has, in part, led to the development of a bewildering number of technology-based approaches to control biofilms. In this section we will, because of the breadth of emerging technologies, cover some biofilm control technologies that have garnered public and scientific recognition and also have the potential to very efficiently treat different polymicrobial biofilms under a variety of environmental conditions. As with the *Biological Strategies* section, this section is by no means all-encompassing and only salient details are given. We suggest that where readers take interest in a given topic area, they consider starting with the assigned references to gain further insight before pursuing more in-depth research, especially if the reader is not familiar with electronic, mechanical, or chemical engineering.

4.1 Photoactivation Technology

A technology that has recently been developed for use in biofilm removal is that of photodynamic therapy. The term "photodynamic therapy" is an all-encompassing one, describing a broad approach to using nontoxic light-sensitive compounds which, when exposed to certain wavelengths of light, become toxic to cells

(Hamblin and Mróz 2008). It is also known by alternate names such as "photodynamic inactivation" or "photo-activated disinfection," though the latter is usually reserved exclusively for dentistry (O'Riordan et al. 2006; Bergmans et al. 2008). Though its application for use against biofilms and bacteria is relatively recent, the use of photodynamic therapies has been around for much longer—in fact, it has been used against cancer cells as early as the 1970s (Dolmans et al. 2003). There are numerous dyes that can be used as photosensitizers, but from a clinically relevant perspective the list is more limited. This is the result of having to choose a photosensitizer that is both nontoxic to humans in its normal state, as well as one that is ideal for its target site. The most common photosensitizer for the purpose of anti-biofilm use appears to be the class of phenothiazines which include toluidine blue O and methylene blue (Usacheva et al. 2001).

Photodynamic therapy works through the introduction of various detrimental particles to a target area, such as a site of infection or a polymicrobial biofilm. Upon excitation at a defined wavelength of light, the photosensitizer is transformed to its active form in which free radicals and reactive oxygen species are produced (Ochsner 1997). These products are the active agents that damage the targeted cells. While photodynamic therapy may be limited in scope due to its inability to treat systemic infection, for more localized infections it may be more advantageous than traditional treatments. As Demidova and Hamblin note, the actual photosensitizers themselves can be chosen based on the target area and the light source can be aimed, thus specifically providing a two-pronged attack as far as local specificity is concerned (Demidova and Hamblin 2004). Further, such ability to localize the treatment can provide a clear advantage over certain systemic therapies such as antibiotic use, which can negatively affect other aspects of the body. This can also be considered a limitation, however, as it is confines photodynamic therapy to localized, accessible infections. Even more promising is the discovery that photodynamic therapy can extend beyond just the killing of cells, by being able to destroy certain virulence factors such as LPS from *E. coli* and proteases from *P. aeruginosa* (Kömerik et al. 2007).

One important aspect of photodynamic therapy is that it does not seem to produce negative effects in mammalian tissue, as suggested from animal studies (Komerik et al. 2002). This is the result of choosing photosensitizers that do not affect mammalian cells, which may also be aided by the extremely short life of the oxidative species (Moan and Berg 1991; Demidova and Hamblin 2004). An additional positive aspect is the ability of photodynamic therapy to affect a wide range of microorganisms. Unlike antibiotics, photodynamic therapy has been shown to be effective when used not only against bacteria but also fungi and viruses as well (Hamblin and Hasan 2004). It should be noted however that different photosensitizers are often needed based on the nature of the target cells—for instance, with Gram-positive and Gram-negative bacteria. Due to the cell wall differences between these two cell types, it was found that Gram-positive species were more susceptible to photosensitizers of virtually any charge, while Gram-negative species were generally limited to cationic ones, unless the cell wall was altered through additional means (Maisch 2009).

The use of photodynamic therapy has a key advantage in that bacterial resistance is unlikely to develop (Tang et al. 2007). One of the largest clinical impacts of this technology seems to be in dental fields, as there are concerns that the use of current antimicrobial formulations can select for bacterial resistance (Roberts and Mullany 2010). This is a very logical application given the tendencies of biofilms to rapidly form in the oral cavity (Diaz et al. 2006; Teles et al. 2012) and, from a technical standpoint, the ease of applying photosensitizers and subsequent light treatment to the affected area. While the clinical data appear to be rather limited at this point, there have been studies illustrating inactivation of dental plaque biofilm species and various pathogens that are known to cause periodontal disease and dental caries (Burns et al. 1995; Rovaldi et al. 2000). Wilson made the connection that the ability to kill *S. mutans* using this technology may result in major innovations for both the prevention and treatment of dental caries (Wilson 2004). The promise of photodynamic therapy is immense and while it is still being successfully used in some clinical applications, its potential in the killing of surface biofilms from the human body is alluring.

4.2 Nonthermal Plasma Technology

One of the more exciting technologies that have recently garnered traction is that of low-temperature or nonthermal plasma, also known as "cool plasma." Plasma is a unique state of matter that results from the rapid ionization of a gas (Laroussi 2012). Plasma is commonly obtained through subjecting gas to extremely high temperatures, but it can also be obtained by passing gas through high-voltage electricity (Rupf et al. 2011; Laroussi 2012; Traba et al. 2013). The latter method allows for the gas to be ionized into plasma at a much lower temperature, one that is cool enough to come in close contact with tissue without risk of thermal damage (Fig. 3). Many different gases have been used for nonthermal plasma with the most common being argon, nitrogen, helium, or some combination of these (Rupf et al. 2011; Gasset et al. 2012; Traba et al. 2013). Nonthermal plasma itself is not a new technology, but its application in inactivating bacteria, even within biofilms, is extremely innovative (Ermolaeva et al. 2011) (Fig. 3). The generation of plasma at mild temperatures facilitates direct contact with many surfaces on the body that might harbor biofilms including teeth, skin, and even open wounds. Fortunately, studies have already shown that nonthermal plasma poses little immediate danger to skin cells, as the amount of UV light that is emitted is well below the threshold for damage and the temperatures that are generated are apparently insufficient to cause human cell damage (Lademann et al. 2011). While this does not provide complete assurance, to our knowledge there do not seem to be any clinical data regarding detrimental effects from exposure to nonthermal plasma and long-term negative effects associated with repeated use.

Nonthermal plasma is an interesting tool because it is capable of eradicating biofilms in two ways. First, there is a direct killing of the cells which make up the biofilm involved, and second a physical removal of the biofilm mass from the surface occurs. Plasma kills microorganisms through the production of highly

Fig. 3 (a) Generation of nonthermal plasma. (b) Photograph showing a zone of inhibition on a plate of *Staphylococcus aureus* after treatment with non-thermal plasma for approximately 45 seconds. Images courtesy of Mr. Luke Raymond and Miss Ella Dolan, University of Michigan

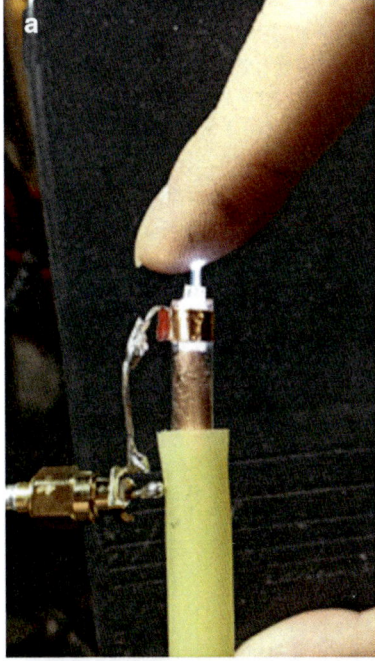

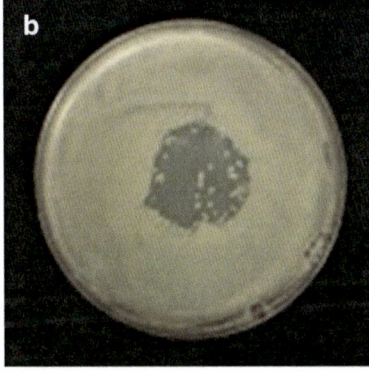

reactive particles that are produced during the ionization of gas including free radicals, UV radiation, and reactive oxygen species (Ehlbeck et al. 2011). Alkawareek et al. suggest that given all the possible cytotoxic products produced by plasma, it is difficult to ascertain whether cell death is derived from a combined effect or individual products of the plasma generation process (Shimizu et al. 2008; Alkawareek et al. 2012). In the second process, the biofilm is removed through plasma exposure in a process called "etching" and is a direct result of the binding of reactive species to outer parts of the cells and subsequent destabilization of the biofilm through cellular de-adhesion (Moisan et al. 2002). Low-powered and high-powered nonthermal plasma both seem to have this ability, but there are advantages and disadvantages to each. Specifically, high-powered nonthermal plasma can kill very effectively and quickly, but the rate at which it etches the biofilms is so fast that often living cells are removed, potentially resulting in contamination of other areas such as a bodily surface or medical devices. Conversely, low power plasma takes a longer exposure time to both kill and remove cells, but because etching happens at a much later stage in treatment, all cells are much more likely to have been killed before they are removed (Traba et al. 2013).

These above-described characteristics make plasma a very alluring candidate for various applications in the medical field. One major potential application is for the cleaning of medical devices. A study done by Rupf and colleagues illustrated the proficiency of plasma to successfully kill biofilms on titanium surfaces (Rupf et al. 2011). This is significant because titanium is often used for medical implants due to its unique surface properties which allows for great biocompatibility. Unfortunately, this same surface also makes it a prime target for biofilm development, and it can be challenging to sterilize without degrading the integrity of the surface (Burgers et al. 2010; Traba et al. 2013). Another exciting avenue is the use in wound treatment. Early clinical trials in humans have indicated that nonthermal plasma can decrease bacterial load in chronic wound infections (Isbary et al. 2010), and similar results have been reported in animal models (Ermolaeva et al. 2011).

4.3 Nanotechnology-Based Technologies

The field of nanotechnology offers a promising avenue of research in the fight against biofilms. An example of such technology is based upon nanoparticles of nanoparticles. Nanoparticles are extremely small bead-like structures with sizes generally ranging between 1 and 100 nm although most are under 50 nm (Haynes et al. 2002). A key advantage to using nanoparticles is that their physicochemical properties (e.g., size, structure, composition, hydrophobicity, charge, etc.) may be "tuned" by varying the precursors and procedures in their construction (Bagwe et al. 2006; Wang et al. 2009; Cheng et al. 2013). As a consequence, nanoparticles have found places for use in biomolecular sensing, microbiological and macro-biological imaging (e.g., quantum dots), drug delivery, and disease therapy (Bhardwaj et al. 2009; Wang et al. 2009). Numerous different types of nanoparticles for use as antimicrobial agents have been documented and often involve the use of metals. While these can be individual metals or combinations, the most notable appears to be silver for microbial control. This is not surprising considering that silver has long been known to have antimicrobial properties (Slawson et al. 1992). Sondi et al. have shown that silver nanoparticles created via the reduction of silver ions using ascorbic acid are able to prevent growth of *E. coli* (Sondi and Salopek-Sondi 2004). Rather than using the nanoparticle itself as the antimicrobial source, there are also other ways of using biologically inert nanoparticles to deliver the antimicrobial agents. In most of these cases, the active agent(s) are attached to the surface of the nanoparticles which acts as a carrier support for localized delivery to species within the polymicrobial biofilm and a release of a much greater concentration of the antimicrobial (Wilkinson et al. 2011). For instance, attaching nitric oxide to silica nanoparticles has proven to be extremely potent at killing Gram-positive and Gram-negative species within biofilms, with *P. aeruginosa* and *E. coli* being most susceptible (Hetrick et al. 2009). Similarly, there are nanoparticles that can carry more than one antibiotic, for simultaneous release, and different combinations of multiple drugs on each particle can be used to overcome the antimicrobial resistance mechanisms observed in biofilms and resident species (Pelgrift and Friedman 2013).

Perhaps one of the more interesting uses of nanoparticles is based upon the use of liposomes. Liposomes are artificial vesicles composed of a phospholipid bilayer that can be used to encapsulate specific molecules of interest (Zhang and Granick 2006). Liposomes can thus be used for the delivery of antimicrobials to bacteria and even bacteria in biofilms (Jones 2005) and have a key benefit that water-soluble or oil-soluble compounds can be delivered to targeted biofilms and the surfaces on which they reside. Unfortunately, liposomes also have the tendency to coalesce (fuse with one another) and have limited shelf life. Work by Zhang and colleagues (2010) demonstrated that nanoparticles could be used to stabilize liposome structure (increasing shelf life) and also prevent them from coalescing.

4.4 Antimicrobial Surfaces and Coatings

One of the most vital components to the development of any biofilm is an appropriate surface to adhere, for without such a surface the complex network of cell–cell interactions can never occur. Indeed, surface properties have been of particular interest for exploitation in the field of marine biofouling (Tribou and Swain 2010; Banerjee et al. 2011; Scardino and de Nys 2011). Conceivably, rendering a surface resilient to the development of a conditioning film and/or toxic to a bacterium will make it non-colonizable and is likley more effective than controlling established recalcitrant biofilms. While there are many examples of variations of this technology, a couple of interesting approaches will be briefly described.

Much effort has been invested into the design of surfaces that can prevent biofilm formation. There is a vast array of surface types that use different strategies to achieve anti-adhesion of bacterial cells. One such option is to coat a surface directly with an antibiotic. For instance, coating surfaces with vancomycin and gentamicin has been shown to be effective for prophylactic purposes (Radin et al. 1997; Alt et al. 2006). Such coatings can be made for controlled release over a relatively short period of time or covalently bonded to the desired surface for long-term protection (Lucke et al. 2003; Edupuganti et al. 2007). Although what could conceivably be considered a generally sound approach, this strategy is inadequate against bacteria that are already resistant to the antimicrobial agent. To overcome this shortcoming, Inbakandan and colleagues found that silver-coated nanoparticles can be an effective solution when used as an antifouling coating (Inbakandan et al. 2013).

Another interesting coating material is polyethylene glycol (PEG). Although not a relatively new technology, PEG has gained notoriety as a frequently used antibiofilm coating agent because it prevents the adsorption of macromolecules and proteinaceous bacterial cell–surface appendages onto the surfaces (Park et al. 1998; Kingshott and Griesser 1999). Indeed, using PEG technology, it has been shown that liposomes containing ciprofloxacin that are sequestered within a PEG-gelatin hydrogel inhibit biofilm formation on catheters (DiTizio et al. 1998; Finelli et al. 2002). Such findings demonstrate how different technologies can be successfully used in concert with one another to prevent biofilm development through inhibiting colonization.

A recent and novel approach takes an entirely different route in surface protection. This form utilizes a liquid-infused surface, rather than traditional solid surfaces. In this method, a liquid coating is held in place by filling in and adhering to a very porous solid base layer (Epstein et al. 2012). This allows for the anti-biofilm properties of the liquid, which is a result of its constant motion, to always remain as a barrier to any potential colonization by bacteria. Remarkably, the liquid surface has the ability to maintain its homogeneity even when submerged in another liquid, provided that the liquid layer chosen for protection is more chemically adherent to the porous surface than the liquid in which it is submerged (Wong et al. 2011). Further, due to the nature of the liquid layer, it is able to self-repair damage due to the tendency of the liquid to naturally flow to the damaged part of the layer (Wong et al. 2011).

5 Concluding Remarks: It's Not What You Have, It's What You Do with It

While this chapter has only scraped the surface of describing novel technologies and approaches that are being developed to control polymicrobial biofilms, it should be noted that many of these concepts are still in their infancy. With the increasing concern of the spread of antimicrobial resistance and the link of numerous chronic illnesses to polymicrobial biofilms (e.g., chronic wounds and periodontal disease), it is clearly evident that research into these novel approaches and technologies should be further accelerated. However, other concerns, such as methods of study and approaches to testing, also need to be addressed. For example, are the laboratory model systems that microbiologists typically use to assess the effectiveness of a treatment against a polymicrobial biofilm appropriate? In 1964, Harry Smith (1964) made the following profound and precocious statement, "While many microbiologists advocate studying microbial behavior under natural conditions, few of them do so. This is because their morale for overcoming the difficulties is constantly sapped by the attractive ease of working with laboratory cultures."

Indeed, the use of growth media is extremely important and bacteria respond very differently to challenges when grown in different growth media. As a case in point, Umland et al. (2012) demonstrated that genes identified as being possible antimicrobial targets in *Acinetobacter baumannii*, using nutrient media, overlapped poorly with those identified by growing the bacterium in human ascites, an ex vivo medium that reflects the infection environment. The authors underscored the importance of using "clinically relevant media and in vivo validation while screening for essential genes for the purpose of developing new antimicrobials" (Umland et al. 2012). Another example demonstrating the importance of environmentally germane conditions was presented by Du and Kolenbrander (2000) who demonstrated that various genes, including those that produce coaggregation adhesins, are upregulated in human saliva compared with brain heart infusion medium. Such issues are likely compounded if (as in polymicrobial biofilms) multiple species are present in the community that is being studied because interactions between component species may be masked (Kolenbrander 2011). In addition, it is becoming clear that not only is growth of species in laboratory media likely to elicit different responses by bacteria as compared to growth in real-world environmental conditions, but repetitive growth in such un-representative laboratory media can lead to the clonal selection of strains with characteristics unlike the originally isolated wild-type progenitor (Sato et al. 2002; Davidson et al. 2008; McLoon et al. 2011). Thus, approaches to directly assess the effectiveness of new technologies to control polymicrobial biofilms need to be performed under conditions representative of the natural environment and with multispecies communities that have not been grown in artificial conditions. Such a notion has begun to make its way into polymicrobial biofilm research studies and model systems that are located in the environment being studied, such as retrievable enamel chip models in dental plaque biofilm studies (Palmer et al. 2001) or using in vivo bioluminescence-based technologies (Chauhan et al. 2012; Vande Velde et al. 2013), are being developed. Alternatively, models are being refined to closely replicate the original environment by using harvested *real-world* milieu and polymicrobial biofilm material. Examples range from dental plaque biofilm microfluidic devices (Nance et al. 2013) to constant depth film fermenter domestic drain microcosms (McBain et al. 2003). With the use of representative model systems and real-world polymicrobial communities, the evaluation of the effectiveness of new technologies to inhibit or control polymicrobial biofilms will likely be more sensitive and revealing.

In conclusion, a thankfully unbridled understanding of the nature as well as approaches to control polymicrobial communities has developed over the last few decades. Instead of being reductionist in our approaches, we have begun to be holistic and consider polymicrobial communities at individual and multispecies levels. This has allowed us to unravel the reasons why bacteria within polymicrobial biofilms possess enhanced abilities, compared to free-floating communities, and why effective biological or technological strategies to control these communities need to take into account the properties of individuals within these communities and the properties of biofilm as a whole. The recent developments in biological and technological approaches to polymicrobial biofilm control offer real prospects for developing new strategies against biofilms of relevance to healthcare or industry alike. Nevertheless, significant hurdles remain. Finding strategies that are both safe and efficacious is not trivial and this may explain why the road to successful biofilm control measures has seen relatively few successes to date. Combinations of mechanical methods with chemical agents are perhaps the most likely candidates for successful biofilm control. However, each biofilm system is distinct and it cannot be expected that one approach will work for all polymicrobial biofilms. Therefore, it is important to keep developing new ideas and to maintain the momentum for research on techniques to combat polymicrobial biofilm communities.

References

Ahmed NA, Petersen FC, Scheie AA (2007) AI-2 quorum sensing affects antibiotic susceptibility in *Streptococcus anginosus*. J Antimicrob Chemother 60:49–53

Ahmed NA, Petersen FC, Scheie AA (2009) AI-2/LuxS is involved in increased biofilm formation by *Streptococcus intermedius* in the presence of antibiotics. Antimicrob Agents Chemother 53:4258–4263

Alanis AJ (2005) Resistance to antibiotics: are we in the post-antibiotic era? Arch Med Res 36:697–705

Alberghini S, Polone E, Corich V, Carlot M, Seno F, Trovato A, Squartini A (2009) Consequences of relative cellular positioning on quorum sensing and bacterial cell-to-cell communication. FEMS Microbiol Lett 292:149–161

Alkawareek MY, Algwari QT, Laverty G, Gorman SP, Graham WG, O'Connell D, Gilmore BF (2012) Eradication of *Pseudomonas aeruginosa* biofilms by atmospheric pressure non-thermal plasma. PloS One 7:e44289

Allukian M Jr, Adekugbe O (2008) The practice and infrastructure of dental public health in the United States. Dent Clin North Am 52:259–280, v

Alt V, Bitschnau A, Osterling J et al (2006) The effects of combined gentamicin–hydroxyapatite coating for cementless joint prostheses on the reduction of infection rates in a rabbit infection prophylaxis model. Biomaterials 27:4627–4634

Anderl JN, Franklin MJ, Stewart PS (2000) Role of antibiotic penetration limitation in *Klebsiella pneumoniae* biofilm resistance to ampicillin and ciprofloxacin. Antimicrob Agents Chemother 44:1818–1824

Anon. (1952) The history of the prevention of fouling. Marine fouling and its prevention (edition USNI), pp 211–223

Bagwe RP, Hilliard LR, Tan W (2006) Surface modification of silica nanoparticles to reduce aggregation and nonspecific binding. Langmuir 22:4357–4362

Bandara HM, Lam OL, Jin LJ, Samaranayake L (2012) Microbial chemical signaling: a current perspective. Crit Rev Microbiol 38:217–249

Banerjee I, Pangule RC, Kane RS (2011) Antifouling coatings: recent developments in the design of surfaces that prevent fouling by proteins, bacteria, and marine organisms. Adv Mater 23:690–718

Bergmans L, Moisiadis P, Huybrechts B, Van Meerbeek B, Quirynen M, Lambrechts P (2008) Effect of photo-activated disinfection on endodontic pathogens ex vivo. Int Endod J 41:227–239

Bernier SP, Beeston AL, Sokol PA (2008) Detection of N-acyl homoserine lactones using a traI-luxCDABE-based biosensor as a high-throughput screening tool. BMC Biotechnol 8:59

Bhardwaj SB, Mehta M, Gauba K (2009) Nanotechnology: role in dental biofilms. Indian J Dent Res 20:511–513

Bigger JW (1944) Treatment of Staphylococcal infections with penicillin. Lancet ii:497–499

Boles BR, Horswill AR (2008) Agr-mediated dispersal of *Staphylococcus aureus* biofilms. PLoS Pathog 4:e1000052

Bos R, van der Mei HC, Busscher HJ (1999) Physico-chemistry of initial microbial adhesive interactions–its mechanisms and methods for study. FEMS Microbiol Rev 23:179–230

Bridier A, Briandet R, Thomas V, Dubois-Brissonnet F (2011) Resistance of bacterial biofilms to disinfectants: a review. Biofouling 27:1017–1032

Brown MR, Allison DG, Gilbert P (1988) Resistance of bacterial biofilms to antibiotics: a growth-rate related effect? J Antimicrob Chemother 22:777–780

Burgers R, Gerlach T, Hahnel S, Schwarz F, Handel G, Gosau M (2010) In vivo and in vitro biofilm formation on two different titanium implant surfaces. Clin Oral Implants Res 21:156–164

Burghard, Leishman, Moynihan, Wright (1915) Office internatonal d'hygiene publique. vii: 946

Burns T, Wilson M, Pearson GJ (1995) Effect of dentine and collagen on the lethal photosensitization of *Streptococcus mutans*. Caries Res 29:192–197

Buroni S, Pasca MR, Flannagan RS et al (2009) Assessment of three resistance-nodulation-cell division drug efflux transporters of *Burkholderia cenocepacia* in intrinsic antibiotic resistance. BMC Microbiol 9:200

Callow ME (1993) A review of fouling in freshwaters. Biofouling 7:313–327

Carrel A, Dehelly G (1917) The treatment of infected wounds. Paul Hoeber, New York

Chaignon P, Sadovskaya I, Ragunah C, Ramasubbu N, Kaplan JB, Jabbouri S (2007) Susceptibility of staphylococcal biofilms to enzymatic treatments depends on their chemical composition. Appl Microbiol Biotechnol 75:125–132

Characklis WG (1990) Biofilm processes. In: Characklis WG, Marshall KC (eds) Biofilms. Wiley, New York, pp 195–231

Chauhan A, Lebeaux D, Decante B, Kriegel I, Escande MC, Ghigo JM, Beloin C (2012) A rat model of central venous catheter to study establishment of long-term bacterial biofilm and related acute and chronic infections. PloS One 7:e37281

Chen X, Schauder S, Potier N, Van Dorsselaer A, Pelczer I, Bassler BL, Hughson FM (2002) Structural identification of a bacterial quorum-sensing signal containing boron. Nature 415:545–549

Cheng J, Gu YJ, Cheng SH, Wong WT (2013) Surface functionalized gold nanoparticles for drug delivery. J Biomed Nanotechnol 9:1362–1369

Chung J, Goo E, Yu S et al (2011) Small-molecule inhibitor binding to an N-acyl-homoserine lactone synthase. Proc Natl Acad Sci USA 108:12089–12094

Cisar JO, Kolenbrander PE, McIntire FC (1979) Specificity of coaggregation reactions between human oral streptococci and strains of *Actinomyces viscosus* or *Actinomyces naeslundii*. Infect Immun 24:742–752

Coenye T, Van Acker H, Peeters E, Sass A, Buroni S, Riccardi G, Mahenthiralingam E (2011) Molecular mechanisms of chlorhexidine tolerance in *Burkholderia cenocepacia* biofilms. Antimicrob Agents Chemother 55:1912–1919

Cosgrove SE (2006) The relationship between antimicrobial resistance and patient outcomes: mortality, length of hospital stay, and health care costs. Clin Infect Dis 42(Suppl 2):S82–S89

Costerton JW (1995) Overview of microbial biofilms. J Ind Microbiol 15:137–140

Costerton JW, Cheng KJ, Geesey GG, Ladd TI, Nickel JC, Dasgupta M, Marrie TJ (1987) Bacterial biofilms in nature and disease. Annu Rev Microbiol 41:435–464

Crist TB, Walsh RA, Alexander RW et al (2003) The research community now and in the future: APM's recommendations for NIH priorities. Am J Med 114:710–713

Das T, Sharma PK, Busscher HJ, van der Mei HC, Krom BP (2010) Role of extracellular DNA in initial bacterial adhesion and surface aggregation. Appl Environ Microbiol 76:3405–3408

Davidson CJ, White AP, Surette MG (2008) Evolutionary loss of the rdar morphotype in *Salmonella* as a result of high mutation rates during laboratory passage. ISME J 2:293–307

Davies D (2003) Understanding biofilm resistance to antibacterial agents. Nat Rev Drug Discov 2:114–122

Davies DG, Marques CN (2009) A fatty acid messenger is responsible for inducing dispersion in microbial biofilms. J Bacteriol 191:1393–1403

De Kievit TR, Parkins MD, Gillis RJ et al (2001) Multidrug efflux pumps: expression patterns and contribution to antibiotic resistance in *Pseudomonas aeruginosa* biofilms. Antimicrob Agents Chemother 45:1761–1770

de Toledo A, Nagata E, Yoshida Y, Oho T (2012) *Streptococcus oralis* coaggregation receptor polysaccharides induce inflammatory responses in human aortic endothelial cells. Mol Oral Microbiol 27:295–307

Declerck P (2010) Biofilms: the environmental playground of *Legionella pneumophila*. Environ Microbiol 12:557–566

Demidova TN, Hamblin MR (2004) Photodynamic therapy targeted to pathogens. Int J Immunopathol Pharmacol 17:245–254

Diaz PI, Chalmers NI, Rickard AH, Kong C, Milburn CL, Palmer RJ Jr, Kolenbrander PE (2006) Molecular characterization of subject-specific oral microflora during initial colonization of enamel. Appl Environ Microbiol 72:2837–2848

Dickschat JS (2010) Quorum sensing and bacterial biofilms. Nat Prod Rep 27:343–369

DiTizio V, Ferguson GW, Mittelman MW, Khoury AE, Bruce AW, DiCosmo F (1998) A liposomal hydrogel for the prevention of bacterial adhesion to catheters. Biomaterials 19:1877–1884

Dobretsov S, Teplitski M, Paul V (2009) Mini-review: quorum sensing in the marine environment and its relationship to biofouling. Biofouling 25:413–427

Dolmans DE, Fukumura D, Jain RK (2003) Photodynamic therapy for cancer. Nat Rev Cancer 3:380–387

Drago L, Gismondo MR, Lombardi A, de Haen C, Gozzini L (1997) Inhibition of in vitro growth of enteropathogens by new *Lactobacillus* isolates of human intestinal origin. FEMS Microbiol Lett 153:455–463

Du LD, Kolenbrander PE (2000) Identification of saliva-regulated genes of *Streptococcus gordonii* DL1 by differential display using random arbitrarily primed PCR. Infect Immun 68:4834–4837

Duan K, Dammel C, Stein J, Rabin H, Surette MG (2003) Modulation of *Pseudomonas aeruginosa* gene expression by host microflora through interspecies communication. Mol Microbiol 50:1477–1491

Eckert R, He J, Yarbrough DK, Qi F, Anderson MH, Shi W (2006) Targeted killing of *Streptococcus mutans* by a pheromone-guided "smart" antimicrobial peptide. Antimicrob Agents Chemother 50:3651–3657

Edupuganti OP, Antoci V Jr, King SB et al (2007) Covalent bonding of vancomycin to Ti6Al4V alloy pins provides long-term inhibition of *Staphylococcus aureus* colonization. Bioorg Med Chem Lett 17:2692–2696

Edwards AM, Grossman TJ, Rudney JD (2006) Fusobacterium nucleatum transports noninvasive *Streptococcus cristatus* into human epithelial cells. Infect Immun 74:654–662

Egland PG, Palmer RJ Jr, Kolenbrander PE (2004) Interspecies communication in *Streptococcus gordonii-Veillonella atypica* biofilms: signaling in flow conditions requires juxtaposition. Proc Natl Acad Sci USA 101:16917–16922

Egwari LO, Rotimi VO, Coker AO (2000) An experimental mouse model to study the pathogenicity of *Prevotella bivia* and investigations of possible virulence. West Indian Med J 49:20–26

Ehlbeck J, Schnabel U, Polak M et al (2011) Low temperature atmospheric pressure plasma sources for microbial decontamination. J Phys D Appl Phys 44:013002

Eke PI, Dye BA, Wei L, Thornton-Evans GO, Genco RJ, Cdc Periodontal Disease Surveillance workgroup: James Beck GDRP (2012) Prevalence of periodontitis in adults in the United States: 2009 and 2010. J Dent Res 91:914–920

Epstein AK, Wong TS, Belisle RA, Boggs EM, Aizenberg J (2012) Liquid-infused structured surfaces with exceptional anti-biofouling performance. Proc Natl Acad Sci USA 109:13182–13187

Ermolaeva SA, Varfolomeev AF, Chernukha MY et al (2011) Bactericidal effects of non-thermal argon plasma in vitro, in biofilms and in the animal model of infected wounds. J Med Microbiol 60:75–83

Evans DJ, Brown MR, Allison DG, Gilbert P (1990a) Susceptibility of bacterial biofilms to tobramycin: role of specific growth rate and phase in the division cycle. J Antimicrob Chemother 25:585–591

Evans DJ, Allison DG, Brown MR, Gilbert P (1990b) Effect of growth-rate on resistance of gram-negative biofilms to cetrimide. J Antimicrob Chemother 26:473–478

Evans K, Passador L, Srikumar R, Tsang E, Nezezon J, Poole K (1998) Influence of the MexAB-OprM multidrug efflux system on quorum sensing in *Pseudomonas aeruginosa*. J Bacteriol 180:5443–5447

Exner M, Kramer A, Lajoie L, Gebel J, Engelhart S, Hartemann P (2005) Prevention and control of health care-associated waterborne infections in health care facilities. Am J Infect Control 33:S26–S40

Finelli A, Burrows LL, DiCosmo FA, DiTizio V, Sinnadurai S, Oreopoulos DG, Khoury AE (2002) Colonization-resistant antimicrobial-coated peritoneal dialysis catheters: evaluation in a newly developed rat model of persistent *Pseudomonas aeruginosa* peritonitis. Perit Dial Int 22:27–31

Fisher ES, Bynum JP, Skinner JS (2009) Slowing the growth of health care costs—lessons from regional variation. N Engl J Med 360:849–852

Fitzgerald JR, Foster TJ, Cox D (2006) The interaction of bacterial pathogens with platelets. Nat Rev Microbiol 4:445–457

Flemming HC, Wingender J (2010) The biofilm matrix. Nat Rev Microbiol 8:623–633

Ganeshnarayan K, Shah SM, Libera MR, Santostefano A, Kaplan JB (2009) Poly-N-acetylglucosamine matrix polysaccharide impedes fluid convection and transport of the cationic surfactant cetylpyridinium chloride through bacterial biofilms. Appl Environ Microbiol 75:1308–1314

Gasset M, Fricke K, Koban I et al (2012) Atmospheric pressure plasma: a high-performance tool for the efficient removal of biofilms. PLoS One 7:e42539

Gibbons RJ, Nygaard M (1970) Interbacterial aggregation of plaque bacteria. Arch Oral Biol 15:1397–1400

Gibson H, Taylor JH, Hall KE, Holah JT (1999) Effectiveness of cleaning techniques used in the food industry in terms of the removal of bacterial biofilms. J Appl Microbiol 87:41–48

Gilbert P, Maira-Litran T, McBain AJ, Rickard AH, Whyte FW (2002) The physiology and collective recalcitrance of microbial biofilm communities. Adv Microb Physiol 46:202–256

Gillis RJ, White KG, Choi KH, Wagner VE, Schweizer HP, Iglewski BH (2005) Molecular basis of azithromycin-resistant *Pseudomonas aeruginosa* biofilms. Antimicrob Agents Chemother 49:3858–3867

Gu H, Hou S, Yongyat C, De Tore S, Ren D (2013) Patterned biofilm formation reveals a mechanism for structural heterogeneity in bacterial biofilms. Langmuir 29:11145–11153

Guo M, Gamby S, Nakayama S, Smith J, Sintim HO (2012) A pro-drug approach for selective modulation of AI-2-mediated bacterial cell-to-cell communication. Sensors 12:3762–3772

Hall-Stoodley L, Stoodley P (2002) Developmental regulation of microbial biofilms. Curr Opin Biotechnol 13:228–233

Hamblin MR, Hasan T (2004) Photodynamic therapy: a new antimicrobial approach to infectious disease? Photochem Photobiol Sci 3:436–450

Hamblin MR, Mróz P (2008) Advances in photodynamic therapy: basic, translational, and clinical. Artech House, Boston, MA

Handely PS, Rickard AH, High NJ, Leach SA (2001) Coaggregation—is it a universal phenomenum? In: Gilbert P, Allison DG, Brading B, Verran J, Walker J (eds) Biofilm community interactions: chance or necessity? Bioline, Cardiff, pp 1–10

Haynes CL, McFarland AD, Smith MT, Hulteen JC, Van Duyne RP (2002) Angle-resolved nanosphere lithography: manipulation of nanoparticle size, shape, and interparticle spacing. J Phys Chem B 106:1898–1902

He XS, Shi WY (2009) Oral microbiology: past, present and future. Int J Oral Sci 1:47–58

Herzberg M, Kaye IK, Peti W, Wood TK (2006) YdgG (TqsA) controls biofilm formation in *Escherichia coli* K-12 through autoinducer 2 transport. J Bacteriol 188:587–598

Hetrick EM, Shin JH, Paul HS, Schoenfisch MH (2009) Anti-biofilm efficacy of nitric oxide-releasing silica nanoparticles. Biomaterials 30:2782–2789

Hill KE, Malic S, McKee R, Rennison T, Harding KG, Williams DW, Thomas DW (2010) An in vitro model of chronic wound biofilms to test wound dressings and assess antimicrobial susceptibilities. J Antimicrob Chemother 65:1195–1206

Hirsch EF (2008) "The Treatment of Infected Wounds," Alexis Carrel's contribution to the care of wounded soldiers during World War I. J Trauma 64:S209–S210

Hoiby N, Bjarnsholt T, Givskov M, Molin S, Ciofu O (2010) Antibiotic resistance of bacterial biofilms. Int J Antimicrob Agents 35:322–332

Hojo K, Nagaoka S, Ohshima T, Maeda N (2009) Bacterial interactions in dental biofilm development. J Dent Res 88:982–990

Howell-Jones RS, Wilson MJ, Hill KE, Howard AJ, Price PE, Thomas DW (2005) A review of the microbiology, antibiotic usage and resistance in chronic skin wounds. J Antimicrob Chemother 55:143–149

Inbakandan D, Kumar C, Abraham LS, Kirubagaran R, Venkatesan R, Khan SA (2013) Silver nanoparticles with anti microfouling effect: a study against marine biofilm forming bacteria. Colloids Surf B Biointerfaces 111C:636–643

Isbary G, Morfill G, Schmidt HU et al (2010) A first prospective randomized controlled trial to decrease bacterial load using cold atmospheric argon plasma on chronic wounds in patients. Br J Dermatol 163:78–82

Itoh Y, Wang X, Hinnebusch BJ, Preston JF 3rd, Romeo T (2005) Depolymerization of beta-1,6-N-acetyl-D-glucosamine disrupts the integrity of diverse bacterial biofilms. J Bacteriol 187:382–387

Jakubovics NS, Shields RC, Rajarajan N, Burgess JG (2013) Life after death: the critical role of extracellular DNA in microbial biofilms. Lett Appl Microbiol 57(6):467–475

Jiang T, Li M (2013) Quorum sensing inhibitors: a patent review. Expert Opin Ther Pat 23:867–894

Jones MN (2005) Use of liposomes to deliver bactericides to bacterial biofilms. Methods Enzymol 391:211–228

Kaplan JB (2010) Biofilm dispersal: mechanisms, clinical implications, and potential therapeutic uses. J Dent Res 89:205–218

Kaplan JB, Ragunath C, Ramasubbu N, Fine DH (2003) Detachment of *Actinobacillus actinomycetemcomitans* biofilm cells by an endogenous beta-hexosaminidase activity. J Bacteriol 185:4693–4698

Kebschull M, Demmer RT, Papapanou PN (2010) "Gum bug, leave my heart alone!"—epidemiologic and mechanistic evidence linking periodontal infections and atherosclerosis. J Dent Res 89:879–902

Keren I, Kaldalu N, Spoering A, Wang Y, Lewis K (2004) Persister cells and tolerance to antimicrobials. FEMS Microbiol Lett 230:13–18

Kingshott P, Griesser HJ (1999) Surfaces that resist bioadhesion. Curr Opin Solid State Mater Sci 4:403–412

Kjelleberg S, Molin S (2002) Is there a role for quorum sensing signals in bacterial biofilms? Curr Opin Microbiol 5:254–258

Klapper I, Gilbert P, Ayati BP, Dockery J, Stewart PS (2007) Senescence can explain microbial persistence. Microbiology 153:3623–3630

Kmet V, Lucchini F (1997) Aggregation-promoting factor in human vaginal *Lactobacillus* strains. FEMS Immunol Med Microbiol 19:111–114

Kmet V, Callegari ML, Bottazzi V, Morelli L (1995) Aggregation-promoting factor in pig intestinal *Lactobacillus* strains. Lett Appl Microbiol 21:351–353

Kolenbrander PE (1988) Intergeneric coaggregation among human oral bacteria and ecology of dental plaque. Annu Rev Microbiol 42:627–656

Kolenbrander PE (2011) Multispecies communities: interspecies interactions influence growth on saliva as sole nutritional source. Int J Oral Sci 3:49–54

Kolenbrander PE, London J (1993) Adhere today, here tomorrow: oral bacterial adherence. J Bacteriol 175:3247–3252

Kolenbrander PE, Palmer RJ Jr, Rickard AH, Jakubovics NS, Chalmers NI, Diaz PI (2006) Bacterial interactions and successions during plaque development. Periodontol 2000 42:47–79

Kolenbrander PE, Palmer RJ Jr, Periasamy S, Jakubovics NS (2010) Oral multispecies biofilm development and the key role of cell-cell distance. Nat Rev Microbiol 8:471–480

Komerik N, Curnow A, MacRobert AJ, Hopper C, Speight PM, Wilson M (2002) Fluorescence biodistribution and photosensitising activity of toluidine blue o on rat buccal mucosa. Lasers Med Sci 17:86–92

Kömerik N, Wilson M, Poole S (2007) The effect of photodynamic action on two virulence factors of Gram-negative bacteria. Photochem Photobiol 72:676–680

Kozarov E, Sweier D, Shelburne C, Progulske-Fox A, Lopatin D (2006) Detection of bacterial DNA in atheromatous plaques by quantitative PCR. Microbes Infect 8:687–693

Lademann O, Kramer A, Richter H et al (2011) Skin disinfection by plasma-tissue interaction: comparison of the effectivity of tissue-tolerable plasma and a standard antiseptic. Skin Pharmacol Physiol 24:284–288

Lamarche MG, Deziel E (2011) MexEF-OprN efflux pump exports the *Pseudomonas* quinolone signal (PQS) precursor HHQ (4-hydroxy-2-heptylquinoline). PLoS One 6:e24310

Lappin-Scott HM, Bass C (2001) Biofilm formation: attachment, growth, and detachment of microbes from surfaces. Am J Infect Control 29:250–251

Laroussi M (2012) Plasma medicine: applications of low-temperature gas plasmas in medicine and biology. Cambridge University Press, Cambridge

Leriche V, Briandet R, Carpentier B (2003) Ecology of mixed biofilms subjected daily to a chlorinated alkaline solution: spatial distribution of bacterial species suggests a protective effect of one species to another. Environ Microbiol 5:64–71

Lewis K (2010) Persister cells. Annu Rev Microbiol 64:357–372

Lu TK, Collins JJ (2007) Dispersing biofilms with engineered enzymatic bacteriophage. Proc Natl Acad Sci USA 104:11197–11202

Lucke M, Schmidmaier G, Sadoni S, Wildemann B, Schiller R, Haas NP, Raschke M (2003) Gentamicin coating of metallic implants reduces implant-related osteomyelitis in rats. Bone 32:521–531

Lyautey E, Jackson CR, Cayrou J, Rols JL, Garabetian F (2005) Bacterial community succession in natural river biofilm assemblages. Microb Ecol 50:589–601

Mah TF, O'Toole GA (2001) Mechanisms of biofilm resistance to antimicrobial agents. Trends Microbiol 9:34–39

Maisch T (2009) A new strategy to destroy antibiotic resistant microorganisms: antimicrobial photodynamic treatment. Mini Rev Med Chem 9:974–983

Malik A, Kakii K (2003) Intergeneric coaggregations among *Oligotropha carboxidovorans* and *Acinetobacter* species present in activated sludge. FEMS Microbiol Lett 224:23–28

Marsh PD, Bradshaw DJ (1995) Dental plaque as a biofilm. J Ind Microbiol 15:169–175

Martiny AC, Jorgensen TM, Albrechtsen HJ, Arvin E, Molin S (2003) Long-term succession of structure and diversity of a biofilm formed in a model drinking water distribution system. Appl Environ Microbiol 69:6899–6907

McBain AJ, Bartolo RG, Catrenich CE et al (2003) Microbial characterization of biofilms in domestic drains and the establishment of stable biofilm microcosms. Appl Environ Microbiol 69:177–185

McBain AJ, Sufya N, Rickard AH (2013) Biofilm recalcitrance: theories and mechanisms. In: Fraise AP, Maillard J-Y, Sattar SA (eds) Russell, Hugo and Ayliffe's principles and practice of disinfection, preservation and sterilization. Wiley, Hoboken, NJ, pp 87–94

McLoon AL, Guttenplan SB, Kearns DB, Kolter R, Losick R (2011) Tracing the domestication of a biofilm-forming bacterium. J Bacteriol 193:2027–2034

Miller ST, Xavier KB, Campagna SR, Taga ME, Semmelhack MF, Bassler BL, Hughson FM (2004) *Salmonella typhimurium* recognizes a chemically distinct form of the bacterial quorum-sensing signal AI-2. Mol Cell 15:677–687

Min KR, Rickard AH (2009) Coaggregation by the freshwater bacterium *Sphingomonas natatoria* alters dual-species biofilm formation. Appl Environ Microbiol 75(12):3987–3997

Moan J, Berg K (1991) The photodegradation of porphyrins in cells can be used to estimate the lifetime of singlet oxygen. Photochem Photobiol 53:549–553

Moisan M, Barbeau J, Crevier M-C, Pelletier J, Philip N, Saoudi B (2002) Plasma sterilization. Methods and mechanisms. Pure Appl Chem 74:349–358

Nance WC, Dowd SE, Samarian D, Chludzinski J, Delli J, Battista J, Rickard AH (2013) A high-throughput microfluidic dental plaque biofilm system to visualize and quantify the effect of antimicrobials. J Antimicrob Chemother 68:2550–2560

National Center for Health Statistics (2012) Health, United States, 2011: With Special Feature on Socioeconomic Status and Health. Hyattsville, MD

Neu TR (1996) Significance of bacterial surface-active compounds in interaction of bacteria with interfaces. Microbiol Rev 60:151–166

O'Riordan K, Sharlin DS, Gross J et al (2006) Photoinactivation of *Mycobacteria* in vitro and in a new murine model of localized *Mycobacterium bovis* BCG-induced granulomatous infection. Antimicrob Agents Chemother 50:1828–1834

Ochsner M (1997) Photophysical and photobiological processes in the photodynamic therapy of tumours. J Photochem Photobiol B Biol 39:1–18

Palmer RJ Jr, White DC (1997) Developmental biology of biofilms: implications for treatment and control. Trends Microbiol 5:435–440

Palmer RJ Jr, Wu R, Gordon S, Bloomquist CG, Liljemark WF, Kilian M, Kolenbrander PE (2001) Retrieval of biofilms from the oral cavity. Methods Enzymol 337:393–403

Palmer RJ Jr, Diaz PI, Kolenbrander PE (2006) Rapid succession within the *Veillonella* population of a developing human oral biofilm in situ. J Bacteriol 188:4117–4124

Paranhos HF, Silva-Lovato CH, Souza RF, Cruz PC, Freitas KM, Peracini A (2007) Effects of mechanical and chemical methods on denture biofilm accumulation. J Oral Rehabil 34:606–612

Park KD, Kim YS, Han DK, Kim YH, Lee EH, Suh H, Choi KS (1998) Bacterial adhesion on PEG modified polyurethane surfaces. Biomaterials 19:851–859

Parsek MR, Greenberg EP (2005) Sociomicrobiology: the connections between quorum sensing and biofilms. Trends Microbiol 13:27–33

Pearson JP, Van Delden C, Iglewski BH (1999) Active efflux and diffusion are involved in transport of *Pseudomonas aeruginosa* cell-to-cell signals. J Bacteriol 181:1203–1210

Pelgrift RY, Friedman AJ (2013) Nanotechnology as a therapeutic tool to combat microbial resistance. Adv Drug Deliv Rev 65(13–14):1803–1815

Percival SL, Thomas JG, Williams DW (2010) Biofilms and bacterial imbalances in chronic wounds: anti-Koch. Int Wound J 7:169–175

Peters BM, Jabra-Rizk MA, O'May GA, Costerton JW, Shirtliff ME (2012) Polymicrobial interactions: impact on pathogenesis and human disease. Clin Microbiol Rev 25:193–213

Peterson BW, van der Mei HC, Sjollema J, Busscher HJ, Sharma PK (2013) A distinguishable role of eDNA in the viscoelastic relaxation of biofilms. MBio 4:e00497–e00513

Radin S, Campbell JT, Ducheyne P, Cuckler JM (1997) Calcium phosphate ceramic coatings as carriers of vancomycin. Biomaterials 18:777–782

Raes J, Bork P (2008) Molecular eco-systems biology: towards an understanding of community function. Nat Rev Microbiol 6:693–699

Raina S, De Vizio D, Odell M, Clements M, Vanhulle S, Keshavarz T (2009) Microbial quorum sensing: a tool or a target for antimicrobial therapy? Biotechnol Appl Biochem 54:65–84

Rao D, Arvanitidou E, Du-Thumm L, Rickard AH (2011) Efficacy of an alcohol-free CPC-containing mouthwash against oral multispecies biofilms. J Clin Dent 22:187–194

Reid G, Burton J (2002) Use of Lactobacillus to prevent infection by pathogenic bacteria. Microbes Infect 4:319–324

Reid G, McGroarty JA, Angotti R, Cook RL (1988) *Lactobacillus* inhibitor production against *Escherichia coli* and coaggregation ability with uropathogens. Can J Microbiol 34:344–351

Rickard AH, Leach SA, Hall LS, Buswell CM, High NJ, Handley PS (2002) Phylogenetic relationships and coaggregation ability of freshwater biofilm bacteria. Appl Environ Microbiol 68:3644–3650

Rickard AH, Gilbert P, High NJ, Kolenbrander PE, Handley PS (2003) Bacterial coaggregation: an integral process in the development of multi-species biofilms. Trends Microbiol 11:94–100

Rivera MF, Lee JY, Aneja M et al (2013) Polymicrobial infection with major periodontal pathogens induced periodontal disease and aortic atherosclerosis in hyperlipidemic ApoE (null) mice. PLoS One 8:e57178

Roberts AP, Mullany P (2010) Oral biofilms: a reservoir of transferable, bacterial, antimicrobial resistance. Expert Rev Anti Infect Ther 8:1441–1450

Roberts ME, Stewart PS (2005) Modelling protection from antimicrobial agents in biofilms through the formation of persister cells. Microbiology 151:75–80

Rogers AH (1976) The oral cavity as a source of potential pathogens in focal infection. Oral Surg Oral Med Oral Pathol 42:245–248

Rovaldi CR, Pievsky A, Sole NA, Friden PM, Rothstein DM, Spacciapoli P (2000) Photoactive porphyrin derivative with broad-spectrum activity against oral pathogens In vitro. Antimicrob Agents Chemother 44:3364–3367

Roy V, Adams BL, Bentley WE (2011) Developing next generation antimicrobials by intercepting AI-2 mediated quorum sensing. Enzyme Microb Technol 49:113–123

Roy V, Meyer MT, Smith JA, Gamby S, Sintim HO, Ghodssi R, Bentley WE (2013) AI-2 analogs and antibiotics: a synergistic approach to reduce bacterial biofilms. Appl Microbiol Biotechnol 97:2627–2638

Rüdiger H, Gabius HJ (2009) The biochemical basis and coding capacity of the sugar code. In: Gabius HJ (ed) The sugar code: fundamentals of glycosciences. Weinheim, Wiley-VCH Verlag

Rupf S, Idlibi AN, Marrawi FA et al (2011) Removing biofilms from microstructured titanium ex vivo: a novel approach using atmospheric plasma technology. PLoS One 6:e25893

Sato Y, Okamoto K, Kizaki H (2002) gbpC and pac gene mutations detected in *Streptococcus mutans* strain GS-5. Oral Microbiol Immunol 17:263–266

Scardino AJ, de Nys R (2011) Mini review: biomimetic models and bioinspired surfaces for fouling control. Biofouling 27:73–86

Schillinger C, Petrich A, Lux R et al (2012) Co-localized or randomly distributed? Pair cross correlation of in vivo grown subgingival biofilm bacteria quantified by digital image analysis. PLoS One 7:e37583

Schwering M, Song J, Louie M, Turner RJ, Ceri H (2013) Multi-species biofilms defined from drinking water microorganisms provide increased protection against chlorine disinfection. Biofouling 29:917–928

Sen CK, Gordillo GM, Roy S et al (2009) Human skin wounds: a major and snowballing threat to public health and the econom. Wound Repair Regen 17:763–771

Seymour GJ, Ford PJ, Cullinan MP, Leishman S, Yamazaki K (2007) Relationship between periodontal infections and systemic disease. Clin Microbiol Infect 13(Suppl 4):3–10

Shakir A, Elbadawey MR, Shields RC, Jakubovics NS, Burgess JG (2012) Removal of biofilms from tracheoesophageal speech valves using a novel marine microbial deoxyribonuclease. Otolaryngol Head Neck Surg 147:509–514

Shank EA, Kolter R (2009) New developments in microbial interspecies signaling. Curr Opin Microbiol 12:205–214

Shapiro JA, Nguyen VL, Chamberlain NR (2011) Evidence for persisters in *Staphylococcus epidermidis* RP62a planktonic cultures and biofilms. J Med Microbiol 60:950–960

Shields RC, Mokhtar N, Ford M, Hall MJ, Burgess JG, ElBadawey MR, Jakubovics NS (2013) Efficacy of a marine bacterial nuclease against biofilm forming microorganisms isolated from chronic rhinosinusitis. PLoS One 8:e55339

Shimizu T, Steffes B, Pompl R et al (2008) Characterization of microwave plasma torch for decontamination. Plasma Processes Polym 5:577–582

Shirtliff ME, Peters BM, Jabra-Rizk MA (2009) Cross-kingdom interactions: *Candida albicans* and bacteria. FEMS Microbiol Lett 299:1–8

Simoes LC, Simoes M, Vieira MJ (2008) Intergeneric coaggregation among drinking water bacteria: evidence of a role for *Acinetobacter calcoaceticus* as a bridging bacterium. Appl Environ Microbiol 74:1259–1263

Slawson RM, Van Dyke MI, Lee H, Trevors JT (1992) Germanium and silver resistance, accumulation, and toxicity in microorganisms. Plasmid 27:72–79

Smith H (1964) Microbial behavior in natural and artificial systems. In: Smith H, Taylor J (eds) Microbial behavior, "in vivo" and "in vitro". Cambridge University Press, Cambridge, pp 1–29

Smith DM, Snow DE, Rees E et al (2010) Evaluation of the bacterial diversity of pressure ulcers using bTEFAP pyrosequencing. BMC Med Genomics 3:41

Smith F, Dryburgh N, Donaldson J, Mitchell M (2013) Debridement for surgical wounds. Cochrane Database Syst Rev 9, CD006214

Sondi I, Salopek-Sondi B (2004) Silver nanoparticles as antimicrobial agent: a case study on *E. coli* as a model for Gram-negative bacteria. J Colloid Interface Sci 275:177–182

Spoering AL, Lewis K (2001) Biofilms and planktonic cells of *Pseudomonas aeruginosa* have similar resistance to killing by antimicrobials. J Bacteriol 183:6746–6751

Stewart PS, Grab L, Diemer JA (1998) Analysis of biocide transport limitation in an artificial biofilm system. J Appl Microbiol 85:495–500

Stickler DJ, Morris NS, McLean RJ, Fuqua C (1998) Biofilms on indwelling urethral catheters produce quorum-sensing signal molecules in situ and in vitro. Appl Environ Microbiol 64:3486–3490

Stoodley P, Sauer K, Davies DG, Costerton JW (2002) Biofilms as complex differentiated communities. Annu Rev Microbiol 56:187–209

Sufya N, Allison DG, Gilbert P (2003) Clonal variation in maximum specific growth rate and susceptibility towards antimicrobials. J Appl Microbiol 95:1261–1267

Suntharalingam P, Cvitkovitch DG (2005) Quorum sensing in streptococcal biofilm formation. Trends Microbiol 13:3–6

Tang HM, Hamblin MR, Yow CM (2007) A comparative in vitro photoinactivation study of clinical isolates of multidrug-resistant pathogens. J Infect Chemother 13:87–91

Teles FR, Teles RP, Uzel NG, Song XQ, Torresyap G, Socransky SS, Haffajee AD (2012) Early microbial succession in redeveloping dental biofilms in periodontal health and disease. J Periodontal Res 47:95–104

Teles R, Teles F, Frias-Lopez J, Paster B, Haffajee A (2013) Lessons learned and unlearned in periodontal microbiology. Periodontol 2000 62:95–162

Thiel V, Vilchez R, Sztajer H, Wagner-Dobler I, Schulz S (2009) Identification, quantification, and determination of the absolute configuration of the bacterial quorum-sensing signal autoinducer-2 by gas chromatography-mass spectrometry. Chembiochem 10:479–485

Tiratsoo JNH (1999) Pipeline pigging technology. Butterworth-Heinemann, Houston, TX

Traba C, Chen L, Liang JF (2013) Low power gas discharge plasma mediated inactivation and removal of biofilms formed on biomaterials. Curr Appl Phys 13:S12–S18

Tribou M, Swain G (2010) The use of proactive in-water grooming to improve the performance of ship hull antifouling coatings. Biofouling 26:47–56

Tsai AG, Williamson DF, Glick HA (2011) Direct medical cost of overweight and obesity in the USA: a quantitative systematic review. Obes Rev 12:50–61

Umland TC, Schultz LW, MacDonald U, Beanan JM, Olson R & Russo TA (2012) In vivo-validated essential genes identified in Acinetobacter baumannii by using human ascites overlap poorly with essential genes detected on laboratory media. mBio 3

Usacheva MN, Teichert MC, Biel MA (2001) Comparison of the methylene blue and toluidine blue photobactericidal efficacy against gram-positive and gram-negative microorganisms. Lasers Surg Med 29:165–173

Vande Velde G, Kucharikova S, Schrevens S, Himmelreich U, Van Dijck P (2014) Towards non-invasive monitoring of pathogen-host interactions during Candida albicans biofilm formation using in vivo bioluminescence. Cell Microbiol 16(1):115–130

Vandevoorde L, Christiaens H, Verstraete W (1992) Prevalence of coaggregation reactions among chicken lactobacilli. J Appl Bacteriol 72:214–219

Vornhagen J, Stevens M, McCormick DW, Dowd SE, Eisenberg JN, Boles BR, Rickard AH (2013) Coaggregation occurs amongst bacteria within and between biofilms in domestic showerheads. Biofouling 29:53–68

Walmsley AD, Laird WR, Lumley PJ (1992) Ultrasound in dentistry. Part 2–periodontology and endodontics. J Dent 20:11–17

Wang X, Wood TK (2011) Toxin-antitoxin systems influence biofilm and persister cell formation and the general stress response. Appl Environ Microbiol 77:5577–5583

Wang X, Liu LH, Ramstrom O, Yan M (2009) Engineering nanomaterial surfaces for biomedical applications. Exp Biol Med 234:1128–1139

Watters C, Everett JA, Haley C, Clinton A, Rumbaugh KP (2013) Insulin treatment modulates the host immune system to enhance *Pseudomonas aeruginosa* wound biofilms. Infect Immun 82 (1):92–100

Whitchurch CB, Tolker-Nielsen T, Ragas PC, Mattick JS (2002) Extracellular DNA required for bacterial biofilm formation. Science 295:1487

Wilkinson LJ, White RJ, Chipman JK (2011) Silver and nanoparticles of silver in wound dressings: a review of efficacy and safety. J Wound Care 20:543–549

Wilson M (2004) Lethal photosensitisation of oral bacteria and its potential application in the photodynamic therapy of oral infections. Photochem Photobiol Sci 3:412–418

Wimpenny J (1996) Ecological determinants of biofilm formation. Biofouling 10:43–63

Wolcott R, Costerton JW, Raoult D, Cutler SJ (2013) The polymicrobial nature of biofilm infection. Clin Microbiol Infect 19:107–112

Wong T-S, Kang SH, Tang SKY, Smythe EJ, Hatton BD, Grinthal A, Aizenberg J (2011) Bioinspired self-repairing slippery surfaces with pressure-stable omniphobicity. Nature 477:443–447

Xavier KB, Bassler BL (2003) LuxS quorum sensing: more than just a numbers game. Curr Opin Microbiol 6:191–197

Xu X, Stewart PS, Chen X (1996) Transport limitation of chlorine disinfection of *Pseudomonas aeruginosa* entrapped in alginate beads. Biotechnol Bioeng 49:93–100

Younes JA, van der Mei HC, van den Heuvel E, Busscher HJ, Reid G (2012) Adhesion forces and coaggregation between vaginal staphylococci and lactobacilli. PloS One 7:e36917

Yu T, de la Rosa C, Lu R (2004) Microsensor measurement of oxygen concentration in biofilms: from one dimension to three dimensions. Water Sci Technol 49:353–358

Zhang L, Granick S (2006) How to stabilize phospholipid liposomes (using nanoparticles). Nano Lett 6:694–698

Zhang L, Pornpattananangku D, Hu CM, Huang CM (2010) Development of nanoparticles for antimicrobial drug delivery. Curr Med Chem 17:585–594

Ziegler S, Dolch K, Geiger K et al (2013) Oxygen-dependent niche formation of a pyrite-dependent acidophilic consortium built by archaea and bacteria. ISME J 7:1725–1737

Antibiofilm Strategies in the Food Industry

Pilar Teixeira and Diana Rodrigues

Abstract Biofilms in food processing plants represent not only a problem to human health but also cause economic losses by technical failure in several systems. In fact, many foodborne outbreaks have been found to be associated with biofilms. Biofilms may be prevented by regular cleaning and disinfection, but this does not completely prevent biofilm formation. Besides, due to their diversity and to the development of specialized phenotypes, it is well known that biofilms are more resistant to cleaning and disinfection than planktonic microorganisms. In recent years, a considerable effort has been made in the prevention of microbial adhesion and biofilm formation on food processing surfaces and novel technologies have been introduced. In this context, this chapter discusses the main conventional and emergent strategies that have been employed to prevent bacterial adhesion to food processing surfaces and thus to efficiently maintain good hygiene throughout the food industries.

1 Introduction

Food processing environments provide a diversity of favorable conditions for biofilm formation such as the presence of nutrients and moisture and the inocula of microorganisms from raw products. Hence, while totally undesirable, biofilms are formed in all food processing surfaces such as plastic, glass, metal, wood, etc. "Dead zones," like cracks, corners, joints, and gaskets, are places where biofilm can remain after cleaning. In addition, biofilms provide a protective environment, in which exopolymeric substances (EPS) lead to a significantly higher tolerance of biofilm cells to many stresses including disinfectants or sanitizers than to freefloating cells or planktonic cells (Gilbert et al. 2001). These biofilms are potential

---

P. Teixeira (✉) · D. Rodrigues
Institute for Biotechnology and Bioengineering, Centre of Biological Engineering,
University of Minho, Campus de Gualtar, 4710-057 Braga, Portugal
e-mail: pilar@deb.uminho.pt

sources of contamination with the consequent spoilage of foods as well as the transmission of foodborne microorganisms. Moreover, when a biofilm detaches from the surface, individual microorganisms can easily be spread, contaminating the surrounding environment and causing cross and post-processing contamination. In addition, biofilms are often responsible for the interference of mechanical locks in the process of heat transfer, as well as for the increased rate of corrosion on surfaces. In drinking water systems, for instance, biofilms can clog pipes, leading to decreases in speed and capacity, which means increased energy usage. Similarly, biofilm formation in heat exchangers and cooling towers can reduce heat transfer and efficiency. Moreover, the ability of bacteria to persist in biofilms on the metal surfaces of processing facilities can also cause corrosion of the surface due to acid production by bacteria. From the above-mentioned, it can be concluded that biofilms in food industries can cause serious health problems and large economic losses.

Many food safety problems can be avoided if good manufacturing practices codified in 21 CFR 110 are followed (FDA 2004). In fact, most of the problems are due to inefficient hygienic practices among employees, language barriers, ineffective training of employees, the existence of biofilms in niche environments, ineffective use of cleaning agents/disinfectants, lack of sanitary equipment design, reactive instead of routine maintenance, ineffective application of sanitation principles, contamination of raw materials with microorganisms, allergens and/or toxins, post-processing contamination microorganisms, allergens and/or toxins, incorrect labeling or packaging, older equipment (more difficult to clean), corrosion of metal containers/equipment/utensils, and contamination with cleaner/sanitizer residues (FDA 2004). However, it is generally accepted that the main problem of the food industry is the survival of foodborne pathogens or microorganisms that cause food spoilage, due to inadequate disinfection of instruments or surfaces that come in contact with food resulting in the formation of biofilms. Biofilms are problematic mainly in food industry sectors such as dairy processing, brewing, fresh produce, poultry processing, and red meat processing (Frank et al. 2003; Jessen and Lammert 2003; Somers and Wong 2004; Chen et al. 2007). These industries are the principal reservoirs for *Salmonella, Campylobacter, Listeria, Yersinia enterocolitica*, and *Staphylococcus aureus* worldwide, which transmit disease to consumers when the contaminated products are inappropriately cooked (Farber and Peterkin 1991; Dewanti and Wong 1995; Kim et al. 2008).

Since biofilms are a great concern in the food industry, many studies have been performed in order to find an efficient strategy to their control and eradication. However, the most important antibiofilm approach will always be to prevent microbial adhesion and biofilm formation by regular cleaning and disinfection of surfaces.

2 Main Foodborne Pathogens

Illnesses caused by ingestion of contaminated food include a broad range of diseases and are a rising global public health concern. The contamination of food can be caused by microorganisms or chemicals, can take place at every step in the process from food production to consumption, and might be a consequence of environmental contamination (such as pollution of soil, air, or water). Although the main clinical presentation of foodborne illnesses consists of gastrointestinal symptoms, they may also have gynecological, neurological, immunological, and other symptoms. Multiorgan breakdown and even cancer can be caused by the intake of contaminated food and is associated with a substantial burden of disability and mortality (WHO 2013). According to the European Food Safety Authority (EFSA 2012), 5,262 foodborne outbreaks were reported in the European Union in 2010, leading to a large amount of human infections and hospitalizations, and causing 25 deaths. The majority of outbreaks that occurred in 2010 were caused by *Salmonella*, viruses, *Campylobacter*, and bacterial toxins. Besides these microorganisms, *Listeria monocytogenes* and *Escherichia coli* are also among the main foodborne pathogens responsible for severe human infections. Moreover, all mentioned bacteria are known to form biofilms on food contact equipment and food surfaces, causing financial losses and severe health problems (Kumar and Anand 1998; Chae and Schraft 2000; Wirtanen et al. 2000).

L. monocytogenes is a facultative intracellular bacterium that is ubiquitous in the environment and pathogenic to humans, since it causes listeriosis—a predominately foodborne illness that has a higher mortality rate in comparison with other foodborne diseases (EFSA 2012). This bacterium is commonly found in diverse foodstuffs as well as in animal feed, soil, water, plants, sewage, and fecal matter (Moltz and Martin 2005; Tompkin 2002). Moreover, ready-to-eat food, uncooked meat products, vegetables, poultry, and soft cheeses have all been reported as vehicles of listeriosis (Teixeira et al. 2007b; Conter et al. 2009; Jadhav et al. 2012), with ingestion of contaminated food being the main route of transmission for humans (Dussurget 2008). Contamination of food by *L. monocytogenes* may happen through several distinct routes, such as staff equipment, uncooked materials, or contact surfaces (Møretrø and Langsrud 2004; Teixeira et al. 2008). Nevertheless, as far as commercial foodstuff is concerned, contamination by these bacteria is not frequently a consequence of flaws in cleaning and disinfection, but it is due to cross-contamination in the post-processing environment (Ryser and Marth 2007; Latorre et al. 2010). This typically takes place in spaces where organic remains accumulate and biocidal compounds have reduced access (slicers, joints, cutting equipment, etc.), which are favorable for continuous biofilm development and provide an opportunity for some strains to become dominant and persevere at the food plant (Verghese et al. 2011).

Salmonella spp. are a group of food contaminant organisms with significant importance in the food industry. Although there are currently more than 2,500 identified serotypes of *Salmonella*, *Salmonella enterica* serovars Enteritidis and

Typhimurium most commonly cause human disease. It is believed that some salmonellosis outbreaks were due to the inexistent or deficient cleaning and disinfection of surfaces and tools (e.g., Ellis et al. 1998; Reij and Aantrekker 2004; Giraudon et al. 2009; Podolak et al. 2010). In fact, several studies have shown that these bacteria are able to colonize various food contact surfaces (e.g., Teixeira et al. 2007a; Oliveira et al. 2006; Rodrigues et al. 2011), and it was also reported that *Salmonella* adhere and form biofilms in food processing facilities (Joseph et al. 2001). Moreover, it has already been well established that the antimicrobial efficiency of diverse biocidal agents is inferior against these biofilms than for their respective planktonic cells. Accordingly, nine disinfectants usually applied in the food industry and efficient against planktonic *Salmonella* cells revealed a variable efficiency against biofilms, with products containing 70 % ethanol being most efficient (Møretrø et al. 2009). Previous studies have also pointed out that, in comparison to *Salmonella* planktonic cells, biofilms were more resistant to trisodium phosphate (Scher et al. 2005), chlorine, and iodine (Joseph et al. 2001).

Campylobacter spp. are foodborne pathogens with the ability to colonize different inert surfaces (Kusumaningrum et al. 2003; Sanders et al. 2007; Shi and Zhu 2009) and are also frequently isolated from poultry and poultry processing. *Campylobacter jejuni* has been the most predominant strain found in such environments (Deming et al. 1987; Sanders et al. 2007) and, consequently, several studies have been performed in order to understand the behavior of this bacterium (Trachoo et al. 2002; Dykes et al. 2003; Hanning et al. 2008). One of the main findings was that, although *C. jejuni* does not readily form a biofilm, it does form mixed biofilms with enterococci (Trachoo and Brooks 2005), within which it gains a higher tolerance to various chemical biocides (Trachoo and Frank 2002). The fact that *C. jejuni* adhesion and colonization of surfaces is eased by a preexisting biofilm (Hanning et al. 2008) highlights the importance of intensifying the control of biofilms, especially in poultry environments where these bacteria are more commonly found.

E. coli O157:H7 is among the most severe foodborne pathogens, with outbreaks related mainly to ingestion of undercooked meat (Proctor et al. 2002), but also with other contamination routes such as drinkable (Swerdlow et al. 1992) and leisure water (Ackman et al. 1997). The adhesion and biofilm formation ability of *E. coli* O157:H7 on diverse food contact surfaces existent in the meat industry has been investigated, and it was observed that these bacteria adhered to and developed biofilms on such materials, even at low temperatures (Dourou et al. 2011). It was also found that the adhesion of these bacteria was affected by the existence of other microbes on the surfaces (Klayman et al. 2009; Marouani-Gadri et al. 2009). As an example, a study conducted by Habimana and coworkers (2010) showed that *E. coli* O157:H7 cells were entrenched and enclosed in an *Acinetobacter calcoaceticus* biofilm, which is in agreement with several other reports that demonstrated multispecies biofilms enhanced the chances for pathogens to flourish in food processing environments (Habimana et al. 2010; Stewart and Franklin 2008).

Bacillus spp. and especially *Bacillus cereus* are associated with food spoilage (Andersson et al. 1995; Janneke et al. 2007). Since *B. cereus* is ubiquitous in the environment, contamination by this bacterium is quite unavoidable in food industry facilities. As an example, over 12 % of the microbial biofilms found in a commercial dairy plant corresponded to *B. cereus* (Sharma and Anand 2002). In addition, this bacterium produces spores that can endure a large range of adverse conditions and promptly attach to food contact surfaces, due to their highly hydrophobic character (Lindsay et al. 2006). *B. cereus* is responsible for two kinds of gastrointestinal diseases, diarrheal and emetic, and the outbreaks associated with this bacterium have been related to the ingestion of several different food items, such as meat, fish, vegetables, rice, milk, cheeses, pasta, and foodstuff with sauces (puddings, roasted, and salads). Moreover, between 1998 and 2008, 1,229 foodborne outbreaks reported in the USA were caused by this bacterium as well as by *Clostridium perfringens* and *S. aureus* (Bennett et al. 2013).

3 Antibiofilm Strategies in the Food Industry

Microbial adhesion to food processing surfaces is a rather fast process, and therefore, cleaning and disinfection of such surfaces is often not sufficient to prevent the adhesion of microorganisms. In fact, cleaning only removes approximately 90 % of bacteria from surfaces and does not kill them (Srey et al. 2013), so disinfection is crucial. Nevertheless, an adequate frequency of disinfection should be carefully determined to avoid accumulation of both particulates and bacterial cells present on abiotic surfaces. The main strategy to prevent biofilm formation is to avoid bacterial adhesion by choosing the correct materials and performing the appropriate cleaning methods. In this context, it is of utmost importance to use materials that do not promote or even suppress biofilm formation. Antimicrobial agents should be applied to walls, ceilings, and floors. Surfaces should have modified physicochemical properties or be impregnated with biocides or antimicrobials to minimize bacterial colonization (Rogers et al. 1995). Hydrophobic surfaces are more prone to biofilm formation than hydrophilic ones. It is also essential that equipment design is smooth and does not contain faults like crevices, corners, cracks, gaskets, valves, and joints, which are vulnerable areas for biofilm accumulation and not easily accessible to sanitizers. Cleaning and disinfection should be performed regularly before bacteria firmly attach to surfaces. To this end, cleaning-in-place (CIP) procedures have been used and sometimes include physical methods, such as mechanical brushing, chemical agents, such as detergents, and biological agents, like enzymes to obtain a biofilm-free industrial environment (Kumar and Anand 1998). Even with these procedures microorganisms can remain on surfaces. Thus, Good Manufacturing Practice (GMP), Good Hygienic Practices (GHPs), Good Agricultural Practices (GAPs), and Hazard Analysis and Critical Control Points (HACCP) have been established for controlling food quality and safety (Myszka and Czaczyk 2011). The HACCP system has the advantage of improving product safety by anticipating and preventing health hazards before they occur. Nevertheless, adhesion and biofilm formation on food processing surfaces and food spoilage and contamination still occur. In recent years several physical and chemical methods have been developed to avoid/control biofilm formation and will be discussed below.

4 Current Approaches

4.1 Chemical Disinfection

To obtain an efficient disinfection, surfaces should be properly cleaned. However, disinfection can be affected by environmental conditions such as temperature, pH, concentration, contact time, soiling and type of surface or medium to be disinfected, and the presence of organic substances including fat, carbohydrates and protein-based materials (Møretrø et al. 2012). Disinfectants may also differ in their ability to kill target microorganisms. There is a wide range of chemical disinfectants, which can be divided according to their mode of action: oxidizing agents including chlorine-based compounds, hydrogen peroxide, ozone and peracetic acid, surface-active compounds including quaternary ammonium compounds (QACs) and acid anionic compounds, and iodophores (van Houdt and Michiels 2010). Chlorine-based compounds, such as hypochlorite, are widely used in the food industry because chlorine has a broad spectrum of activity, acts fast, and is usually cheap. This compound has been shown to be highly effective against biofilms (Toté et al. 2010; da Silva et al. 2011) and has greater efficacy in low pH than alkaline pH environments (Araújo et al. 2011).

Disinfectants containing hydrogen peroxide or peracetic acid are regarded as environmentally friendly because they decompose into oxygen and water (or acetic acid). Hydrogen peroxide affects the biofilm matrix, has been found to be effective against biofilm cells and is widely used in disinfectants (Robbins et al 2005; Shikongo-Nambabi et al. 2010). Hydrogen peroxide-based disinfectives also have a broad spectrum of activity and act fast. Peracetic acid has the advantage of being relatively stable in the presence of organic compounds compared to other disinfectant types. Several studies have reported its efficacy against biofilms. For instance, Cabeça et al. (2008) showed that 0.50 % w/v peracetic acid reduced 24 h-old *L. monocytogenes* biofilms by 5 log. Similarly, Frank and coworkers (2003) demonstrated that 2.0 mL/L peracetic acid reduced *L. monocytogenes* biofilms more than 6 log on stainless steel in the presence of fat, protein, and soil after 10 min of exposure.

Ozone is regarded as an environmentally friendly disinfectant as it rapidly disintegrates into water and oxygen. Unfortunately, its instability can cause it to react and disintegrate before reaching the target organism. However, ozone is a potent antimicrobial agent, which can be used against bacteria, fungi, viruses, protozoa, and bacterial and fungal spores (Khardre et al. 2001).

Quaternary ammonium compounds (QACs) are active against a range of vegetative bacteria and can be used over a wide temperature so they are widely used in the food industry. However, they are usually not used in CIP because of foaming and their activity is reduced in the presence of hard water. Also, their degradability in the environment is slow and residues may contribute to resistance development in bacteria.

Disinfectants based on alcohols are effective against a wide range of microorganisms and are relatively robust in the presence of organic material. However, their use is limited due to safety reasons (health and flammability) and their relatively high price. Alcohols are therefore mainly used for hand disinfection and on equipment that does not stand in water (Møretrø et al. 2012).

Due to the abovementioned reasons, a disinfectant must be carefully chosen according to the type of application and some aspects must be taken into account: the disinfectant must be environmentally friendly and economical; should be safe to use (nontoxic and nonallergenic), have no negative impact on surface materials (corrosiveness, staining and reactivity), be stable during storage and over a wide range of pH and temperatures, be robust to environmental factors (soil, hard water, and dilution), and have a broad spectrum of activity (Møretrø et al. 2012). Furthermore, it is of the outmost importance to know the mode of growth of the target organisms (i.e., planktonic, adhered, or biofilm). The efficacy of the disinfectant is strongly dependent on this factor because cells within a biofilm are more tolerant to antimicrobial agents than their planktonic counterparts. Wirtanen and Mattila-Sandholm (1992) showed that increased biofilm age may also lead to enhanced resistance against disinfectants and biocides. Usually, to obtain a good sanitary effect, when there is a biofilm present, it is necessary to combine an extensive mechanical action, such as scrubbing or scraping, with the use of cleaning and sanitizing agents. Chemical disinfectants react with the exopolymeric matrix of biofilms, which enhances the mechanical biofilm removal. Otherwise, chemical disinfectants can kill planktonic bacterial cells, while the exopolymeric matrix remains unaffected. Thus, chemical and mechanical treatment can have a synergistic effect in biofilm removal.

4.2 Physical Methods

The most commonly used physical method to remove biofilms is the manual cleaning of surfaces using scrubbers. Pressure washing is another approach currently being used that consists of rinsing surfaces with hot or cold water, the application of a detergent for the required contact time, and rinsing the surface before the application of a disinfectant. Usually, water is applied at 125 °C for 30 min and this method is considered as very effective in eliminating microbial communities. However, Wirtanen and Matilla-Sandholm (1993) verified that 3-day-old biofilms were difficult to completely remove even at this temperature. Kiskó and Szabó-Szabó (2011) also observed that hot water was not sufficient to eliminate *Pseudomonas aeruginosa* and *Pseudomonas stutzeri* biofilms from surfaces. The disadvantage of this method is that hot water denatures proteins and increases the adhesion properties of equipment, which can aid in the formation of biofilms, so it is not advisable. In order to be more efficient in biofilm removal, this method should be combined with chemical disinfection.

Ultrasounds, the application of electrical fields and super-high magnetic fields have been identified as newer physical methods for biofilm control. These approaches will be addressed below.

5 Emergent Approaches

5.1 Ultrasons

Ultrasonication has been reported as an efficient biofilm removal method. This technique is particularly useful in surface decontamination where the inrush of fluid that accompanies cavitational collapse near a surface is nonsymmetric (Chemat et al. 2011). The particular advantage of ultrasonic cleaning in this context is that it can reach crevices that are not easily reached by conventional cleaning methods. The use of ultrasound allows the destruction of a variety of fungi, bacteria, and viruses in a much reduced processing time when compared to thermal treatment at similar temperatures (Chemat et al. 2011). However, by itself, this technique doesn't eliminate all the bacteria in food industries and thus it is recommended to be used in combination with other treatment techniques (Srey et al. 2013). In fact, it has been postulated that ultrasound induces cavitation within the biofilm, which increases transport of solutes, as antimicrobial agents, through the biofilm or outer bacterial membranes (Carmen et al. 2005). Thus, there is a synergistic effect between ultrasound and other antimicrobial agents. For instance, the combination of ultrasound and ethylenediaminetetraacetic acid (EDTA), and ultrasound and enzymes showed a higher efficacy in removing biofilms. Baumann and coworkers (2009) also showed a significant effect on biofilm removal on stainless steel food contact surfaces by combining the use of ozonation and sonication.

5.2 Electrical Methods

Electrical methods for controlling bacterial adhesion have received special attention and are regarded to be environmental friendly because they use "electrons" as the nontoxic reaction mediator. These methods can be divided into current and potential applications, and each application can be conducted in the cathodic, anodic, and block (or alternating) modes (Hong et al. 2008). Electrical methods have been applied in some studies to prevent bacterial adhesion and to detach adhered bacteria, but it was verified that the removed bacteria could again accumulate on the surface and thus the problem of surface contamination continues. Besides, according to Wagnera et al. (2004), when an anodic current or potential is applied, the inactivated bacteria tend to remain on the surface providing new sites for bacterial adhesion. Thus, the control of bacterial adhesion through the exclusive application of anodic current is still limited. In order to try to overcome these limitations, Hong and colleagues (2008) investigated the specific role of electric currents in bacterial detachment and inactivation when a constant current was applied in the cathodic, anodic, and block modes. These authors observed that the application of cathodic current promoted the detachment of adhered bacteria by electrorepulsive forces, but bacteria remaining on the surface were still viable. On the other hand, the anodic current inactivates most of the remaining bacteria. Thus, these authors concluded that the best electrical strategy for reducing bacterial adhesion consists of the application of a block current.

Flint and coauthors (2000) observed that it may be possible to disrupt the attachment of thermo-resistant streptococci to stainless steel by applying a small voltage. In fact, when a voltage of 9 V and a current of 40 mA were applied to a suspension of *S. thermophilus* held between stainless steel electrodes, attachment to the cathode was reduced, whereas attachment to the anode was inhibited. This may result from the disruption of the electrical bilayer on the substrate.

An approach using electrical current to enhance the activity of antimicrobials against established biofilms has also been proposed. Blenkinsopp et al. (1992) found that three common industrial biocides (glutaraldehyde, a quaternary ammonium compound and kathon) exhibited enhanced action when applied against *P. aeruginosa* biofilms within a low strength electric field with a low current density.

Concerning its mode of action, it has been suggested that the mechanism of antibacterial activity of electrical current results from the oxidation of enzymes and coenzymes, membrane damage leading to the leakage of essential cytoplasmic constituents, and toxic substances (e.g., H_2O_2, oxidizing radicals, and chlorine molecules) produced as a result of electrolysis and/or a decreased bacterial respiratory rate (del Pozo et al. 2009).

5.3 Electrolyzed Water

Electrolyzed water (EW) has been used in the food industry as a novel disinfecting agent. This process was shown to be more efficient than water and chlorine solutions as a sanitizer of meats, some fresh products, cutting boards, and utensils. EW is generated in a cell containing inert positively charged and negatively charged electrodes separated by a septum (membrane or diaphragm) (Al-Haq et al. 2005). By electrolysis, a dilute sodium chloride solution dissociates into acidic electrolysed water (AEW; pH between 2 and 3, oxidation–reduction potential of N1100 mV, and an active chlorine content of 10–90 mg/L), and basic electrolyzed water (BEW; pH between 10 and 13 and oxidation–reduction potential of −800 to −900 mV) (Hricova et al. 2008). Neutral electrolyzed water (NEW; pH 7–8) is produced by adding hydroxyl ions to AEW or by using a single chamber (Hricova et al. 2008). AEW has been determined to have a strong bactericidal effect on several pathogenic food bacteria such as *L. monocytogenes* (Park et al. 2004), *C. jejuni* (Park et al. 2002), *E. coli* O157:H7 (Park et al. 2004), *S.* Enteritidis (Koseki et al. 2003) and others, having more antimicrobial effect than BEW. Thus, according to Møretrø et al. (2012), a combination of BEW and AEW is more efficient than AEW alone. AEW has also been demonstrated to have an antibiofilm effect, namely, to inactivate *L. monocytogenes* biofilms on stainless steel surfaces. Treatment with acidic EO water for 30–120 s reduced the bacteria population by 4.3–5.2 log CFU/coupon (Ayebah et al. 2005). NEW is advantageous because it does not promote corrosion of processing equipment or irritation of skin and is stable because chlorine loss is significantly reduced at pH values of 6–9 (Len et al. 2002). In general, electrolysed water is considered environmental friendly because it is generated from water and a dilute salt solution and reverts to water after use.

5.4 Antimicrobial Materials

Numerous efforts have been made in order to impede microbial adhesion and biofilm development by altering surface physicochemical properties (Rodriguez et al. 2007), integrating antimicrobial compounds into materials, and/or coating surfaces with biocides (Gottenbos et al. 2001). As a result, a large variety of materials and products are now available to be applied in the food industry, household, and for personal use (e.g., conveyor belts, refrigerators, cutting boards, and boxes for transport of food). Nevertheless, it is highly important to notice that all these materials and products must be seen as an extra contamination obstacle and not as a substitute for correct sanitary procedures (Kampmann et al. 2008; Møretrø et al. 2006).

One of the main biocidal agents incorporated in materials is triclosan, which can be applied in plastic polymers and has Microban® as a trade name (http://www.microban.com). Although a vast amount of products available nowadays contain this antimicrobial agent, there is evidence that its efficacy may not be satisfactory. Accordingly, although a plastic enclosing 1.5 g/kg triclosan had restrained *S. typhimurium* growth in an agar plate assay, when beef was vacuum sealed using the same material, no effect was observed on *S. typhimurium* development on meat compared to the control after up to 14 days incubation at different temperatures (Cutter 1999). Moreover, when Rodrigues et al. (2011) compared *Salmonella* Enteritidis adhesion on silestones (quartz surfaces incorporating Microban®, used as kitchen bench stones) and on other food contact surfaces without antimicrobial treatment, no significant effect was found. Although the results concerning biofilm formation highlighted a potential bacteriostatic activity of this antibacterial agent, all materials tested did not support food safety, revealing that these surfaces imply a cautious use and a correct sanitation when applied in food processing areas (Rodrigues et al. 2011). Furthermore, some worries have been associated with the wholesale application of triclosan in the household area, mainly because of the concern about expansion of resistant bacteria (Levy 2001; Webber et al. 2008).

5.5 Surface Coatings and Surface Modifications

Since stainless steel is one of the most commonly used materials in the food industry and food processing areas, several modifications have been made in order to prevent microbial colonization: coating with antimicrobial compounds; implantation of ions to lower surface energy; creation of bioactive surfaces (e.g., immobilized enzymes); production of diamond-like carbon surfaces; coating with a molecular brush (steric hindrance); development of silica surfaces to create either a hard glass-like surface or a hydrophilic anionic surface; or integration of polytetrafluoroethylene (PTFE) into the surfaces. Zhao and coworkers (2005a) reported a decrease of 94–98 % in *E. coli* adhesion to Ag-PTFE-coated stainless steel, in comparison to titanium surfaces, silver coating, or uncoated stainless steel. Moreover, these same researchers also produced surfaces with particular energies known to avoid biofouling by using coatings of PTFE, nickel, copper, and phosphorus (Zhao et al. 2005b; Zhao and Liu 2006).

Titanium dioxide (TiO_2) and, more recently, nitrogen-doped titanium dioxide (N-TiO_2) coatings are other possible forms to enhance food contact surface performance in terms of better hygiene and easier sanitation. When Rodrigues et al. (2013) compared *L. monocytogenes* viability on N-TiO_2 coated and uncoated stainless steel and glass, satisfactory results were found on the coated surfaces since, for most conditions tested, survival rates decreased below 50 %. Nevertheless, no successful disinfection was accomplished, since the required bacterial reduction of at least 3 log was not achieved (Rodrigues et al. 2013). Thus, N-TiO_2 coating still requires more investigation and enhancement in order to become a really useful tool against microbial contamination of food contact surfaces. In fact, new surface coatings and different disinfectant agents are regularly investigated worldwide, but these data have yet to be transferred to the industry due to several reasons, such as process consistency, charges, product quality and safety, and maintenance (Goode et al. 2013).

In work dealing with biofilm control, microparticles ($CaCO_3$) coated with benzyldimethyldodecylammonium chloride have successfully repelled biofilm formation (Ferreira et al. 2010), and various researchers have shown that silver coatings prevented biofilm formation (Hashimoto 2001; Knetsch and Koole 2011). Furthermore, passive coatings of organic polymers are also a promising approach to prevent microbial contamination. Due to the propensity of some plastics to microbial degradation, efforts have been made to integrate inhibitors into these materials. Price and coworkers (1991) have shown that, compared to a control polymer, a significant decrease of attachment and viability of *Klebsiella pneumoniae, S. aureus*, and *P. aeruginosa* was achieved on an ethylene vinyl acetate/low-density polyethylene product containing a low-solubility commercial quaternary amine complex. Although further studies are needed, this seems to be a promising application to control microbial contamination on food contact surfaces. Nevertheless, it is also important to note that not all antimicrobial coatings tested so far have shown efficacy. For example, a polystyrene surface coated with antimicrobial fullerene-based nanoparticles was created aiming to prevent biofilm formation by *Pseudomonas mendocina*, but it actually enhanced biofilm development (Lyon et al. 2008). This demonstrates that antibacterial nanomaterials can lose their efficacy when applied as coatings.

Another possible way to avoid biofilm formation is by steric hindrance, or blocking, of bacterial adhesion by means of a "molecular brush," which involves coating a surface with an inert material that physically prevents bacterial adhesion. Namely, polyethylene glycol (PEG) is the most investigated molecular brush that controls protein adsorption to materials (Jönsson and Johansson 2004). Although the prevention of protein adsorption by a molecular brush has generally been established, its usefulness in preventing microbial attachment is somehow controversial. In fact, Wei and coworkers (2003) have reported that stainless steel coated with PEG inhibited the adsorption of b-lactoglobulin, but did not inhibit the adhesion of *Pseudomonas* sp. and *L. monocytogenes* cells. A possible explanation for these observations may be related with the particular nature of the PEG layer used, as well as the complexity of bacterial adhesion, since protein interactions are not the only aspect that influences it.

5.6 Natural Compounds

Recently, the emergence of antibiotic-resistant strains and the reluctance of consumers toward the use of chemical products, such as biocides, have led to a search for natural alternative products. The use of biocides as sanitizers in the food industry has associated concerns such as biocide biodegradability, their risk to human health, and their environmental impact (Cappitelli et al. 2006). The use of substances obtained from plants is preferred since they may have been used in traditional medicine for a long time, they are generally considered to be safe by consumers, and are not known to cause harm to the environment (Leonard et al. 2010). Essential oils (EOs) or their constituents are one of the more promising and natural alternative antimicrobial agents. EOs are volatile, natural, complex compounds characterized by a strong odor and are obtained from plant material (flowers, buds, seeds, leaves, twigs, bark, herbs, wood, fruits, and roots). Concerning their mode of action, they pass through the bacterial cell wall and cytoplasmic membrane, disrupt the structure of the different layers of polysaccharides, fatty acids, and phospholipids, and permeabilize them (Bakkali et al. 2008).

Oliveira et al. (2012) evaluated the antibacterial potential of EOs from *Cinnamomum cassia* bark and *Melaleuca alternifolia* and *Cymbopogon flexuosus* leaves against planktonic and sessile cells of *E. coli* (EPEC) and *L. monocytogenes*. These authors observed that all of the EOs and combinations tested possessed antibacterial activity against planktonic cells; however, the EO of *C. cassia* was the most effective antibiofilm agent. Jadhav et al. (2013) also observed the inhibitory effect of the essential oil obtained from yarrow (*Achillea millefolium*) against planktonic cells and biofilms of *L. monocytogenes* and *Listeria innocua* isolates obtained from food processing environments.

Other natural compounds are biosurfactants that are surface-active compounds of microbial origin and have attracted attention due to their low toxicity and high biodegradability, when compared to synthetic surfactants (Nitschke et al. 2005; Banat et al. 2010). The adsorption of biosurfactants to a solid surface can modify its hydrophobicity and thus bacterial adhesion and consequently biofilm formation. One study investigated whether surfactin from *Bacillus subtilis* and rhamnolipids from *P. aeruginosa* could reduce the adhesion and/or disrupt the biofilms of some foodborne pathogenic bacteria (Gomes and Nitschke 2012). It was observed that after 2 h contact with surfactin at 0.1 % concentration, the preformed biofilms of *S. aureus* were reduced by 63.7 %, *L. monocytogenes* by 95.9 %, *S.* Enteritidis by 35.5 %, and the mixed culture biofilm by 58.5 %. Concerning the effect of rhamnolipids, it was observed that, at a concentration of 0.25 %, they removed 58.5 % of the *S. aureus* biofilm, 26.5 % of *L. monocytogenes*, 23.0 % of *S.* Enteritidis, and 24.0 % of the mixed species biofilm. Nevertheless, although the replacement of synthetic surfactants by biosurfactants would provide advantages such as biodegradability and low toxicity, their use has been limited by their relatively high production cost, as well as scarce information on their toxicity in humans (Rodrigues 2011).

5.7 *Enzymes*

Enzymes are biological catalysts, i.e., substances that increase the rate of chemical reactions without being used up. In other words, enzymes are proteins capable of lowering the activation energy of a chemical reaction; their action relies on the possibility of interacting with the substrate to be transformed, via its active site (Glinel et al. 2012).

Concerning their mode of action, enzymes immobilized on a material surface and in contact with a biological environment may act against biofilm in various ways. Enzymes may impair the initial step of surface colonization by microorganisms by cleavage of proteins and carbohydrates; these types of enzymes are called adhesive-degrading enzymes. Enzymes may also have a biocidal effect when they compromise the viability of living organisms growing on surfaces. In the first category, enzymes such as proteases can impede microbial adhesion by hydrolyzing peptidic bonds (Rawlings et al. 2006), while glycosidases specifically break ester bonds of polysaccharides, which are the main constituents of microbial adhesives (Moss 2006). van Speybroeck et al. (1996) reported the use of an enzymatic preparation comprised of exopolysaccharide-degrading enzymes, particularly the colanic acid-degrading enzymes, derived from a strain of *Streptomyces* for the removal and/or prevention of biofilm formation on surfaces.

Molobela et al. (2010) tested proteases (savinase, everlase, and polarzyme) and amylase (amyloglucosidase and bacterial amylase novo) activity on biofilms formed by *P. fluorescens* and on extracted EPS. They observed that everlase and savinase were the most effective enzymatic treatments for removing biofilms and degrading the EPS.

Enzymes have also been used as antibiofilm coatings. In this case, they can be either covalently grafted onto solid substrates or incorporated into polymer matrices to produce antibacterial coatings and it is thought that enzymes impair one or several "bricks" of the biofilm construction (Glinel et al. 2012). Yuan and coworkers (2011) tested a coating composed by coupling lysozyme on a PEG layer against two different bacterial species, Gram-negative *E. coli* and Gram-positive *S. aureus*. These authors observed that more than 90 % of *S. aureus* and ~80 % of *E. coli* that adhered to lysozyme-functionalized surfaces were damaged within 4 h. In addition, these coatings showed long-term activity since the antibacterial effect against *S. aureus* was retained after a contact time of ~36 h. However the effect faded over time for *E. coli*. This result was probably due to the fact that lysozyme is more active toward peptidoglycans present in the Gram-positive bacterial wall than toward the double membrane of the Gram-negative cell wall. It can be concluded that, as the structural composition of EPS varies even among bacteria of the same species, the mode of action and the consequent efficiency of enzymes will also be variable.

Therefore, enzymes constitute an important alternative for biofilm removal in the food industry. Though, it must be noted that enzymes, as coatings, may contribute to the unwanted degradation of substances surrounding the surface coating. In addition, enzymes that produce biocidal substances have to be approved by the appropriate legislative body before being implemented.

5.8 Quorum-Sensing Interfering Molecules

Quorum sensing (QS) or cell-to-cell communication is employed by a diverse group of bacteria, including those commonly associated with food. Through the mechanisms of QS, bacteria communicate with each other by producing the signaling molecules known as autoinducers and are consequentially able to express specific genes in response to population density. Since several types of signaling molecules have been detected in different spoiled food products, disrupting the QS circuit can potentially play a major role in controlling microbial gene expression related to human infection and food spoilage (Bai and Rai 2011). QS inhibitors can be developed in order to target synthesis of the cell signaling molecules themselves or to block these signaling systems (Bai and Rai 2011).

QS systems appear to be involved in all phases of biofilm formation. They regulate population density and the metabolic activity within the mature biofilm to fit the nutritional demands and resources available. Furthermore, bacteria within biofilms have markedly different transcriptional programs from planktonic bacteria of the same strain (Asad and Opal 2008).

The relation between QS and biofilm formation in food-related bacteria has been observed by several authors. However, according to Bai and Rai (2011), though signaling molecules have been detected in biofilms, their precise role in the different stages of biofilm formation is still not clear.

Kerekes and coauthors (2013) investigated the effect of clary sage, juniper, lemon, and marjoram essential oils and their major components on the formation of bacterial and yeast biofilms and on the inhibition of AHL mediated QS and verified that the compounds tested seemed to be good candidates for prevention of biofilm formation and inhibition of the AHL-mediated QS mechanism.

Furanones are one of the most studied QS inhibitors and it was demonstrated that they were able to control multicellular behavior induced by autoinducer-1 (Manefield et al. 2002) and autoinducer-2 (Ren et al. 2004) in Gram-negative microorganisms.

5.9 Bacteriophages

Bacteriophages (phages) are viruses that infect bacteria and can be found in the same biosphere niches as their bacterial hosts (Kutter and Sulakvelidze 2005). They were originally found by Harkin in 1896 and were applied in the cure of microbial infections previous to antibiotic discovery. The application of phages to control biofilms can be a practicable, natural, harmless, and greatly specific way to deal with numerous microorganisms implicated in biofilm formation (Kudva et al. 1999). In fact, phages and their endolysins have already been used to stop biofilm development by *L. monocytogenes* and *E. coli* (Gaeng et al. 2009; Sharma et al. 2005). Accordingly, a *L. monocytogenes* phage (ATCC 23074-B1) was effectively used for biofilm eradication (Hibma et al. 1997), and a synergistic effect of an alkaline disinfectant and a phage has been described for the eradication of *E. coli* O157:H7 biofilms grown on stainless steel (Sharma et al. 2005). Moreover, Lu and Collins (2007) produced a phage that expresses a biofilm-degrading enzyme, which attacked both biofilm bacteria and matrix, leading to more than 99.9 % elimination of the biofilm cells.

A study conducted by Sillankorva and coworkers (2008) showed that the phage phiIBB-PF7A can be an outstanding natural agent regarding its ability to lyse *P. fluorescens* biofilm cells in a very short period of time. This same phage was also applied to control a *P. fluorescens* and *Staphylococcus lentus* mixed biofilm and led to a remarkable decline in the attached bacterial cells (*P. fluorescens*).

Moreover, it was also shown that phages can be effective in both monoculture and mixed-culture biofilms and competently reach and lyse their specific host, despite the coresidence of a nonvulnerable species (Sillankorva et al. 2010). When Briandet and coworkers (2008) investigated the dispersion and response of phages within biofilms, it was observed that phages were able to penetrate distinct biofilm complexes. In addition, these authors found that, in general, phages within biofilms are immobilized, reproduced, and released by a lytic cycle, connecting with their specific binding sites on the hosts. Moreover, Tait et al. (2002) reported that phages and bacteria were able to progressively coexist in biofilms, and therefore recommended a combination of phages and polysaccharide depolymerases and disinfectant for improved biofilm control. On the other hand, Brooks and Flint (2008) have suggested that it may be productive to look for phages in biofilm samples from food industry facilities and to apply them against microbial communities found in the same environment. Moreover, since phages are likely to be highly host specific, this approach should not represent any danger to other fractions of the production, even though the application of a phage mixture would likely to be required due to arising host resistance.

Although it is already known that infection of biofilm cells by phages is highly dependent on several factors, such as their chemical composition, phage concentration, temperature, media, and growth stage (Sillankorva et al. 2004; Chaignon et al. 2007), there is much more to explore and explain. Since dairy foodstuffs are highly vulnerable to contamination by bacterial biofilms, the dairy industry has become the leader of exploiting phages as an antibacterial approach (Thallinger et al. 2013), and it is expected that the development of highly efficient and inexpensive methods of genetic material treatment and DNA sequencing will accelerate the finding and creation of engineered phages.

6 Conclusions

Due to the ability of foodborne pathogens to form biofilms on diverse food contact surfaces, leading to a continuous contamination of food, prevention and elimination of biofilms are significant concerns for the food industry. Currently, the best practical ways to prevent biofilm development consists of a successful application of hygienic and sanitation compounds, appropriated sanitation, and a good operation of the process line. Although much progress has been made in this area, out-of-date prevention means are still being applied. Nevertheless, given the ability of bacteria to become resistant and consequently to endure approaches that used to be efficient, new methods of elimination for these microbial communities are continuously required. However, a lot more is still left to discover about the effect of antibacterial compounds on biofilms and their subsequent recovery reaction. This, together with an improved knowledge about the mechanisms involved in biofilm formation on food contact surfaces is of utmost importance towards the goal of achieving a novel, highly effective, cheaper, and ecological tactic to assure food safety.

Acknowledgments P. Teixeira and D. Rodrigues acknowledge the financial support of the Portuguese Foundation for Science and Technology through the grants SFRH/BPD/86732/2012 and SFRH/BPD/72632/2010, respectively.

References

Ackman D, Marks S, Mack P, Caldwell M, Root T, Birkhead G (1997) Swimming-associated haemorrhagic colitis due to *Escherichia coli* O157:H7 infection: evidence of prolonged contamination of a fresh water lake. Epidemiol Infect 119:1–8

Al-Haq MI, Sugiyama J, Isobe S (2005) Applications of electrolyzed water in agriculture & food industries. Food Sci Technol Res 11(2):135–150

Andersson A, Ronner U, Granum PE (1995) What problems does the food industry have with the spore-forming pathogens *Bacillus cereus* and *Clostridium perfringens*? Int J Food Microbiol 28:145–155

Araújo P, Lemos M, Mergulhão F, Melo L, Simões M (2011) Antimicrobial resistance to disinfectants in biofilms. In: Méndez-Vilas A (ed) Science against microbial pathogens: communicating current research and technological advances. Formatex, Spain, pp 826–834

Asad S, Opal SM (2008) Bench-to-bedside review: quorum sensing and the role of cell-to-cell communication during invasive bacterial infection. Crit Care 12:236–246

Ayebah B, Hung YC, Frank JF (2005) Enhancing the bactericidal effect of electrolyzed water on *Listeria monocytogenes* biofilms formed on stainless steel. J Food Prot 68:1375–1380

Bai AJ, Rai VR (2011) Bacterial quorum sensing and food industry. Comp Rev Food Sci Food Safety 10:184–194

Bakkali F, Averbeck S, Averbeck D, Idaomar M (2008) Biological effects of essential oils: a review. Food Chem Toxicol 46:446–475

Banat IM, Franzetti A, Gandolfi I, Bestetti G, Martinotti MG, Fracchia L et al (2010) Microbial biosurfactants production, applications and future potential. Appl Microbiol Biotechnol 87:427–444

Baumann AR, Martin SE, Feng H (2009) Removal of *Listeria monocytogenes* biofilms from stainless steel by use of ultrasound and ozone. J Food Prot 72(6):1306–1309

Bennett SD, Walsh KA, Gould LH (2013) Foodborne disease outbreaks caused by *Bacillus cereus*, *Clostridium perfringens*, and *Staphylococcus aureus*-United States, 1998–2008. Clin Infect Dis 57(3):425–433

Blenkinsopp SA, Khoury AE, Costerton JW (1992) Electrical enhancement of biocide efficacy against *Pseudomonas aeruginosa* biofilms. Appl Environ Microbiol 58(11):3770–3773

Briandet R, Lacroix-Gueu P, Renault M, Lecart S, Meylheuc T, Bidnenko E et al (2008) Fluorescence correlation spectroscopy to study diffusion and reaction of bacteriophages inside biofilms. Appl Environ Microbiol 74(7):2135–2143

Brooks JD, Flint SH (2008) Biofilms in the food industry: problems and potential solutions. Int J Food Sci Technol 43(12):2163–2176

Cabeça TK, Pizzolitto AC, Pizzolitto EL (2008) Assessment of action of disinfectants against *Listeria monocytogenes* biofilms. J Food Nut 17(2):121–125

Cappitelli F, Principi P, Sorlini C (2006) Biodeterioration of modern materials in contemporary collections: can biotechnology help? Trends Biotechnol 24:350–354

Carmen JC, Roeder BL, Nelson JL, Ogilvie RL, Robison RA, Schaalje GB, Pitt WG (2005) Treatment of biofilm infections on implants with low-frequency ultrasound and antibiotics. Am J Infect Control 33:78–82

Chae MS, Schraft H (2000) Comparative evaluation of adhesion and biofilms formation of different *Listeria monocytogenes* strains. Int J Food Microbiol 62:103–111

Chaignon P, Sadovskaya I, Ragunah C, Ramasubbu N, Kaplan JB, Jabbouri S (2007) Susceptibility of staphylococcal biofilms to enzymatic treatments depends on their chemical composition. Appl Microbiol Biotechnol 75:125–132

Chemat F, Zill-e-Huma, Khan MK (2011) Applications of ultrasound in food technology: processing, preservation and extraction. Ultrason Sonochem 18:813–835

Chen J, Rossman ML, Pawar DM (2007) Attachment of enterohemorragic *Escherichia coli* to the surface of beef and a culture medium. LWT- Food Sci Technol 40:249–254

Conter M, Vergara A, Di Ciccio P, Zanardi E, Ghidini S, Ianieri A (2009) Polymorphism of *actA* gene is not related to in vitro virulence of *Listeria monocytogenes*. Int J Food Microbiol 137:100–105

Cutter CN (1999) The effectiveness of triclosan-incorporated plastic against bacteria on beef surfaces. J Food Prot 62(5):474–479

da Silva PMB, Acosta EJR, Pinto R, Graeff L, Spolidorio MDMP, Almeida RS et al (2011) Microscopical analysis of *Candida albicans* biofilms on heat-polymerised acrylic resin after chlorhexidine gluconate and sodium hypochlorite treatments. Mycoses 54(6):712–717

del Pozo JL, Rouse MS, Mandrekar JN, Steckelberg JM, Patel R (2009) The electricidal effect: reduction of *Staphylococcus* and *Pseudomonas* biofilms by prolonged exposure to low-intensity electrical current. Antimicrob Agents Chemother 53:41–45

Deming MS, Tauxe RV, Blake PA, Dixon SE, Fowler BS, Jones TS et al (1987) *Campylobacter enteritis* at a university: transmission from eating chicken and from cats. Am J Epidemiol 126(3):526–534

Dewanti R, Wong AC (1995) Influence of culture conditions on biofilm formation by *Escherichia coli* O157:H7. Int J Food Microbiol 26(2):147–164

Dourou D, Beauchamp CS, Yoon Y, Geornaras I, Belk KE, Smith GC et al (2011) Attachment and biofilm formation by *Escherichia coli* O157:H7 at different temperatures, on various food-contact surfaces encountered in beef processing. Int J Food Microbiol 149(3):262–268

Dussurget O (2008) New insight into determinants of *Listeria monocytogenes* virulence. Int Rev Cell Mol Biol 270:1–38

Dykes GA, Sampathkumar B, Korber DR (2003) Planktonic or biofilm growth affects survival, hydrophobicity and protein expression patterns of a pathogenic *Campylobacter jejuni* strain. Int J Food Microbiol 89(1):1–10

EFSA (2012) European Food Safety Authority, European Centre for Disease Prevention and Control; The European Union summary report on trends and sources of zoonoses, zoonotic agents and food-borne outbreaks in 2010. EFSA J 10(3):2597, [442pp]. Available at http://www.efsa.europa.eu/efsajournal

Ellis A, Preston M, Borczyk A, Miller B, Stone P, Hatton B et al (1998) A community outbreak of *Salmonella berta* associated with a soft cheese product. Epidemiol Inf 120(1):29–35

Farber JM, Peterkin PI (1991) *Listeria monocytogenes*, a food-borne pathogen. Microbiol Rev 55(3):476–511

FDA/ERG (2004) GMPs—section two: literature review of common food safety problems and applicable controls. August 2004. http://www.fda.gov/Food/GuidanceRegulation/CGMP/ucm110911.htm

Ferreira C, Pereira AM, Melo LF (2010) Advances in industrial biofilm control with micro-nanotechnology. Appl Microbiol:845–854

Flint SH, Brooks JD, Bremer PJ (2000) Properties of the stainless steel substrate influencing the adhesion of thermo-resistant *streptococci*. J Food Eng 43:235–242

Frank JF, Ehlers J, Wicker L (2003) Removal of *Listeria monocytogenes* and poultry soil-containing biofilms using chemical cleaning and sanitizing agents under static conditions. Food Prot Trends 23:654–663

Gaeng S, Scherer S, Neve H, Loessner MJ (2009) Gene cloning and expression and secretion of *Listeria monocytogenes* bacteriophagelytic enzymes in *Lactococcus lactis*. Appl Environ Microbiol 66:951–2958

Gilbert P, Das JR, Jones MV, Allison DG (2001) Assessment of resistance towards biocides following the attachment of micro-organisms to, and growth on, surfaces. J Appl Microbiol 91 (2):248–254

Giraudon I, Cathcart S, Blomqvist S, Littleton A, Surman-Lee S, Mifsud A et al (2009) Large outbreak of *Salmonella* phage type 1 infection with high infection rate and severe illness associated with fast food premises. Public Health 123(6):444–447

Glinel K, Thebault P, Humblot V, Pradier CM, Jouenne T (2012) Antibacterial surfaces developed from bio-inspired approaches. Acta Biomater 8(5):1670–1684

Gomes M, Nitschke M (2012) Evaluation of rhamnolipid and surfactin to reduce the adhesion and remove biofilms of individual and mixed cultures of food pathogenic bacteria. Food Control 25:441–447

Goode KR, Konstantia A, Robbins PT, Fryer PJ (2013) Fouling and cleaning studies in the food and beverage industry classified by cleaning type. Comp Rev Food Sci Food Safety 12(2): 121–143

Gottenbos B, van der Mei HC, Klatter F, Nieuwenhuis P, Busscher HJ (2001) In vitro and in vivo antimicrobial activity of covalently coupled quaternary ammonium silane coatings on silicone rubber. Biomaterials 23:1417–1423

Habimana O, Heir E, Langsrud S, Asli AW, Møretr T (2010) Enhanced surface colonization by *Escherichia coli* O157:H7 in biofilms formed by an *Acinetobacter calcoaceticus* isolate from meat-processing environments. Appl Environ Microbiol 76(13):4557–4559

Hanning I, Jarquin R, Slavik M (2008) *Campylobacter jejuni* as a secondary colonizer of poultry biofilms. J Appl Microbiol 105(4):1199–1208

Hashimoto H (2001) Evaluation of the anti-biofilm effect of a new anti-bacterial silver citrate/lecithin coating in an in-vitro experimental system using a modified Robbins device. Kansenshogaku Zasshi 75(8):678–685

Hibma AM, Jassim SA, Griffiths MW (1997) Infection and removal of L-forms of *Listeria monocytogenes* with bred bacteriophage. Int J Food Microbiol 4:197–207

Hong SH, Jeong J, Shim S, Kang H, Kwon S, Ahn KH, Yoon J (2008) Effect of electric currents on bacterial detachment and inactivation. Biotechnol Bioeng 100(2):379–386

Hricova D, Stephan R, Zweifel C (2008) Electrolyzed water and its application in the food industry. J Food Prot 71(9):1934–1947

Jadhav S, Bhave M, Palombo EA (2012) Methods used for the detection and subtyping of *Listeria monocytogenes*. J Microbiol Methods 88:327–341

Jadhav S, Shah R, Bhave M, Palombo EA (2013) Inhibitory activity of yarrow essential oil on *Listeria* planktonic cells and biofilms. Food Control 29:125–130

Janneke G, Wijman E, Patrick P, de Leeuw LA, Mooezelaar R, Zwietering MH et al (2007) Air–liquid interface biofilms of *Bacillus cereus*: formation, sporulation, and dispersion. Appl Environ Microbiol 73:1481–1488

Jessen B, Lammert L (2003) Biofilm and disinfection in meat processing plants. Int Biodeterior Biodegradation 51:265–269

Jönsson M, Johansson HO (2004) Effect of surface grafted polymers on the adsorption of different model proteins. Colloids Surf B 37:71–81

Joseph B, Otta SK, Karunasagar I, Karunasagar I (2001) Biofilm formation by *Salmonella* spp. on food contact surfaces and their sensitivity to sanitizers. Int J Food Microbiol 64:367–372

Kampmann Y, De Clerck E, Kohn S, Patchala DK, Langerock R, Kreyenschmidt J (2008) Study on the antimicrobial effect of silver-containing inner liners in refrigerators. J Appl Microbiol 104(6):1808–1814

Kerekes EB, Deák É, Takó M, Tserennadmid R, Petkovits T, Vágvölgyi C, Krisch J (2013) Anti-biofilm forming and anti-quorum sensing activity of selected essential oils and their main components on food-related micro-organisms. J Appl Microbiol 115(4):933–942

Khardre MA, Yousef AE, Kim JG (2001) Microbiological aspects of ozone applications in food: a review. J Food Sci 66(9):1242–1252

Kim TJ, Young BM, Young GM (2008) Effect of flagellar mutations on *Yersinia enterocolitica* biofilm formation. Appl Environ Microbiol 74(17):5466–5474

Kiskó G, Szabó-Szabó O (2011) Biofilm removal of *Pseudomonas* strains using hot water sanitation. Acta Univ Sapientiae Alimentaria 4:69–79

Klayman BJ, Volden PA, Stewart PS, Camper AK (2009) *Escherichia coli* O157:H7 requires colonizing partner to adhere and persist in a capillary flow cell. Environ Sci Technol 43(6): 2105–2111

Knetsch MLW, Koole LH (2011) New strategies in the development of antimicrobial coatings: the example of increasing usage of silver and silver nanoparticles. Polymers 3(1):340–366

Koseki S, Yoshida K, Kamitani Y, Itoh K (2003) Influence of inoculation method, spot inoculation site, and inoculation size on the efficacy of acidic electrolyzed water against pathogens on lettuce. J Plant Prot 66:2010–2016

Kudva IT, Jelacic S, Tarr PI, Youderian P, Hovde CJ (1999) Biocontrol of *Escherichia coli* O157 with O157-specific bacteriophages. Appl Environ Microbiol 65:3767

Kumar CG, Anand SK (1998) Significance of microbial biofilms in food industry: a review. Int J Food Microbiol 42(1–2):9–27

Kusumaningrum HD, Riboldi G, Hazeleger WC, Beumer RR (2003) Survival of foodborne pathogens on stainless steel surfaces and crosscontamination to foods. Int J Food Microbiol 85:227–236

Kutter E, Sulakvelidze A (2005) Introduction. In: Kutter E, Sulakvelidze A (eds) Bacteriophages biology and applications. CRC, Boca Raton, FL, pp 1–4

Latorre AA, Van Kessel JS, Karns JS, Zurakowski MJ, Pradhan AK, Boor KJ et al (2010) Biofilm in milking equipment on a dairy farm as a potential source of bulk tank milk contamination with *Listeria monocytogenes*. J Dairy Sci 93:2792–2802

Len SV, Hung YC, Chung D, Andreson JL, Erickson MC, Morita K (2002) Effects of storage conditions and pH on chlorine loss in electrolyzed oxidizing (EO) water. J Agric Food Chem 50:209–212

Leonard CM, Virijevic S, Regnier T, Combrinck S (2010) Bioactivity of selected essential oils and some components on *Listeria monocytogenes* biofilms. S Afr J Bot 76:676–680

Levy SB (2001) Antibacterial household products: cause for concern. Emerg Infect Dis 7(3): 512–515

Lindsay D, Brözel VS, von Holy A (2006) Biofilm-spore response in *Bacillus cereus* and *Bacillus subtilis* during nutrient limitation. J Food Prot 69:1168–1172

Lu TK, Collins JJ (2007) Dispersing biofilms with engineered enzymatic bacteriophage. Proc Natl Acad Sci USA 104:11197–11202

Lyon DY, Brown D, Sundstrom ER, Alvarez PJJ (2008) Assessing the antibiofouling potential of a fullerene-coated surface. Int Biodeterior Biodegradation 62(4):475–478

Manefield M, Rasmussen TB, Henzter M, Andersen JB, Steinberg P, Kjelleberg S, Givskov M (2002) Halogenated furanones inhibit quorum sensing through accelerated LuxR turnover. Microbiology 148:1119–1127

Marouani-Gadri N, Augier G, Carpentier B (2009) Characterization of bacterial strains isolated from a beef-processing plant following cleaning and disinfection: influence of isolated strains on biofilm formation by *Sakaï* and EDL 933 *E. coli* O157:H7. Int J Food Microbiol 133(1–2): 62–67

Molobela P, Cloete TE, Beukes M (2010) Protease and amylase enzymes for biofilm removal and degradation of extracellular polymeric substances (EPS) produced by *Pseudomonas fluorescens* bacteria. Afr J Microbiol Res 4(14):1515–1524

Moltz AG, Martin SE (2005) Formation of biofilms by Listeria monocytogenes under various growth conditions. J Food Prot 68:92–97

Møretrø T, Langsrud S (2004) *Listeria monocytogenes*: biofilm formation and persistence in food-processing environments. Biofilms 1:107–121

Møretrø T, Sonerud T, Mangelrød E, Langsrud S (2006) Evaluation of the antimicrobial effect of a triclosan-containing industrial floor used in the food industry. J Food Prot 69(3):627–633

Møretrø T, Vestby LK, Nesse LL, Storheim SE, Kotlarz K, Langsrud S (2009) Evaluation of efficacy of disinfectants against *Salmonella* from the feed industry. J Appl Microbiol 106: 1005–1012

Møretrø T, Heir E, Nesse LL, Vestby LK, Langsrud S (2012) Control of *Salmonella* in food related environments by chemical disinfection. Food Res Int 45:532–544

Moss G (2006) In: Typton KF, Boyce S (eds) Enzyme nomenclature. London (online edition)

Myszka K, Czaczyk K (2011) Bacterial biofilms on food contact surfaces: a review. Pol J Food Nutr Sci 61(3):173–180

Nitschke M, Costa SGVAO, Haddad R, Gonçalves LAG, Eberlin MN, Contiero J (2005) Oil wastes as unconventional substrates for rhamnolipid biosurfactant production by *Pseudomonas aeruginosa* LBI. Biotechnol Prog 21:1562–1566

Oliveira K, Oliveira T, Teixeira P, Azeredo J, Henriques M, Oliveira R (2006) Comparison of the adhesion ability of different *Salmonella* enteritidis serotypes to materials used in kitchens. J Food Prot 69(10):2352–2356

Oliveira M, Brugnera D, Nascimento J, Piccoli R (2012) Control of planktonic and sessile bacterial cells by essential oils. Food Bioprod Proc 90:809–818

Park H, Hung YC, Kim C (2002) Effectiveness of electrolyzed water as a sanitizer for treating different surfaces. J Food Prot 65:1276–1280

Park H, Hung YC, Chung D (2004) Effect of chlorine and pH on efficacy of electrolyzed water for inactivating *Escherichia coli* O157:H7 and *Listeria monocytogenes*. Int J Food Microbiol 91:13–18

Podolak R, Enache E, Stone W, Black DG, Elliott PH (2010) Sources and risk factors for contamination, survival, persistence, and heat resistance of *Salmonella* in low-moisture foods. J Food Prot 73(10):1919–1936

Price DL, Sawant AD, Aheam DG (1991) Activity of an insoluble antimicrobial quaternary amine complex in plastics. J Ind Microbiol 8:83–90

Proctor ME, Kurzynski T, Koschmann C, Archer JR, Davis JP (2002) Four strains of *Escherichia coli* O157:H7 isolated from patients during an outbreak of disease associated with ground beef: importance of evaluating multiple colonies from an outbreak-associated product. J Clin Microbiol 40:1530–1533

Rawlings ND, Morton FR, Barett AJ (2006) EROPS: the peptidase database. Nucleic Acids Res 34:D270–D272

Reij MW, Aantrekker ED (2004) Recontamination as a source of pathogens in processed food. Int J Food Microbiol 91(1):1–11

Ren D, Bedzyk LA, Ye RW, Thomas SM, Wood TK (2004) Differential gene expression shows natural brominated furanones interfere with the autoinducer-2 bacterial signalling system of *Escherichia coli*. Biotechnol Bioeng 88:630–642

Robbins JB, Fisher CW, Moltz AG, Martin SE (2005) Elimination of *Listeria monocytogenes* biofilms by ozone, chlorine, and hydrogen peroxide. J Food Prot 68(3):494–498

Rodrigues LR (2011) Inhibition of bacterial adhesion on medical devices. Adv Exp Med Biol 715 (351–67):715

Rodrigues D, Teixeira P, Oliveira R, Azeredo J (2011) *Salmonella enterica* enteritidis biofilm formation and viability on regular and triclosan incorporated bench cover materials. J Food Protect 74(6):32–37

Rodrigues D, Teixeira P, Tavares CJ, Oliveira R, Azeredo J (2013) Food contact surfaces coated with nitrogen-doped titanium dioxide: effect on *Listeria monocytogenes* survival under different light sources. Appl Surf Sci 270:1–5

Rodriguez A, Autio WR, McLandsborough LA (2007) Effects of inoculation level, material hydration, and stainless steel surface roughness on the transfer of *Listeria monocytogenes* from inoculated bologna to stainless steel and high-density polyethylene. J Food Protect 70:1423–1428

Rogers J, Dowsett AB, Keevil CW (1995) A paint incorporating silver to control mixed biofilms containing *Legionella pneumophila*. J Ind Microbiol 15(4):377–383

Ryser ET, Marth EH (2007) *Listeria*, listeriosis and food safety, 3rd edn. CRC, Boca Raton, FL

Sanders SQ, Boothe DH, Frank JF, Arnold JW (2007) Culture and detection of *Campylobacter jejuni* within mixed microbial populations of biofilms on stainless steel. J Food Protect 70(6): 1379–1385

Scher K, Romling U, Yaron S (2005) Effect of heat, acidification, and chlorination on *Salmonella enterica* serovar Typhimurium cells in a biofilm formed at the air-liquid interface. Appl Environ Microbiol 71:1163–1168

Sharma M, Anand SK (2002) Biofilms evaluation as an essential component of HACCP for food/dairy industry: a case. Food Control 13:469–477

Sharma M, Ryu J, Beuchat LR (2005) Inactivation of *Escherichia coli* O157:H7 in biofilm on stainless steel by treatment with an alkaline cleaner and a bacteriophage. J Appl Microbiol 99:449–459

Shi X, Zhu X (2009) Biofilm formation and food safety in food industries. Trends Food Sci Technol 20:407–413

Shikongo-Nambabi M, Kachigunda B, Venter SN (2010) Evaluation of oxidising disinfectants to control *Vibrio* biofilms in treated seawater used for fish processing. Water SA 36(3):215–220

Sillankorva S, Oliveira R, Vieira MJ, Sutherland I, Azeredo J (2004) *Pseudomonas fluorescens* infection by bacteriophage PhiS1: the influence of temperature, host growth phase and media. FEMS Microbiol Lett 241(1):13–20

Sillankorva S, Neubauer P, Azeredo J (2008) *Pseudomonas fluorescens* biofilms subjected to phage phiIBB-PF7A. BMC Biotechnol 8:79

Sillankorva S, Neubauer P, Azeredo J (2010) Phage control of dual species biofilms of *Pseudomonas fluorescens* and *Staphylococcus lentus*. Biofouling 26(5):567–575

Somers EB, Wong AC (2004) Efficacy of two cleaning and sanitizing combinations on *Listeria monocytogenes* biofilms formed at low temperature on a variety of materials in the presence of ready-to-eat-meat residue. J Food Prot 67:2218–2229

Srey S, Jahid IK, Há SD (2013) Biofilm formation in food industries: a food safety concern. Food Control 31:572–585

Stewart PS, Franklin MJ (2008) Physiological heterogeneity in biofilms. Nat Rev Microbiol 6(3): 199–210

Swerdlow DL, Woodruff BA, Brady RC, Griffin PM, Tippen S, Donnell HD Jr, Geldreich E, Payne BJ, Meyer A Jr, Wells JG et al (1992) A waterborne outbreak in Missouri of *Escherichia coli* O157:H7 associated with bloody diarrhea and death. Ann Intern Med 117:812–819

Tait K, Skillman LC, Sutherland IW (2002) The efficacy of bacteriophage as a method of biofilm eradication. Biofouling 18(4):305–311

Teixeira P, Lima JC, Azeredo J, Oliveira R (2007a) Note. Colonisation of bench cover materials by *Salmonella typhimurium*. Food Sci Technol Int 13(1):5–10

Teixeira PC, Leite GM, Domingues RJ, Silva J, Gibbs PA, Ferreira JP (2007b) Antimicrobial effects of a microemulsion and a nanoemulsion on enteric and other pathogens and biofilms. Int J Food Microbiol 118:15–19

Teixeira P, Lima J, Azeredo J, Oliveira R (2008) Adhesion of *Listeria monocytogenes* to materials commonly found in domestic kitchens. Int J Food Sci Technol 43(7):1239–1244

Thallinger B, Prasetyo EN, Nyanhongo GS, Guebitz GM (2013) Antimicrobial enzymes: an emerging strategy to fight microbes and microbial biofilms. Biotechnol J 8:97–109

Tompkin RB (2002) Control of Listeria in the food-processing environment. J Food Prot 65(4): 709–723

Toté K, Horemans T, Vanden Berghe D, Maes L, Cos P (2010) Inhibitory effect of biocides on the viable masses and matrices of *Staphylococcus aureus* and *Pseudomonas aeruginosa* biofilms. Appl Environ Microbiol 76(10):3135–3142

Trachoo N, Brooks JD (2005) Attachment and heat resistance of *Campylobacter jejuni* on *Enterococcus faecium* biofilm. Pak J Biol Sci 8:599–605

Trachoo N, Frank JF (2002) Effectiveness of chemical sanitizers against *Campylobacter jejuni*-containing biofilms. J Food Prot 65:117–1121

Trachoo N, Frank JF, Stern NJ (2002) Survival of *Campylobacter jejuni* in biofilms isolated from chicken houses. J Food Prot 65(7):1110–1116

Van Houdt R, Michiels CW (2010) Biofilm formation and the food industry, a focus on the bacterial outer surface. J Appl Microbiol 109(4):1117–1131

van Speybroeck MMP, Bruggeman G, van Poele J, van Peel KLI, van Damme EJ (1996) Exopolysaccharide-degrading enzyme and use of the same PCT Patent Appl. WO 9631610

Verghese B, Lok M, Wen J, Alessandria V, Chen Y, Kathariou S et al (2011) comK prophage junction fragments as markers for *Listeria monocytogenes* genotypes unique to individual meat and poultry processing plants and a model for rapid niche-specific adaptation, biofilm formation, and persistence. Appl Environ Microbiol 77:3279–3292

Wagnera VE, Kobersteinb JT, Bryer JD (2004) Protein and bacterial fouling characteristics of peptide and antibody decorated surfaces of PEGpoly(acrylic acid) co-polymers. Biomaterials 25:2247–2263

Webber MA, Randall LP, Cooles S, Woodward MJ, Piddock LJV (2008) Triclosan resistance in *Salmonella enterica* serovar Typhimurium. J Antimicrob Chemother 62(1):83–91

Wei J, Ravn DB, Gram L, Kingshott P (2003) Stainless steel modified with poly(ethylene glycol) can prevent protein adsorption but not bacterial adhesion. Colloids Surf B 32:275–291

Wirtanen G, Matilla-Sandholm T (1993) Epifluorescence image analysis and cultivation of foodborne biofilm bacteria grown on stainless steel surfaces. J Food Prot 56:678–683

Wirtanen G, Mattila-Sandholm T (1992) Effect of the growth phase of foodborne biofilms on their resistance to a chloride sanitizer. Part II. Lebensm Wiss Technol 25:50–54

Wirtanen G, Saarela M, Mattila-Sandholm T (2000) Biofilms-impact on hygiene in food industries. In: Bryers JD (ed) Biofilms II: process analysis and applications. Wiley-Liss, New York, pp 327–372

World Health Organization (WHO) (2013) Foodborne diseases. http://www.who.int/topics/foodborne_diseases/en/

Yuan S, Wan D, Liang B, Pehkonen SO, Ting YP, Neoh KG et al (2011) Lysozyme-coupled (poly (ethylene glycol) methacrylate)-stainless steel hybrids and their antifouling and antibacterial surfaces. Langmuir 27:2761–2774

Zhao Q, Liu Y (2006) Modification of stainless steel surfaces by electroless Ni-P and small amount of PTFE to minimize bacterial adhesion. J Food Eng 72:266–272

Zhao Q, Liu Y, Wang C (2005a) Development and evaluation of electroless Ag-PTFE composite coatings with anti-microbial and anti-corrosion properties. Appl Surf Sci 252:1620–1627

Zhao Q, Liu Y, Wang C, Wang S, Müller-Steinhagen H (2005b) Effect of surface free energy on the adhesion of biofouling and crystalline fouling. Chem Eng Sci 60:4858–4865

Part III
The Future of Antibiofilm Agents

Biofilm Inhibition by Nanoparticles

D. Bakkiyaraj and S.K. Pandian

Abstract Infectious diseases are of immediate concern due to their high rate of morbidity and mortality. Infectious diseases are life threatening in the current scenario as the causative agents are resistant to almost all the drugs in use. Apart from well-known factors like efflux pumps, receptor modifications, and drug inactivation, formation of biofilms attributes to broad-spectrum resistance toward antimicrobials. This necessitates the search for novel therapeutics that effectively control drug-resistant pathogens. Targeting biofilm formation is one such strategy to combat infectious diseases much more effectively. For over a decade diverse sources of synthetic to semisynthetic agents derived from microbes to plants have been tested for their antibiofilm potential with limited success. The birth of nanotechnology provided new insights into antibiofilm research as these nanoparticles are highly reactive and effective in penetrating the biofilm matrix. This chapter comprehensively summarizes the synthesis, application, weakness, and antibiofilm potential of nanoparticles.

1 Introduction

Infectious diseases are of major concern as they can result in high mortality and morbidity. Bacteria and fungi are the major pathogens that cause infections in humans and development of resistance by these organisms contributes to the severity of infections. Diseases caused by multidrug-resistant (MDR) pathogens are extremely difficult to treat and have a major impact on the economy. Even though numerous mechanisms like activation of efflux pumps, decreased permeability to antagonists, production of enzymes (e.g., beta-lactamase to inactivate antimicrobials), and mutation in target proteins facilitate resistance, biofilm

---

D. Bakkiyaraj • S.K. Pandian (✉)
Department of Biotechnology, Alagappa University, Science Campus, Karaikudi 630004, TN, India
e-mail: sk_pandian@rediffmail.com

formation has been the predominant mechanism of broad-spectrum tolerance. Biofilm are complexes made up of microbes surrounded by a hydrated matrix that is secreted by these indwelling microbes to protect or facilitate their growth in hostile environments. Biofilms are characterized by their extracellular polymeric substances (EPS) which contain polysaccharides, proteins, lipids, and nucleic acids. Apart from conferring resistance to the inhabitants, biofilms also facilitate various other functions like aggregation, retention of water and nutrients, absorption of nutrients, protection against host immune responses, and horizontal gene transfer. Studies have even suggested that the multicellular behavior of biofilm inhabitants is similar to higher multicellular organisms.

In theory, antibiofilm agents are less likely to cause selective pressure for the evolution of resistance because they do not kill pathogens as do antibiotics. Successful antibiofilm agents can either inhibit the formation of biofilms or disrupt mature biofilms. Antibiofilm agents are preferred over antibiotics in some instances as they prevent or disrupt biofilms, facilitating their clearance by the host immune system. Numerous sources from soil to sea and herbs to plants have been screened for antibiofilm activity. Even, some antibiotics have been shown to possess antibiofilm activity at sublethal concentrations. In addition, synthetic agents with antibiofilm properties are of interest because of their feasibility and availability for application.

Various agents such as synthetic chemicals, microbial secondary metabolites, phenolic compounds and other phytochemicals from plants, antibiotics at their sublethal concentrations, nucleases, proteases and other enzymes, peptides, etc., were shown to have the potential to inhibit biofilm formation and/or disrupt mature biofilms. Numerous synthetic chemicals like thiazolidinone derivatives (Pan et al. 2010; Rane et al. 2012), aminoimidazoles (Furlani et al. 2012), diazopyrazole derivatives (Raimondi et al. 2012), bromopyrrole alkaloids (Rane et al. 2013), etc., were shown to have antibiofilm potential against Gram-positive bacterial pathogens like *Staphylococcus aureus*, *Streptococcus epidermidis*, and *Enterococcus faecalis*. Other synthetics like niclosamide (Imperi et al. 2013), esomeprazole (Singh et al. 2012), chlorogenic acid (Karunanidhi et al. 2013), and zinc (Wu et al. 2013) showed promising results against various Gram negative bacteria, especially *Pseudomonas aeruginosa*. Other chemicals such as caspofungin (Bink et al. 2012) and farnesol (Ramage et al. 2002) were shown to be active against the biofilms of *Candida albicans*. Lastly, antibiotics at sublethal concentrations were shown to inhibit biofilms, which is of interest, as these sublethal concentrations are less likely to induce the development of resistance (Balaji et al. 2013; Gilbert et al. 2002; Latimer et al. 2012).

Various microbial extracts (Bakkiyaraj and Pandian 2010; Bakkiyaraj et al. 2012; Nithya et al. 2010b, 2011; Nithya and Pandian 2010) and their secondary metabolites like usnic acid and atranorin (Pompilio et al. 2013), glycolipid biosurfactants (Kiran et al. 2010), phenylacetic acid (Musthafa et al. 2012b), ophiobolins (Arai et al. 2013), and piperazinedione (Musthafa et al. 2012a) were reported to have antibiofilm properties against bacterial and fungal pathogens. Apart from the microbial metabolites, enzymes were also shown to inhibit the formation of biofilms. Alpha amylase produced by *Bacillus subtilis* (Kalpana et al. 2012), acylase produced by *B. pumilus* (Nithya et al. 2010a), alginate lyase (Lamppa and Griswold 2013), and protease produced by *P. aeruginosa* and actinomycetes (Park et al. 2012a, b) were also shown to have antibiofilm potential against human bacterial pathogens.

Numerous plants have been reported to display antibiofilm activities against bacterial and fungal pathogens. Cinnamaldehyde (Brackman et al. 2008), methyl eugenol (Packiavathy et al. 2012), casbane diterpene (Cardoso Sa et al. 2012), curcumin (Packiavathy et al. 2013), taxodione derivatives (Kuzma et al. 2012), gallic acid and ferulic acid (Borges et al. 2012), and ellagic acid (Sarabhai et al. 2013) are the notable plant products with potential antibiofilm activity.

The latest developments in the field of antibiofilm research employ novel agents like peptides (Amer et al. 2010; Choi and Lee 2012; Reymond et al. 2013; Zhang et al. 2010) and nanoparticles (Anghel et al. 2012; Hernandez-Delgadillo et al. 2012, 2013; Lellouche et al. 2012b; Durmus and Webster 2013; Martinez-Gutierrez et al. 2013; Sawant et al. 2013) as antibiofilm agents. The synthesis, properties, and the application of nanoparticles as antibiofilm agents will be discussed in detail in this chapter.

2 Properties and Synthesis of Nanoparticles

Among the antibiofilm technologies that have recently emerged, nanotechnology is one of the most promising. Nanotechnology can be defined as "a technology of engineering functional systems at molecular scale." Nanotechnology can also be defined as technology involving design, synthesis, and application of materials and devices whose size and shape have been engineered at nanoscale. Particles produced through nanotechnology are called "nanoparticles" and are typically sized less than 100 nm. Nanoparticles are highly reactive and preferred over other bioactive agents because of their higher surface area in contrast to their size. For example, 1 μg of particles of 1 nm^3 size have the same surface area as 1 g of particles of size 1 mm^3. Huge surface area of these nanoparticles facilitates their use as drug carriers.

Even though diverse chemicals like chitosan (Du et al. 2008), carboxymethyl chitosan (Zhao et al. 2013b), poly-gamma-glutamic acid (Liu et al. 2013b), cellulose (Raghavendra et al. 2013), zinc oxide (ZnO) (Dutta et al. 2013; Jones et al. 2008), magnesium fluoride (Lellouche et al. 2009, 2012b, c), polyethyleneimine (Beyth et al. 2010), hydroxyapatite (Evliyaoglu et al. 2011), fullerene (Patel et al. 2013), lipids (terpinen-4-ol) (Sun et al. 2012), and silica (Besinis et al. 2014; Li and Wang 2013) were shown to be useful, metals are the prime component of most nanoparticles. Derivatives of metals, like their oxides, form the base material for synthesis of many nanoparticles. Silver (Antony et al. 2013; Apte et al. 2013b; Besinis et al. 2014; Chernousova and Epple 2013; Jain and Pradeep 2005; Mohanty et al. 2012), gold (Annamalai et al. 2013; Geethalakshmi and

Sarada 2012; Khan et al. 2012; Naz et al. 2013; Pender et al. 2013; Ramamurthy et al. 2013), copper (Eshed et al. 2012; Kim et al. 2006; Pandiyarajan et al. 2013; Pramanik et al. 2012; Singh et al. 2013; Thekkae Padil and Cernik 2013), titanium (Besinis et al. 2014; Jayaseelan et al. 2013; Li et al. 2013), and iron (Das et al. 2013; Grumezescu et al. 2011; Leuba et al. 2013) are the predominant members of metal oxide nanoparticles and other metals like bismuth (Hernandez-Delgadillo et al. 2012, 2013) and cerium oxide (Shah et al. 2012) are other metal nanoparticles shown to possess bioactive potential.

Nanoparticles are synthesized either through a scale-up process, where atoms are grouped together, or a scale-down process, where larger molecules are minced to nanoscale. Irrespective of the methods used, synthesis of nanoparticles involves evaporation/dissolution, nucleation, and growth.

The synthesis of nanoparticles by scale-down or sizing-down processes can be achieved either by attrition or milling, followed by size-dependent grouping and selection. Scale-up processes can be broadly classified into three groups: gas phase fabrication; liquid phase fabrication; and biosynthesis or green synthesis of nanoparticles.

2.1 Gas or Vapor Phase Nanoparticle Fabrication

This process involves the evaporation of solid and liquid precursors to gaseous precursors followed by supersaturation, producing an intermediate product. Nucleation or condensation of these intermediate products results in primary particles. These primary particles, upon grain growth and agglomeration, produce nanoparticles and nanoclusters, respectively. Methods that employ gas phase fabrication are as follows:

1. Methods using solid precursors (Iskandar 2009)

   - Inert gas condensation
   - Pulsed laser ablation
   - Spark discharge generation
   - Ion sputtering

2. Methods using liquid or vapor precursors (Suciu et al. 2003)

   - Chemical vapor synthesis
   - Spray pyrolysis
   - Laser pyrolysis/photochemical synthesis
   - Thermal plasma synthesis
   - Flame synthesis
   - Flame spray pyrolysis
   - Low-temperature reactive synthesis

2.2 Liquid Phase Nanoparticle Fabrication

Liquid phase fabrication involves wet chemistry and the general process includes the surface reaction of solid and liquid precursors to produce corresponding intermediate products. Such intermediate products are converted to primary particles either by nucleation or condensation similar to gas phase fabrication, followed by growth or agglomeration to produce nanoparticles or nanoclusters, respectively. Methods that employ liquid phase fabrication are:

1. Co-precipitation (Murray et al. 2000)
2. Solvothermal methods (Yang et al. 2006)
3. Sol–gel methods (Yu et al. 2004)
4. Synthesis in structure media (e.g., Microemulsion) (Capek 2004)
5. Microwave synthesis (Tsuji et al. 2005)
6. Sonochemical synthesis (Zhang and Yu 2003)

2.3 Biological Synthesis of Nanoparticles

Synthesis of nanoparticles catalyzed by bacteria or fungi or their products is of considerable interest as it employs cleaner and greener technology. Numerous fungi and bacteria have been utilized for the bioconversion of raw chemicals into nanoparticles. For instance, the ability of the marine yeast *Yarrowia lipolytica* to catalyze the synthesis of gold nanoparticles has been reported (Agnihotri et al. 2009; Apte et al. 2013a, b). Biosynthesis of silver, gold, and bimetallic nanoparticles by fungi like *Phanerochaete chrysosporium*, *Penicillium* sp., and *Neurospora crassa* has also been reported (Castro-Longoria et al. 2011; Du et al. 2010; Vigneshwaran et al. 2006). Similarly, the synthesis of silver nanoparticles with antimicrobial potential by psychrophilic bacteria such as *Pseudomonas antarctica* and *Arthrobacter kerguelensis* has also been reported (Shivaji et al. 2011). *Lactobacillus fermentum* (Sintubin et al. 2009) and *Shewanella oneidensis* (Suresh et al. 2010) were also shown to catalyze the production of silver nanoparticles with antimicrobial potential.

Though there are numerous reports on the microbe-mediated synthesis of nanoparticles, very few studies have described the biomolecules involved in this synthesis. For example, nitrate reductase along with a protein from *Aspergillus niger* and nitrate reductase along with rhamnolipids from *P. aeruginosa* were shown to be indispensable for the synthesis of nanoparticles (Gade et al. 2008; Kumar and Mamidyala 2011). Similarly, the role of cell-bound melanin produced by the yeast *Y. lipolytica* and certain proteins produced by marine fungi *A. tubingensis* and *Bionectria ochroleuca* in the synthesis of silver nanoparticles with antibiofilm activity have been reported recently (Apte et al. 2013a, b; Rodrigues et al. 2013).

3 Diverse Applications of Nanoparticles

Nanoparticles or nanomaterials in general have diverse applications in various fields.

3.1 Industrial Applications

Many microelectronic instruments such as transistors have adapted nanotechnology (Thompson and Parthasarathy 2006). Carbon nanotubes are reported to be the nanoscale alternatives to conventional semiconductor crystals because of their diverse electronic properties from metallic to semiconducting (Jacoby 2002) or superconducting (Cristina and Kevin 2005). Carbon nanotubes have been shown to be useful in making low-voltage field-emission displays (Carey 2003). Nanomaterials like aerogel intercalation electrode materials, nanocrystalline alloys, nanosized composite materials, carbon nanotubes, and nanosized transition metal oxides have shown promise in the development of lithium-ion batteries with increased capacity and lifecycle over their conventional counterparts (Liu et al. 2006a; Scott et al. 2011).

Nanocrystalline materials synthesized by the sol–gel technique exhibit foam-like structures called "aerogel" which find application as insulation material in industries because of their negligible thermal conductivity (Hrubesh and Poco 1995). Paints that have incorporated nanoparticles (Titanium oxide) demonstrate enhanced mechanical properties, such as scratch resistance. For example, the wear resistance of paint-nanocomposite coatings is claimed to be ten times higher than that of conventional acrylic paints (Mochizuki et al. 2013).

In the automobile industry, nanoparticles of carbon black act as filler in the polymer matrix of tires and are used for mechanical reinforcement. Nanocomposites containing the flakes of clay and plastics and nanosized clay are used in manufacturing the exteriors of cars with superior properties like scratch resistance compared to traditional materials.

Nanoparticles have found their way into the food industry due to their antimicrobial properties. For example, silver-montmorillonite (Ag-MMT) nanoparticles were used in the prevention of food spoilage (Costa et al. 2011). In addition to preventing the growth of food-spoiling microbes, Ag-MMT nanoparticles also preserved color, odor, and firmness of the food (Costa et al. 2011).

Nanoparticles also have potential in controlling pollution because of their ability to catalyze the conversion of toxic gases (carbon monoxide and nitrogen oxide) from the exhaust of vehicles and power generators. Iron nanoparticles, along with palladium, converted detrimental products in groundwater to inert or less harmful products (He and Zhao 2005). The nanoparticles were also shown to be effective in removing organic chlorine (a carcinogen) from water contaminated with the chlorine-based organic solvents (used in dry cleaners).

3.2 Nanoparticles in Biotechnology and Medicine

Carbon nanotubes have been used as probe tips in atomic force microscopy (AFM) which is used for high-resolution imaging of nucleic acids, immunoglobulins, etc. (Hafner et al. 2001). Molecular recognition and the chemical forces between the interacting molecules can be studied by attaching AFM tips bearing these biomolecules (Hafner et al. 2001).

Nanofiber scaffolds have been employed in the regeneration of cells and organs. Experiments on a hamster with a detached optic tract demonstrated that a peptide nanofiber scaffold could facilitate the regeneration of axonal tissue (Ellis-Behnke et al. 2006). Titanium dioxide and zinc oxide are used in sunscreens and cosmetics to absorb and reflect UV light.

Nanotube membranes can act as channels for highly selective transport of molecules and ions between solutions that are present on both sides of the membrane (Jirage et al. 1997). For instance, membranes containing nanotubes with small inner dimensions (less than 1 nm) were useful for the separation of small molecules on the basis of molecular size, while the nanotubes with larger inner diameters (20–60 nm) were used to separate proteins (Martin and Kohli 2003).

The ability of nanoparticles to target and penetrate specific cells and organs has also been explored in nanomedicine. Nanospheres made of biodegradable (facilitating timely release) polymers and drugs have potential applications in acidic microenvironments as in the case of tumor tissues or sites of inflammation (Kamaly et al. 2012). Nanoparticles acted as drug carriers for the targeted release of a conjugate containing chlorotoxin (a peptide that selectively binds to glioblastoma cells) and liposomes encapsulating antisense oligonucleotides or small interfering RNAs for effective treatment of glioblastoma (Costa et al. 2013). Similarly, numerous other studies have independently demonstrated the utility of nanoparticles as drug carriers in different tumor types (Amoozgar et al. 2013; Leifert et al. 2013; Liu et al. 2013a; Shi et al. 2013; Vivek et al. 2013).

In addition, surface-functionalized nanoparticles can be used to infuse cell membranes at a much higher level than nanoparticles without a functionalized surface, which can be employed for transfer of genetic material into living cells (Lewin et al. 2000). Silica nanospheres coated with ammonium groups (cation) can bind to DNA (anion) through electrostatic interactions, which could be used to deliver the latter into the cells (Kneuer et al. 2000).

Nanospheres can act as carriers for antigens and toxoids for potential use in vaccination. Studies involving antigen-coated polystyrene nanospheres as vaccine carriers targeting human dendritic cells have been under trial for nasal vaccination (Matsusaki et al. 2005). Studies have also unveiled the potential of nanoparticles in the diagnosis and treatment of various cancers. For instance, a study by Yin et al. (2013) showed enhanced anticancer action of curcumin upon coupling it with nanoparticles made from methoxy poly(ethylene glycol)-polycaprolactone (PCL) block copolymers (Yin et al. 2013). Similarly, the silver nanoparticles were shown to inhibit lung cancer cells in a concentration-dependent manner

(Sankar et al. 2013). Iron nanoparticles coupled with high-resolution MRI detected lymph node metastases in patients with prostate cancer at a stage undetectable by any other method (Harisinghani et al. 2003) and the gold nanoparticles were employed for the accurate detection of matriptase—a cancer biomarker protein overexpressed in all types of cancer (Deng et al. 2013). Lastly, nanoparticles made of compounds with oxygen vacancies (CeO_2 and Y_2O_3) (Schubert et al. 2006) have been demonstrated to possess neuroprotective and anti-apoptotic properties.

3.3 Antimicrobial Activity of Nanoparticles

Nanoparticles have been considered to be some of the most effectual bioactive agents mainly because of their large surface area to volume ratio (Hamouda 2012). Nanopowders possess antimicrobial properties against various bacterial, fungal and viral human pathogens (Koper et al. 2002; Bosi et al. 2003) and can rapidly kill bacterial cells (90 % in 1 h). The antibacterial properties of silver and titanium dioxide nanoparticles have been assessed as coatings for surgical masks (Li et al. 2006), in addition to many other clinical uses.

Nanoparticles shown to have antimicrobial effects include silver (Lara et al. 2010; Lok et al. 2006), titanium dioxide (Li et al. 2006), fullerenes (Bosi et al. 2003), zinc oxide (Brayner et al. 2006), and magnesium fluoride (Lellouche et al. 2012c). The antibacterial activity of fullerenes was reported against *Escherichia coli*, Salmonella and *Streptococcus* spp. (Bosi et al. 2003). The ability of zinc oxide nanoparticles to disturb the membrane permeability of *E. coli* has also been reported (Brayner et al. 2006). The wide spectrum antimicrobial activity of silver nanoparticles has been attributed to their ability to destabilize the bacterial outer membrane and deplete adenosine triphosphate (principal form of energy) in bacteria (Lara et al. 2010; Lok et al. 2006).

Fullerenes have also been shown to have neuroprotective, anti-apoptotic, and anti-HIV activities (Bosi et al. 2003). Size-dependent interactions of silver nanoparticles and HIV-1 virus were reported, which resulted in the inhibition of host–viral interactions (Elechiguerra et al. 2005). Numerous other studies have demonstrated the antimicrobial potential of various nanoparticles and drug–nanoparticle conjugates against bacterial, fungal, and viral pathogens (Zheng et al. 2013; Zhao et al. 2013a; Zhang et al. 2013c; Xiong et al. 2013; Westendorf 2013; Wang and Lim 2013; Wang et al. 2013; Vidic et al. 2013; Tavassoli Hojati et al. 2013; Su et al. 2013; Shimizu et al. 2013; Mohanty et al. 2012; Mallick et al. 2012; Lellouche et al. 2012a; Costa et al. 2011; Mukhopadhyay et al. 2010; Huda et al. 2010; Sanpui et al. 2008; Pinto et al. 2013; Hernandez-Delgadillo et al. 2013; Monteiro et al. 2012; Khan et al. 2012; Lara et al. 2010).

4 Nanoparticles as Antibiofilm Agents

The application of nanoparticles is an emerging area of antibiofilm or antipathogenic research. Nanoparticles are preferred over other agents due to their acute ability to penetrate EPS and cell membranes. Nanoparticles were found to be efficient drug carriers, effectively transporting drugs across the biofilm matrix. Silver, iron and zinc nanoparticles have received the most attention as antibiofilm agents. Silver nanoparticles are the predominant ones with antibiofilm activity against Gram-positive bacteria like *S. aureus*, *S. epidermidis*, *E. faecalis*, and *Streptococcus mutans*; Gram-negative bacteria like *P. aeruginosa*, *Salmonella paratyphi*, *E. coli*, and *Acinetobacter baumannii*; and fungal pathogens like *C. albicans* and *C. glabrata*. Details of nanoparticles with antibiofilm activities, along with their target pathogens, are given in Table 1.

The use of nanoparticles in combination with other antibiotics or drugs was found to have superior action than when alone. Chitosan nanoparticles loaded with Tamoxifen were effective in controlling tumor development in breast cancer cell lines (Vivek et al. 2013). Similarly, the side effects of daunorubicin were reduced significantly when combined with titanium oxide nanoparticles, which increased the target specificity and anticancer activity in leukemia cells (Zhang et al. 2012). Silver nanoparticles in combination with conventional antibiotics like ampicillin, chloramphenicol, and kanamycin have shown antibiofilm activity against Gram-positive and negative bacterial pathogens including *E. faecium*, *S. aureus*, *E. coli*, and *P. aeruginosa* (Hwang et al. 2012).

5 Demerits of Nanoparticles

Even though nanoparticles have historically been considered inert, they are actually highly reactive. The large surface area of nanoparticles can be both a pro and a con to their application in biology. Nanoparticles are commonly found in dust and aerosols. Inhaled nanoparticles deposited in the lungs are cleared through host processes such as mucociliary escalation into the gastrointestinal tract (from where they are eliminated through the feces) (Semmler et al. 2004), lymphatic system (Liu et al. 2006b), and circulatory systems (Oberdorster et al. 2005). Failure to clear these nanoparticles results in accumulation in lungs, subsequently increasing the risk of lung cancer (Borm et al. 2004). Accumulation of nanoparticles in lungs also elicits an inflammatory response that damages host tissues (Oberdorster et al. 1994). Adverse effects to nanoparticles include impaired phagocytosis, inflammation, epithelial cell proliferation followed by fibrosis, emphysema, and the initiation of tumors (Ferin 1994; Oberdorster et al. 1994; Nikula et al. 1995; Dasenbrock et al. 1996; Driscoll et al. 1996; Borm et al. 2004).

Inhalation of nanoparticles can also result in immune suppression and reduction in the ability of the immune system to combat infections (Lucarelli et al. 2004).

Table 1 Antibiofilm activity of nanoparticles and their target pathogens

Nanoparticle	Target organism	References
Silver nanoparticles	S. paratyphi, P. aeruginosa, S. epidermidis	Apte et al. (2013b), Kalishwaralal et al. (2010)
Bismuth oxide aqueous colloidal nanoparticles	C. albicans, S. mutans	Hernandez-Delgadillo et al. (2012, 2013)
Nano-oil formulation from *Mentha piperita* L.	Staphylococcus sp.	Anghel and Grumezescu (2013)
Nanoemulsion (detergent, oil, and water) in combination with cetylpyridinium chloride	A. baumannii	Hwang et al. (2013)
Silver and gold incorporated polyurethane, polycaprolactam, polycarbonate, and polymethylmethacrylate	E. coli	Sawant et al. (2013)
Silver nanoparticles in combination with nystatin and chlorhexidine	C. albicans, C. glabrata	Monteiro et al. (2012, 2013)
Silver nanoparticle and 12-methacryloyloxydodecylpyridinium bromide (MDPB)	Dental plaque microcosm biofilms	Zhang et al. (2013a, b)
Zinc	Actinobacillus pleuropneumoniae, S. typhimurium, Haemophilus parasuis, E. coli, S. aureus, S. suis	Wu et al. (2013)
Magnetite nanoparticles	C. albicans	Anghel et al. (2012)
Eugenia carryophyllata essential oil stabilized by iron oxide/oleic acid core/shell nanostructures	S. aureus	Grumezescu et al. (2011, 2012)
Zinc and copper oxide nanoparticles	S. mutans	Eshed et al. (2012)
Zerovalent bismuth nanoparticle	S. mutans	Hernandez-Delgadillo et al. (2012)
Silver nanoparticles in combination with ampicillin, chloramphenicol, and kanamycin	Enterococcus faecium, S. aureus, E. coli, P. aeruginosa	Hwang et al. (2012)
Dextran sulfate nanoparticle complex containing ofloxacin and levofloxacin	P. aeruginosa	Cheow and Hadinoto (2012)
PEG-stabilized lipid nanoparticles loaded with terpinen-4-ol	C. albicans	Sun et al. (2012)
Magnesium fluoride nanoparticles	S. aureus, E. coli	Lellouche et al. (2009, 2012b, c)
Yttrium fluoride nanoparticles	S. aureus, E. coli	Lellouche et al. (2012a)
Iron oxide/oleic acid in combination with essential oil from *Rosmarinus officinalis*	C. albicans, C. tropicalis	Chifiriuc et al. (2012)
Gold nanoparticles and methylene blue	C. albicans	Khan et al. (2012)
Starch-stabilized silver nanoparticles	S. aureus, P. aeruginosa	Mohanty et al. (2012)

(continued)

Table 1 (continued)

Nanoparticle	Target organism	References
Iron oxide–oleic acid nanofluid	S. aureus	Grumezescu et al. (2011)
Chitosan, zinc oxide, nitric oxide nanoparticles	E. faecalis	Shrestha et al. (2010)
Quaternary ammonium polyethylenimine nanoparticles	Oral biofilms	Beyth et al. (2010)
Zinc oxide nanoparticles, chitosan nanoparticles, and combination of both	E. faecalis	Kishen et al. (2008)
Polyurethane nanocomposite	S. epidermidis	Styan et al. (2007)

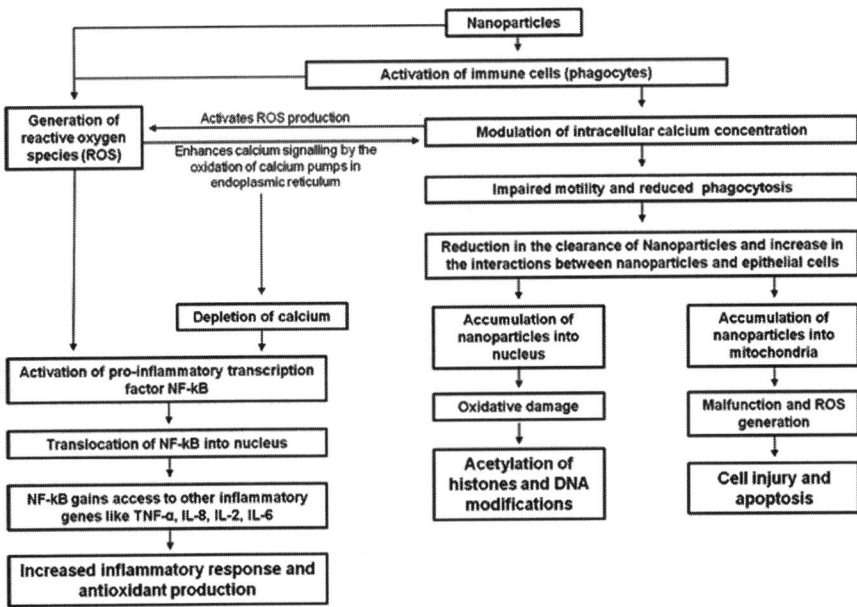

Fig. 1 Molecular mechanisms involved in nanoparticle-induced cellular toxicity (adopted and modified from Buzea et al. 2007)

Exposure to nanoparticles like zirconium dioxide (ZrO_2) induces overexpression of viral receptors and in turn results in hyper-reaction of the immune system and subsequent unwarranted inflammation (Lucarelli et al. 2004). In vivo and in vitro studies have shown the ability of nanoparticles (fullerenes, carbon nanotubes, quantum dots, and automobile exhaust) to initiate the production of reactive oxygen species (ROS) (Oberdorster et al. 2005), which has been shown to play a key role in cell damage by peroxidizing lipids, damaging proteins and nucleic acids, interfering with signaling functions, and modulating gene expression (Brown et al. 2004; Risom et al. 2005; Peters et al. 2006; Mehta et al. 2008). Malfunction of mitochondria has also been observed upon nanoparticle treatment as they effectively enter these organelles and contribute to oxidative stress and damage (Li et al. 2003; Xia et al. 2006; Sioutas et al. 2005). There is also evidence of the adverse effects (appearance of thrombi) of nanoparticles on the cardiovascular system (Schulz et al. 2005; Nemmar et al. 2002; Vermylen et al. 2005; Hoet et al. 2004). Uptake of nanoparticles through skin results in their accumulation in the lymphatic system causing podoconiosis (Corachan 1988; Blundell et al. 1989) and Kaposi's sarcoma (Montella et al. 1997; Mott et al. 2002). Molecular mechanisms involved in nanoparticle-mediated cellular toxicity are schematically represented (Fig. 1).

6 Conclusions

Nanotechnology is a nascent field of science with promising potential in many fields including physics, chemistry, biology, pharmacology, and medicine. As discussed in this chapter, nanoparticles can be our friend or foe. Although there are reports which state nanoparticles are toxic, there is always the potential for improvement and development of safe and effective novel nanoparticles or nanocomposites. The utility of nanoparticles as drug carriers appears to be an important tool for targeted tumor therapy, and enhancing the efficacy of drugs could be another attractive application for nanomaterials. Even though the use of nanoparticles in vivo is debatable for now, their use on inanimate objects is effective. Without any doubt, the future will witness increasing use of nanoparticles in many fields, hopefully for the improvement of mankind.

References

Agnihotri M, Joshi S, Kumar AR, Zinjarde S, Kulkarni S (2009) Biosynthesis of gold nanoparticles by the tropical marine yeast *Yarrowia lipolytica* NCIM 3589. Mater Lett 63 (15):1231–1234

Amer LS, Bishop BM, van Hoek ML (2010) Antimicrobial and antibiofilm activity of cathelicidins and short, synthetic peptides against Francisella. Biochem Biophys Res Commun 396(2):246–251

Amoozgar Z, Park J, Lin Q, Weidle JH 3rd, Yeo Y (2013) Development of quinic acid-conjugated nanoparticles as a drug carrier to solid tumors. Biomacromolecules 14(7):2389–2395

Anghel I, Grumezescu AM (2013) Hybrid nanostructured coating for increased resistance of prosthetic devices to staphylococcal colonization. Nanoscale Res Lett 8(1):6

Anghel I, Grumezescu AM, Andronescu E, Anghel AG, Ficai A, Saviuc C, Grumezescu V, Vasile BS, Chifiriuc MC (2012) Magnetite nanoparticles for functionalized textile dressing to prevent fungal biofilms development. Nanoscale Res Lett 7(1):501

Annamalai A, Christina VL, Sudha D, Kalpana M, Lakshmi PT (2013) Green synthesis, characterization and antimicrobial activity of Au NPs using *Euphorbia hirta* L. leaf extract. Colloids Surf B Biointerfaces 108:60–65

Antony JJ, Nivedheetha M, Siva D, Pradeepha G, Kokilavani P, Kalaiselvi S, Sankarganesh A, Balasundaram A, Masilamani V, Achiraman S (2013) Antimicrobial activity of *Leucas aspera* engineered silver nanoparticles against *Aeromonas hydrophila* in infected *Catla catla*. Colloids Surf B Biointerfaces 109:20–24

Apte M, Girme G, Bankar A, Ravikumar A, Zinjarde S (2013a) 3, 4-dihydroxy-L-phenylalanine-derived melanin from *Yarrowia lipolytica* mediates the synthesis of silver and gold nanostructures. J Nanobiotechnol 11:2

Apte M, Sambre D, Gaikawad S, Joshi S, Bankar A, Kumar AR, Zinjarde S (2013b) Psychrotrophic yeast *Yarrowia lipolytica* NCYC 789 mediates the synthesis of antimicrobial silver nanoparticles via cell-associated melanin. AMB Express 3(1):32

Arai M, Niikawa H, Kobayashi M (2013) Marine-derived fungal sesterterpenes, ophiobolins, inhibit biofilm formation of *Mycobacterium* species. J Nat Med 67(2):271–275

Bakkiyaraj D, Pandian SK (2010) *In vitro* and *in vivo* antibiofilm activity of a coral associated actinomycete against drug resistant *Staphylococcus aureus* biofilms. Biofouling 26(6):711–717

Bakkiyaraj D, Sivasankar C, Pandian SK (2012) Inhibition of quorum sensing regulated biofilm formation in *Serratia marcescens* causing nosocomial infections. Bioorg Med Chem Lett 22 (9):3089–3094

Balaji K, Thenmozhi R, Pandian SK (2013) Effect of subinhibitory concentrations of fluoroquinolones on biofilm production by clinical isolates of *Streptococcus pyogenes*. Indian J Med Res 137(5):963–971

Besinis A, De Peralta T, Handy RD (2014) The antibacterial effects of silver, titanium dioxide and silica dioxide nanoparticles compared to the dental disinfectant chlorhexidine on *Streptococcus mutans* using a suite of bioassays. Nanotoxicology 8(1):1–16

Beyth N, Yudovin-Farber I, Perez-Davidi M, Domb AJ, Weiss EI (2010) Polyethyleneimine nanoparticles incorporated into resin composite cause cell death and trigger biofilm stress in vivo. Proc Natl Acad Sci USA 107(51):22038–22043

Bink A, Kucharikova S, Neirinck B, Vleugels J, Van Dijck P, Cammue BP, Thevissen K (2012) The nonsteroidal antiinflammatory drug diclofenac potentiates the in vivo activity of caspofungin against *Candida albicans* biofilms. J Infect Dis 206(11):1790–1797

Blundell G, Henderson WJ, Price EW (1989) Soil particles in the tissues of the foot in endemic elephantiasis of the lower legs. Ann Trop Med Parasitol 83(4):381–385

Borges A, Saavedra MJ, Simoes M (2012) The activity of ferulic and gallic acids in biofilm prevention and control of pathogenic bacteria. Biofouling 28(7):755–767

Borm PJ, Schins RP, Albrecht C (2004) Inhaled particles and lung cancer, part B: paradigms and risk assessment. Int J Cancer 110(1):3–14

Bosi S, Da Ros T, Spalluto G, Prato M (2003) Fullerene derivatives: an attractive tool for biological applications. Eur J Med Chem 38(11–12):913–923

Brackman G, Defoirdt T, Miyamoto C, Bossier P, Van Calenbergh S, Nelis H, Coenye T (2008) Cinnamaldehyde and cinnamaldehyde derivatives reduce virulence in *Vibrio* spp. by decreasing the DNA-binding activity of the quorum sensing response regulator LuxR. BMC Microbiol 8:149

Brayner R, Ferrari-Iliou R, Brivois N, Djediat S, Benedetti MF, Fievet F (2006) Toxicological impact studies based on *Escherichia coli* bacteria in ultrafine ZnO nanoparticles colloidal medium. Nano Lett 6(4):866–870

Brown DM, Donaldson K, Borm PJ, Schins RP, Dehnhardt M, Gilmour P, Jimenez LA, Stone V (2004) Calcium and ROS-mediated activation of transcription factors and TNF-alpha cytokine gene expression in macrophages exposed to ultrafine particles. Am J Physiol Lung Cell Mol Physiol 286(2):L344–L353

Buzea C, Blandino IIP, Robbie K (2007) Nanomaterials and nanoparticles: sources and toxicity. Biointerphases 2(4):MR17–MR172

Capek I (2004) Preparation of metal nanoparticles in water-in-oil (w/o) microemulsions. Adv Colloid Interface Sci 110(1–2):49–74

Cardoso Sa N, Cavalcante TT, Araujo AX, dos Santos HS, Albuquerque MR, Bandeira PN, da Cunha RM, Cavada BS, Teixeira EH (2012) Antimicrobial and antibiofilm action of Casbane Diterpene from *Croton nepetaefolius* against oral bacteria. Arch Oral Biol 57(5):550–555

Carey JD (2003) Engineering the next generation of large-area displays: prospects and pitfalls. Philos Trans A Math Phys Eng Sci 361(1813):2891–2907

Castro-Longoria E, Vilchis-Nestor AR, Avalos-Borja M (2011) Biosynthesis of silver, gold and bimetallic nanoparticles using the filamentous fungus *Neurospora crassa*. Colloids Surf B Biointerfaces 83(1):42–48

Cheow WS, Hadinoto K (2012) Green preparation of antibiotic nanoparticle complex as potential anti-biofilm therapeutics via self-assembly amphiphile-polyelectrolyte complexation with dextran sulfate. Colloids Surf B Biointerfaces 92:55–63

Chernousova S, Epple M (2013) Silver as antibacterial agent: ion, nanoparticle, and metal. Angew Chem Int Ed Engl 52(6):1636–1653

Chifiriuc C, Grumezescu V, Grumezescu AM, Saviuc C, Lazar V, Andronescu E (2012) Hybrid magnetite nanoparticles/*Rosmarinus officinalis* essential oil nanobiosystem with antibiofilm activity. Nanoscale Res Lett 7:209

Choi H, Lee DG (2012) Antimicrobial peptide pleurocidin synergizes with antibiotics through hydroxyl radical formation and membrane damage, and exerts antibiofilm activity. Biochim Biophys Acta 1820(12):1831–1838

Corachan M (1988) Endemic non-filarial elephantiasis of the lower limbs: podoconiosis. Med Clin (Barc) 91(3):97–100

Costa C, Conte A, Buonocore GG, Del Nobile MA (2011) Antimicrobial silver-montmorillonite nanoparticles to prolong the shelf life of fresh fruit salad. Int J Food Microbiol 148(3):164–167

Costa PM, Cardoso AL, Mendonca LS, Serani A, Custodia C, Conceicao M, Simoes S, Moreira JN, Pereira de Almeida L, Pedroso de Lima MC (2013) Tumor-targeted chlorotoxin-coupled nanoparticles for nucleic acid delivery to glioblastoma cells: a promising system for glioblastoma treatment. Mol Ther Nucleic Acids 2:e100

Cristina B, Kevin R (2005) Assembling the puzzle of superconducting elements: a review. Supercond Sci Technol 18(1):R1

Das B, Mandal M, Upadhyay A, Chattopadhyay P, Karak N (2013) Bio-based hyperbranched polyurethane/Fe_3O_4 nanocomposites: smart antibacterial biomaterials for biomedical devices and implants. Biomed Mater 8(3):035003

Dasenbrock C, Peters L, Creutzenberg O, Heinrich U (1996) The carcinogenic potency of carbon particles with and without PAH after repeated intratracheal administration in the rat. Toxicol Lett 88(1–3):15–21

Deng D, Zhang D, Li Y, Achilefu S, Gu Y (2013) Gold nanoparticles based molecular beacons for *in vitro* and in vivo detection of the matriptase expression on tumor. Biosens Bioelectron 49C:216–221

Driscoll KE, Carter JM, Howard BW, Hassenbein DG, Pepelko W, Baggs RB, Oberdorster G (1996) Pulmonary inflammatory, chemokine, and mutagenic responses in rats after subchronic inhalation of carbon black. Toxicol Appl Pharmacol 136(2):372–380

Du WL, Xu YL, Xu ZR, Fan CL (2008) Preparation, characterization and antibacterial properties against *E. coli* K(88) of chitosan nanoparticle loaded copper ions. Nanotechnology 19(8):085707

Du L, Xian L, Feng J-X (2010) Rapid extra-/intracellular biosynthesis of gold nanoparticles by the fungus *Penicillium* sp. J Nanopart Res 13(3):921–930

Durmus NG, Webster TJ (2013) Eradicating antibiotic-resistant biofilms with silver-conjugated superparamagnetic iron oxide nanoparticles. Adv Healthc Mater 2(1):165–171

Dutta RK, Nenavathu BP, Gangishetty MK, Reddy AV (2013) Antibacterial effect of chronic exposure of low concentration ZnO nanoparticles on *E. coli*. J Environ Sci Health A Tox Hazard Subst Environ Eng 48(8):871–878

Elechiguerra JL, Burt JL, Morones JR, Camacho-Bragado A, Gao X, Lara HH, Yacaman MJ (2005) Interaction of silver nanoparticles with HIV-1. J Nanobiotechnol 3:6

Ellis-Behnke RG, Liang YX, You SW, Tay DK, Zhang S, So KF, Schneider GE (2006) Nano neuro knitting: peptide nanofiber scaffold for brain repair and axon regeneration with functional return of vision. Proc Natl Acad Sci USA 103(13):5054–5059

Eshed M, Lellouche J, Matalon S, Gedanken A, Banin E (2012) Sonochemical coatings of ZnO and CuO nanoparticles inhibit *Streptococcus mutans* biofilm formation on teeth model. Langmuir 28(33):12288–12295

Evliyaoglu Y, Kobaner M, Celebi H, Yelsel K, Dogan A (2011) The efficacy of a novel antibacterial hydroxyapatite nanoparticle-coated indwelling urinary catheter in preventing biofilm formation and catheter-associated urinary tract infection in rabbits. Urol Res 39(6):443–449

Ferin J (1994) Pulmonary retention and clearance of particles. Toxicol Lett 72(1–3):121–125

Furlani RE, Yeagley AA, Melander C (2012) A flexible approach to 1,4-di-substituted 2-aminoimidazoles that inhibit and disperse biofilms and potentiate the effects of beta-lactams against multi-drug resistant bacteria. Eur J Med Chem 62C:59–70

Gade AK, Bonde P, Ingle AP, Marcato PD, Duran N, Rai MK (2008) Exploitation of *Aspergillus niger* for synthesis of silver nanoparticles. J Biobased Mater Bioenergy 2:243–247

Geethalakshmi R, Sarada DV (2012) Gold and silver nanoparticles from *Trianthema decandra*: synthesis, characterization, and antimicrobial properties. Int J Nanomedicine 7:5375–5384

Gilbert P, Allison DG, McBain AJ (2002) Biofilms in vitro and in vivo: do singular mechanisms imply cross-resistance? J Appl Microbiol 92(Suppl):98S–110S

Grumezescu AM, Saviuc C, Chifiriuc MC, Hristu R, Mihaiescu DE, Balaure P, Stanciu G, Lazar V (2011) Inhibitory activity of Fe(3) O(4)/oleic acid/usnic acid-core/shell/extra-shell nanofluid on *S. aureus* biofilm development. IEEE Trans Nanobiosci 10(4):269–274

Grumezescu AM, Chifiriuc MC, Saviuc C, Grumezescu V, Hristu R, Mihaiescu DE, Stanciu GA, Andronescu E (2012) Hybrid nanomaterial for stabilizing the antibiofilm activity of *Eugenia carryophyllata* essential oil. IEEE Trans Nanobiosci 11(4):360–365

Hafner JH, Cheung CL, Woolley AT, Lieber CM (2001) Structural and functional imaging with carbon nanotube AFM probes. Prog Biophys Mol Biol 77(1):73–110

Hamouda IM (2012) Current perspectives of nanoparticles in medical and dental biomaterials. J Biomed Res 26(3):143–151

Harisinghani MG, Barentsz J, Hahn PF, Deserno WM, Tabatabaei S, van de Kaa CH, de la Rosette J, Weissleder R (2003) Noninvasive detection of clinically occult lymph-node metastases in prostate cancer. N Engl J Med 348(25):2491–2499

He F, Zhao D (2005) Preparation and characterization of a new class of starch-stabilized bimetallic nanoparticles for degradation of chlorinated hydrocarbons in water. Environ Sci Technol 39(9):3314–3320

Hernandez-Delgadillo R, Velasco-Arias D, Diaz D, Arevalo-Nino K, Garza-Enriquez M, De la Garza-Ramos MA, Cabral-Romero C (2012) Zerovalent bismuth nanoparticles inhibit *Streptococcus mutans* growth and formation of biofilm. Int J Nanomedicine 7:2109–2113

Hernandez-Delgadillo R, Velasco-Arias D, Martinez-Sanmiguel JJ, Diaz D, Zumeta-Dube I, Arevalo-Nino K, Cabral-Romero C (2013) Bismuth oxide aqueous colloidal nanoparticles inhibit *Candida albicans* growth and biofilm formation. Int J Nanomedicine 8:1645–1652

Hoet PH, Bruske-Hohlfeld I, Salata OV (2004) Nanoparticles: known and unknown health risks. J Nanobiotechnol 2(1):12

Hrubesh LW, Poco JF (1995) Thin aerogel films for optical, thermal, acoustic and electronic applications. J Non Cryst Solids 188((1–2)):46–53

Huda S, Smoukov SK, Nakanishi H, Kowalczyk B, Bishop K, Grzybowski BA (2010) Antibacterial nanoparticle monolayers prepared on chemically inert surfaces by cooperative electrostatic adsorption (CELA). ACS Appl Mater Interfaces 2(4):1206–1210

Hwang IS, Hwang JH, Choi H, Kim KJ, Lee DG (2012) Synergistic effects between silver nanoparticles and antibiotics and the mechanisms involved. J Med Microbiol 61(Pt 12):1719–1726

Hwang YY, Ramalingam K, Bienek DR, Lee V, You T, Alvarez R (2013) Antimicrobial activity of nanoemulsion in combination with cetylpyridinium chloride on multidrug-resistant *Acinetobacter baumannii*. Antimicrob Agents Chemother 57(8):3568–3575

Imperi F, Massai F, Ramachandran Pillai C, Longo F, Zennaro E, Rampioni G, Visca P, Leoni L (2013) New life for an old drug: the anthelmintic drug niclosamide inhibits *Pseudomonas aeruginosa* quorum sensing. Antimicrob Agents Chemother 57(2):996–1005

Iskandar F (2009) Nanoparticle processing for optical applications: a review. Adv Powder Technol 20(4):283–292

Jacoby M (2002) Nanoscale electronics. Chem Eng News Arch 80(39):38–43

Jain P, Pradeep T (2005) Potential of silver nanoparticle-coated polyurethane foam as an antibacterial water filter. Biotechnol Bioeng 90(1):59–63

Jayaseelan C, Rahuman AA, Roopan SM, Kirthi AV, Venkatesan J, Kim SK, Iyappan M, Siva C (2013) Biological approach to synthesize TiO2 nanoparticles using *Aeromonas hydrophila* and its antibacterial activity. Spectrochim Acta A Mol Biomol Spectrosc 107:82–89

Jirage KB, Hulteen JC, Martin CR (1997) Nanotubule-based molecular-filtration membranes. Science 278(5338):655–658

Jones N, Ray B, Ranjit KT, Manna AC (2008) Antibacterial activity of ZnO nanoparticle suspensions on a broad spectrum of microorganisms. FEMS Microbiol Lett 279(1):71–76

Kalishwaralal K, BarathManiKanth S, Pandian SR, Deepak V, Gurunathan S (2010) Silver nanoparticles impede the biofilm formation by *Pseudomonas aeruginosa* and *Staphylococcus epidermidis*. Colloids Surf B Biointerfaces 79(2):340–344

Kalpana BJ, Aarthy S, Pandian SK (2012) Antibiofilm activity of alpha-amylase from *Bacillus subtilis* S8-18 against biofilm forming human bacterial pathogens. Appl Biochem Biotechnol 167(6):1778–1794

Kamaly N, Xiao Z, Valencia PM, Radovic-Moreno AF, Farokhzad OC (2012) Targeted polymeric therapeutic nanoparticles: design, development and clinical translation. Chem Soc Rev 41 (7):2971–3010

Karunanidhi A, Thomas R, van Belkum A, Neela V (2013) *In vitro* antibacterial and antibiofilm activities of chlorogenic acid against clinical isolates of *Stenotrophomonas maltophilia* including the trimethoprim/sulfamethoxazole resistant strain. Biomed Res Int 2013:392058

Khan S, Alam F, Azam A, Khan AU (2012) Gold nanoparticles enhance methylene blue-induced photodynamic therapy: a novel therapeutic approach to inhibit *Candida albicans* biofilm. Int J Nanomedicine 7:3245–3257

Kim YH, Lee DK, Cha HG, Kim CW, Kang YC, Kang YS (2006) Preparation and characterization of the antibacterial Cu nanoparticle formed on the surface of SiO_2 nanoparticles. J Phys Chem B 110(49):24923–24928

Kiran GS, Sabarathnam B, Selvin J (2010) Biofilm disruption potential of a glycolipid biosurfactant from marine *Brevibacterium casei*. FEMS Immunol Med Microbiol 59(3):432–438

Kishen A, Shi Z, Shrestha A, Neoh KG (2008) An investigation on the antibacterial and antibiofilm efficacy of cationic nanoparticulates for root canal disinfection. J Endod 34(12):1515–1520

Kneuer C, Sameti M, Bakowsky U, Schiestel T, Schirra H, Schmidt H, Lehr CM (2000) A nonviral DNA delivery system based on surface modified silica-nanoparticles can efficiently transfect cells in vitro. Bioconjug Chem 11(6):926–932

Koper OB, Klabunde JS, Marchin GL, Klabunde KJ, Stoimenov P, Bohra L (2002) Nanoscale powders and formulations with biocidal activity toward spores and vegetative cells of bacillus species, viruses, and toxins. Curr Microbiol 44(1):49–55

Kumar CG, Mamidyala SK (2011) Extracellular synthesis of silver nanoparticles using culture supernatant of *Pseudomonas aeruginosa*. Colloids Surf B Biointerfaces 84(2):462–466

Kuzma L, Wysokinska H, Rozalski M, Budzynska A, Wieckowska-Szakiel M, Sadowska B, Paszkiewicz M, Kisiel W, Rozalska B (2012) Antimicrobial and anti-biofilm properties of new taxodione derivative from hairy roots of *Salvia austriaca*. Phytomedicine 19(14):1285–1287

Lamppa JW, Griswold KE (2013) Alginate lyase exhibits catalysis-independent biofilm dispersion and antibiotic synergy. Antimicrob Agents Chemother 57(1):137–145

Lara HH, Garza-Trevino EN, Ixtepan-Turrent L, Singh DK (2010) Silver nanoparticles are broad-spectrum bactericidal and virucidal compounds. J Nanobiotechnol 9:30

Latimer J, Forbes S, McBain AJ (2012) Attenuated virulence and biofilm formation in *Staphylococcus aureus* following sublethal exposure to triclosan. Antimicrob Agents Chemother 56(6):3092–3100

Leifert A, Pan-Bartnek Y, Simon U, Jahnen-Dechent W (2013) Molecularly stabilised ultrasmall gold nanoparticles: synthesis, characterization and bioactivity. Nanoscale 5(14):6224–6242

Lellouche J, Kahana E, Elias S, Gedanken A, Banin E (2009) Antibiofilm activity of nanosized magnesium fluoride. Biomaterials 30(30):5969–5978

Lellouche J, Friedman A, Gedanken A, Banin E (2012a) Antibacterial and antibiofilm properties of yttrium fluoride nanoparticles. Int J Nanomedicine 7:5611–5624

Lellouche J, Friedman A, Lahmi R, Gedanken A, Banin E (2012b) Antibiofilm surface functionalization of catheters by magnesium fluoride nanoparticles. Int J Nanomedicine 7:1175–1188

Lellouche J, Friedman A, Lellouche JP, Gedanken A, Banin E (2012c) Improved antibacterial and antibiofilm activity of magnesium fluoride nanoparticles obtained by water-based ultrasound chemistry. Nanomedicine 8(5):702–711

Leuba KD, Durmus NG, Taylor EN, Webster TJ (2013) Carboxylate functionalized superparamagnetic iron oxide nanoparticles (SPION) for the reduction of *S. aureus* growth post biofilm formation. Int J Nanomedicine 8:731–736

Lewin M, Carlesso N, Tung CH, Tang XW, Cory D, Scadden DT, Weissleder R (2000) Tat peptide-derivatized magnetic nanoparticles allow *in vivo* tracking and recovery of progenitor cells. Nat Biotechnol 18(4):410–414

Li LL, Wang H (2013) Enzyme-coated mesoporous silica nanoparticles as efficient antibacterial agents in vivo. Adv Healthc Mater 2(10):1351–1360

Li N, Sioutas C, Cho A, Schmitz D, Misra C, Sempf J, Wang M, Oberley T, Froines J, Nel A (2003) Ultrafine particulate pollutants induce oxidative stress and mitochondrial damage. Environ Health Perspect 111(4):455–460

Li Y, Leung P, Yao L, Song QW, Newton E (2006) Antimicrobial effect of surgical masks coated with nanoparticles. J Hosp Infect 62(1):58–63

Li D, Cui F, Zhao Z, Liu D, Xu Y, Li H, Yang X (2013) The impact of titanium dioxide nanoparticles on biological nitrogen removal from wastewater and bacterial community shifts in activated sludge. Biodegradation. doi:10.1007/s10532-013-9648-z

Liu HK, Wang GX, Guo Z, Wang J, Konstantinov K (2006a) Nanomaterials for lithium-ion rechargeable batteries. J Nanosci Nanotechnol 6(1):1–15

Liu J, Wong HL, Moselhy J, Bowen B, Wu XY, Johnston MR (2006b) Targeting colloidal particulates to thoracic lymph nodes. Lung Cancer 51(3):377–386

Liu G, Mao J, Jiang Z, Sun T, Hu Y, Zhang C, Dong J, Huang Q, Lan Q (2013a) Transferrin-modified doxorubicin-loaded biodegradable nanoparticles exhibit enhanced efficacy in treating brain glioma-bearing rats. Cancer Biother Radiopharm 28(9):691–696

Liu Y, Sun Y, Xu Y, Feng H, Fu S, Tang J, Liu W, Sun D, Jiang H, Xu S (2013b) Preparation and evaluation of lysozyme-loaded nanoparticles coated with poly-gamma-glutamic acid and chitosan. Int J Biol Macromol 59:201–207

Lok CN, Ho CM, Chen R, He QY, Yu WY, Sun H, Tam PK, Chiu JF, Che CM (2006) Proteomic analysis of the mode of antibacterial action of silver nanoparticles. J Proteome Res 5(4):916–924

Lucarelli M, Gatti AM, Savarino G, Quattroni P, Martinelli L, Monari E, Boraschi D (2004) Innate defence functions of macrophages can be biased by nano-sized ceramic and metallic particles. Eur Cytokine Netw 15(4):339–346

Mallick S, Sharma S, Banerjee M, Ghosh SS, Chattopadhyay A, Paul A (2012) Iodine-stabilized Cu nanoparticle chitosan composite for antibacterial applications. ACS Appl Mater Interfaces 4(3):1313–1323

Martin CR, Kohli P (2003) The emerging field of nanotube biotechnology. Nat Rev Drug Discov 2(1):29–37

Martinez-Gutierrez F, Boegli L, Agostinho A, Sanchez EM, Bach H, Ruiz F, James G (2013) Antibiofilm activity of silver nanoparticles against different microorganisms. Biofouling 29(6):651–660

Matsusaki M, Larsson K, Akagi T, Lindstedt M, Akashi M, Borrebaeck CA (2005) Nanosphere induced gene expression in human dendritic cells. Nano Lett 5(11):2168–2173

Mehta M, Chen LC, Gordon T, Rom W, Tang MS (2008) Particulate matter inhibits DNA repair and enhances mutagenesis. Mutat Res 657(2):116–121

Mochizuki D, Tamura S, Yasutake H, Kataoka T, Mitsuo K, Wada Y (2013) A photostable bi-luminophore pressure-sensitive paint measurement system developed with mesoporous silica nanoparticles. J Nanosci Nanotechnol 13(4):2777–2781

Mohanty S, Mishra S, Jena P, Jacob B, Sarkar B, Sonawane A (2012) An investigation on the antibacterial, cytotoxic, and antibiofilm efficacy of starch-stabilized silver nanoparticles. Nanomedicine 8(6):916–924

Monteiro DR, Silva S, Negri M, Gorup LF, de Camargo ER, Oliveira R, Barbosa DB, Henriques M (2012) Silver nanoparticles: influence of stabilizing agent and diameter on antifungal activity against *Candida albicans* and *Candida glabrata* biofilms. Lett Appl Microbiol 54(5):383–391

Monteiro DR, Silva S, Negri M, Gorup LF, de Camargo ER, Oliveira R, Barbosa DB, Henriques M (2013) Antifungal activity of silver nanoparticles in combination with nystatin and chlorhexidine digluconate against *Candida albicans* and *Candida glabrata* biofilms. Mycoses 56(6):672–680

Montella M, Franceschi S, Geddes da Filicaia M, De Macro M, Arniani S, Balzi D, Delfino M, Iannuzzo M, Buonanno M, Satriano RA (1997) Classical Kaposi sarcoma and volcanic soil in southern Italy: a case-control study. Epidemiol Prev 21(2):114–117

Mott JA, Meyer P, Mannino D, Redd SC, Smith EM, Gotway-Crawford C, Chase E (2002) Wildland forest fire smoke: health effects and intervention evaluation, Hoopa, California, 1999. West J Med 176(3):157–162

Mukhopadhyay A, Basak S, Das JK, Medda SK, Chattopadhyay K, De G (2010) Ag-TiO$_2$ nanoparticle codoped SiO$_2$ films on ZrO$_2$ barrier-coated glass substrates with antibacterial activity in ambient condition. ACS Appl Mater Interfaces 2(9):2540–2546

Murray CB, Kagan CR, Bawendi MG (2000) Synthesis and characterization of monodisperse nanocrystals and close-packed nanocrystal assemblies. Annu Rev Mater Sci 30(1):545–610

Musthafa KS, Balamurugan K, Pandian SK, Ravi AV (2012a) 2,5-Piperazinedione inhibits quorum sensing-dependent factor production in *Pseudomonas aeruginosa* PAO1. J Basic Microbiol 52(6):679–686

Musthafa KS, Sivamaruthi BS, Pandian SK, Ravi AV (2012b) Quorum sensing inhibition in *Pseudomonas aeruginosa* PAO1 by antagonistic compound phenylacetic acid. Curr Microbiol 65(5):475–480

Naz SS, Islam NU, Shah MR, Alam SS, Iqbal Z, Bertino M, Franzel L, Ahmed A (2013) Enhanced biocidal activity of Au nanoparticles synthesized in one pot using 2, 4-dihydroxybenzene carbodithioic acid as a reducing and stabilizing agent. J Nanobiotechnol 11(1):13

Nemmar A, Hoylaerts MF, Hoet PH, Dinsdale D, Smith T, Xu H, Vermylen J, Nemery B (2002) Ultrafine particles affect experimental thrombosis in an in vivo hamster model. Am J Respir Crit Care Med 166(7):998–1004

Nikula KJ, Snipes MB, Barr EB, Griffith WC, Henderson RF, Mauderly JL (1995) Comparative pulmonary toxicities and carcinogenicities of chronically inhaled diesel exhaust and carbon black in F344 rats. Fundam Appl Toxicol 25(1):80–94

Nithya C, Pandian SK (2010) The *in vitro* antibiofilm activity of selected marine bacterial culture supernatants against *Vibrio* spp. Arch Microbiol 192(10):843–854

Nithya C, Aravindraja C, Pandian SK (2010a) *Bacillus pumilus* of Palk Bay origin inhibits quorum-sensing-mediated virulence factors in Gram-negative bacteria. Res Microbiol 161(4):293–304

Nithya C, Begum MF, Pandian SK (2010b) Marine bacterial isolates inhibit biofilm formation and disrupt mature biofilms of *Pseudomonas aeruginosa* PAO1. Appl Microbiol Biotechnol 88(1):341–358

Nithya C, Devi MG, Pandian SK (2011) A novel compound from the marine bacterium *Bacillus pumilus* S6-15 inhibits biofilm formation in gram-positive and gram-negative species. Biofouling 27(5):519–528

Oberdorster G, Ferin J, Lehnert BE (1994) Correlation between particle size, *in vivo* particle persistence, and lung injury. Environ Health Perspect 102(Suppl 5):173–179

Oberdorster G, Oberdorster E, Oberdorster J (2005) Nanotoxicology: an emerging discipline evolving from studies of ultrafine particles. Environ Health Perspect 113(7):823–839

Packiavathy IASV, Agilandeswari P, Musthafa KS, Pandian SK, Ravi AV (2012) Antibiofilm and quorum sensing inhibitory potential of *Cuminum cyminum* and its secondary metabolite methyl eugenol against Gram negative bacterial pathogens. Food Res Int 45(1):85–92

Packiavathy IA, Sasikumar P, Pandian SK, Ravi A (2013) Prevention of quorum-sensing-mediated biofilm development and virulence factors production in *Vibrio* spp. by curcumin. Appl Microbiol Biotechnol 97(23):10177–10187

Pan B, Huang RZ, Han SQ, Qu D, Zhu ML, Wei P, Ying HJ (2010) Design, synthesis, and antibiofilm activity of 2-arylimino-3-aryl-thiazolidine-4-ones. Bioorg Med Chem Lett 20(8):2461–2464

Pandiyarajan T, Udayabhaskar R, Vignesh S, James RA, Karthikeyan B (2013) Synthesis and concentration dependent antibacterial activities of CuO nanoflakes. Mater Sci Eng C Mater Biol Appl 33(4):2020–2024

Park JH, Lee JH, Cho MH, Herzberg M, Lee J (2012a) Acceleration of protease effect on *Staphylococcus aureus* biofilm dispersal. FEMS Microbiol Lett 335(1):31–38

Park JH, Lee JH, Kim CJ, Lee JC, Cho MH, Lee J (2012b) Extracellular protease in Actinomycetes culture supernatants inhibits and detaches *Staphylococcus aureus* biofilm formation. Biotechnol Lett 34(4):655–661

Patel MB, Harikrishnan U, Valand NN, Modi NR, Menon SK (2013) Novel cationic quinazolin-4 (3H)-one conjugated fullerene nanoparticles as antimycobacterial and antimicrobial agents. Arch Pharm (Weinheim) 346(3):210–220

Pender DS, Vangala LM, Badwaik VD, Willis CB, Aguilar ZP, Sangoi TN, Paripelly R, Dakshinamurthy R (2013) Bactericidal activity of starch-encapsulated gold nanoparticles. Front Biosci 18:993–1002

Peters A, Veronesi B, Calderon-Garciduenas L, Gehr P, Chen LC, Geiser M, Reed W, Rothen-Rutishauser B, Schurch S, Schulz H (2006) Translocation and potential neurological effects of fine and ultrafine particles a critical update. Part Fibre Toxicol 3:13

Pinto RJ, Almeida A, Fernandes SC, Freire CS, Silvestre AJ, Neto CP, Trindade T (2013) Antifungal activity of transparent nanocomposite thin films of pullulan and silver against *Aspergillus niger*. Colloids Surf B Biointerfaces 103:143–148

Pompilio A, Pomponio S, Di Vincenzo V, Crocetta V, Nicoletti M, Piovano M, Garbarino JA, Di Bonaventura G (2013) Antimicrobial and antibiofilm activity of secondary metabolites of lichens against methicillin-resistant *Staphylococcus aureus* strains from cystic fibrosis patients. Future Microbiol 8(2):281–292

Pramanik A, Laha D, Bhattacharya D, Pramanik P, Karmakar P (2012) A novel study of antibacterial activity of copper iodide nanoparticle mediated by DNA and membrane damage. Colloids Surf B Biointerfaces 96:50–55

Raghavendra GM, Jayaramudu T, Varaprasad K, Sadiku R, Ray SS, Mohana Raju K (2013) Cellulose-polymer-Ag nanocomposite fibers for antibacterial fabrics/skin scaffolds. Carbohydr Polym 93(2):553–560

Raimondi MV, Maggio B, Raffa D, Plescia F, Cascioferro S, Cancemi G, Schillaci D, Cusimano MG, Vitale M, Daidone G (2012) Synthesis and anti-staphylococcal activity of new 4-diazopyrazole derivatives. Eur J Med Chem 58:64–71

Ramage G, Saville SP, Wickes BL, Lopez-Ribot JL (2002) Inhibition of *Candida albicans* biofilm formation by farnesol, a quorum-sensing molecule. Appl Environ Microbiol 68(11):5459–5463

Ramamurthy CH, Padma M, Samadanam ID, Mareeswaran R, Suyavaran A, Kumar MS, Premkumar K, Thirunavukkarasu C (2013) The extra cellular synthesis of gold and silver nanoparticles and their free radical scavenging and antibacterial properties. Colloids Surf B Biointerfaces 102:808–815

Rane RA, Sahu NU, Shah CP (2012) Synthesis and antibiofilm activity of marine natural product-based 4-thiazolidinones derivatives. Bioorg Med Chem Lett 22(23):7131–7134

Rane RA, Sahu NU, Shah CP, Shah NK (2013) Design, synthesis and anti-staphylococcal activity of marine pyrrole alkaloid derivatives. J Enzyme Inhib Med Chem. doi:10.3109/14756366.2013.793183

Reymond JL, Bergmann M, Darbre T (2013) Glycopeptide dendrimers as *Pseudomonas aeruginosa* biofilm inhibitors. Chem Soc Rev 42:4814–4822

Risom L, Moller P, Loft S (2005) Oxidative stress-induced DNA damage by particulate air pollution. Mutat Res 592(1–2):119–137

Rodrigues AG, Ping LY, Marcato PD, Alves OL, Silva MC, Ruiz RC, Melo IS, Tasic L, De Souza AO (2013) Biogenic antimicrobial silver nanoparticles produced by fungi. Appl Microbiol Biotechnol 97(2):775–782

Sankar R, Karthik A, Prabu A, Karthik S, Shivashangari KS, Ravikumar V (2013) *Origanum vulgare* mediated biosynthesis of silver nanoparticles for its antibacterial and anticancer activity. Colloids Surf B Biointerfaces 108:80–84

Sanpui P, Murugadoss A, Prasad PV, Ghosh SS, Chattopadhyay A (2008) The antibacterial properties of a novel chitosan-Ag-nanoparticle composite. Int J Food Microbiol 124(2):142–146

Sarabhai S, Sharma P, Capalash N (2013) Ellagic acid derivatives from *Terminalia chebula* Retz. Downregulate the expression of quorum sensing genes to attenuate *Pseudomonas aeruginosa* PAO1 virulence. PLoS ONE 8(1):e53441

Sawant SN, Selvaraj V, Prabhawathi V, Doble M (2013) Antibiofilm properties of silver and gold incorporated PU, PCLm, PC and PMMA nanocomposites under two shear conditions. PLoS ONE 8(5):e63311

Schubert D, Dargusch R, Raitano J, Chan SW (2006) Cerium and yttrium oxide nanoparticles are neuroprotective. Biochem Biophys Res Commun 342(1):86–91

Schulz H, Harder V, Ibald-Mulli A, Khandoga A, Koenig W, Krombach F, Radykewicz R, Stampfl A, Thorand B, Peters A (2005) Cardiovascular effects of fine and ultrafine particles. J Aerosol Med 18(1):1–22

Scott ID, Jung YS, Cavanagh AS, Yan Y, Dillon AC, George SM, Lee SH (2011) Ultrathin coatings on nano-LiCoO2 for Li-ion vehicular applications. Nano Lett 11(2):414–418

Semmler M, Seitz J, Erbe F, Mayer P, Heyder J, Oberdorster G, Kreyling WG (2004) Long-term clearance kinetics of inhaled ultrafine insoluble iridium particles from the rat lung, including transient translocation into secondary organs. Inhal Toxicol 16(6–7):453–459

Shah V, Shah S, Shah H, Rispoli FJ, McDonnell KT, Workeneh S, Karakoti A, Kumar A, Seal S (2012) Antibacterial activity of polymer coated cerium oxide nanoparticles. PLoS ONE 7(10):e47827

Shi P, Aluri S, Lin YA, Shah M, Edman M, Dhandhukia J, Cui H, Mackay JA (2013) Elastin-based protein polymer nanoparticles carrying drug at both corona and core suppress tumor growth in vivo. J Control Release 171(3):330–338

Shimizu N, Otsuka K, Sawada H, Maejima T, Shirotake S (2013) Bacteriolysis by vancomycin-conjugated acryl nanoparticles and morphological component analysis. Drug Dev Ind Pharm. doi:10.3109/03639045.2013.788012

Shivaji S, Madhu S, Singh S (2011) Extracellular synthesis of antibacterial silver nanoparticles using psychrophilic bacteria. Process Biochem 46(9):1800–1807

Shrestha A, Shi Z, Neoh KG, Kishen A (2010) Nanoparticulates for antibiofilm treatment and effect of aging on its antibacterial activity. J Endod 36(6):1030–1035

Singh V, Arora V, Alam MJ, Garey KW (2012) Inhibition of biofilm formation by esomeprazole in *Pseudomonas aeruginosa* and *Staphylococcus aureus*. Antimicrob Agents Chemother 56 (8):4360–4364

Singh S, Ashfaq M, Singh RK, Joshi HC, Srivastava A, Sharma A, Verma N (2013) Preparation of surfactant-mediated silver and copper nanoparticles dispersed in hierarchical carbon micro-nanofibers for antibacterial applications. New Biotechnol 30:656–665

Sintubin L, De Windt W, Dick J, Mast J, van der Ha D, Verstraete W, Boon N (2009) Lactic acid bacteria as reducing and capping agent for the fast and efficient production of silver nanoparticles. Appl Microbiol Biotechnol 84(4):741–749

Sioutas C, Delfino RJ, Singh M (2005) Exposure assessment for atmospheric ultrafine particles (UFPs) and implications in epidemiologic research. Environ Health Perspect 113(8):947–955

Styan K, Abrahamian M, Hume E, Poole-Warren LA (2007) Antibacterial polyurethane organosilicate nanocomposites. Key Eng Mat 342:757–760

Su R, Jin Y, Liu Y, Tong M, Kim H (2013) Bactericidal activity of Ag-doped multi-walled carbon nanotubes and the effects of extracellular polymeric substances and natural organic matter. Colloids Surf B Biointerfaces 104:133–139

Suciu CV, Iwatsubo T, Deki S (2003) Investigation of a colloidal damper. J Colloid Interface Sci 259(1):62–80

Sun LM, Zhang CL, Li P (2012) Characterization, antibiofilm, and mechanism of action of novel PEG-stabilized lipid nanoparticles loaded with terpinen-4-ol. J Agric Food Chem 60 (24):6150–6156

Suresh AK, Pelletier DA, Wang W, Moon JW, Gu B, Mortensen NP, Allison DP, Joy DC, Phelps TJ, Doktycz MJ (2010) Silver nanocrystallites: biofabrication using *Shewanella oneidensis*, and an evaluation of their comparative toxicity on gram-negative and gram-positive bacteria. Environ Sci Technol 44(13):5210–5215

Tavassoli Hojati S, Alaghemand H, Hamze F, Ahmadian Babaki F, Rajab-Nia R, Rezvani MB, Kaviani M, Atai M (2013) Antibacterial, physical and mechanical properties of flowable resin composites containing zinc oxide nanoparticles. Dent Mater 29(5):495–505

Thekkae Padil VV, Cernik M (2013) Green synthesis of copper oxide nanoparticles using gum karaya as a biotemplate and their antibacterial application. Int J Nanomedicine 8:889–898

Thompson S, Parthasarathy S (2006) Moore's law: the future of Si microelectronics. Mater Today 9(6):20–25

Tsuji M, Hashimoto M, Nishizawa Y, Kubokawa M, Tsuji T (2005) Microwave-assisted synthesis of metallic nanostructures in solution. Chem Eur J 11(2):440–452

Vermylen J, Nemmar A, Nemery B, Hoylaerts MF (2005) Ambient air pollution and acute myocardial infarction. J Thromb Haemost 3(9):1955–1961

Vidic J, Stankic S, Haque F, Ciric D, Le Goffic R, Vidy A, Jupille J, Delmas B (2013) Selective antibacterial effects of mixed ZnMgO nanoparticles. J Nanopart Res 15(5):1595

Vigneshwaran N, Kathe AA, Varadarajan PV, Nachane RP, Balasubramanya RH (2006) Biomimetics of silver nanoparticles by white rot fungus, *Phaenerochaete chrysosporium*. Colloids Surf B Biointerfaces 53(1):55–59

Vivek R, Nipun Babu V, Thangam R, Subramanian KS, Kannan S (2013) pH-responsive drug delivery of chitosan nanoparticles as Tamoxifen carriers for effective anti-tumor activity in breast cancer cells. Colloids Surf B Biointerfaces 111C:117–123

Wang X, Lim TT (2013) Highly efficient and stable Ag-AgBr/TiO2 composites for destruction of *Escherichia coli* under visible light irradiation. Water Res 47(12):4148–4158

Wang H, Liu J, Wu X, Tong Z, Deng Z (2013) Tailor-made Au@Ag core-shell nanoparticle 2D arrays on protein-coated graphene oxide with assembly enhanced antibacterial activity. Nanotechnology 24(20):205102

Westendorf AM (2013) Applications of nanoparticles for treating cutaneous infection. J Invest Dermatol 133(5):1133–1135

Wu C, Labrie J, Tremblay YD, Haine D, Mourez M, Jacques M (2013) Zinc as an agent for the prevention of biofilm formation by pathogenic bacteria. J Appl Microbiol 115(1):30–40

Xia T, Kovochich M, Brant J, Hotze M, Sempf J, Oberley T, Sioutas C, Yeh JI, Wiesner MR, Nel AE (2006) Comparison of the abilities of ambient and manufactured nanoparticles to induce cellular toxicity according to an oxidative stress paradigm. Nano Lett 6(8):1794–1807

Xiong R, Lu C, Zhang W, Zhou Z, Zhang X (2013) Facile synthesis of tunable silver nanostructures for antibacterial application using cellulose nanocrystals. Carbohyd Polym 95(1):214–219

Yang X, Konishi H, Xu H, Wu M (2006) Comparative sol–hydro(Solvo)thermal synthesis of TiO_2 nanocrystals. Eur J Inorg Chem 2006(11):2229–2235

Yin H, Zhang H, Liu B (2013) Superior anticancer efficacy of curcumin-loaded nanoparticles against lung cancer. Acta Biochim Biophys Sin (Shanghai) 45(8):634–640

Yu JC, Wang X, Fu X (2004) Pore-wall chemistry and photocatalytic activity of mesoporous titania molecular sieve films. Chem Mater 16(8):1523–1530

Zhang L, Yu JC (2003) A sonochemical approach to hierarchical porous titania spheres with enhanced photocatalytic activity. Chem Commun 9(16):2078–2079

Zhang R, Zhou M, Wang L, McGrath S, Chen T, Chen X, Shaw C (2010) Phylloseptin-1 (PSN-1) from *Phyllomedusa sauvagei* skin secretion: a novel broad-spectrum antimicrobial peptide with antibiofilm activity. Mol Immunol 47(11–12):2030–2037

Zhang H, Wang C, Chen B, Wang X (2012) Daunorubicin-TiO_2 nanocomposites as a "smart" pH-responsive drug delivery system. Int J Nanomedicine 7:235–242

Zhang K, Cheng L, Imazato S, Antonucci JM, Lin NJ, Lin-Gibson S, Bai Y, Xu HH (2013a) Effects of dual antibacterial agents MDPB and nano-silver in primer on microcosm biofilm, cytotoxicity and dentine bond properties. J Dent 41(5):464–474

Zhang K, Li F, Imazato S, Cheng L, Liu H, Arola DD, Bai Y, Xu HH (2013b) Dual antibacterial agents of nano-silver and 12-methacryloyloxydodecylpyridinium bromide in dental adhesive to inhibit caries. J Biomed Mater Res B Appl Biomater 101(6):929–938

Zhang X, Li Z, Yuan X, Cui Z, Bao H, Li X, Liu Y, Yang X (2013c) Cytotoxicity and antibacterial property of titanium alloy coated with silver nanoparticle-containing polyelectrolyte multi-layer. Mater Sci Eng C Mater Biol Appl 33(5):2816–2820

Zhao J, Wang Z, Dai Y, Xing B (2013a) Mitigation of CuO nanoparticle-induced bacterial membrane damage by dissolved organic matter. Water Res 47(12):4169–4178

Zhao L, Zhu B, Jia Y, Hou W, Su C (2013b) Preparation of biocompatible carboxymethyl chitosan nanoparticles for delivery of antibiotic drug. Biomed Res Int 2013:236469

Zheng F, Wang S, Wen S, Shen M, Zhu M, Shi X (2013) Characterization and antibacterial activity of amoxicillin-loaded electrospun nano-hydroxyapatite/poly(lactic-co-glycolic acid) composite nanofibers. Biomaterials 34(4):1402–1412

Drug Delivery Systems That Eradicate and/or Prevent Biofilm Formation

Mohammad Sajid, Mohd Sajjad Ahmad Khan, Swaranjit Singh Cameotra, and Iqbal Ahmad

Abstract The capability to form biofilms contributes significantly to the pathogenesis of microbial infections by various mechanisms including a decrease in susceptibility to antimicrobial agents. Over the past few years therapy against biofilm has undergone a revolutionary shift to effectively kill biofilm-producing microorganisms. With the advancement of biotechnology, emphasis has been made to effectively deliver antimicrobial agents against biofilm. In this regard, particulate materials have attracted enormous attention as drug delivery systems, not only for the controlled release of drugs but also because of the rapid development of synthetic methods for controlling morphology and particle size from the micro to the nanoscale. For targeted drug delivery a number of constituents in the process of biofilm formation have been studied as targets for novel drug delivery technologies. In this chapter, the contribution of various drug delivery systems made up of amphiphilic molecules (liposome, niosomes), polymer (PLGA), chitin (chitosan), and dendrimer, and their potential to deliver antimicrobial agents against biofilm is discussed.

1 Introduction

The age of nanostructural delivery systems began with the development of liposome by Bangham et al. (1965). Since then a large number of nanoparticulate systems have been developed and as of now, the sheer number and types of nanoparticulate structures that have been already developed or are being researched

---

M. Sajid • M.S.A. Khan • S.S. Cameotra
Environmental Biotechnology and Microbial Biochemistry Laboratory, Institute of Microbial Technology, Chandigarh, 160036, India
e-mail: sajid.zilli387@gmail.com; khanmsa@hotmail.com; swaranjitsingh@yahoo.com

I. Ahmad (✉)
Department of Agricultural Microbiology, Aligarh Muslim University, Aligarh, 202002, India
e-mail: ahmadiqbal8@yahoo.co.in

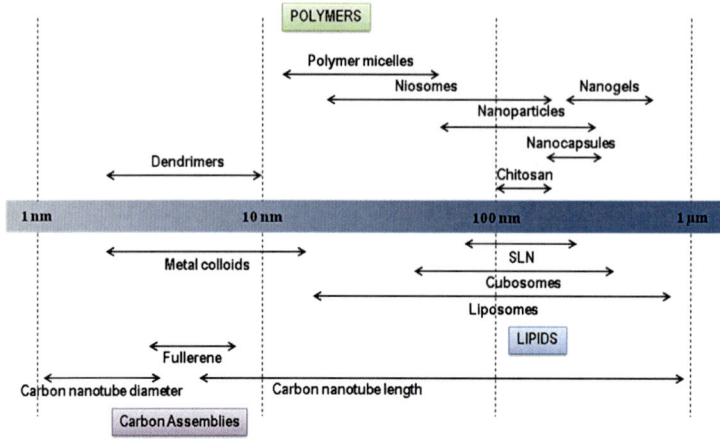

Fig. 1 Size ranges of various drug delivery systems

is tremendous. The concept of the 'magic bullet' proposed a century ago by Nobel laureate Paul Ehrlich came to reality with the recent appearance of several approved forms of drug-targeting systems for the treatment of certain cancers and serious infectious diseases.

The lack of newer antibiotics and the emergence of multiple drug resistant to the conventional antibiotics have shifted research to the optimization of existing drugs. A wide range of materials, such as natural or synthetic polymers, lipids, surfactants, and dendrimers (Fig. 1), have been employed as drug carriers which significantly affect the pharmokinetics and phamacodynamics of the drug (Duncan 2003, 2006; Sampathkumar and Yarema 2005; Torchilin 2008). Therapeutic and preventive strategies, in order to be successful, depend not only on the appropriate choice of the active principle but also to a large extent on the use of an appropriate delivery system. This holds true for difficult to deliver compounds; in particular drugs that have poor solubility (hydrophobic) and poor permeability. In fact, the quest to find the ideal therapeutic strategy will continue until a drug with maximum efficacy and no side effects is found.

Many present day drugs are plagued by narrow therapeutic windows and toxicity. These bottlenecks can be circumvented, and the therapeutic effectiveness of existing drugs can be improved, through the use of an appropriately designed delivery system, which can modify the distribution of the drug in the body by targeting it to desired site and by controlling its release. These ideas are being realized through the use of nanotechnology to develop nanoparticulate delivery systems for drug and antigen delivery. For example, these drug delivery systems could diffuse into the mucus environment surrounding a biofilm where local and controlled release of the antibiofilm ingredient increases its efficacy (Meers et al. 2008; Suk et al. 2009; Tang et al. 2009).

2 Vesicular Systems

Vesicular systems are highly ordered assemblies of one or several concentric bilayers that are formed when certain amphiphilic building blocks are dispersed in water. A wide variety of lipids and surfactants can be used to prepare vesicular carriers (Crommelin and Schreier 1994; Mullertz et al. 2010). The composition of the vesicles influences their physicochemical characteristics such as size, charge, thermodynamic phase, lamellarity, and bilayer elasticity. These physicochemical characteristics have a significant effect on the behavior of the vesicles and hence on their effectiveness as a drug delivery system. Vesicular systems have been able to address the problems of drug insolubility, instability, and rapid degradation. These systems delay elimination of rapidly metabolizable drugs and function as sustained release systems. Encapsulation of a drug in vesicular structures can be predicted to prolong the existence of the drug in systemic circulation, and reduce the toxicity if selective uptake can be achieved (Todd et al. 1982). Vesicular systems can incorporate both hydrophilic and lipophilic drugs. Hydrophilic drugs can be entrapped into the internal aqueous compartment, whereas amphiphilic, lipophilic, and charged hydrophilic drugs can be associated with the vesicle bilayer by hydrophobic and/or electrostatic interactions (Martin and Lloyd 1992).

The applications of vesicles in drug delivery are based on physicochemical and colloidal characterization such as composition, size, and loading efficiency and the stability of the carrier, as well as their biological interactions with the cells. A major interaction is lipid exchange whereby liposomal lipids are exchanged with the lipids of various cell membranes. This depends on the mechanical stability of the bilayer and can be reduced by the addition of cholesterol (which gives rise to greatly improve mechanical properties, such as increased stretching elastic modulus, resulting in stronger membranes and reduced permeability). The second major interaction is adsorption into cells, which occurs when the attractive forces (electrostatic, electrodynamic, van der walls, hydrophobic interaction, hydrogen bonding, etc.) exceed repulsive forces (electrostatic, steric, hydration, undulation, protrusion, etc.) (Lasic 1993; Lipowsky 1995; Israelachvili 1991).

2.1 Liposome-Based Drug Delivery Systems

Liposomes are globular lipid vesicles with a bilayer membrane consisting of amphiphilic lipid molecules (Zhang and Granick 2006). The liposome or lipid vesicles are self-forming, enclosed lipid bilayers upon hydration; liposomal drug delivery systems have played a considerable role in the transport of potent drugs to improve therapeutics. Liposomal drug delivery system can be made of either natural or synthetic lipids. One of the most frequently used lipid moieties in liposome preparations is phosphotidylcholine, which is an electrically neutral phospholipid that contains fatty acyl chains of varying degrees of saturation and

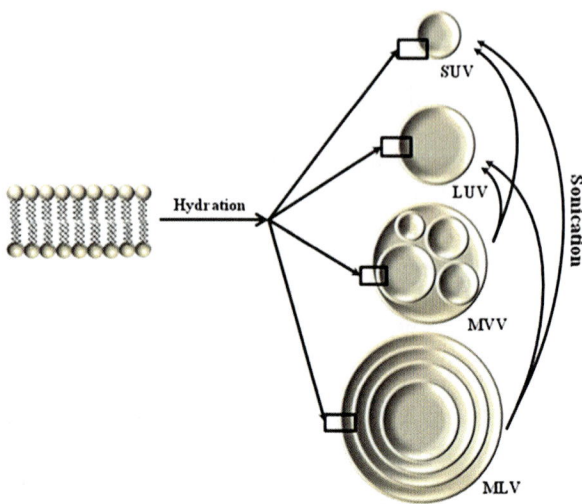

Fig. 2 Relative sizes of liposomal systems

length. Essentially, cholesterol is incorporated into the formulation to regulate membrane rigidity and stability. Structurally, liposomes can be classified into multilamellar vesicles (MLVs), which consist of manifold phospholipid bilayer membranes, and unilamellar vesicles (ULVs), which have a single lipid bilayer (Fig. 2). ULVs can be further classified into small unilamellar vesicles (SUVs) and large unilamellar vesicles (LUVs) according to their size range (Vemuri and Rhodes 1995).

Types and sizes of vesicular systems

Type	Specifications	Diameter (µm)
MLV	Multilamellar large vesicles	>0.5
MVV	Multivesicular vesicles	0.1–20
OLV	Oligolamellar vesicles	0.1–1.0
SUV	Small unilamellar vesicles	20–100
LUV	Large unilamellar vesicles	>100
GUV	Giant unilamellar vesicles	>1

Recently liposomal formulations have been designed to reduce toxicity and increase accumulation at the target site. There are several methods of liposome preparation based on lipid drug interaction and liposome disposition mechanisms including the inhibition of rapid clearance of liposome by controlling particle size, charge, and surface hydration. Most clinical applications of liposomal drug delivery are targeted to tissue with or without expression of target identification molecules on the lipid membrane. Liposomes are characterized with respect to physical, chemical, and biological parameters. This approach of drug delivery provides safer and more efficacious administration of several classes of drugs including antiviral, antifungal, antimicrobials, vaccines, anti-tubercular drugs, and gene therapeutics (Duzgunes et al. 1999; Johnson et al. 1998; Alipour et al. 2008; Watson et al. 2012; Pandey et al. 2004; Nakase et al. 2005), while also inhibiting biofilm formation (Table 2).

One of the characteristic features of liposomes are their lipid bilayer structure, which resembles cell membranes and can readily fuse with microbes. By directly fusing with bacterial membranes, the drug loaded in liposomes can be released in the cell membranes or the interior of the bacteria. The unique structure of liposomes, a lipid membrane surrounding an aqueous cavity, allows them to carry both hydrophobic and hydrophilic compounds without chemical modification. In addition, the liposome surface can be easily functionalized with "stealth" material to increase their in vivo stability or target ligands via preferential delivery by liposomes. For example, polyethylene glycol (PEG) has been frequently conjugated to liposome surfaces to create a stealth layer that prolongs the circulation lifetime of liposomes in the blood stream. On the other hand, by attaching targeting ligands such as antibodies, antibody segments, aptamers, peptides, or small molecule ligands to the surface of the liposomes, they can selectively bind to microorganisms or infected cells and then release the drug payloads to kill or inhibit the growth of the microorganisms (Zhang et al. 2010; Alphandary et al. 2000; Maruyama et al. 1990).

Liposomes are appealing drug carrier systems, especially against colonizing microorganisms, due to several factors such as good biocompatibility; their ability to carry drugs with very different characteristics (hydrophobic to hydrophilic); and encapsulation of the drug protects it from the biological milieu and facilitates drug transport to a specific target site (Vyas et al. 2007; Tamilvanan et al. 2008).

2.1.1 Liposomal Formulation for Delivery of Drugs to Biofilms

Several studies have been performed investigating the interaction between liposomes and bacterial biofilms (Kim and Jones 2004; Jones 2009). Halwani et al. (2008) showed that this new strategy can be used to deliver two agents at the same time in order to prevent *Pseudomonas aeruginosa* biofilm formation and resistance in vitro. DiTizio et al. (1998) developed a liposomal hydrogel to deliver ciprofloxacin that reduced bacterial adhesion to a silicone catheter in a rat model of persistent *P. aeruginosa* peritonitis. This technique opened new avenues for the development of novel antimicrobial peritoneal dialysis catheters as well as other types of catheters (Finelli et al. 2002). Buckler et al. (2008) reported that liposomal antifungal lock therapy could be a possible alternative to catheter removal. In fact, this therapy was previously tested with success in an animal model of *C. albicans* biofilm-associated CVC infection (Schinabeck et al. 2004). A different study used liposomal amphotericin B (LAMB) at the minimal inhibitory concentration in a catheter continuous flow model for *Candida* and showed that hyphae growth diminished 20 % in comparison with the traditional antifungal therapy after 24 h of treatment with LAMB; additionally, the extracellular matrix was undetectable (Seidler et al. 2010). Among various liposomal delivery systems approved for clinical use, four of them target fungi (Table 1).

Table 1 Liposomal formulation of antifungal agents in clinical trial

Product name	Active drug	Route of injection	Approved indication	Trial phase	Reference
Nyotran	Nystatin	Intravenous	Systemic fungal infection	Phase I/II	Immordino et al. (2006)
AmBisome	Amphotericin-B	Intravenous	Fungal infections	Phase II/III	Immordino et al. (2006), Meunier et al. (1991)
Abelcet	Amphotericin-B	*Intravenous*	Sever fungal infections	Phase I/II	Enzon Pharmaceuticals, Wasan and Lopez-Berestein (1996)
Amphotec	Amphotericin-B	*Intravenous*	Sever fungal infections	Phase II/III	Denning et al. (1994), Three Rivers Pharmaceuticals

Between the cationic and anionic liposomes used to deliver the bactericide Triclosan to *Streptococcus salivarius* DBD and *Streptococcus sanguis* C104 biofilms and their mixed biofilms, anionic liposomes were most effective in inhibiting the growth of *S. sanguis* C104 biofilms, whereas growth of *S. salivarius* DBD could not be effectively inhibited by liposomal Triclosan. Growth inhibition of mixed biofilms by liposomal Triclosan reflected the effects found on the single species biofilms, whereas anionic liposomes inhibited the growth of biofilms with a high content of *S. sanguis* C104 (Robinson et al. 2001). Additionally, confocal laser scanning microscopy has been used to visualize the adsorption of fluorescently labeled liposomes on immobilized biofilms of the bacterium *Staphylococcus aureus* (Ahmed et al. 2002).

Liposomes can target matrix or biofilm cells by specific attachment, allowing the drug to be released in the vicinity of the microorganisms; although, in the case of adhesion of yeast cells to human cells, further study is needed on the ability of these systems to prevent adhesion but not affect adhered cells. So, this nanotechnology is indeed a promising research area, but requires more studies to fully understand the mechanism behind the antimicrobial activity.

2.2 Niosome-Based Drug Delivery Systems

Niosomes are nonionic surfactant vesicles constructed by the hydration of synthetic nonionic surfactants, with or without amalgamation of cholesterol or other lipids. They are also called nonionic surfactant vesicles (NISV) or novasomes (Brewer and Alexander 1994; Gupta et al. 1996). They are vesicular systems similar to liposomes that can be used as carriers of amphiphilic and lipophilic molecules but have a better stability and release profile (Mukherjee et al. 2007), lower cost, and less

Fig. 3 Niosomal drug delivery system

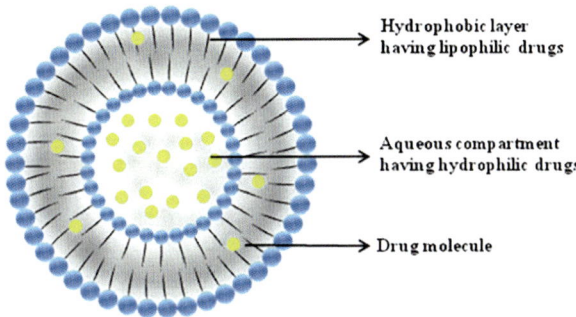

variation in purity in their manufacturing with phospholipids as compared to liposomes (Fang et al. 2001). Niosomes are promising vehicles for drug delivery because they are nonionic, resulting in less toxicity, and they are more specific (Alam et al. 2009). The hydrophilic core of the molecule provides an ideal domain for hydrophilic drugs. In addition, lipophilic compounds can incorporate into their hydrophobic domain (Fig. 3). In addition, modification of niosomes can direct them to specific targets. Niosomal systems can target antibiotics to the surface of bacterial biofilms. Based on the vesicle size, niosomes can be divided into three groups. These are small unilamellar vesicles (SUV, size = 0.025–0.05 μm), multilamellar vesicles (MLV, size= >0.05 μm), and large unilamellar vesicles (LUV, size= >0.10 μm).

Opinions about the usefulness of niosomes in the delivery of drugs vary depending on the use, e.g., encapsulating toxic drugs such as anti-AIDS drugs, anticancer drugs, antiviral drugs, and antimicrobial drugs. However, they do provide a promising carrier system in comparison with ionic drug carriers, which are relatively toxic. However, the technology utilized in niosomes is still in its infancy. Hence, future research is required to develop an appropriate technology for large production (Kazi et al. 2010).

3 Polymer-Based Drug Delivery Systems

Polymer-based drug delivery systems consist of hydrophilic and hydrophobic regions. The hydrophobic part forms a polymeric core containing the drugs, while the hydrophilic segment protects the core from degradation. A variety of biodegradable polymers have been used to form the hydrophobic polymeric core, including poly(lactic acid) (PLA), poly(glycolic acid) (PGA), poly(lactide-co-glycolide) (PLGA), poly(carprolactone) (PCL), and poly(cyanoacrylate) (PCA), whereas polyethylene glycol (PEG) has been commonly used for the hydrophilic segment. Polymer-based drug delivery systems possess several unique characteristics for antimicrobial drug delivery. They are structurally stable and can be synthesized

Table 2 Liposomal formulations for antibiofilm drug delivery

Formulation	Encapsulated drug	Biofilm-producing target organism	Reference
Liposome	Penicillin G	*Staphylococcus aureus*	Kim and Jones (2004)
Liposome	Gentamicin	*Pseudomonas aeruginosa*	Halwani et al. (2008)
Solid supported liposome	Triclosan	*Streptococcus oralis*	Catuogno and Jones (2003)
Liposome	Vancomycin	*Staphylococcus aureus*	Kim et al. (1999)
Liposome	Amikacin	*Pseudomonas aeruginosa*	Meers et al. (2008)
Cationic nanoliposomes	Rifampin	*Staphylococcus epidermidis*	Sharif et al. (2012a, b)
Liposome	Amphotericin-B	*Candida albicans*	Ramage et al. (2013)
Liposome	Bismuth–thiol and tobramycin	*Pseudomonas aeruginosa*	Halwani et al. (2009)
Liposome	Clarithromycin	*Pseudomonas aeruginosa*	Alhajlan et al. (2013)
Liposomes	Bismuth–ethanedithiol with tobramycin	*Pseudomonas aeruginosa*	Alipour et al. (2010)
Liposome combined beta-TCP scaffold	Gentamicin	*Staphylococcus aureus*	Zhu et al. (2010)
Mannosylated liposome	Metronidazole	*Staphylococcus aureus*	Vyas et al. (2007)

with a controlled size distribution. Particle properties such as size, zeta potentials, and drug release profiles can be precisely tuned by selecting different polymer lengths, surfactants, and organic solvents during the synthesis. Functional groups present on the surface of polymeric nanoparticles can be chemically modified with either drug moieties or targeting ligands (Tanihara et al. 1999).

In the past few years, the use of biocompatible or biodegradable polymers has gained importance in the medicinal field as antimicrobial carriers against biofilms (Table 2). Although molecules that undergo bio-erosion are also used interchangeably with biodegradation, the two differ; erosion occurs by the dissolution of chain fragments in non-cross-linked systems without chemical alterations to the molecular structure, while biodegradation occurs through the covalent bond cleavage by a chemical reaction. Both biodegradation and erosion can occur as a surface or bulk process. In surface degradation the polymeric matrix is progressively removed from the surface, but the polymer volume fraction remains almost unchanged. In contrast, during bulk degradation no significant changes occur in the physical size of the polymer until it is nearly completely degraded or eroded. Among these products, polymeric microspheres, polymer micelles, and hydrogel-type materials have been shown to be effective nanocarriers in enhancing drug targeting specificity, lowering systemic drug toxicity, improving treatment absorption rates, and providing protection for pharmaceuticals against biochemical degradation (Sanli et al. 2008).

3.1 Poly(DL-lactide-co-glycolide) Nanocapsule-Based Drug Delivery Systems

PLGA is a clinically approved (FDA), biodegradable, and biocompatible polymer considered safe for controlled release formulations (Lu et al. 2009). PLGA is polyester consisting of one or more different hydroxy acid monomers, D-lactic, L-lactic, and/or glycolic acids. In general, the polymer, can be made to be highly crystalline [e.g., poly(L-lactic acid)], or completely amorphous [e.g., poly(DL-lactic-co-glycolic acid)], can be processed into most any shape and size (down to $\leq$200 nm), and can encapsulate molecules of virtually any size. PLGA is one of the most effectively used polymers for the development of a drug delivery system because it undergoes hydrolysis in the body to produce the biodegradable metabolite monomers, lactic acid and glycolic acid, which are assimilated by the body, resulting in minimal systemic toxicity (Wu 1995). Many approaches have been proposed for the preparation of PLGA particles. The emulsification–evaporation method (Sahoo et al. 2004), the spontaneous emulsification–solvent diffusion method (SESD) (Bilati et al. 2005), the nanoprecipitation method (Govender et al. 1999; Rivera et al. 2004), and the spray-drying method (Takashima et al. 2007; Cheng et al. 2008) are all widely used in preparing PLGA nano/microparticles of various size.

PLGA polymers have been used to trap several antibiotics in nanoparticle formulations, demonstrating improved delivery and antibiotic efficacy (Pillai et al. 2008; Mohammadi et al. 2010; Kashi et al. 2012). Notably, Cheow et al. (2010) reported the preparation of levofloxacin loaded poly(DL-lactide-co-glycolide) (PLGA) and poly(caprolactone) (PLC) nanoparticles and showed that, to be effective against *E. coli* biofilm cells, the formulation required a biphasic extended release profile. This release profile consisted of an initial fast release, providing high antibiotic concentrations for biofilm eradication, followed by slower extended release that minimized biofilm growth and infection exacerbation. Katherine et al. (2012) verified that cinnamaldehyde (CA) and carvacrol (CARV) in solution significantly impaired bacterial growth, and therefore biofilm formation, by *E. coli, P. aeruginosa*, and *S. aureus* at low concentrations. A PLGA formulation of gentamicin also demonstrated improved antimicrobial effects on peritoneal infection caused by *P. aeruginosa* biofilms in vivo (Sharif et al. 2012a, b).

3.2 Others Polymer-Based Drug Delivery Systems

The University of Washington Engineered Biomaterials group has developed a novel drug delivery polymer matrix consisting of a poly(2-hydroxyethyl methacrylate) hydrogel coated with ordered methylene chains that form an ultrasound-responsive coating. This system was able to retain the drug ciprofloxacin inside the polymer in the absence of ultrasound, but showed significant drug release when low-intensity ultrasound was applied (Norris et al. 2005).

4 Dendrimer-Based Drug Delivery Systems

Dendrimers are highly ordered and repeatedly branched globular macromolecules produced by stepwise iterative approaches first described by Buhleier et al. (1978). The structure of dendrimers consists of three distinctive architectural regions: a core, layers of branched repeat units emerging from the core, and functional end groups on the outer layer of repeat units (Grayson and Frechet 2001). Tomalia et al. (1984) reported the synthesis and characterization of the first family of polyamidoamine (PAMAM) dendrimers, which has developed into one of the most accepted dendrimers since. Two synthetic approaches, divergent and convergent, have been developed to synthesize dendrimer systems for delivering various types of drugs. In the divergent approach, synthesis initiates from a core and emanates outward through a repetition of coupling and activation steps. During the first coupling reaction, the peripheral functional groups of the core react with the complementary reactive groups to form new latent branch points at the coupling sites and increase the number of peripheral functional groups. These latent functional groups are then activated to couple with additional monomers. The activation of the latent functional groups can be achieved by removal of protecting groups, coupling with secondary molecules, or reactive functionalization (Fig. 4). Large excess of reagents is required to drive the activation step to completion. The resulting dendrimer products can then be separated from the excess reagents by distillation, precipitation, or ultrafiltration.

In contrast, the convergent approach initiates the synthesis from the periphery and progresses inward (Fig. 5). This approach starts with coupling end groups to each branch of the monomer, followed by the activation of a single functional group located at the focal point of the first wedge-shape dendritic fragment or dendron. Higher generation dendron is synthesized by the coupling of the activated dendron to an additional monomer. After repetition of the coupling and activation steps, a globular dendrimer is formed by attaching a number of dendrons to a polyfunctional core. Dendrimers thus synthesized can be effectively purified. However, synthesis of large dendrimers above the sixth generation is difficult (Grayson and Frechet 2001).

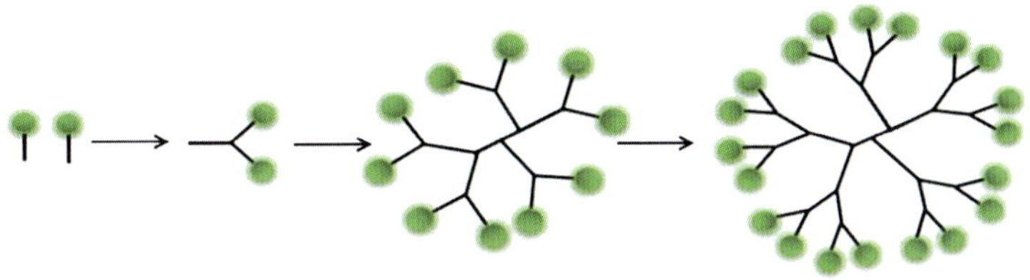

Fig. 4 Divergent synthesis of dendrimer: from core to surface

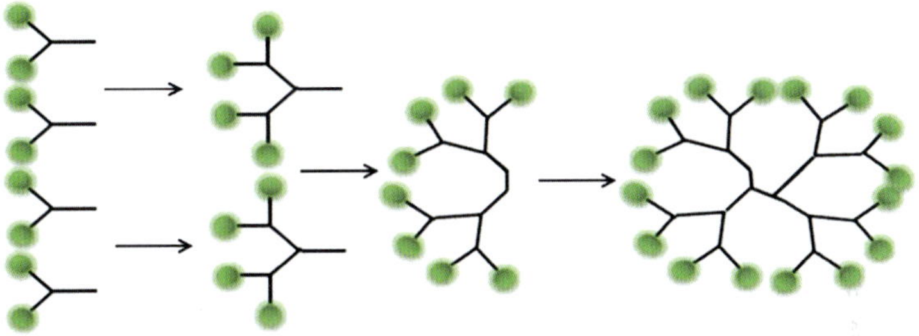

Fig. 5 Convergent synthesis of dendrimer: fragment condensation

Dendrimers acquire several unique properties that make them an excellent platform for antimicrobial drug delivery. The highly branched nature of dendrimers provides enormous surface area to size ratio and allows great reactivity with microorganisms in vivo. In addition, both hydrophobic and hydrophilic agents can be loaded into dendrimers. Hydrophobic drugs can be trapped inside the cavity in the hydrophobic core and hydrophilic drugs can be linked to the multivalent surfaces of dendrimers through covalent conjugation or electrostatic interaction (Florence 2005; Gillies and Frechet 2005). The polycationic nature of dendrimer biocides facilitates the initial electrostatic adsorption to negatively charged bacteria. Glycopeptide dendrimers (branched oligopeptides) have great potential to block *P. aeruginosa* biofilm formation and induce biofilm dispersal in vitro (Emma et al. 2008) due to their high affinity toward the *P. aeruginosa* lectins LecB and LecA, which are responsible for formation of antibiotic-resistant biofilms. The dendrimeric peptide (RW) 4D has been shown to be active against multidrug-resistant (MDR) *S. aureus* and *E. coli* (D31) as well as *Acinetobacter baumannii* and *E. coli RP437* (Hou et al. 2009).

5 Chitosan-Based Drug Delivery Systems

Chitosan is a natural cationic polymer that consists of D-glucosamine and N-actyl-glucosamine linked via β-(1,4)-glycosidic bonds and is obtained by the alkaline deacetylation of chitin. It is nontoxic, biocompatible, biodegradable, and has

Table 3 Polymer-based drug delivery systems with antibiofilm activity

Delivery system	Encapsulated drug	Biofilm-producing target organism	Reference
Poly(DL-lactic-co-glycolic acid) microparticles	Rifampin, Clindamycin hydrochloride	*Staphylococcus aureus*	Daniel et al. (2012)
Poly(DL-lactide-co-glycolide (PLGA) nanocapsules	Carvacrol	*Staphylococcus epidermidis*	Iannitelli et al. (2011)
Polymer nano composite	Silver and gold nanoparticles	*Escherichia coli*	Sawant et al. (2013)
Poly(DL-lactic acid)	Vancomycin	*Staphylococcus aureus* (MRSA)	Ravelingien et al. (2010)
Poly(L-lactic acid)	Ciprofloxacin	*Pseudomonas aeruginosa*	Owusu-Ababio et al. (1995)

Table 4 Chitosan-based formulations for antibiofilm drug delivery

Delivery system	Encapsulated drug	Biofilm-producing target organism	Reference
n-HA/CS/KGM	Vancomycin	*Staphylococcus aureus*	Ma et al. (2011)
Chitosan	Rifampin	*Staphylococcus epidermidis, Staphylococcus aureus*	Zhengbing and Sun (2009)
Chitosan microspheres	Tetracyclin	*Pseudomonas aeruginosa*	Nahla et al. (2012)
Chitosan	Gentamicin	*Staphylococcus* spp. *clinical isolates*	Tunney et al. (2008)

mucusoadhesive properties due to its electrostatic interactions with the sialic groups of mucin (Huang and Khort 2004). The chemical modification of chitosan imparts amphiphilicity, which is an important characteristic for the formation of self-assembled nanoparticles and biofilm delivery (Tables 3 and 4). The hydrophobic cores of the nanoparticles can act as reservoirs or microcontainers for various bioactive substances such as antimicrobial or anticancer agents, genes, and vaccines. Due to its biocompatibility, biodegradability, lack of toxicity, and adsorption properties, chitosan is also used as a stabilizing agent to prepare silver, gold, and platinum nanoparticles (Namasivayam and Roy 2013). Due to the cationic nature of chitosan, it strongly binds to anionic antibiotics through the formation of ionic complexes (Zhengbing and Sun 2009).

6 Conclusions

The biofilm matrix helps microbes to aggregate and adhere to various biological and nonbiological surfaces. Drugs against biofilm components and the machinery used by bacteria to establish biofilms (e.g., quorum sensing) will undoubtedly help to combat biofilm infections. While natural compounds and synthetic analogues have been used effectively to inhibit biofilm formation by quorum quenching, most are still in the preclinical phase, often due to cytotoxicity, low solubility, rapid degradation, and/or clearance in the blood stream. Drug delivery systems such as liposomes, niosomes, polymer-based particles, and dendrimers should be able to overcome these issues and facilitate effective delivery of antimicrobial agents to microbial niches including the biofilm. Liposomal and niosomal materials can be easily manipulated for the design of a new ideal temporary store to enhance therapeutic efficacy or improve the drug release profile is of a great interest. In addition to this, multifunctional liposome and niosome formulations with targeted moieties such as antibodies, peptides, glycoproteins, polysaccharides, and receptors may increase liposomal drug accumulation in the biofilm. Drug delivery using PLGA or PLGA-based polymers is an attractive research area with the primary goal of increasing the therapeutic effects of drugs while minimizing side effects and should provide limitless opportunities. The therapeutic advantages of PLGA as a drug carrier are becoming apparent and will soon be associated with every route of drug administration, making them feasible candidates for drug delivery systems. The medical management of malignancies has been impacted by PLGA-based drug delivery systems, but soon, other medical specialties will also use these novel forms of drug delivery to achieve optimal treatment success. As more clinical data become available, our knowledge of their pharmacology and long-term health effects will expand, leading to the development of more rationally designed and optimized drug-loaded PLGA-based delivery systems. Ideally, these systems will have improved selectivity, efficacy, and safety and will be fully accepted by the market. Dendrimer and chitosan have also proved to be suitable carriers for delivery of a variety of drugs. Surface engineering will further improve the effectiveness of dendrimer and chitosan-based drug delivery systems against biofilm eradication. In summary, the majority of antimicrobial drug delivery systems are presently in preclinical development, but several have been approved for clinical use. With the ongoing efforts in this field, there is no doubt that these drug delivery systems will continue to improve management of bacterial infections, especially in life-threatening diseases.

References

ABELCET-(Amphotericin-b, Dimyristoylphosphatidylcholine, dl- and Dimyristoylphosphatidyl-glycerol, dl- injection). Enzon Pharmaceuticals, Bridgewater, NJ

Ahmed K, Gribbon PN, Jones MN (2002) The application of confocal microscopy to the study of liposome adsorption onto bacterial biofilms. J Liposome Res 12(4):285–300

Alam M, Dwivedi V, Khan AA, Mohammad O (2009) Efficacy of niosomal formulation of diallyl sulfide against experimental candidiasis in Swiss albino mice. Nanomedicine 4(7):713–724

Alhajlan M, Alhariri M, Omri A (2013) Efficacy and safety of liposomal clarithromycin and its effect on *Pseudomonas aeruginosa* virulence factors. Antimicrob Agents Chemother 57(6): 2694–2704

Alipour M, Halwani M, Omri A, Suntres ZE (2008) Antimicrobial effectiveness of liposomal polymyxin B against resistant Gram-negative bacterial strains. Int J Pharm 355(1–2):293–298

Alipour M, Suntres ZE, Lafrenie RM, Omri A (2010) Attenuation of *Pseudomonas aeruginosa* virulence factors and biofilms by co-encapsulation of bismuth–ethanedithiol with tobramycin in liposomes. J Antimicrob Chemother 65:684–693

Alphandary HP, Andremont A, Couvreur P (2000) Targeted delivery of antibiotics using liposomes and nanoparticles: research and applications. Int J Antimicrob Agents 13:155–168

AMPHOTEC-Distributed by Three Rivers Pharmaceuticals, LLC, Warrendale, PA. (US Patent Numbers 4,822,777; 5,032,582; 5,194,266; 5,077,057)

Bangham AD, Standish MM, Watkins JC (1965) Diffusion of univalent ions across the lamellae of swollen phospholipids. J Mol Biol 13:238–252

Bilati U, Allemann E, Doelker E (2005) Development of a nanoprecipitation method intended for the entrapment of hydrophilic drugs into nanoparticles. Eur J Pharm Sci 24:67–75

Brewer JM, Alexander J (1994) Studies on the adjuvant activity of non-ionic surfactant vesicles: adjuvant-driven IgG2a production independent of MHC control. Vaccine 12(7):613–619

Buckler BS, Sams RN, Goei VL, Krishnan KR, Bemis MJ, Parker DP, Murray DL (2008) Treatment of central venous catheter fungal infection using liposomal amphotericin-B lock therapy. Pediatr Infect Dis J 27:762–764

Buhleier E, Wehner W, Vogtle F (1978) Cascade and nonskid-chainlike synthesis of molecular cavity topologies. Synthesis 1978:155–158

Catuogno C, Jones MN (2003) The antibacterial properties of solid supported liposomes on *Streptococcus oralis* biofilms. Int J Pharm 257:125–140

Cheng FY, Wang SP, Su CH, Tsai TL, Wu PC, Shieh DB, Chen JH, Hsieh PC, Yeh CS (2008) Stabilizer-free poly (lactide-co-glycolide) nanoparticles for multimodal biomedical probes. Biomaterials 29:2104–2112

Cheow WS, Chang MW, Hadinoto K (2010) Antibacterial efficacy of inhalable levofloxacin-loaded polymeric nanoparticles against *E. coli* biofilm cells: the effect of antibiotic release profile. Pharm Res 27:1597–1609

Crommelin DJA, Schreier H (1994) Liposomes. In: Kreuter J (ed) Colloidal drug delivery systems. Dekker, New York, pp 73–190

Daniel M, Chessman R, Al-Zahid S, Richards B, Rahman C, Ashraf W, McLaren J, Cox H, Qutachi O, Fortnum H, Fergie N, Shakesheff K, Birchall JP, Bayston RR (2012) Biofilm eradication with biodegradable modified-release antibiotic pellets: a potential treatment for glue ear. Arch Otolaryngol Head Neck Surg 138(10):942–949

Denning DW, Lee JY, Hostetler JS, Pappas P, Kauffman CA et al (1994) NIAID Mycoses Study Group multicenter trial of oral itraconazole therapy for invasive aspergillosis. Am J Med 97:135–144

DiTizio V, Ferguson GW, Mittelman MW, Khoury AE, Bruce AW, DiCosmo FA (1998) Liposomal hydrogel for the prevention of bacterial adhesion to catheters. Biomaterials 19:1877–1884

Duncan R (2003) The dawning era of polymer therapeutics. Nat Rev Drug Discov 2:347–360

Duncan R (2006) Polymer conjugates for drug targeting: from inspired to inspiration. J Drug Target 14:333–335

Duzgunes N, Pretzer E, Simoes S, Slepushkin V, Konopka K, Flasher D, de Lima MC (1999) Liposome-mediated delivery of antiviral agents to human immunodeficiency virus-infected cells. Mol Membr Biol 16(1):111–118

Emma MVJ, Shanika AC, Elena K, Lieven B, Rameshwar UK, Martina C, Kai-Malte B, Stephen PD, Miguel C, Paul W, Remy L, Cristina N, Frank R, Karl-Erich J, Tamis D, Jean-Louis R (2008) Inhibition and dispersion of *Pseudomonas aeruginosa* biofilms by glycopeptide dendrimers targeting the fucose-specific lectin LecB. Chem Biol 15:1249–1257

Fang JY, Hong CT, Chiu WT (2001) Effect of liposomes and niosomes on skin permeation of enoxacin. Int J Pharm 219:61–72

Finelli A, Burrows LL, DiCosmo FA, DiTizio V, Sinnadurai S, Oreopoulos DG, Khoury AE (2002) Colonization-resistant antimicrobial-coated peritoneal dialysis catheters: evaluation in a newly developed rat model of persistent *Pseudomonas aeruginosa* peritonitis. Perit Dial Int 22:27–31

Florence AT (2005) Dendrimers: a versatile targeting platform. Adv Drug Deliv 57:2104–2105

Gillies ER, Frechet JM (2005) Dendrimers and dendritic polymers in drug delivery. Drug Discov Today 10:35–43

Govender T, Stolnik S, Garnett MC, Illum L, Davis SS (1999) PLGA nanoparticles prepared by nanoprecipitation: drug loading and release studies of a water soluble drug. J Control Release 52:171–185

Grayson SM, Frechet JM (2001) Convergent dendrons and dendrimers: from synthesis to applications. Chem Rev 101:3819–3868

Gupta RK, Varanelli CL, Griffin P, Wallach DFH, Siber GR (1996) Adjuvant properties of non-phospholipid liposomes (Novasomes®) in experimental animals for human vaccine antigens. Vaccine 14(3):219–225

Halwani M, Yebio B, Suntres ZE, Alipour M, Azghani AO, Omri A (2008) Co-encapsulation of gallium with gentamicin in liposomes enhances antimicrobial activity of gentamicin against *Pseudomonas aeruginosa*. J Antimicrob Chemother 62:1291–1297

Halwani M, Hebert S, Suntres ZE, Lafrenie RM, Azghani AO, Omri A (2009) Bismuth–thiol incorporation enhances biological activities of liposomal tobramycin against bacterial biofilm and quorum sensing molecules production by *Pseudomonas aeruginosa*. Int J Pharm 373(1–2):141–146

Hou S, Zhou C, Liu Z, Young AW, Shi Z, Ren D, Kallenbach NR (2009) Antimicrobial dendrimer active against *Escherichia coli* biofilms. Bioorg Med Chem Lett 19:5478–5481

Huang M, Khort LLY (2004) Uptake and cytotoxicity of chitosan molecule and nanoparticles: effect of molecular weight and degree of deacetylation. Pharm Res 21:344–353

Iannitelli A, Grande R, Di Stefano A, Di Giulio M, Sozio P, Bessa LJ, Laserra S, Paolini C, Protasi F, Cellini L (2011) Potential antibacterial activity of carvacrol-loaded poly(DL-lactide-co-glycolide) (PLGA) nanoparticles against microbial biofilm. Int J Mol Sci 12(8):5039–5051

Immordino ML, Dosio F, Cattel L (2006) Stealth liposomes: review of the basic science, rationale, and clinical applications, existing and potential. Int J Nanomedicine 1:297–315

Israelachvili JN (1991) Intermolecular and surface forces. Academic, London

Johnson EM, Ojwang JO, Szekely A, Wallace TL, Warnock DW (1998) Comparison of in vitro antifungal activities of free and liposome-encapsulated nystatin with those of four amphotericin B formulations. Antimicrob Agents Chemother 42(6):1412–1416

Jones MN (2009) Use of liposomes to deliver bactericides to bacterial biofilms. Methods Enzymol 391:211–228

Kashi TS, Eskandarion S, Esfandyari-Manesh M (2012) Improved drug loading and antibacterial activity of minocycline-loaded PLGA nanoparticles prepared by solid/oil/water ion pairing method. Int J Nanomedicine 7:221–234

Katherine RZ, Jessica DS, Menachem E (2012) Biodegradable polymer (PLGA) coatings featuring cinnamaldehyde and carvacrol mitigate biofilm formation. Langmuir 28:13993–13999

Kazi KM, Mandal AS, Biswas N, Guha A, Chatterjee S, Behera M, Kuotsu K (2010) Niosome: a future of targeted drug delivery systems. J Adv Pharm Technol Res 1(4):374–380

Kim HJ, Jones MN (2004) The delivery of benzyl penicillin to *Staphylococcus aureus* biofilms by use of liposomes. J Liposome Res 14:123–139

Kim HJ, Michael Gias EL, Jones MN (1999) The adsorption of cationic liposomes to *Staphylococcus aureus* biofilms. Colloids Surf A Physicochem Eng Aspects 149:561–570

Lasic DD (1993) Liposomes, from physics to applications. Elsevier, Amsterdam

Lipowsky R (1995) Generic interactions of flexible membranes. Elsevier, Amsterdam

Lu JM, Wang X, Marin-Muller C (2009) Current advances in research and clinical applications of PLGA-based nanotechnology. Expert Rev Mol Diagn 9:325–341

Ma T, Shang BC, Tang H, Zhou TH, Xu GL, Li HL, Chen QH, Xu YQ (2011) Nano-hydroxyapatite/chitosan/konjac glucomannan scaffolds loaded with cationic liposomal vancomycin: preparation, in vitro release and activity against *Staphylococcus aureus* biofilms. J Biomater Sci Polym Ed 22(12):1669–1681

Martin GP, Lloyd AW (1992) Basic principles of liposomes for drug use. In: Braun-Falco O et al (eds) Liposome dermatics, Griesbach conference. Springer, Berlin, pp 20–26

Maruyama K, Kennel SJ, Huang L (1990) Lipid composition is important for highly efficient target binding and retention of immunoliposomes. Proc Natl Acad Sci USA 87:5744–5748

Meers P, Neville M, Malinin V, Scotto AW, Sardaryan G, Kurumunda R, Mackinson GJ, Fisher S, Perkins WR (2008) Biofilm penetration, triggered release and in vivo activity of inhaled liposomal amikacin in chronic *Pseudomonas aeruginosa* lung infections. J Antimicrob Chemother 61:859–868

Meunier F, Prentice HG, Ringdén O (1991) Liposomal amphotericin B (AmBisome): safety data from a phase II/III clinical trial. J Antimicrob Chemother 28:B83–B91

Mohammadi G, Valizadeh H, Barzegar-Jalali M (2010) Development of azithromycin-PLGA nanoparticles: physicochemical characterization and antibacterial effect against *Salmonella typhi*. Colloids Surf B: Biointerfaces 80:34–39

Mukherjee B, Patra B, Layek B, Mukherjee A (2007) Sustained release of acyclovir from nano-liposomes and nano-niosomes: an in vitro study. Int J Nanomedicine 2(2):213–225

Mullertz A, Ogbonna A, Ren S, Rades T (2010) New perspectives on lipid and surfactant based drug delivery systems for oral delivery of poorly soluble drugs. J Pharm Pharmacol 62(11): 1622–1636

Nahla AM, Hanaa AM, Mona TA (2012) Bactericidal activity of various antibiotics versus tetracycline-loaded chitosan microspheres against *Pseudomonas aeruginosa* biofilms. Afr J Microbiol Res 6(25):5387–5398

Nakase M, Inui M, Okumura K, Kamei T, Nakamura S, Tagawa T (2005) p53 gene therapy of human osteosarcoma using a transferrin-modified cationic liposome. Mol Cancer Ther 4(4): 625–631

Namasivayam SKR, Roy EA (2013) Enhanced antibiofilm activity of chitosan stabilized chemogenic silver nanoparticles against *Escherichia coli*. Int J Sci Res Publ 3:1–9

Norris P, Noble M, Francolini I, Vinogradov AM, Stewart PS, Ratner BD, Costerton JW, Stoodley P (2005) Ultrasonically controlled release of ciprofloxacin from self-assembled coatings on poly (2-hydroxyethyl methacrylate) hydrogels for *Pseudomonas aeruginosa* biofilm prevention. Antimicrob Agents Chemother 49(10):4272–4279

Owusu-Ababio G, Rogers JA, Morck DW, Olson ME (1995) Efficacy of sustained release ciprofloxacin microspheres against device-associated *Pseudomonas aeruginosa* biofilm infection in a rabbit peritoneal model. J Med Microbiol 43(5):368–376

Pandey R, Sharma S, Khuller GK (2004) Liposome-based anti-tubercular drug therapy in a guinea pig model of tuberculosis. Int J Antimicrob Agents 23(4):414–415

Pillai RR, Somayaji SN, Rabinovich M, Hudson MC, Gonsalves KE (2008) Nafcillin loaded PLGA nanoparticles for treatment of osteomyelitis. Biomed Mater 3:034114

Ramage G, Jose A, Sherry L, Lappin DF, Jones B, Williams C (2013) Liposomal amphotericin B displays rapid dose dependant activity against Candida albicans biofilms. Antimicrob Agents Chemother 57(5):2369–2371

Ravelingien M, Mullens S, Luyten J, D'Hondt M, Boonen J, Spiegeleer BD, Coenye T, Vervaet C, Remon JP (2010) Vancomycin release from poly(DL-lactic acid) spray-coated hydroxyapatite fibers. Eur J Pharm Biopharm 76(3):366–370

Rivera PA, Martinez-Oharriz MC, Rubio M, Irache JM, Espuelas S (2004) Fluconazole encapsulation in PLGA microspheres by spray-drying. J Microencapsul 21:203–211

Robinson AM, Bannister M, Creeth JE, Jones MN (2001) The interaction of phospholipid liposomes with mixed bacterial biofilms and their use in the delivery of bactericide. Colloids Surf A Physicochem Eng Aspects 186:43–53

Sahoo SK, Ma W, Labhasetwar V (2004) Efficacy of transferrin-conjugated paclitaxelloaded nanoparticles in a murine model of prostate cancer. Int J Cancer 112:335–340

Sampathkumar SG, Yarema KJ (2005) Targeting cancer cells with dendrimers. Chem Biol 12:5–6

Sanli O, Biçer E, Isiklan N (2008) In vitro release study of diltiazem hydrochloride from poly (vinyl pyrrolidone)/sodium alginate blend microspheres. JAPS 107:1973–1980

Sawant SN, Selvaraj V, Prabhawathi V, Doble M (2013) Antibiofilm properties of silver and gold incorporated PU, PCLm, PC and PMMA nanocomposites under two shear conditions. PLoS ONE 8(5):e63311

Schinabeck MK, Long LA, Hossain MA, Chandra J, Mukherjee PK, Mohamed S, Ghannoum MA (2004) Rabbit model of *Candida albicans* biofilm infection: liposomal amphotericin B antifungal lock therapy. Antimicrob Agents Chemother 48:1727–1732

Seidler M, Salvenmoser S, Muller FM (2010) Liposomal amphotericin B eradicates *Candida albicans* biofilm in a continuous catheter flow model. FEMS Yeast Res 10:492–495

Sharif MA, Derek JQ, Rebecca JI, Brendan FG, Ryan FD, Clifford CT, Christopher JS (2012a) Gentamicin-loaded nanoparticles show improved antimicrobial effects towards *Pseudomonas aeruginosa* infection. Int J Nanomedicine 7:4053–4063

Sharif NM, Bazzaz BF, Malaekeh-Nikouei B (2012b) The effect of cationic nanoliposomes containing rifampin on eradication of *Staphylococcus epidermidis* biofilm. Res Pharm Sci 7 (5):S275

Suk JS, Lai SK, Wang YY, Ensign LM, Zeitlin PL, Boyle MP, Hanes J (2009) The penetration of fresh undiluted sputum expectorated by cystic fibrosis patients by non-adhesive polymer nanoparticles. Biomaterials 30:2591–2597

Takashima Y, Saito R, Nakajima A, Oda M, Kimura A, Kanazawa T, Okada H (2007) Spray-drying preparation of microparticles containing cationic PLGA nanospheres as gene carriers for avoiding aggregation of nanospheres. Int J Pharm 343:262–269

Tamilvanan S, Venkateshan N, Ludwig A (2008) The potential of lipid- and polymer-based drug delivery carriers for eradicating biofilm consortia on device-related nosocomial infections. J Control Release 128:2–22

Tang BC, Dawson M, Lai SK, Wang YY, Suk JS, Yang M, Zeitlin P, Boyle MP, Fu J, Hanes J (2009) Biodegradable polymer nanoparticles that rapidly penetrate the human mucus barrier. Proc Natl Acad Sci USA 106:19268–19273

Tanihara M, Suzuki Y, Nishimura Y, Suzuki K, Kakimaru Y, Fukunishi Y (1999) A novel microbial infection-responsive drug release system. J Pharm Sci 88(5):510–514

Todd JA, Modest EJ, Rossow PW, Tokes ZA (1982) Liposome encapsulation enhancement of methotrexate sensitivity in a transport resistant human leukemic cell line. Biochem Pharmacol 31(4):541–546

Tomalia DA, Dewald JR, Hall MJ, Martin SJ, Smith PB (1984) Preprints of the 1st SPSJ international polymer conference. Society of polymer science, Kyoto, Japan

Torchilin V (2008) Antibody-modified liposomes for cancer chemotherapy. Expert Opin Drug Deliv 5:1003–1025

Tunney MM, Brady AJ, Buchanan F, Newe C, Dunne NJ (2008) Incorporation of chitosan in acrylic bone cement: effect on antibiotic release, bacterial biofilm formation and mechanical properties. J Mater Sci Mater Med 19:1609–1615

Vemuri S, Rhodes CT (1995) Preparation and characterization of liposomes as therapeutic delivery systems: a review. Pharma Acta Helvetiae 70:95–111

Vyas SP, Sihorkar V, Jain S (2007) Mannosylated liposomes for bio-film targeting. Int J Pharm 330(1–2):6–13

Wasan KM, Lopez-Berestein G (1996) Characteristics of lipid-based formulations that influence their biological behavior in the plasma of patients. Clin Infect Dis 23:1126–1138

Watson DS, Endsley AN, Huang L (2012) Design considerations for liposomal vaccines: influence of formulation parameters on antibody and cell-mediated immune responses to liposome associated antigens. Vaccine 30(13):2256–2272

Wu XS (1995) Synthesis and properties of biodegradable lactic/glycolic acid polymers. In: Wise DL et al (eds) Encyclopedic handbook of biomaterials and bioengineering. Dekker, New York, pp 1015–1054

Zhang L, Granick S (2006) How to stabilize phospholipid liposomes (using nanoparticles). Nano Lett 6:694–698

Zhang L, Pornpattananangkul D, Hu CM, Huang CM (2010) Development of nanoparticles for antimicrobial drug delivery. Curr Med Chem 17:585–594

Zhengbing C, Sun Y (2009) Chitosan-based rechargeable long-term antimicrobial and biofilm-controlling systems. J Biomed Mater Res Part A 89:960–967

Zhu CT, Xu YQ, Shi J, Li J, Ding J (2010) Liposome combined porous beta-TCP scaffold: preparation, characterization, and anti-biofilm activity. Drug Deliv 17(6):391–398

Eradication of Wound Biofilms by Electrical Stimulation

Chase Watters and Matt Kay

Abstract Chronic wounds are a major clinical health problem, costing billions of dollars and plaguing millions of people worldwide with increased morbidity and mortality. Treatment of these wound infections is complex and can take months to years. The length and difficulty of treating these wounds are largely attributed to the presence of biofilms created by common microbiological contaminants in the wounded area. Since multidrug-resistant bacterial biofilms persist in these wounds, the biofilms are able to dodge the "magic" antibiotic bullet. Without the aid of antimicrobials to fight off these infections, novel antimicrobials are essential. Recently, the physical sciences have been mined for alternatives to antibiotics. One such promising alternative therapy is the use of electrical stimulation devices to speed the wound healing process. There are three forms of electrical stimulation predominately utilized to treat chronic wound infections: low-intensity direct current, high-voltage pulsed current, and alternating current. The use of these various forms of electrical stimulation is proposed to enhance wound healing via the stimulation of host cells and by inhibiting bacterial biofilm growth. This chapter focuses on studies examining electrical stimulation in conjunction with host and bacterial cells, along with relevant clinical trial studies.

C. Watters (✉)
Department of Surgery, MS 8312, Texas Tech University Health Sciences Center,
3601 4th Street, Lubbock, TX 79430, USA

Department of Immunology and Molecular Microbiology, Texas Tech University
Health Sciences Center, Lubbock, TX 79430, USA
e-mail: chase.m.watters@gmail.com

M. Kay
Department of Immunology and Molecular Microbiology, Texas Tech University
Health Sciences Center, Lubbock, TX 79430, USA

1 Introduction

1.1 Background on Chronic Wound Infections

Chronic wounds are clinically defined as any wound which fails to heal within 30 days and include diabetic foot ulcers, venous leg ulcers, arterial ulcers, and pressure ulcers (Bradley et al. 1999; Falanga 1998). Ulcer wounds all possess one similarity, which is that they penetrate deep into the epidermis of the skin (Robinson and Lynn 2008). Wound infections are usually staged with inclination to severity based on the depth and appearance of the ulcer (The National Pressure Ulcer Advisory Panel 1989). These wounds are resistant to natural healing and can require long-term medical care (Beckrich and Aronovitch 1999). The chronic wound environment supports a variety of microorganisms, including a mixture of gram-negative, gram-positive, aerobic, and anaerobic bacteria, plus fungal cells (Leake et al. 2009). Some of the primary bacterial pathogens most commonly isolated from wound infections are *Staphylococcus aureus, Enterococcus faecalis, Finegoldia magna,* and *Pseudomonas aeruginosa* (Dowd et al. 2008). Indeed, *S. aureus* and *P. aeruginosa* are reported to infect up to 88 % and 33 % of leg ulcers respectively (Hansson et al. 1995; Schmidt et al. 2000).

1.2 Biofilm Pathogenesis in Chronic Wounds

Microorganisms are thought to delay wound healing by causing tissue damage and are difficult to treat because of multidrug resistance and the formation of biofilms (Percival et al. 2010). Biofilms are a conglomerate of bacterial cells, DNA, and proteins intercalated in a sugary slime matrix (Fuxman Bass et al. 2010). Biofilms have been implicated to cause many chronic diseases, and when 50 clinical chronic wounds were evaluated for the presence of biofilm, 60 % of these wounds were biofilm positive (James et al. 2008). The continued presence of bacterial biofilms in wounds is thought to delay wound healing for multiple reasons. The outer coating of the biofilm, called the exopolysaccharide (EPS) matrix, acts as a mechanical barrier to antibodies and complement, preventing biofilm penetration and protecting bacterial cells (Thomson 2011; Jacobsen et al. 2011). The EPS may also prevent fibroblasts, epithelial cells, and keratinocytes from migrating into the wound bed. As long as the biofilm remains, the immune system will try and remove it, causing prolonged inflammation and collateral damage to host tissue (Wolcott et al. 2008). In addition, oxygen is depleted rapidly in biofilms, supporting the growth of anaerobic bacteria and impairing host cell growth and repair. In biofilm-associated chronic wound infections, topical and systemic antibiotics have reduced efficacy. The bacteria within the biofilms are 500–5,000 times more tolerant to antimicrobials than planktonic bacteria, making these infections difficult to treat (Anwar et al. 1990). The tolerance that biofilms display is distinct from the classical idea of antibiotic resistance. Resistance to antibiotics refers to planktonic or biofilm cells that exhibit a transferable genetic mutation or acquire a plasmid or transposon, conferring protection against antimicrobials (Lewis 2007). However, multidrug tolerance observed in biofilms implies a transient, non-heritable phenotypic change proposed to be caused by numerous mechanisms: delayed penetration, stress responses to unfavorable environmental conditions, development of persister cells, and altered microenvironment (Gefen and Balaban 2009). These hypothesized mechanisms enable biofilms to dodge the "magic" antibiotic bullet. Without the aid of antimicrobials, clinicians are left with the physical removal of biofilms as their primary recourse. In chronic wound infections, debridement is an essential therapy, but it is not without a cost (Game 2008). The multiple forms of wound debridement are riddled with problems including patient pain, collateral tissue damage, and the length and cost of treatment (Falabella 2006). Consequently, the physical sciences have been mined for alternatives to standard wound care (SWC). One such promising physical science strategy is electrical stimulation which will be the focus of this chapter.

1.3 Electrical Stimulation of Chronic Wounds

Electrotherapy is already utilized by physical therapists for a wide array of treatments ranging from pain management, iontophoresis, and the stimulation of tissue repair in chronic wounds (American Physical Therapy Association 2005). After lengthy clinical trials and review, the use of electrical stimulation for tissue healing and repair (ESTHR) was approved in the USA as an insured treatment of chronic wounds in 2002 (Centers for Medicare and Medicaid Services 2002). However, insurance companies will only cover ESTHR for wounds which have been treated with SWC for 30 days and display no quantifiable improvements in wound healing. There are three forms of ESTHR (Fig. 1) predominately utilized to treat chronic wound infections: low-intensity direct current (LIDC), high-voltage pulsed current (HVPC), and alternating current (AC). Devices that produce ESTHR are typically made up of a component that produces the electricity and an anode and cathode, which deliver the electrical therapy to the wound site. Regardless of the specified form of electrotherapy, clinical trials have found that ESTHR is an effective adjunctive therapy for treating wound infections. Due to the lack of standardization amongst ESTHR devices, some healthcare providers remain skeptical. And although many promising clinical studies have been performed with ESTHR, it still remains unapproved as a wound healing therapy by the Food and Drug Administration. The contraindications reported from ESTHR treatment include enhanced pain to the patient (Sussman and BatesJensen 1998), potential disruption of pacemakers to those with cardiac irregularities, electrophoretic effects on ions in some medications (zinc, mercury, silver), and collateral tissue damage (Robinson and Snyder-Mackler 2007).

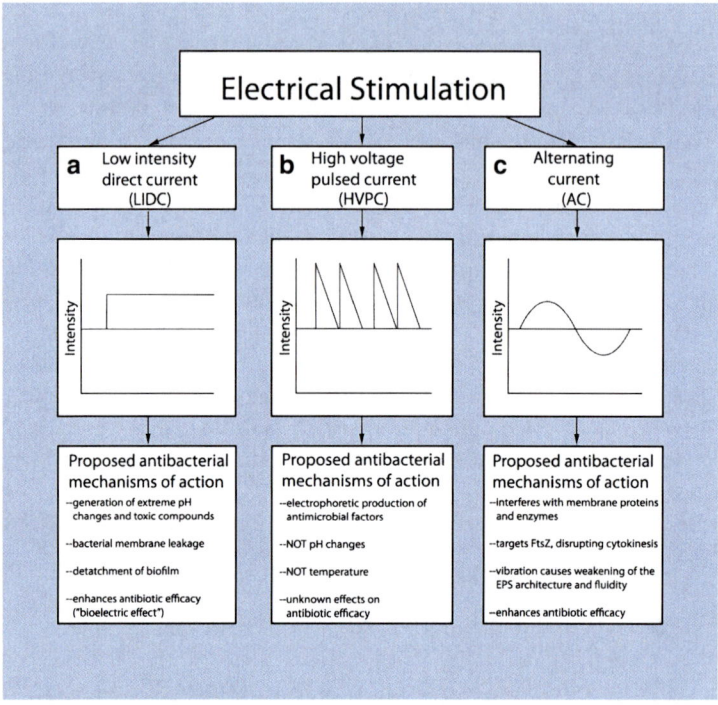

Fig. 1 Types of electrical stimulation used to treat wound infections. (A) Constant LIDC has a monophasic waveform and is applied at low intensities ranging from 20 µA to 32 mA. (B) HVPC has a twin-spike monophasic waveform and is thus a form of DC. HVPC devices deliver electrical stimulation at intensities from 0 to 250 V. (C) AC refers to the continuous application of biphasic waves. In this chapter, we focus on radio frequency alternating current, given at kHz–MHz frequencies and low intensities

2 Current of Injury

Measurable electrical currents that direct wound healing are found in intact and damaged skin of amphibians, mammals, and humans (Borgens et al. 1977; Illingsworth and Barker 1980; Barker et al. 1982; Foulds and Barker 1983; Vanable 1989; Cunliffe-Barnes 1945). Many investigators have measured electronegative voltages from the surface of intact skin and electropositive voltages from the dermal tissue of wounds (Illingsworth and Barker 1980; Barker et al. 1982; Cunliffe-Barnes 1945). These measurable trans-epithelial potentials (TEPs) occur as a result of Na^+ channels in the apical membrane of the skin's mucosal surface that allow extracellular Na^+ to diffuse between epidermal cells (Vanable 1989). Foulds and Barker demonstrated the presence of a "skin battery," with TEPs ranging from 10 mV to almost 60 mV, and an average negative potential of 23.4 mV (Barker et al. 1982). This skin battery voltage effect was discovered to be produced by charged activity in exocrine sweat glands (Wolcott et al. 1969) and can be blocked by applying amiloride (a compound which blocks Na^+ channels in outer epidermal membrane) to mammalian skin (Eltinge et al. 1986).

When wounding occurs in the skin, electrical leaks are produced that short-circuit the skin battery, allowing current to flow out (Barker et al. 1982; Vanable 1989). This effect was demonstrated as early as 1847 (Matteucci 1847), but more recent experiments have displayed currents from wounds of humans and mammals (Barker et al. 1982; DuBois-Reymond 1843). This phenomenon of "current of injury" is measurable in amphibians and mammals and can be sustained in moist environments, but is shut off when a wound dries out (McGinnis and Vanable 1986; Stump and Robinson 1986; Jaffe and Vanable 1984). In a porcine model system, occlusive dressings applied to wounds sustained the injury current at a higher voltage for 4 days as opposed to a sixfold lower voltage from wounds exposed to air (Cheng et al. 1995). Wound current of injury may thereby be sustained with occlusive, moisture-retentive bandages that contribute to the faster healing rate (Alvarez et al. 1983; Wikter 1972). When wounds do occur in the skin, a measurable positive current emerges from the wound. This current can range from 140 mV/mm at wound edge to 0 mV/mm at 3 mm lateral to wound edge, which gradually becomes reduced to nonexistence from regenerated epithelium (McGinnis and Vanable 1986; Jaffe and Vanable 1984). Maintaining or enhancing a "current of injury" response can evidently play a role in accelerating wound healing.

3 Impact of Electrical Stimulation on Wound Healing

3.1 The Wound Healing Process

The normal process of wound healing is observed as occurring in three overlapping phases: *inflammatory, proliferative,* and *remodeling* (or maturation) (Kirsner and Bogensberger 2002). The *inflammatory phase* is characterized by the body's primary response to a wound. This includes hemostasis, autolysis, and phagocytosis, occurring 2–7 days after onset of the initial wound. The phase also stimulates formation of connective tissues and angiogenesis by endothelial cells (Kirsner and Bogensberger 2002). The *proliferative phase* spans 2–3 days to 3 weeks after initial wound trauma. This event is characterized by fibroplasia (new granulation tissue), re-epithelialization, neovascularization, and wound contraction. The new tissue is formed by migration of dermal fibroblasts into the wound, which then proliferate into the different tissue types (Kirsner and Bogensberger 2002). The *remodeling phase* can last from 3 weeks to 3 years. The last phase is characterized by degradation of the initial collagen from the first two phases being replaced by new adult collagen. The type III (fetal) collagen is replaced by type I (adult) collagen, with stronger tissue tension and support overall (Kirsner and Bogensberger 2002).

3.2 In Vitro Studies Examining Electrical Stimulation and Wound Healing

Many studies have been performed elucidating the various methods by which ES enhances wound healing, such as direct cellular responses to variable electrical currents. These include studies on changes in cell synthesis and metabolism and effects on cell migration (Kloth 2005). For example, a continuous electrostatic field of 1,000 V/cm through fibroblast cultures resulted in a 20 % increase in DNA and collagen synthesis after 14 days (Bassett and Herrmann 1968). Another study involving stimulation of human fibroblasts in cell cultures with HVPC demonstrated fibroblast induction of increased rates of DNA and protein synthesis (up to 160 % greater in protein synthesis) (Bourguignon and Bourguignon 1987; Bourguignon et al. 1986, 1989). In addition, the same group of researchers reported that within the first minute of in vitro HPVC stimulation, fibroblasts had an increase in Ca^{2+} uptake, followed by upregulation of insulin receptors on fibroblast membranes by the second minute (Bourguignon et al. 1989). Transforming growth factor-β receptors have also been shown to be upregulated by sixfold in human dermal fibroblasts stimulated by ES (Falanga et al. 1987). Another study involving LIDC stimulation on rat skin reported a fivefold increase in ATP concentrations and 30–40 % increased amino acid uptake above controls (Cheng et al. 1982).

Many investigators have also reported wound healing-related cells migrating toward the anode or cathode of an electric field run through cell cultures (Orida and Feldman 1982; Monguio 1933; Fukushima et al. 1953; Dineur 1891; Canaday and Lee 1991; Erikson and Nuccitelli 1984; Yang et al. 1984; Nishimura et al. 1996; Sheridan et al. 1996). This ability, called galvanotaxis, is the attraction of charged cells moving toward an electric field of opposite polarity (Kloth 2005). Macrophages display this ability by their ability to migrate towards the anode (Orida and Feldman 1982), while neutrophils migrate to the anode or cathode (Monguio 1933; Fukushima et al. 1953). Others have reported that leukocytes migrate toward the cathode in infected areas, suggesting a link between chemical and electrical responses (Monguio 1933; Dineur 1891). Fibroblasts are known to migrate toward the cathode (Bourguignon and Bourguignon 1987; Bourguignon et al. 1986; Canaday and Lee 1991; Erikson and Nuccitelli 1984; Yang et al. 1984). Researchers have also shown that exogenously applied electric fields of the same magnitude as those found in wounded tissue can direct the migration of keratinocytes towards the cathode (Nishimura et al. 1996; Sheridan et al. 1996; Stromberg 1988; Cooper and Schliwa 1985).

3.3 In Vivo Studies Examining Electrical Stimulation and Wound Healing

In vivo studies on the effects on ES on cell migration have discovered significant benefits: a study by Eberardt et al. found a 24 % increase in neutrophils from human wounds stimulated by exogenous ES currents (Eberhart and Korytowski 1986), while a study by Mertz et al. (1993) on epidermal cell migration in response to ES demonstrated 20 % greater wound epithelialization in an ovine model. Several models have been proposed as to how ES aids cell migration in vivo by galvanotaxis. Movement of proteins embedded in the plasma membrane may occur when cells are exposed to an electric field; for example, epidermal growth factor receptors have been shown to move to the cathode side of keratinocytes in a LIDC electric field (Fang et al. 1999). Other affected regions that ES may stimulate to initiate galvanotaxis include membrane depolarizations due to calcium ion flow (Bedlack et al. 1992), changes in cell form and cytoskeleton (Soong et al. 1990; Onuma and Hui 1985, 1986; Luther et al. 1983), and protein kinase activity (Baker and Peng 1993; Peng et al. 1993). Weak electric fields used for galvanotaxis of cells in culture or clinically used currents for enhancement of healing of chronic wounds may replicate the natural electric fields found in mammalian wounds (Nishimura et al. 1996; Sheridan et al. 1996). These differential galvanic effects should be used to plan the therapeutic treatment based on selection of the anode or cathode at various time points for the greatest ES benefit. Although the effect of ESTHR can be explained to some extent by stimulating host cells, there appear to be other mechanisms important to enhance wound healing. The remainder of this chapter will focus on the three primary forms of ESTHR, first including relevant clinical studies and then examining studies on the specified electrical stimulation and bacterial biofilms.

4 Low-Intensity Direct Current Background

Direct current (DC) (sometimes referred to as galvanic or continuous current) is a form of electrical current which constantly flows in only one direction (monophasic) and is generated by thermocouples, batteries, and solar cells. LIDC refers to the continuous form of DC (Fig. 1A) and is applied at intensities ranging from 20 µA to 32 mA (Feedar et al. 1991). LIDC was the first form of electrotherapy explored as a wound care adjunctive therapy (Carey and Lepley 1962) and, not surprisingly, remains as the most commonly researched low-intensity current (Balakatounis and Angoules 2008). In 1962 LIDC was explored in an animal wound model and was found to enhance wound healing and stimulate the immune response (Carey and Lepley 1962). Shortly thereafter, clinical trials further demonstrated LIDC's potency to significantly enhance wound healing.

5 Low-Intensity Direct Current Clinical Studies

The first in-depth electrotherapy clinical trial was performed in 1969 by Wolcott et al. (1969), in which SWC plus LIDC was applied directly to 75 ischemic skin ulcers over the course of a year and a half. In this nonrandom clinical trial, LIDC treatment entailed three daily courses of LIDC, given for 2 h at an intensity range of 20 µA–1 mA. The authors reported a mean healing rate of 13.4 % per week for the electrical treated 75 ischemic skin ulcers. A major limitation of this study was that the LIDC-treated ischemic ulcers lacked a proper control group for comparison. However, the authors also reported that eight patients suffered from bilateral ulcers which were similar in size and location and thus could be utilized as ideal control for the LIDC-treated groups. From these eight patients it was observed that the untreated control ulcers healed at a rate of 5 % per week in comparison to 27 % per week for the LIDC-treated group. In 1985 Carley and Wainapel conducted an unknown blind, randomized clinical trial (RCT), treating 15 inpatient wounds with SWC and 15 with SWC + LIDC for 5 weeks (Carley and Wainapel 1985). For this study LIDC was given twice daily, for 2 h at an intensity of 300–700 µA. Following 5 weeks of therapy the mean healing rate of the control group was 45 % as compared to the 89 % in the LIDC-treated group. The authors also observed wounds that opened up and became infected in the control group following the 5-week study. However, in the LIDC-treated group, the wounds closed and did not experience infection. More recently, Adunsky et al. conducted a double-blind RCT on 63 patients suffering from pressure ulcers with 8 weeks of treatment and 12 weeks of follow-up (Adunsky and Ohry 2005). Interestingly, in this study the authors developed their own ESTHR device termed the "decubitus direct current treatment" (DDCT), which distributed a mixture of direct and alternating current into the wound and was able to determine wound size and record patient data and the current of electricity in the wound before and after treatment. All patients received SWC (debridement and colloidal dressing), while 35 patients were randomly selected to receive DDCT, and 28 other patients received a placebo (sham) treatment. The course of DCCT began with three 20-min sessions daily, but was reduced to two daily sessions after 14 days for undeclared reasons. The electrical parameters of DCCT were not reported. On day 45, the mean healing rate observed in the DDCT group was 44 %, as compared to 14 % in the control group. When the authors conducted a follow-up, they observed no significant differences in wound closure at days 57 and 145. A clear limitation and potential counter indication for this device are that 25 patients dropped out of the study for unstated reasons. These studies support the use of LIDC to treat non-healing wound infections, though further studies were done to examine what effect LIDC had on the bacterial wound burden.

6 Effect of Low-Intensity Current Alone Against Bacterial Biofilms

6.1 In Vitro Studies with Low-Intensity Direct Current and Biofilms

The antimicrobial potential of LIDC alone was first studied and shown to have a bactericidal effect against both planktonic (Rowley 1972; Davis et al. 1989) and bacterial biofilm cells. In 1974 Barranco et al. studied the effect of 0.4–400 µA against *S. aureus* biofilms grown on agar plates. These biofilms were treated for 24 h, with the anode and cathode coated in four different materials: silver, platinum, gold, and stainless steel. The authors reported maximal bactericidal effects at 400 µA, with both electrodes and with all four materials. However, the silver cathode exhibited the maximal bactericidal effects at the lower intensities (0.4 µA, 4 µA). Liu et al. (1997) went on to examine *S. aureus* and *Staphylococcus epidermidis* biofilms treated with LIDC. Biofilms grown on agar plates were treated with 10 µA for 16 h, and significant zones of inhibition were observed around the cathode. Merriman et al. (2004) treated *S. aureus* biofilms grown on agar plates with LIDC for 3 consecutive days and observed significant zones of inhibition. The LIDC treatment was applied for 1 h and generated with stainless steel electrodes producing continuous 500 µA LIDC. Del Pozo et al. (2009a) examined the antimicrobial effect of LIDC against mature biofilms. *S. aureus*, *S. epidermidis*, and *P. aeruginosa* biofilms were grown on Teflon disks for 48 h and then treated continuously for 2–7 days with 0.2–2 mA LIDC. When *Staphylococcus* biofilms were treated for 2 days with 2 mA LIDC, a 4–6 log reduction in bacterial cells was observed. However, *P. aeruginosa* biofilms had to be treated for 7 days with 2 mA LIDC in order to cause only a 3.5–5 log reduction of bacteria. In this study the authors coined the term, the "bioelectric effect," to describe the bactericidal effect of LIDC against bacterial biofilms.

6.2 In Vivo Studies with Low-Intensity Direct Current and Biofilms

In vivo studies examining LIDC against biofilms began in 1974 when Rowley et al. (1974) reported that 200–1,000 µA cathodic LIDC inhibited the growth of 24 h-old *P. aeruginosa* biofilms in rabbit dermal wounds. Van der Borden et al. (2007) devised an interesting wound infection model where three stainless steel pins were inserted into goat tibias and inoculated with *S. epidermidis*. Biofilms were grown on two of the pins, with one receiving 100 µA LIDC and the other used as a negative control, while the third pin acted as structural support. Electricity was given continuously for 21 days immediately following *S. epidermidis* infection, and current was delivered to the pins via a platinum anode. The pin site wounds were scored for infection at day 21, with infection rates of 89 % in the control group, and only 11 % in the LIDC group. Del Pozo et al. (2009b) later tested LIDC therapy versus antibiotic therapy in a chronic foreign body osteomyelitis rabbit model. Osteomyelitis is very serious complication of chronic wounds and lengthens the chronicity and difficulty in treating these infections. In this study, the tibias of rabbits were infected with *S. epidermidis* and these biofilm-associated infections went 4 weeks before therapy was applied. After 4 weeks of infection, the rabbits were divided into three groups: a control group receiving no treatment, an antibiotic group receiving daily intravenous doxycline, and an electrical current group receiving 200 µA of continuous LIDC. Following 4 weeks of treatment the bacterial load was reduced by 38 % in the doxycline- and 73 % in the electrical current-treated groups as compared to the control. These in vivo studies with LIDC appear promising with clinically relevant electrical intensities, although the extensive duration of electrical treatment (21 continuous days) would be clinically difficult.

6.3 Human Studies with Low-Intensity Direct Current and Biofilms

In the first extensive clinical trial treating ischemic ulcers with LIDC (as discussed above), Wolcott et al. (1969) reported that most initial wound swab cultures detected the presence of the *Pseudomonas* and *Proteus* genera. However, after several days of ES the majority of secondary wound cultures were free of pathogens. Bolton et al. (1980) conducted a human study investigating what effect LIDC had on 24 h *S. epidermidis* biofilms inoculated on the epidermis. A 24 h treatment with 100 µA cathodic LIDC resulted in complete eradication of *S. epidermidis* on the human volunteer's skin. An additional human study by Fakhri and Amin examined LIDC treatment on 20 non-healing burn wound infections (Fakhri and Amin 1987). For the study, antibiotic treatment was stopped during the electrical stimulation therapy. At the beginning of the study, patient wound swabs cultured positive for *P. aeruginosa*, *S. aureus*, and *Escherichia coli*. However, 10 min, biweekly applications of 25 mA LIDC resulted in wound swabs that came back as sterile or with a lowered bacterial burden. In summary these findings support the idea that LIDC treatment alone stimulates wound healing via the eradication of the bacterial biofilm burden in wounds.

7 The Antibacterial Mechanisms of Low-Intensity Direct Current

7.1 Antibacterial Mechanisms of Low-Intensity Direct Current Alone

Numerous mechanisms have been proposed to describe the antimicrobial effect of LIDC alone. The most popular proposed mechanism by which LIDC is able to kill pathogens is that it causes extremely basic and acidic conditions near the electrode capable of killing pathogens (Barranco et al. 1974; Merriman et al. 2004). Lui et al. (1997) reported another mechanism of the bactericidal activity of LIDC, which is the generation of toxic compounds (H_2O_2 and chlorine) as a result of electrolysis. Yet, some assert that electrophysical changes and toxic compounds do not completely explain the effect of LIDC on bacterial cells. Laatsch et al. (1995) suggested two alternative mechanisms: firstly, that LIDC is able to disrupt intracellular bacterial metabolism, and, secondly, that LIDC ES overwhelms microorganisms with electrons which continually excites cell membranes, causing membrane leakage of essential cellular components. Van der Borden el al. clearly elucidated another potential mechanism, which is the dispersion and detachment of *Staphylococcus* strains when biofilms were treated with 25–125 µA, causing these cells to switch to a planktonic phenotype and become more sensitive to the immune system and antibacterial compounds (van der Borden et al. 2004).

7.2 Antibacterial Mechanisms of Low-Intensity Direct Current Combined with Antibiotics ("Bioelectric Effect")

Weak direct electric currents have also been shown to enhance the effectiveness of antibacterial agents against biofilms. This scientific peculiarity (first described with LIDC) has been termed "the bioelectric effect" (Costerton et al. 1994). The bioelectric phenomenon has been studied extensively in vitro with LIDC against various bacterial biofilms: *S. epidermis* treated with tobramycin (Khoury et al. 1992), *P. aeruginosa* with tobramycin (Jass and Lappin-Scott 1996), and *S. aureus* with vancomycin (del Pozo et al. 2009a), along with a host of biofilms made by other bacterial species. For a more comprehensive examination of the bioelectric effect, there is a thorough review on the subject (Del Pozo et al. 2008). Some of the proposed mechanisms explaining the bioelectric effect include alteration of the EPS's negative charge, which reduces the ability of biofilms to bind to antibiotics, and modification of EPS integrity (Jass and Lappin-Scott 1996; Del Pozo et al. 2008). Another very interesting mechanism was reported by Niepa et al. (2012) in which LIDC was shown to enhance the efficacy of tobramycin against *P. aeruginosa* biofilm cells, via the reduction of persister cells in the population. In this study, 70 mA/cm^2 LIDC was applied for 1 h in order to target the persister cells.

8 Low-Intensity Direct Current Contraindications

One of the primary problems associated with using DC against biofilms includes the potential production of toxic compounds at the electrode interface causing collateral tissue damage. Since DC cannot be generated by insulated electrodes, the production of these toxic derivatives remains a long-term problem (Giladi et al. 2010). Additionally, such currents may stimulate nerve or muscle cells resulting in pain and muscular contractions in the patient. Due to these inherent problems, it comes as no surprise that only 10 % of polled physical therapists use LIDC on a weekly basis (Robinson and Snyder-Mackler 1988). Other electrical stimulation modalities, with HVPC on the top of the list, are utilized by clinicians and are the subject of research efforts.

9 High-Voltage Pulse Current Background

HVPC devices create pulsed current with maximum capacity ranging from 100 to 250 V. HVPC electricity displays a twin-peaked wave form (Fig. 1B) and lasts for a very short time (5–100 μs). The electricity generated in HVPC can be either DC (monophasic) or AC (biphasic), although DC is the primary electrical source utilized. The idea behind HVPC is that it allows higher voltages to be utilized in such fast pulses that nerve or other human cells are not excited and damaged. HVPC was explored in the 1980s following the successful clinical application of LIDC to wound infections in the early 1970s. The initial popularity of HVPC was linked to the finding that electrodes from HVPC devices did not cause thermal or chemical reactions and thus eliminated many of the unwarranted side effects associated with LIDC (Alon and De Do-minico 1987). Pulsed current has been the most frequently utilized form of ESTHR by physical therapists in the USA, with a reported 2/3 of the polled therapists utilizing HVPC on a daily basis (Robinson and Snyder-Mackler 1988).

10 High-Voltage Pulse Current Clinical Studies

One of the initial HVPC studies was conducted by Feeder and Kloth in 1988 where they conducted a single-blinded RCT with 16 patients suffering from Stage IV decubitus ulcers (Kloth and Feedar 1988). The patients were randomly assigned into a SWC + ESTHR-treated group ($n = 9$) or a control group only receiving

SWC ($n = 7$). Patients in the treatment group received 45 min of HVPC daily, 5 days a week. The HVPC device generated monophasic twin pulsed waves, with an interphase interval of 50 µs, an intensity of 100–175 V, and a total pulse charge accumulation of 342 µC/s. Over a mean length of seven and a half weeks of therapy the wounds of the treated group healed at a mean rate of 44.8 % a week, as compared to −11.6 % in the control group. A decade later, in 1999, the same group performed a larger study comparing the healing of stage II, III, and IV ulcers treated with SWC + HVPC (treated) to those given SWC + sham electrotherapy (control) (Feedar et al. 1991). In this randomized double-blind multicenter study, 47 patients with 50 chronic wounds were randomly assigned to the treatment ($n = 26$) or control group ($n = 24$). Treated wounds were given two daily doses of 30 min HVPC at a pulse frequency of 128 pulses per second (pps), with a maximal intensity of 29.2 mA. Following 4 weeks of therapy, the mean healing rate in the treated group was 14 % with a wound closure rate of 67 %. Conversely in the control group, the mean healing rate observed was 8.25 % with wound closure rate of 44 %.

More recently, in 2010, a single-blinded RCT was conducted with 34 spinal injury patients suffering from stage II, III, IV, and V pressure ulcers (Houghton et al. 2010). These ulcers were examined for 3 months to compare wound size and appearance when given SWC (control, $n = 18$) or SWC + HVPC (treated, $n = 16$). HVPC was given daily for an hour, via the Micro Z device which delivered a twin-peaked monophasic pulsed current with 50 s pulse duration and an intensity of 50 and 150 V. Following the 3-month intervention, the authors observed a significant decrease in wound closure in the SWC + HVPC group (70 %) as compared to the SWC control (36 %).

11 Effect of High-Voltage Pulse Current Alone Against Bacterial Biofilms

The three discussed RCTs clearly demonstrate that HVPC can enhance wound healing in patients, and this could be accomplished by HVPC reducing the bacterial burden in these wounds. This idea was first tested in 1989 by Kincaid and Lavoie who reported that in vitro, HVPC could inhibit the bacterial biofilm growth of three commonly isolated wound pathogens (*S. aureus, E. coli,* and *P. aeruginosa*) (Kincaid and Lavoie 1989). Utilizing stainless steel wires connected to the sides of plastic petri dishes, the authors observed that exposing all three pathogens to HVPC for 2 h at 250 V with 120 pps and a 55 µs interpulse interval caused a bacteriostatic effect at both the cathode and the anode. That same year, Guffey and Asmussen (1989) reported that when HVPC was applied for 30 min at voltages below 160, no effect was observed on *S. aureus* bacterial growth. However, numerous authors have implied that the shorter treatment time (Merriman et al. 2004) and use of controlled current flow (Szuminsky et al. 1994) contributed to their observations. Szuminsky et al. (1994) examined the impact of HVPC on bacterial biofilm growth by four wound pathogens (*E. coli, Klebsiella, P. aeruginosa,* and *S. aureus*). The authors reported growth inhibition at both of the stainless steel electrodes, when biofilms on agarose plates were treated for 30 min with 500 V HVPC and a 70 μs interpulse interval. The clear limitation of this study is that the authors used an electrical intensity almost double of what is applied clinically, which would cause immense pain/discomfort in patients. However, the authors state that there is a significantly lower current flow in the petri plates as compared to human skin, suggesting that utilizing lower settings clinically could also generate bactericidal activity against the tested microorganisms (Szuminsky et al. 1994). This study was repeated in 2004, comparing the antibacterial effectiveness of HVPC to other three electrical modalities (Merriman et al. 2004). The authors treated *S. aureus* biofilms with HVPC for 3 consecutive days and observed significant zones of inhibition. The HVPC treatment was applied for 1 h and was generated with stainless steel electrodes producing 250 V HVPC, with a 70 μs interphase interval. The dose of 250 V seems excessive, but the authors state that this intensity is tolerable to patients if the interpulse interval is long enough (Merriman et al. 2004). Interestingly, the authors observed physical changes from HVPC application, specifically gas formation near the cathode and corrosive discoloration of the plates at the anode, suggesting that physical changes occur in the area directly around the electrodes. Also, in 2004 an abstract reported the successful inhibition of *Streptococcus A* biofilms in vitro with the HVPC-negative polarity (Kuykendall et al. 2004). These make up the majority of studies examining the antimicrobial effect of HVPC in vitro, and the only in vivo study comes from a dissertation by Mark Campolo, who conducted the study at Seton Hall University (Campolo 1999). Campolo utilized a rabbit wound model in which *S. aureus* infection was either left untreated or treated for 1 h for 6 consecutive days with 100 V HVPC and a 70 μs interphase interval. In this study, HVPC treatment did not significantly reduce the bacterial biofilm burden in rabbit wounds, possibly due to a low group number and high variability (Campolo 1999).

12 The Antibacterial Mechanisms of High-Voltage Pulse Current

A clear gap in knowledge is the mechanism by which HVPC inhibits bacterial growth. Szuminsky et al. (1994) tried to determine the mechanism by which 500 V HVPC killed bacteria in vitro and came up with numerous possibilities: the direct action of electricity on the bacteria, local heat generation, electrophoretic production of antimicrobial factors in the media, or pH changes. The authors directly tested temperature increases and observed minimal (<1 °C) temperature rises, discounting heat from the equation. In addition, the authors found some evidence suggesting that extreme pH changes cause antimicrobial effects, but overall they concluded that pH changes caused by HVPC are not the predominant reason for the observed bacterial inhibition. What also negates pH change as the proposed mechanism is that when HVPC was applied to human tissue for 30 min it was not shown to cause electrochemical changes at either electrode (Newton and Karselis 1983). The reason for this is that HVPC produces a minimal average current in the tissue, and thus there are negligible chemical reactions in tissue near the electrode (unlike LIDC) (Campolo 1999). More in-depth studies are required to determine the exact biofilm inhibitory mechanism of action of HVPC, and with minimal in vivo studies, the in vitro results should be taken with caution. Other drawbacks of the current literature examining HVPC and biofilm inhibition are that none of the studies examined the effect of HVPC on mature biofilms and biofilm-specific experiments were not conducted (e.g., crystal violet assays, antimicrobial tolerance, electron microscopy). The HVPC studies used biofilms that were grown on solid, rather than in liquid, media, or directly on the electrodes. One final gap in the current literature is that no study to date has examined the potential of HVPC therapy to enhance antibiotic efficacy against bacterial biofilms. This effect has clearly been established with DC (the bioelectric effect), and hopefully future studies will shed more light on this.

13 Alternating Current Background

AC refers to the uninterrupted bidirectional flow of charged particles which oscillate from a positive charge to a negative charge (Fig. 1C), based on the frequency of the AC. Much like the "War of the Currents" in the early 1900s, DC was the first established form of electricity used in ESTHR, but in this war AC eventually supplanted the use of DC as the standard form of electricity used in the USA. At present, however, there still have not been any clinical trials examining the effectiveness of AC in promoting wound healing. Some authors have incorrectly reported that the electricity supplied by transcutaneous electrical nerve stimulation (TENS) devices is AC (Kloth 2005). Yet, TENS devices produce pulsed monophasic pulsed current (DC) or biphasic pulsed current (AC), not continuous AC. The studies examining this form of pulsed AC are limited and report minimal (Maadi et al. 2010) to no effect (Merriman et al. 2004) on bacterial killing. This lack of antibiofilm activity does not seem to only be associated with low-intensity, low-frequency (LILF) pulsed AC, since low-intensity pulsed DC is also reported to be similarly ineffective (Merriman et al. 2004). In addition, there are few in vitro studies suggesting that LILF constant AC can inhibit biofilm formation (Gabi et al. 2011; Kang et al. 2011) so this form of AC could show promise in the future.

14 Radio Frequency Alternating Current Background

An interesting alternative to the typical LILF pulsed AC is constant radio frequency alternating electric current (RFAC). RFAC is very clinically attractive because at radio frequencies (hundreds of kHz/MHz) constant AC can pass through the human body without nerve or muscle excitation, because the electricity is switching from positive to negative so quickly that human cells are not affected. Therefore, even though the patient has an electric current running though his/her body, no discomfort is felt (d'Arsonval 1894). At certain specifications RFAC can deposit heat in the tissues and be used therapeutically. In fact, RFAC is an FDA-approved treatment of tumor cells, in the form of radio frequency tumor ablation (Locklin and Wood 2005). Radio frequency tumor ablation uses low to medium frequencies of 450 kHz and high intensity of 80–100 W to heat up tumors to the desired temperature of 60–100° centigrade, resulting in tumor ablation. Many hard to reach and inoperable tumors can be treated with radio frequency tumor ablation including lung, liver, and bone tumors (Sharma et al. 2011). Another novel use of RFAC against tumors has recently been demonstrated using low-intensity RFAC at frequencies of 100–200 kHz (Kirson et al. 2004). Using this low-intensity, low-frequency, heat-free form of RFAC resulted in the inhibition of cancer cell growth, both in vitro and in vivo with no effect on normal mammalian cells. This particular use of RFAC is termed tumor treating fields (TTFields). Unlike tumor ablation, where heat causes tumor destruction, TTFields' proposed mechanism of action is that the radio fields interfere with microtubule production in cancer cells as they multiply. By interfering with the tumor cell's microtubules, which are needed for chromosome alignment and separation, they cause mitotic arrest, and cell destruction ensues (Kirson et al. 2004). The reason that normal mammalian cells are not affected is because they do not proliferate as quickly as cancer cells and thus are minimally affected (Kirson et al. 2007). A therapeutic form of RFAC which combines the use of current (with minimal temperature increases) and RF radiation would be ideal because it would not affect the host but would have potential bactericidal activity.

15 Effect of Biofilms Radio Frequency Alternating Current Against Bacterial Biofilms

The most clinically relevant studies concerning RFAC's ability to inhibit bacteria combine RFAC ablation and TTFields. This RFAC chimera features low intensity, high frequencies, and minimal temperature increases. The effect of RFAC at varying parameters in the presence and absence of chloramphenicol was tested on planktonic *P. aeruginosa* and *S. aureus* (Giladi et al. 2008). The results of this in vitro study showed that the RFAC inhibition of planktonic bacteria was intensity (voltage) and frequency (cycles per second of AC field) dependent (Robinson and

Snyder-Mackler 2007). Frequencies between 100 kHz and 50 MHz were examined, and the frequency of 10 MHz was found to inhibit planktonic bacteria most effectively. Low intensities (0–4 V/cm) were utilized and as the intensity increased so did the inhibition of bacterial growth. RFAC was also observed to have an additive inhibitory effect with chloramphenicol. Planktonic *P. aeruginosa* growth was also shown to be reduced by RFAC in vivo (Giladi et al. 2010). Some of the proposed mechanisms of action for RFAC's planktonic cell reduction are as follows: RFAC interferes with membrane proteins and enzymes, RFAC targets FtsZ (a tubulin homologue), and lastly, RFAC affects dividing bacteria during cytokinesis (Giladi et al. 2008). Some of these proposed bacterial targets may translate over to RFAC treatment of biofilms. However, the key therapeutic target to disrupt in the biofilm is ideally the EPS. *E. coli* biofilms were pitted against RFAC with gentamicin and oxytetracycline in vitro (Caubet et al. 2004). Using similar conditions from the planktonic study (10 MHz, low intensity) RFAC decreased the amount of bacteria in a biofilm by more than 60 %, even in the absence of antibiotics. RFAC was also found to synergize effectively with both gentamicin and oxytetracycline. This dramatic eradication of *E. coli* biofilms was proposed to be from EPS matrix alterations. As stated above, the EPS is a negatively charged mechanical barrier surrounding the biofilm. The RFAC oscillates from negative to positive and can vibrate polar/charged molecules. RFAC vibration increases the fluidity of polar membranes. So the RFAC could be increasing fluidity and thus weakening the EPS structure. In a more recent study, authors combined low intensity 1.25 V/cm, 10 MHz RFAC with DC to enhance the efficacy of gentamicin against *E. coli* biofilms in vitro (Kim et al. 2012). The authors did not observe significant reductions in bacteria numbers with ES alone, but when combined with gentamicin they observed a 56 % reduction in bacterial cell viability, as compared to gentamicin treatment alone. Either way, for both planktonic bacteria and biofilms, RFAC has been shown to inhibit growth at high frequencies/low intensities and has the potential to be an interesting alternative ESTHR modality. However, much more work must be pursued to find the optimal parameters and methods to inhibit bacterial biofilms with RFAC in the presence and absence of antibiotics.

16 Conclusions

Chronic wounds continue to be a large source of morbidity and mortality, usually resulting in increased hospital costs during extended medical care. Largely this effect is believed to be due to the presence of biofilms, which display resistance to antibiotics and necessitate painful debridement. Alternative therapies for the treatment of chronic wounds exacerbated by biofilm development include the use of ESTHR devices to speed the healing process. The various forms of ESTHR include LIDC, HVPC, and AC which can enhance wound healing by either stimulating the current of injury or providing bactericidal effects. The normal healing process of wounds in skin involves the body's innate "current of injury" electrical stimulation to provide galvanotaxis for migrating immune cells towards wounded tissue and stimulating fibroblasts to repair and replace dead tissue. This "current of injury" effect has been notably increased with the use of ESTHR, specifically for the treatment of chronic wounds which display complications in normal healing. LIDC has been shown to have an extensive bactericidal effect against both planktonic and biofilm cells. In a trial treating ischemic ulcers with LIDC, secondary wound cultures were cleared of pathogens. LIDC appears to have a direct effect on the physiology of bacterial cells, from disruption of the bacterial metabolism to excitation of the cell membrane, leading to leakage of cellular components. Clinically, HVPC is a much safer option than LIDC, avoiding many potential side effects. HVPC has been shown in vitro to inhibit the bacterial biofilm growth of common wound pathogens: *S. aureus, E. coli,* and *P. aeruginosa*. Though the actual mechanism of bacterial disruption in HVPC treatment of wounds is unknown, HVPC is the preferred treatment by many clinicians. RFAC has been used in treatment of tumors in hard to reach areas of the body. The mechanism of action is believed to be interference with microtubule production, which stops the replication of rapidly growing cancer cells. RFAC is thought to affect bacterial cells in a similar fashion, stopping cell division of planktonic cells and disrupting the EPS architecture of biofilm cells. In conclusion, electrical stimulation provides clinicians with a valuable tool to treat non-healing wound infections. The mechanisms by which electricity can speed wound healing appear linked to increasing the immune response to the wound site, while reducing the overall bacterial burden. Hopefully future studies will focus on defining more standardized parameters, with a special focus on HVPC and RFAC.

References

Adunsky A, Ohry A (2005) Decubitus direct current treatment (DDCT) of pressure ulcers: results of a randomized double-blinded placebo controlled study. Arch Gerontol Geriatr 41(3):261–269

Alon G, De Do-minico G (1987) High voltage stimulation: an integrated approach to clinical electrotherapy, vol 1. Chatanooga Corporation, Hixton, TN

Alvarez O, Mertz PM, Eaglstein W (1983) The effect of occlusive dressing on collagen synthesis and re-epthelialization in superficial wounds. J Surg Res 35:142–148

American Physical Therapy Association (2005) Electrotherapeutic terminology in physical therapy; section on clinical electrophysiology. American Physical Therapy Association, Alexandria, VA

Anwar H, Dasgupta MK, Costerton JW (1990) Testing the susceptibility of bacteria in biofilms to antibacterial agents. Antimicrob Agents Chemother 34(11):2043–2046

Baker L, Peng HB (1993) Tyrosine phosphorylation and acetylcholine receptor cluster formation in cultured Xenopus muscle cells. J Cell Biol 120(1):185–195

Balakatounis KC, Angoules AG (2008) Low-intensity electrical stimulation in wound healing: review of the efficacy of externally applied currents resembling the current of injury. Eplasty 8:e28

Barker A, Jaffe LF, Vanable JW Jr (1982) The glabrous epidermis of cavies contains a powerful battery. Am J Physiol 242:R258–R366

Barranco SD et al (1974) In vitro effect of weak direct current on *Staphylococcus aureus*. Clin Orthop Relat Res 100:250–255

Bassett C, Herrmann I (1968) The effect of electrostatic fields on macromolecular synthesis by fibroblasts in vitro [abstract]. J Cell Biol 39:9

Beckrich K, Aronovitch SA (1999) Hospital-acquired pressure ulcers: a comparison of costs in medical vs. surgical patients. Nurs Econ 17(5):263–271

Bedlack R Jr, Wei M, Loew L (1992) Localized membrane depolarization and localized calcium influx during electric field-guided neurite growth. Neuron 9(3):393–403

Bolton L et al (1980) Direct-current bactericidal effect on intact skin. Antimicrob Agents Chemother 18(1):137–141

Borgens R, Borgens R Jr, Jaffee LF (1977) Bioelectricity and regeneration: I. Initiation of frog limb regeneration by minute currents. J Exp Zool 200:402–417

Bourguignon G, Bourguignon LY (1987) Electric stimulation of protein and DNA synthesis in human fibroblasts. FASEB J 1(5):398–402

Bourguignon G, Bergouignan M, Khorshed A et al (1986) Effect of high voltage pulsed galvanic stimulation on human fibroblasts in cell culture [abstract]. J Cell Biol 103:344a

Bourguignon G, Jy W, Bourguignon L (1989) Electrical stimulation of human fibroblasts cause an increase in Ca2+ influx and the exposure of additional insulin receptors. J Cell Physiol 140 (2):379–385

Bradley M, Cullum N, Sheldon T (1999) The debridement of chronic wounds: a systematic review. Health Technol Assess 3(17 Pt 1): iii–iv, 1–78.

Campolo M (1999) Anti-bacterial effect of Hvpc in vivo. Dissertation and Thesis, paper 242

Canaday D, Lee R (1991) Scientific basis for clinical application of electric fields in soft tissue repair. In: Brighton C, Pollack S (eds) Electromagnetics in biology and medicine. San Francisco Press, San Francisco, CA

Carey LC, Lepley D Jr (1962) Effect of continuous direct electric current on healing wounds. Surg Forum 13:33–35

Carley PJ, Wainapel SF (1985) Electrotherapy for acceleration of wound healing: low intensity direct current. Arch Phys Med Rehabil 66(7):443–446

Caubet R et al (2004) A radio frequency electric current enhances antibiotic efficacy against bacterial biofilms. Antimicrob Agents Chemother 48(12):4662–4664

Centers for Medicare and Medicaid Services (2002) National coverage determination for electrical stimulation(es) and electromagnetic therapy for the treatment of wounds. NCD 270.1. Centers for Medicare and Medicaid Services, Washington, DC

Cheng N, Van Hoof H, Bockx E et al (1982) The effects of electric currents on ATP generation, protein synthesis, and membrane transport in rat skin. Clin Orthop 171:264–272

Cheng K, Tarjan P, Oliveira-Gandia F et al (1995) An occlusive dressing can sustain natural electrical potential of wounds. J Invest Dermatol 104(4):662–665

Cooper M, Schliwa M (1985) Electrical and ionic controls of tissue cell locomotion in DC electrical fields. J Cell Physiol 103:363

Costerton JW et al (1994) Mechanism of electrical enhancement of efficacy of antibiotics in killing biofilm bacteria. Antimicrob Agents Chemother 38(12):2803–2809

Cunliffe-Barnes T (1945) Healing rate of human skin determined by measurements of electrical potential of experimental abrasions: study of treatment with petrolatum and with petrolatum containing yeast and liver abstracts. Am J Surg 69:82–87

d'Arsonval A (1894) Action de l'électricité sur les êtres vivants Exposé des titres et traveaux scientifiques du Dr A. d'Arsonval. Imprimerie de la cour d'appel, Paris, pp 37–77

Davis CP et al (1989) Effects of microamperage, medium, and bacterial concentration on iontophoretic killing of bacteria in fluid. Antimicrob Agents Chemother 33(4):442–447

Del Pozo JL, Rouse MS, Patel R (2008) Bioelectric effect and bacterial biofilms. A systematic review. Int J Artif Organs 31(9):786–795

del Pozo JL et al (2009a) The electricidal effect: reduction of Staphylococcus and pseudomonas biofilms by prolonged exposure to low-intensity electrical current. Antimicrob Agents Chemother 53(1):41–45

Del Pozo JL et al (2009b) The electricidal effect is active in an experimental model of *Staphylococcus epidermidis* chronic foreign body osteomyelitis. Antimicrob Agents Chemother 53(10): 4064–4068

Dineur E (1891) Note sur la sensibilities des leukocytes a l'electricite. Bull Seances Soc Belge Microsc (Bruxelles) 18:113–118

Dowd SE et al (2008) Survey of bacterial diversity in chronic wounds using pyrosequencing, DGGE, and full ribosome shotgun sequencing. BMC Microbiol 8:43

DuBois-Reymond E (1843) Vorlaufiger abrifs einer untersuchung uber densogenanten froschstrom und die electromotorischen fische. Ann Phys U Chem 58:1–4

Eberhart SP, Korytowski G (1986) Effect of transcutaneous electrostimulation on cell composition of skin exudates. Acta Physiol Pol 37:41–46

Eltinge E, Cragoe EJ Jr, Vanable JW Jr (1986) Effects of amiloride analogues on adult *Notophtalmus viridescens* limb stump currents. Comp Biochem Physiol 89A:39–44

Erikson C, Nuccitelli R (1984) Embryonic fibroblast motility and orientation can be influenced by physiological electric fields. J Cell Biol 98:296–307

Fakhri O, Amin MA (1987) The effect of low-voltage electric therapy on the healing of resistant skin burns. J Burn Care Rehabil 8(1):15–18

Falabella AF (2006) Debridement and wound bed preparation. Dermatol Ther 19(6):317–325

Falanga V (1998) Wound healing and chronic wounds. J Cutan Med Surg 3(Suppl 1):S1-1–S1-5

Falanga V, Bourguignon GJ, Bourguignon L (1987) Electrical stimulation increases the expression of fibroblast receptors for transforming growth factor-beta. J Invest Dermatol 88:488–492

Fang K, Ionides E, Oster G et al (1999) Epidermal growth factor receptor relocalization and kinase activity are necessary for directional migration of keratinocytes in DC electric fields. J Cell Sci 112:1967–1978

Feedar JA, Kloth LC, Gentzkow GD (1991) Chronic dermal ulcer healing enhanced with monophasic pulsed electrical stimulation. Phys Ther 71(9):639–649

Foulds L, Barker AT (1983) Human skin battery potentials and their possible role in wound healing. Br J Dermatol 109:515–522

Fukushima K, Sena N, Inui H et al (1953) Studies of galvanotaxis of leukocytes. Med J Osaka Univ 4(2–3):195–208

Fuxman Bass JI et al (2010) Extracellular DNA: a major proinflammatory component of *Pseudomonas aeruginosa* biofilms. J Immunol 184(11):6386–6395

Gabi M et al (2011) Electrical microcurrent to prevent conditioning film and bacterial adhesion to urological stents. Urol Res 39(2):81–88

Game F (2008) The advantages and disadvantages of non-surgical management of the diabetic foot. Diabetes Metab Res Rev 24(Suppl 1):S72–S75

Gefen O, Balaban NQ (2009) The importance of being persistent: heterogeneity of bacterial populations under antibiotic stress. FEMS Microbiol Rev 33(4):704–717

Giladi M et al (2008) Microbial growth inhibition by alternating electric fields. Antimicrob Agents Chemother 52(10):3517–3522

Giladi M et al (2010) Microbial growth inhibition by alternating electric fields in mice with *Pseudomonas aeruginosa* lung infection. Antimicrob Agents Chemother 54(8):3212–3218

Guffey JS, Asmussen MD (1989) In vitro bactericidal effects of high voltage pulsed current versus direct current against *Staphylococcus aureus*. Clin Electrophysiol 1:5–9

Hansson C et al (1995) The microbial-flora in venous leg ulcers without clinical signs of infection—repeated culture using a validated standardized microbiological technique. Acta Derm Venereol 75(1):24–30

Houghton PE et al (2010) Electrical stimulation therapy increases rate of healing of pressure ulcers in community-dwelling people with spinal cord injury. Arch Phys Med Rehabil 91(5):669–678

Illingsworth C, Barker AT (1980) Measurement of electrical currents emerging during the regneration of amputated finger tips in children. Clin Phys Physiol Meas 1:87–89

Jacobsen JN et al (2011) Investigating the humoral immune response in chronic venous leg ulcer patients colonised with *Pseudomonas aeruginosa*. Int Wound J 8(1):33–43

Jaffe L, Vanable JW Jr (1984) Electrical fields and wound healing. Clin Dermatol 2(3):33–44

James GA et al (2008) Biofilms in chronic wounds. Wound Repair Regen 16(1):37–44

Jass J, Lappin-Scott HM (1996) The efficacy of antibiotics enhanced by electrical currents against *Pseudomonas aeruginosa* biofilms. J Antimicrob Chemother 38(6):987–1000

Kang H et al (2011) Bacterial translational motion on the electrode surface under anodic electric field. Environ Sci Technol 45(13):5769–5774

Khoury AE et al (1992) Prevention and control of bacterial infections associated with medical devices. ASAIO J 38(3):M174–M178

Kim YW, Mosteller MP, Meyer MT, Ben-Yoav H, Bentley W, Ghodssi R (2012) Microfluidic biofilm observation, analysis, and treatment (micro-boat) platform. In: Solid-state sensors, actuators and microsystems workshop, Hilton Head, South Carolina

Kincaid CB, Lavoie KH (1989) Inhibition of bacterial growth in vitro following stimulation with high voltage, monophasic, pulsed current. Phys Ther 69(8):651–655

Kirsner RS, Bogensberger G (2002) The normal process of healing. In: Kloth LC, McCulloch JM (eds) Wound healing alternatives in management. F. A. Davis, Philadelphia, PA, pp 3–29

Kirson ED et al (2004) Disruption of cancer cell replication by alternating electric fields. Cancer Res 64(9):3288–3295

Kirson ED et al (2007) Alternating electric fields arrest cell proliferation in animal tumor models and human brain tumors. Proc Natl Acad Sci USA 104(24):10152–10157

Kloth LC (2005) Electrical stimulation for wound healing: a review of evidence from in vitro studies, animal experiments and clinical trials. Int J Low Extrem Wounds 4:23–44

Kloth LC, Feedar JA (1988) Acceleration of wound healing with high voltage, monophasic, pulsed current. Phys Ther 68(4):503–508

Kuykendall C et al (2004) Effects of high voltage pulsed current on bacterial viability: an in vitro study. Wound Repair Regen 12(2):A29

Laatsch L, Ong P, Kloth L (1995) In vitro effects of two silver electrodes on select wound pathogens. J Clin Electrophysiol 7:10–15

Leake JL et al (2009) Identification of yeast in chronic wounds using new pathogen-detection technologies. J Wound Care 18(3):103–104, 106, 108

Lewis K (2007) Persister cells, dormancy and infectious disease. Nat Rev Microbiol 5(1):48–56

Liu WK, Brown MR, Elliott TS (1997) Mechanisms of the bactericidal activity of low amperage electric current (DC). J Antimicrob Chemother 39(6):687–695

Locklin JK, Wood BJ (2005) Radiofrequency ablation: a nursing perspective. Clin J Oncol Nurs 9 (3):346–349

Luther P, Peng H, Lin J (1983) Changes in cell shape and actin distribution induced by constant electric fields. Nature 303(5912):61–64

Maadi H et al (2010) Effect of alternating and direct currents on *Pseudomonas aeruginosa* growth in vitro. Afr J Biotechnol 9(38):6373–6379

Matteucci C (1847) Lectures on the physical phenomena of living beings. In: Pereira J (ed) Carlo Matteucci, 1811-1868. Longman, Brown, Green and Longmans, London, p 435

McGinnis M, Vanable JW Jr (1986) Voltage gradients in newt limb stumps. Prog Clin Biol Res 210:231–238

Merriman HL et al (2004) A comparison of four electrical stimulation types on *Staphylococcus aureus* growth in vitro. J Rehabil Res Dev 41(2):139–146

Mertz P, Davis S, Cazzaniga A et al (1993) Electrical stimulation: acceleration of soft tissue repair by varying the polarity. Wounds 5:153–159

Monguio J (1933) Uber die polare wirkung des galvanischen stromes auf leukozyten. Z Biol 93:553–559

Newton RA, Karselis TC (1983) Skin pH following high voltage pulsed galvanic stimulation. Phys Ther 63(10):1593–1596

Niepa THR, Gilbert JL, Ren DC (2012) Controlling *Pseudomonas aeruginosa* persister cells by weak electrochemical currents and synergistic effects with tobramycin. Biomaterials 33(30): 7356–7365

Nishimura K, Isseroff RR, Nuccitelli R (1996) Human keratinocytes migrate to the negative pole in direct current electrical fields comparable to those measured in mammalian wounds. J Cell Sci 109:199–207

Onuma E, Hui SW (1985) A calcium requirement for electric field induced cell shape changes and preferential orientation. Cell Calcium 6(3):281–292

Onuma E, Hui SW (1986) The effects of calcium on electric field-induced cell shape changes and preferential orientation. Prog Clin Biol Res 210:319–327

Orida N, Feldman JD (1982) Directional protrusive pseudopodial activity and motility in macrophages induced by extra-cellular electric fields. Cell Motil 2:243–255

Peng H, Baker L, Dai Z (1993) A role of tyrosine phosphorylation in the formation of receptor clusters induced by electric fields in cultured Xenopus muscle cells. J Cell Biol 120(1): 194–204

Percival SL, Thomas JG, Williams DW (2010) Biofilms and bacterial imbalances in chronic wounds: anti-Koch. Int Wound J 7(3):169–175

Robinson A, Lynn SM (2008) Clinical electrophysiology, electrotherapy and electrophysiologic testing, 3rd edn. Lippincott Williams and Wilkins, Baltimore, MD, pp 275–299

Robinson AJ, Snyder-Mackler L (1988) Clinical application of electrotherapeutic modalities. Phys Ther 68(8):1235–1238

Robinson AJ, Snyder-Mackler L (2007) Clinical electrophysiology: electrotherapy and electrophysiologic testing. Lippincott Williams and Wilkins, Philadelphia, PA

Rowley BA (1972) Electrical current effects on *E. coli* growth rates. Proc Soc Exp Biol Med 139 (3):929–934

Rowley BA et al (1974) The influence of electrical current on an infecting microorganism in wounds. Ann N Y Acad Sci 238:543–551

Schmidt K et al (2000) Bacterial population of chronic crural ulcers: is there a difference between the diabetic, the venous, and the arterial ulcer? VASA 29(1):62–70

Sharma A et al (2011) How I do it: radiofrequency ablation and cryoablation of lung tumors. J Thorac Imaging 26(2):162–174

Sheridan D, Isseroff RR, Nuccitelli R (1996) Imposition of a physiologic DC electric field alters the migratory response of human keratinocytes on extracellular matrix molecules. J Invest Dermatol 106(4):642–646

Soong H, Parkinson WC, Sulik G et al (1990) Effects of electric fields on cytoskeleton of corneal stromal fibroblasts. Curr Eye Res 9(9):893–901

Stromberg B (1988) Effects of electrical currents on wound contraction. Ann Plast Surg 21(2): 121–123

Stump R, Robinson KR (1986) Ionic current in Xenopus embryos during neuroulation and wound healing. Prog Clin Biol Res 210:2223–2230

Sussman C, BatesJensen BM (1998) Stimulation for wound healing, wound care collaborative practice manual for physical therapists and nurses. Aspen Publishers, Gaithersburg, MD

Szuminsky NJ et al (1994) Effect of narrow, pulsed high voltages on bacterial viability. Phys Ther 74(7):660–667

The National Pressure Ulcer Advisory Panel (1989) Pressure ulcer prevalence, cost, and risk assessment: consensus development conference statement. Decubitus 2:24

Thomson CH (2011) Biofilms: do they affect wound healing? Int Wound J 8(1):63–67

van der Borden AJ, Mei HC, Busscher HJ (2004) Electric-current-induced detachment of *Staphylococcus epidermidis* strains from surgical stainless steel. J Biomed Mater Res B Appl Biomater 68(2):160–164 van der Borden AJ et al (2007) Prevention of pin tract infection in external stainless steel fixator frames using electric current in a goat model. Biomaterials 28(12):2122–2126

Vanable J Jr (1989) Integumentary potentials and wound healing. In: Borgans R et al (eds) Electric fields in vertebrate repair. Alan R Liss, New York, NY, p 183

Wikter G (1972) Epidermal regeneration studies in the domestic pig. In: Maibach H, Rovee D (eds) Epidermal wound healing. Year Book Medical Publishers, Chicago, IL, pp 71–112

Wolcott LE, Wheeler P, Hardwicke HM et al (1969) Accelerated healing of skin ulcers by electrotherapy. South Med J 62:795–801

Wolcott RD, Rhoads DD, Dowd SE (2008) Biofilms and chronic wound inflammation. J Wound Care 17(8):333–341

Yang W, Onuma EK, Hui A (1984) Response of C3H/10T1/2 fibroblasts to an external steady electric field stimulation. Exp Cell Res 155:92–97

The Effects of Photodynamic Therapy in Oral Biofilms

Michelle Peneluppi Silva, Juliana Campos Junqueira, and Antonio Olavo Cardoso Jorge

Abstract Increased drug resistance in pathogenic microorganisms, leading to a decrease in the effectiveness of antibiotic, antiviral, antiparasitic and antifungal therapy, has generated international concern. This has triggered numerous studies seeking alternative antimicrobial technologies independent of pharmacology, which resulted in success in therapeutic protocols. One of the most important examples of these innovations is antimicrobial photodynamic therapy. Since the beginning of the last century, the combined use of light and dyes has been used to eliminate microorganisms. However, this therapy was forgotten after the discovery of antibiotics in the 1950s and is returning to the research field today. Numerous studies have shown that photodynamic therapy is an effective way to eliminate microorganisms, especially those that cause infections, including those in the oral cavity. However, it is important to realise that many infectious diseases will continue to need systemic therapy. The rapid growth in the use of photodynamic therapy during recent years in dentistry illustrates the number of infections that can be treated medically in this manner in the future.

1 Historical Considerations of Photodynamic Therapy

The rapid rise in antibiotic resistance among several species of pathogenic bacteria characterises the end of a period that extends over the past 50 years referred to as the "antibiotic era" (Bell 2003; Hamblin and Hasan 2004). Rapid bacterial reproduction and mutation have contributed to the prevalence of the survival of microorganisms in the presence of antibiotics. In addition, genetic elements, such as enzymes encoded by resistance plasmids and efflux pumps, can be transferred between species. The inappropriate prescribing of antimicrobials, especially for

M.P. Silva (✉) • J.C. Junqueira • A.O.C. Jorge
Department of Biosciences and Oral Diagnosis, Institute of Science and Technology, UNESP – Univ Estadual Paulista, Francisco Jose Longo 777, São Dimas, São José dos Campos, Brazil
e-mail: mipeneluppi@ig.com.br

viral diseases, the lack of commitment of some patients to complete the treatment and the widespread use of antibiotics in animal feed only exacerbate the problem by repeatedly selecting more resistant strains (Harrison and Svec 1998). Worldwide, the inexorable growth of multiresistant bacteria has resulted in a great effort by researchers to find alternative therapies for which microorganisms do not easily develop resistance (Dai et al. 2009).

Reports of photodynamic therapy (PDT) date from the time of the Egyptian civilisation through ingestion of plants (containing psoralens, furo [3,2-g] coumarin or 6-hydroxy-5-benzofuran-acrylic acid δ-lactone) and sunlight to treat diseases such as vitiligo (Simplicio et al. 2002) and psoriasis. Ancient documents found in India and China also describe the use of PDT (Simplicio et al. 2002; Ochsner 1997a). In 1900, Oscar Raab, Ludwig-Maximillian University, Munich, described the lethal action of the dye acridine and light on *Paramecium*, a unicellular organism that causes malaria (Sternberg and Dolphin 1998; Kübler et al. 2001; Pervaiz 2001; Malik et al. 2010).

The German physician Friedrich Meyer-Betz pioneered the development of studies on phototherapy irradiation (PRT) with porphyrin in 1913. He observed that skin inoculation with 200 mg haematoporphyrin had no effect, but when the inoculated skin was exposed to light, the subjects developed photosensitivity that lasted for a few months (Simplicio et al. 2002).

John Toth, product manager of Medical Devices Corp Cooper/Cooper Lasersonics, confirmed the chemical effect of therapy using photosensitisers and an argon laser; he wrote the first study to rename this therapy photodynamic therapy. Currently, photodynamic therapy is used to treat various cancers and infectious diseases (Dougherty and Marcus 1992; Lui and Anderson 1992; Ackroyd et al. 2001).

2 Photodynamic Therapy

Photodynamic therapy is an effective alternative treatment for localised microbial infections but also for oral ulcers and chronic infections (Wainwright and Crossley 2004). In dentistry, there is ongoing research into the treatment of periodontal disease with photodynamic therapy as a beneficial method and complementary treatment to conventional methods (Walker 1996; Manch-Citron et al. 2000; Feres et al. 2002; Malik et al. 2010). Photodynamic therapy has emerged as a new non-invasive therapeutic method to treat infections caused by various bacteria, fungi and viruses (Jori et al. 2006).

Combining the use of a photosensitising substance and light to antibiotic treatment is referred to as antimicrobial photodynamic therapy (APT), as illustrated in Fig. 1. This therapeutic modality illustrates an important fact: without repeated applications, there is no selective pressure for resistant bacteria, so microbial resistance is not developed (Wainwright and Crossley 2004). Due to the

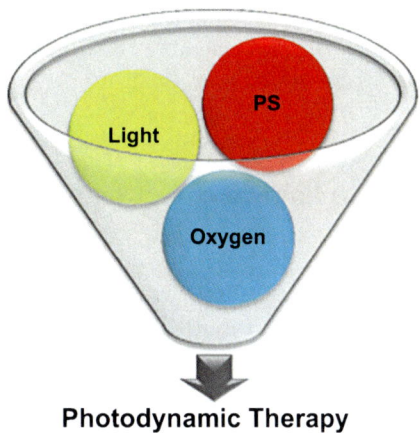

Fig. 1 Triad formative elements of PDT

antimicrobial action, PDT has other names, such as photodynamic inactivation (PI) or photoactivated disinfection (PAD) (Bonsor et al. 2006; Nagata et al. 2012).

Compared with other cytotoxic therapies, photodynamic therapy has the potential to act in the infected tissue or the target cell, and the light source may relate directly to the injury site (Demidova and Hamblin 2004). Thus, an important feature of photodynamic therapy is the dual selectivity, first through concentration of the photosensitiser by specific binding to target tissue, and second by restricting the irradiation to a specific volume. In photodynamic antibacterial therapy, the photodestruction is primarily caused by damage to the cell membrane and DNA (Schafer et al. 1998; Bertoloni et al. 2000; Romanova et al. 2003; Soukos and Goodson 2011).

A fundamental difference in susceptibility to PDT between Gram-positive and Gram-negative bacteria was recognised in the 1990s (Nitzan et al. 1992). Generally, neutral, anionic and cationic photosensitising molecules can efficiently destroy Gram-positive bacteria, whereas only cationic photosensitisers or strategies, which alter the permeability of Gram-negative bacteria, in combination with non-cationic photosensitisers, are capable of inactivating Gram-negative bacteria. This is due to the presence of a cytoplasmic membrane that is surrounded by a relatively porous cell wall composed of peptidoglycan and lipoteichoic acid in Gram-positive bacteria; this structure allows for the diffusion of the photosensitiser (Nagata et al. 2012). The cell wall of Gram-negative bacteria comprises the inner cytoplasmic membrane and the outer membrane, a periplasmic space that is interspersed with peptidoglycan. The outer membrane forms an effective permeability barrier between the external environment and the cell, binding and limiting the penetration of the photosensitiser (Minnock et al. 2000). Studies have shown that Gram-positive bacteria are more susceptible to photoinactivation (Malik et al. 1992; Bertoloni et al. 1992; Soukos et al. 1998) than Gram-negative bacteria (Nitzan et al. 1995; Soukos et al. 1998).

The fungal cell wall has a relatively thick layer of chitin and beta-glucan that has intermediate permeability between Gram-positive and Gram-negative bacteria (Dai et al. 2009).

3 Mechanism of Action of PDT

Photodynamic therapy is based on the concept that a particular compound, or photosensitiser curing unit, can be preferably located in certain tissues and subsequently activated by light of the appropriate wavelength to generate singlet oxygen and free radicals that are cytotoxic to the microorganisms in the target tissue (Soukos and Goodson 2011). Photodynamic therapy involves two stages: the first stage involves the application of a photosensitising agent, and the second stage involves the application of light directly to the treated area. When light is combined with the photosensitising agent, phototoxic reactions are induced to destroy the microbial cells (Malik et al. 2010).

In this process, photon absorption only occurs when the wavelength of irradiated light belongs to the absorption spectrum of the photosensitive substance. After absorption of light, the photosensitiser ground state goes to the excited singlet state with a short half-life. The compound excited singlet can return to the ground state by emitting light in the form of fluorescence or can pass to a triplet excited state with a long half-life through a process referred to as crossing between systems (Ochsner 1997b; Henderson and Dougherty 1992; Malik et al. 2010). In the excited triplet state, the photosensitiser can undergo two types of reactions: type I and type II. In a type I reaction, there is a transfer of a proton or an electron from the photosensitiser excited triplet state to the substrate or solvent molecules generating an ion radical anion or cation, which reacts with the oxygen in the ground state to form reactive species oxygen. In type II reactions, energy transfer occurs from the photosensitiser triplet state directly to molecular oxygen, generating highly cytotoxic singlet oxygen (Ochsner 1997b; Castano et al. 2004; Henderson and Dougherty 1992). Reactions can occur simultaneously and depend on the type of photosensitiser used and the concentration of the substrate and oxygen (Castano et al. 2004; Soukos and Goodson 2011), as shown in Fig. 2.

There are two basic mechanisms proposed to explain the lethal damage caused by PDT in bacteria: DNA damage and cytoplasmic membrane damage, causing leakage of cellular contents or inactivation of membrane transport systems and enzymes. Studies have reported that treatment of bacteria with different photosensitisers and light causes DNA damage. However, while DNA damage occurs, this is not a major cause of bacterial cell death (Bertoloni et al. 2000; Hamblin and Hasan 2004).

Several microbial cells are susceptible to the photooxidation effect caused by singlet oxygen. The photooxidation effect includes inactivation of enzymes and other proteins and lipid peroxidation, leading to lysis of cell membranes, mitochondria and lysosomes (Gonzales and Maisch 2012).

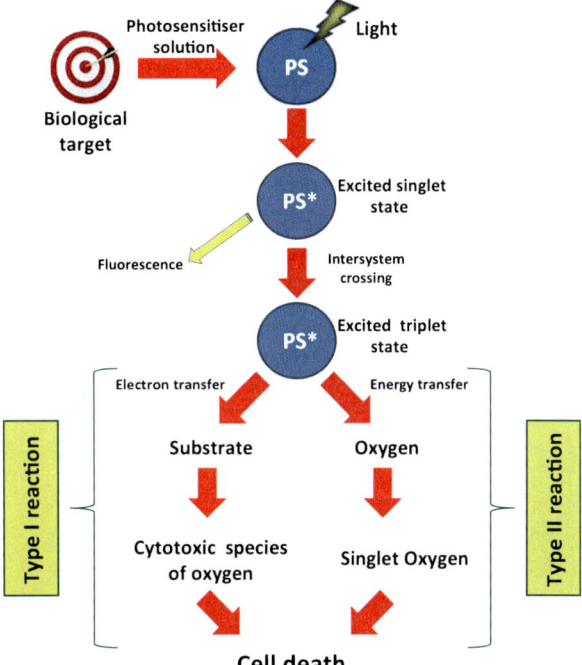

Fig. 2 Mechanism of action of PDT

The mechanism of photodynamic damage in fungal cells results when reactive oxygen species cross and pierce cell walls and membranes, thereby allowing for displacement of the photosensitiser into the cell. Afterward, oxidising species generated by light excitation induce photodestruction of internal organelles and cell death. Thus, singlet oxygen generated by the excitation of the photosensitiser is an oxidising agent that the cell has no specific cellular defence against (Donnelly et al. 2008; Gonzales and Maisch 2012). Antioxidant enzymes, such as catalase and peroxidise, protect against certain reactive oxygen species, but not against singlet oxygen, which inactivates certain antioxidant enzymes, such as catalase (Kim et al. 2001; Gonzales and Maisch 2012).

Thus, the photosensitive inactivation of microorganisms is a complex phenomenon and is dependent on several parameters, including the probability of absorption of the photosensitiser in the outer membrane of the cell, the cell concentration and target location (Canete et al. 1993), the microbial species and the type of photosensitiser, light source and pre-irradiation.

3.1 Singlet Oxygen

Singlet oxygen corresponds to three electronically excited states that are higher than molecular oxygen in the ground state, 3Σ. Thus, the oxygen molecule in its ground state is important for photochemical processes due to its high chemical potential and unique characteristics of reactivity (Turro 1991). According to Molecular Orbital Theory, the electron configuration of oxygen in the ground state has two unpaired electrons in molecular orbital degenerates πx and πy. These electrons tend to have the same maximum spin multiplicity and produce a state of low energy. This is why the ground state of oxygen is a triplet. The lifetime of singlet oxygen in solution is deeply influenced by the nature of the solvent: in water, for example, the lifetime of singlet oxygen is approximately 4.0 μs (Ochsner 1997b; Dougherty et al. 1998). In biological systems, singlet oxygen lifetimes are extremely low, less than 0.04 μs. Therefore, their range is very low, <0.02 μm (Malik et al. 2010).

3.2 Photosensitiser

Initial preparations of photosensitisers for photodynamic therapy were based on a complex mixture of porphyrins, referred to as haematoporphyrin derivatives. Extensive chemical and biological studies have been performed in the past 20 years to identify new photosensitisers that belong to different classes of compounds (Brown et al. 2004; Tardivo et al. 2005).

A photosensitiser is capable of absorbing light of a specific wavelength and transforming it into useful energy. In PDT, the photosensitiser is responsible for the production of cytotoxic agents, their main function being to induce a desired biological effect (Sharman et al. 1999; Lee et al. 2012). The ideal photosensitiser should not be toxic by itself but should exert toxicity only after activation by irradiation, not cause allergic reactions or hypotension, be water soluble, exhibit rapid excretion and be biologically stable. Moreover, such a photosensitiser should be photochemically effective and selective (Meisel and Kocher 2005), have a strong binding affinity for microorganisms, have a low affinity for mammalian cells (avoiding photodestruction of host tissues), have a minimal risk of promoting mutagenic processes and have low chemical toxicity (Jori et al. 2006; Soukos and Goodson 2011).

Over 400 compounds have known photosensitising properties, including dyes, medicines, cosmetics, chemicals and many natural substances (Malik et al. 2010). Antimicrobial photosensitisers such as porphyrins, phthalocyanines and phenothiazines, such as toluidine blue O and methylene blue, have a positive charge and may act directly on both Gram-negative and Gram-positive bacteria (Merchat et al. 1996a; Minnock et al. 1996; Wilson et al. 1995). The positive charge appears to promote binding of the photosensitiser to the bacterial outer membrane, inducing localised damage and favouring penetration (Merchat et al. 1996b). Toluidine blue and methylene blue are commonly used in oral antimicrobial photodynamic therapy. Toluidine blue is vital for staining the mucus abnormalities in the cervix and oral cavity, and, furthermore, defines the extent of injury before excision

(Lingen et al. 2008). Also, it has been shown to be a potent photosensitiser for destruction of oral bacteria (Wilson et al. 1995; Soukos and Goodson 2011).

Methylene blue has been used as a photosensitising agent since 1920 (Wainwright et al. 2007), for the detection of premalignant mucosa (Ojetti et al. 2007) and as a marker dye in surgery (Creagh et al. 1995). The hydrophobicity of methylene blue (Wainwright et al. 1997) along with its low molecular weight and positive charge allows for its passage between the porin protein channels in the outer membrane of Gram-negative bacteria (Usacheva et al. 2003).

Some photosensitisers, such as toluidine blue and methylene blue, were tested in association with red laser low intensity to promote bactericidal effects in vivo (Wong et al. 2005; Komerik et al. 2003). Currently, erythrosine is used clinically as a PDT plaque disclosing agent that induces bacterial cell death (1.5 $\log_{10}$) in *Streptococcus mutans* biofilms in vitro (Wood et al. 2006; Metcalf et al. 2006). Erythrosine has several advantages compared to other photosensitisers, including having no toxicity to the host and being approved for use in food products (Lee et al. 2012).

In a search for effective photosensitisers with absorption in the red band, certain methods were developed with the malachite green dye. This dye is a member of the triarylmethane family, along with crystal violet, and shows strong absorption of red light (Prates et al. 2007). Malachite green has been used in dental practice to visualise dental biofilms and as a colourimetric test to evaluate dental erosions (Attin et al. 2005). This photosensitiser effectively reduced the cell viability of various microorganisms, including *Aggregatibacter actinomycetemcomitans* (Prates et al. 2007), *Staphylococcus*, *Enterobacteriaceae* and *Candida* (Junqueira et al. 2010; Souza et al. 2010). However, all the studies listed above were carried out with planktonic cultures.

Low concentrations of photosensitisers are more suitable for providing low toxicity and high solubility and are unlikely to stain teeth (Nagata et al. 2012).

3.3 Light Source

The light source also influences the effects of PDT, and the high absorption coefficient, concentration of the photosensitiser and light energy flux incident determine the effectiveness of phototherapy (Malik et al. 2010). In the past, the activation of the photosensitiser was obtained through a variety of light sources, such as argon-pumped dye laser, potassium titanyl phosphate (KTP)—or neodymium: yttrium aluminium garnet (Nd/YAG)-pumped dye lasers and gold vapour- or copper vapour-pumped dye lasers. These laser systems were complex and expensive (Kübler 2005).

Currently, the literature presents three main classes of light sources in clinical PDT: laser, LED (light-emitting diode) and halogen lamps. Lasers have certain advantages, such as monochromaticity, high efficiency (>90 %), high power and interstitial light delivery devices; however, they are also expensive. Activation of photosensitisers is also performed by diode laser-emitting light with a specific wavelength. These devices are portable and their cost is much lower than argon lasers. Gallium and aluminium and helium–neon have been used for PDT with diode lasers. Diode lasers are very convenient and reliable, but possess only one wavelength and require a separate drive for each photosensitiser because of the different absorption wavelengths (Soukos and Goodson 2011).

Sources of non-laser light, such as LEDs, have also been applied in PDT (Allison et al. 2004; Juzeniene et al. 2004; Pieslinger et al. 2006; Steiner 2006). LED has become a viable technology for PDT in recent years, particularly for irradiating easily accessible tissue surfaces. LEDs allow a greater irradiation time, are easy to handle and have the convenience of being small and lightweight with lower costs than lasers (Konopka and Goslinski 2007; Chen et al. 2002). The main advantages of LED over the laser sources or laser diode are its low cost and ease of configuration matrices in different irradiation geometries. As with laser diodes, LED has a wavelength of fixed output, but the cost per watt is significantly smaller, so different units can be built for each photosensitiser (Wilson and Patterson 2008).

Halogen lamps may be spectrally filtered to match any photosensitiser but cannot be efficiently coupled into fibre optic bundles and cause heating. As a broadband source, the output power of a halogen lamp is less effective compared to the peak power laser activation of the photosensitiser (Nagata et al. 2012). It is not an ideal light source due to the low-density light power and low-energy fluence of the light. However, this type of lamp is already widely used as a curing unit in dental clinics where other devices are not needed for use as a light source in PDT (Lee et al. 2012).

The conventional lamp has an emission wavelength between 400 and 520 nm, similar to the region of absorption of the photosensitiser erythrosine, 500–550 nm (Vahabi et al. 2011; Bolean et al. 2010).

4 Mechanism of Action of PDT in Oral Biofilm

The oral cavity is colonised by a complex community of microorganisms and specific relatively highly interrelated, aerobic and anaerobic bacteria, including Gram-positive and Gram-negative bacteria, fungi, mycoplasma, protozoa and viruses (Konopka and Goslinski 2007). Thus, oral biofilm, previously referred to as bacterial plaque, consists of complex microbial communities embedded in a polymer matrix of bacteria and saliva and is formed on the surface of the teeth and soft tissues of the oral cavity (Costerton et al. 1999). Biofilm-related diseases such as caries, periodontal disease and chronic oral infections are prevalent, placing oral and general health at risk (Wilson et al. 1996; Meisel and Kocher 2005; Pereira et al. 2011). Biofilm formation is a multistep process involving initial attachment to the surface, cell growth, the formation of microcolonies and finally maturation, resulting in a three-dimensional biofilm (Mang et al. 2012).

The antimicrobial activity of photosensitisers is mediated by singlet oxygen, which, due to its high chemical reactivity, directly affects extracellular molecules. The polysaccharides present in the extracellular matrix of the biofilm are also susceptible to photodamage. This dual activity is not exhibited by antibiotics and is a significant advantage of antimicrobial photodynamic therapy. The disintegration of the biofilm can inhibit the exchange of plasmids involved in the transfer of antibiotic resistance (Konopka and Goslinski 2007).

One of the main pathogenic bacteria and contributor to cariogenic biofilm formation is *Streptococcus mutans* (Kreth et al. 2004). Studies have shown that different species of yeasts and bacteria are associated with biofilm, including *Candida* spp., *Staphylococcus* spp., *Streptococcus* spp., *Lactobacillus* spp., *Pseudomonas* spp., *Enterobacter* spp., and *Actinomyces* spp. (Glass et al. 2001; Ribeiro et al. 2009). The initial colonisation of the tooth or repair of surfaces for various initial colonising bacteria is replaced by later colonisers, such as S. *mutans*, and an increase in the number of anaerobic bacteria that adhere to the tooth enamel or dentin surfaces is then restored (Wei et al. 2006; Mang et al. 2012) (Fig. 3).

The microbial species that comprise the biofilm are highly interactive and use the system of intercellular signalling or *quorum sensing*. This phenomenon promotes collective behaviour in a microbial population by improving access to nutrients and niches as well as promoting collective defence against other competitive organisms (Williams 2007).

A related antifungal resistance and recurrence of infection are seen by *Candida* spp. that also form biofilms (White et al. 1998; Chandra et al. 2001). The main species responsible for most cases of candidiasis is *C. albicans*. Furthermore, this phenotype is less susceptible to antifungal agents compared to their planktonic counterparts (Ramage et al. 2001). For this reason, alternatives are needed for effective treatment of superficial infections caused mainly by cells in biofilms.

Although biofilms formed by fungi exhibit more complex structures than those formed by bacteria, studies indicate that this factor does not affect the photodynamic processes in such cells (Paardekooper et al. 1995; Donnelly et al. 2008). For this reason, the *Candida* cell can be destroyed by anionic photosensitisers (Bliss et al. 2004).

Fungal cells may be inactivated using photosensitisers under the same conditions used for other microbial classes. The phenothiazine photosensitiser toluidine blue has been shown to be highly effective in photodynamic inactivation of *C. albicans*. In contradiction to bacteria, yeasts appear to be susceptible to photosensitisers regardless of load, most likely reflecting differences in the outer cell architecture. However, similarities of yeast with mammalian cells should be considered, especially given that cationic photosensitisers have better uptake by mammalian cells (Wainwright and Crossley 2004).

Studies analysed the effect of antimicrobial photodynamic therapy in biofilms of *C. albicans* mediated by the photosensitiser toluidine blue and lighting with a Paterson lamp (635 nm). Higher concentrations of the photosensitiser associated with increased incubation time were required for the biofilms to achieve microbial reduction equivalent to planktonic cells. The low susceptibility of biofilms

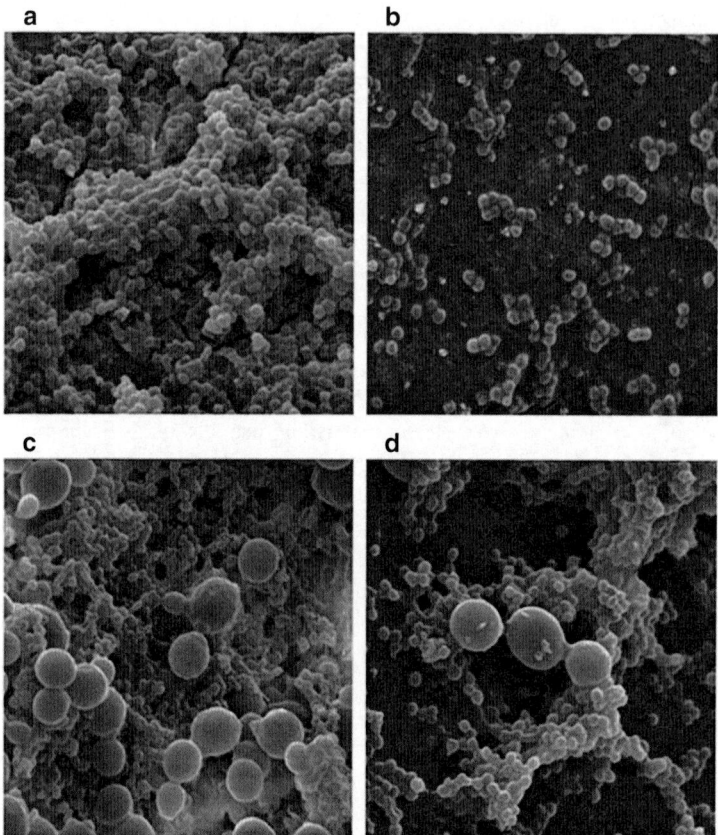

Fig. 3 Scanning electron microscope of different biofilms. Image (**a**) refers to control biofilms of *Staphylococcus aureus* not sensitised with methylene blue (MB) and not exposed to laser light and (**b**) shows biofilms sensitised with MB for 5 min and exposed to laser for 98 s. Image (**c**) refers to control biofilms of *Candida albicans*, *Staphylococcus aureus* and *Streptococcus mutans* not sensitised with MB and not exposed to laser light and (**d**) shows sensitised biofilms with MB for 5 min and exposed to laser for 98 s. Magnification = ×10,000. Partially reproduced from Pereira et al. (2011). Lasers in medical science by European Laser Association, Reproduced with permission of Springer-Verlag, London Ltd

presented by *C. albicans* can be explained by the structural differences between biofilm bacteria and yeast or failure of light to penetrate thick biofilms formed by *Candida* (Lee et al. 2004).

There are differences in physiological and environmental bacteria growing in biofilms compared to planktonic counterparts. In biofilms, bacteria respond best to the antimicrobial methods of nutrient deprivation, pH changes, oxygen radicals, and antibiotics compared to bacteria in the planktonic form (Wei et al. 2006).

Some phenotypic changes may render the bacteria more resistant to environmental changes as they move from the planktonic to the sessile state of biofilm, including, for example, enhanced resistance to acid generated in the formation of dental caries (Welin et al. 2003). During the last decade, there has been interest in the possibility of replacing or supplementing conventional mechanical therapeutic measures to remove biofilms with chemical agents, such as antiseptics or antibiotics. Due to various limitations of antibacterial measures, including antibiotic resistance, attempts are being made to introduce photodynamic therapy as an alternative to antibacterial and mechanical measures (Mang et al. 2012).

Bacteria isolated from the oral cavity, including pathogenic, periodontal and cariogenic bacteria associated with endodontic lesions, are sensitive to PDT (Lee et al. 2012; Giusti et al. 2008; Garcez et al. 2008), showing susceptibility to different therapeutic protocols applied in vitro and in vivo (Bliss et al. 2004; Foschi et al. 2007). Several studies have concluded that photosensitisers combined with suitable light sources are capable of destroying a variety of microorganisms in localised infections, experimental models and planktonic cultures (Meisel and Kocher 2005; Hamblin and Hasan 2004; Araujo et al. 2010).

Bacteria in biofilms are less accessible to antibiotics due to their protection within the polymer matrix and the bacterial adhesion to teeth or epithelia (Vitkov et al. 2002). In vitro susceptibility tests in model biofilms revealed significant microorganism survival after treatment with antibiotics (Biel 2010). The absorption of photosensitisers into the matrix is prevented in the same way as antibiotics. Device ultrasonics or photomechanical waves can improve the absorption efficiency of these substances (Qian et al. 1997). The photodynamic treatment also influences the structure of the biofilm, decreasing layer thickness and biomass (Malik et al. 2010).

The application of PDT in the maintenance treatment of periodontal disease removes biofilms in residual pockets and makes mechanical treatment, by means of scraping the root, no longer necessary. Thus, the patient may experience less tooth hypersensitivity. PDT may decrease the risk of bacteraemia, which usually occurs after periodontal treatment procedures. On the other hand, there is unequivocal evidence demonstrating the risk of periodontal systemic diseases, such as cardiovascular disease and diabetes (Greenwell and Bissada 2002). If antibiotic resistance continues to increase, PDT could be a valuable alternative for most indications in which antibiotics have been administered previously without satisfactory results. The number of immunocompromised patients can generate new challenges for treatment strategies (Malik et al. 2010).

In the future, difficulties with antibiotic therapy may arise due to increased resistance to antibiotics commonly used in periodontics, an increase in the number of immunocompromised patients (Ryder 2002) and periodontal infections caused by various pathogens requiring different antibiotics that cause adverse reactions (Muller et al. 2002).

5 PDT in the Treatment of Oral Biofilms

As one of the leading causes of bacterial infections in humans, biofilms pose a serious problem for health care (Donlan 2001). Professor Michael Wilson and colleagues (Wilson 1993), Eastman Dental Institute, University College London, UK, pioneered the application of photodynamic therapy as an alternative to mechanical and antimicrobial regimens in eliminating microbial species present in the oral biofilm (Soukos and Goodson 2011).

In the literature, there are a number of studies showing a variety of protocols for the use of PDT, but it is necessary to fully analyse the properties of photosensitisers and light sources used in dentistry to develop a successful treatment.

Lee and colleagues (2012) confirmed the positive effect of PDT in the reduction of biofilms formed by *S. mutans* using erythrosine and halogen light. Thus, a significant reduction of biofilm formation of *S. mutans* in response to PDT could be obtained in most dental offices for no expense given that erythrosine and halogen light are conventionally used in dental offices. Four treatment conditions were established: no photosensitiser or irradiation (control), photosensitiser alone, photosensitiser and irradiation and irradiation alone. It was observed that only the combination treatment resulted in significant increases in microbial destruction, with rates of 75 % and 55 % after 8 h of incubation and 74 % and 42 % after 12 h of incubation for biofilms in a brain heart infusion broth supplemented with 0 % or 0.1 % sucrose, respectively.

Mang et al. (2012) evaluated the effect of PDT using 25–125 µg/mL porfimer sodium with a photosensitiser laser (light source) at 630 nm for 5 min in the treatment of localised infections caused by *S. mutans* biofilms. The authors demonstrated that there was a significant reduction of *S. mutans*. Maximum efficiency was observed when the biofilms were exposed to a combination of the photosensitiser and light. Porfimer sodium at 25 µg/mL with an incubation time shorter than 5 min (30 J/cm^2) resulted in a significant reduction in the viability of bacteria in biofilms. Optimal parameters were obtained at a concentration of 125 µg/mL with an incubation of 5 min (60 J/cm^2). From the results of this study, it was concluded that the microbial reduction was significant even when the bacteria were incorporated into an extracellular matrix because the photosensitiser was combined with the appropriate wavelength of emitted light.

Studies performed by Li et al. (2013) evaluated the efficacy of 5-aminolevulinic acid (ALA) associated with the laser (0, 100, 200 and 300 J/cm^2) in photodynamic therapy in biofilms formed by *S. aureus* and *S. epidermidis* resistant to methicillin. The treatment showed great potential for elimination of biofilm strains resistant to methicillin, dependent on the density of light energy. Also, as a natural precursor of protoporphyrin IX [PpIX], ALA can be used in the treatment of infectious diseases through local, systemic and oral administration.

Garcez et al. (2007) developed a real-time method using bioluminescent bacteria and a camera that provided low light images to evaluate the antimicrobial effects in the treatment of root canal infections caused by *Proteus mirabilis* and

P. aeruginosa biofilms. These authors quantitatively compared conventional endodontic treatment with PDT combination. For the PDT treatment, conjugated polyethylenimine and chlorine (e6) were used as photosensitisers and a 660 nm laser was emitted into the root canal by a 200 µm optical fibre. The PDT treatment was compared and combined with standard endodontic treatment, mechanical debridement and antiseptic irrigation. Endodontic therapy alone reduced bacterial bioluminescence by 90 % and PDT treatment alone reduced it by 95 %. The combination of these treatments caused greater than 98 % reduction and showed lower bacterial growth after 24 h compared to either treatment alone. The results suggested efficacy in the use of PDT as an adjunct to conventional endodontic treatment.

Eick et al. (2013) studied the effect of photoactivated disinfection (PAD) using toluidine blue and LED (625–635 nm) in species of microorganisms associated with periodontal, peri-implant and periodontopathic biofilm-forming bacteria. Sixteen microbial species, including *P. gingivalis* and *A. actinomycetemcomitans*, and a mixture consisting of 12 species suspended in saline, with or without 25 % human serum, were exposed to photoactivation. In addition, monotypic biofilms, consisting of *P. gingivalis* and *A. actinomycetemcomitans*, and heterotypic biofilms were grown on titanium discs in 24-well plates, and artificial periodontal pockets were exposed to PAD with or without pre-treatment with 0.25 % hydrogen peroxide. Analysing the results together, the authors concluded that the photoactivated disinfection with LED is effective against periodontopathic microbial species, even in the presence of serum. PAD with or without hydrogen peroxide reduced the viability of monotypic biofilms, while heterotypic biofilms were less sensitive. In this study, we observed that complete elimination of multi-species biofilms with PAD is unlikely, highlighting the importance of prior mechanical removal. We also observed increased antimicrobial activity after photoactivated disinfection with hydrogen peroxide, indicating the relevant potential for this method in adjunctive antimicrobial treatment of periodontal infections and peri-implants.

Andrade et al. (2013) evaluated the effects of pre-irradiation time (PIT) in photodynamic therapy mediated by curcumin (CUR) in planktonic and biofilm cultures of strains of *C. albicans*, *C. glabrata* and *C. dublinienses*. Suspensions and biofilms of *Candida* spp. remain in contact with different concentrations of CUR at time intervals of 1, 5, 10 and 20 min prior to irradiation and activation by LED. Control samples received no light or CUR. After PDT, the suspensions were plated on Sabouraud dextrose agar, and the plaque results were obtained using the XTT reduction method. Different PIT showed no significant differences in PDT-mediated CUR suspensions of *Candida* spp. There was complete inactivation of the three species of *Candida* with the combination of 20.0 mM CUR after 5, 10 and 20 min of PIT. The biofilms showed a significant reduction in cell viability after PDT. Generally, the three *Candida* species evaluated showed greater reductions in cell viability with 40.0 mM CUR and 20 min of PIT.

Vilela and colleagues (2012) compared the action of malachite green with the phenothiazine photosensitisers methylene blue and toluidine blue (concentrations ranging from 37.5 to 3,000 µM) in biofilms of *S. aureus* and *Escherichia coli* with a

660 nm laser diode. The authors concluded that the most significant microbial reduction was achieved with the photosensitiser malachite green at higher concentrations than those used for the phenothiazine dyes. However, to establish the safety of malachite green as a photosensitiser for PDT, studies in human cells are necessary.

The environment of the oral cavity is completely different from laboratory culture or an in vitro environment, making it difficult to provide an ideal condition for the study of PDT. However, despite these limitations, in general, the research shows promising results in this field.

6 Advantages of PDT

The use of antimicrobial photodynamic therapy has advantages such as the following:

- The lack of genotoxicity and mutagenicity to microorganisms and human cells, favouring long-term safety (Gonzales and Maisch 2012).
- Photosensitiser and subsequent localised reactions do not damage the surrounding tissue or harm the resident microbiota in the tissue (Alvarez et al. 2012).
- Possibility of reduced treatment time, even in individuals more susceptible to infections and particularly prone to develop resistance (Gonzales and Maisch 2012).
- Not only inactivation of microorganisms inhabiting the biofilm but also the biofilm structures (Kishen et al. 2010; Collins et al. 2010; Saino et al. 2010).
- Direct action on extracellular molecules, such as polysaccharides present in the extracellular polymeric substances, inhibits the exchange of plasmids and, thus, the transfer of antibiotic resistance, avoiding new colonisation and preventing recurrence of infection due to the high chemical reactivity of singlet oxygen and other reactive oxygen species (Wainwright and Crossley 2004).
- Do not induce resistance, so repeated applications of PDT can be performed if treatment is not sufficient to disrupt biofilm structures and inactivate cells (Nagata et al. 2012).
- Easy access to superficial infections (Gonzales and Maisch 2012).
- Ability of reactive oxygen species in inactivating virulence factors secreted by microorganisms, especially proteins that can cause tissue damage and may remain even after efficient microbial reduction (Yordanov et al. 2008).

7 Conclusion

This chapter sought to provide the reader with a description of the use of photodynamic therapy in oral biofilms by presenting an overview of the concept and mechanism of the action of PDT. There have been studies reported on oral biofilms for the development of therapeutic agents that block their formation or promote the disintegration of the biofilm microbial community to control the growth of microorganisms in the oral cavity. Thus, PDT may be an alternative to conventional periodontal therapeutic methods. Although PDT is still in the experimental stage of development and testing, the method can be an adjunct to conventional antibacterial measures in periodontology in places of difficult access for mechanical treatment (Malik et al. 2010). PDT is a safe, minimally invasive and non-toxic treatment to control biofilm formation (Lee et al. 2012; Soukos and Goodson 2011).

It is indeed hoped that this therapy will be applied for future clinical use in dentistry. The success of PDT depends on the choice of the most appropriate dose of the photosensitiser and the optimal administration time of the photosensitiser after light activation.

References

Ackroyd R, Kelty C, Brown N, Reed M (2001) The history of photodetection and photodynamic therapy. Photochem Photobiol 74:656–669

Allison RR, Mota HC, Sibata CH (2004) Clinical PD/PDT in North America: an historical review. Photodiagnosis Photodyn Ther 1:263–277

Alvarez MG, Gómez ML, Mora SJ, Milanesio ME, Durantini EM (2012) Photodynamic inactivation of *Candida albicans* using bridged polysilsesquioxane films doped with porphyrin. Bioorg Med Chem 20:4032–4039

Andrade MC, Ribeiro AP, Dovigo LN, Brunetti IL, Giampaolo ET, Bagnato VS, Pavarina AC (2013) Effect of different pre-irradiation times on curcumin-mediated photodynamic therapy against planktonic cultures and biofilms of *Candida* spp. Arch Oral Biol 58:200–210

Araujo PV, Cortes ME, de Abreu Poletto LT (2010) Photodynamic therapy of cariogenic agents: a systematic review. J Laser Appl 22:13–24

Attin T, Becker K, Hannig C, Buchalla W, Wiegand A (2005) Suitability of a malachite green procedure to detect minimal amounts of phosphate dissolved in acidic solutions. Clin Oral Investig 9:203–207

Bell SG (2003) Antibiotic resistance: is the end of an era near? Neonatal Netw 22:47–54

Bertoloni G, Rossi F, Valduga G, Jori G, Ali H, van Lier JE (1992) Photosensitizing activity of water- and lipid-soluble phthalocyanines on prokaryotic and eukaryotic microbial cells. Microbios 71:33–46

Bertoloni G, Lauro FM, Cortella G, Merchat M (2000) Photosensitizing activity of hematoporphyrin on *Staphylococcus aureus* cells. Biochim Biophys Acta 1475:169–174

Biel MA (2010) Photodynamic therapy of bacterial and fungal biofilm infections. Methods Mol Biol 635:175–194

Bliss JM, Bigelow CE, Foster TH, Haidaris CG (2004) Susceptibility of Candida species to photodynamic effects of photofrin. Antimicrob Agents Chemother 48:2000–2006

Bolean M, Paulino TP, Thedei G Jr, Ciancaglini P (2010) Photodynamic therapy with rose bengal induces GroEL expression in *Streptococcus mutans*. Photomed Laser Surg 28:S79–S84

Bonsor SJ, Nichol R, Reid TM, Pearson GJ (2006) An alternative regimen for root canal disinfection. Br Dent J 201:101–105

Brown SB, Brown EA, Walker I (2004) The present and the future role of photodynamic therapy in cancer treatment. Lancet Oncol 5:497–508

Canete M, Villanueva A, Juarranz A (1993) Uptake and photoeffectiveness of two thiazines in HeLa cells. Anticancer Drug Des 8:471–477

Castano AP, Demidova TN, Hamblin MR (2004) Mechanisms in photodynamic therapy: part one—photosensitizers, photochemistry and cellular localization. Photodiagnosis Photodyn Ther 1:279–293

Chandra J, Kuhn DM, Mukherjee PK, Hoyer LL, McCormick T, Ghannoum MA (2001) Biofilm formation by the fungal pathogen *Candida albicans*: development, architecture, and drug resistance. J Bacteriol 183:5385–5394

Chen J, Keltner L, Christophersen J, Zheng F, Krouse M, Singhal A, Wang SS (2002) New technology for deep light distribution in tissue for phototherapy. Cancer J 8:154–163

Collins TL, Markus EA, Hassett DJ, Robinson JB (2010) The effect of a cationic porphyrin on *Pseudomonas aeruginosa* biofilms. Curr Microbiol 61:411–416

Costerton JW, Stewart PS, Greenberg EP (1999) Bacterial biofilms: a common cause of persistent infections. Science 284:1318–1322

Creagh TA, Gleeson M, Travis D, Grainger R, McDermott TE, Butler MR (1995) Is there a role for in vivo methylene blue staining in the prediction of bladder tumour recurrence? Br J Urol 75:477–479

Dai T, Huang YY, Hamblin MR (2009) Photodynamic therapy for localized infections—state of the art. Photodiagnosis Photodyn Ther 6:170–188

Demidova TN, Hamblin MR (2004) Photodynamic therapy targeted to pathogens. Int J Immunopathol Pharmacol 17:245–254

Donlan RM (2001) Biofilm formation: a clinically relevant microbiological process. Clin Infect Dis 33:1387–1392

Donnelly RF, McCarron PA, Tunney MM (2008) Antifungal photodynamic therapy. Microbiol Res 163:1–12

Dougherty TJ, Marcus SL (1992) Photodynamic therapy. Eur J Cancer 28A:1734–1742

Dougherty TJ, Gomer CJ, Henderson BW, Jori G, Kessel D, Korbelik M, Moan J, Peng Q (1998) Photodynamic therapy. J Natl Cancer Inst 90:889–905

Eick S, Markauskaite G, Nietzsche S, Laugisch O, Salvi GE, Sculean A (2013) Effect of photoactivated disinfection with a light-emitting diode on bacterial species and biofilms associated with periodontitis and peri-implantitis. Photodiagnosis Photodyn Ther 10:156–167, http://dx.doi.org/10.1016/j.pdpdt.2012.12.001

Feres M, Haffajee AD, Allard K, Som S, Goodson JM, Socransky SS (2002) Antibiotic resistance of subgingival species during and after antibiotic therapy. J Clin Periodontol 29:724–735

Foschi F, Fontana CR, Ruggiero K, Riahi R, Vera A, Doukas AG, Pagonis TC, Kent R, Stashenko PP, Soukos NS (2007) Photodynamic inactivation of *Enterococcus faecalis* in dental root canals in vitro. Lasers Surg Med 39:782–787

Garcez AS, Ribeiro MS, Tegos GP, Nunez SC, Jorge AO, Hamblin MR (2007) Antimicrobial photodynamic therapy combined with conventional endodontic treatment to eliminate root canal biofilm infection. Lasers Surg Med 39:59–66

Garcez AS, Nunez SC, Hamblin MR, Ribeiro MS (2008) Antimicrobial effects of photodynamic therapy on patients with necrotic pulps and periapical lesion. J Endod 34:138–142

Giusti JS, Santos-Pinto L, Pizzolito AC, Helmerson K, Carvalho-Filho E, Kurachi C, Bagnato VS (2008) Antimicrobial photodynamic action on dentin using a light-emitting diode light source. Photomed Laser Surg 26:281–287

Glass RT, Bullard JW, Hadley CS, Mix EW, Conrad RS (2001) Partial spectrum of microorganisms found in dentures and possible disease implications. J Am Osteopath Assoc 101:92–94

Gonzales FP, Maisch T (2012) Photodynamic inactivation for controlling *Candida albicans* infections. Fungal Biol 116:1–10

Greenwell H, Bissada NF (2002) Emerging concepts in periodontal therapy. Drugs 62:2581–2587

Hamblin MR, Hasan T (2004) Photodynamic therapy: a new antimicrobial approach to infectious disease? Photochem Photobiol Sci 3:436–450

Harrison JW, Svec TA (1998) The beginning of the end of the antibiotic era? Part II. Proposed solutions to antibiotic abuse. Quintessence Int 29:223–229

Henderson BW, Dougherty TJ (1992) How does photodynamic therapy work? Photochem Photobiol 55:145–157

Jori G, Fabris C, Soncin M, Ferro S, Coppellotti O, Dei D, Fantetti L, Chiti G, Roncucci G (2006) Photodynamic therapy in the treatment of microbial infections: basic principles and perspective applications. Lasers Surg Med 38:468–481

Junqueira JC, Ribeiro MA, Rossoni RD, Barbosa JO, Querido SM, Jorge AO (2010) Antimicrobial photodynamic therapy: photodynamic antimicrobial effects of malachite green on *Staphylococcus*, *Enterobacteriaceae*, and *Candida*. Photomed Laser Surg 28:S67–S72

Juzeniene A, Juzenas P, Ma LW, Iani V, Moan J (2004) Effectiveness of different light sources for 5-aminolevulinic acid photodynamic therapy. Lasers Med Sci 19:139–149

Kim SY, Kwon OJ, Park JW (2001) Inactivation of catalase and superoxide dismutase by singlet oxygen derived from photoactivated dye. Biochimie 83:437–444

Kishen A, Upadya M, Tegos GP, Hamblin MR (2010) Efflux pump inhibitor potentiates antimicrobial photodynamic inactivation of *Enterococcus faecalis* biofilm. Photochem Photobiol 86:1343–1349

Komerik N, Nakanishi H, MacRobert AJ, Henderson B, Speight P, Wilson M (2003) In vivo killing of *Porphyromonas gingivalis* by toluidine blue-mediated photosensitization in an animal model. Antimicrob Agents Chemother 47:932–940

Konopka K, Goslinski T (2007) Photodynamic therapy in dentistry. J Dent Res 86:694–707

Kreth J, Hagerman E, Tam K, Merritt J, Wong DT, Wu BM, Myung MV, Shi W, Qi F (2004) Quantitative analyses of *Streptococcus mutans* biofilms with quartz crystal microbalance, microjet impingement and confocal microscopy. Biofilms 1:277–284

Kübler AC (2005) Photodynamic therapy. Med Laser Appl 20:37–45

Kübler AC, Scheer M, Zöller JE (2001) Photodynamic therapy of head and neck cancer. Onkologie 224:230–237

Lee CF, Lee CJ, Chen CT, Huang CT (2004) delta-Aminolaevulinic acid mediated photodynamic antimicrobial chemotherapy on *Pseudomonas aeruginosa* planktonic and biofilm cultures. J Photochem Photobiol B 75:21–25

Lee YH, Park HW, Lee JH, Seo HW, Lee SY (2012) The photodynamic therapy on *Streptococcus mutans* biofilms using erythrosine and dental halogen curing unit. Int J Oral Sci 4:196–201

Li X, Guo H, Tian Q, Zheng G, Hu Y, Fu Y, Tan H (2013) Effects of 5-aminolevulinic acid-mediated photodynamic therapy on antibiotic-resistant staphylococcal biofilm: an in vitro study. J Surg Res. doi:10.1016/j.jss.2013.03.094

Lingen MW, Kalmar JR, Karrison T, Speight PM (2008) Critical evaluation of diagnostic aids for the detection of oral cancer. Oral Oncol 44:10–22

Lui H, Anderson RR (1992) Photodynamic therapy in dermatology: shedding a different light on skin disease. Arch Dermatol 128:1631–1636

Malik Z, Ladan H, Nitzan Y (1992) Photodynamic inactivation of gram-negative bacteria: problems and possible solutions. J Photochem Photobiol B 14:262–266

Malik R, Manocha A, Suresh DK (2010) Photodynamic therapy—a strategic review. Indian J Dent Res 21:285–291

Manch-Citron JN, Lopez GH, Dey A, Rapley JW, MacNeill SR, Cobb CM (2000) PCR monitoring for tetracycline resistance genes in subgingival plaque following site-specific periodontal therapy: a preliminary report. J Clin Periodontol 27:437–446

Mang TS, Tayal DP, Baier R (2012) Photodynamic therapy as an alternative treatment for disinfection of bacteria in oral biofilms. Lasers Surg Med 44:588–596

Meisel P, Kocher T (2005) Photodynamic therapy for periodontal diseases: state of the art. J Photochem Photobiol B 79:159–170

Merchat M, Bertolini G, Giacomini P, Villanueva A, Jori G (1996a) Meso-substituted cationic porphyrins as efficient photosensitizers of gram-positive and gram-negative bacteria. J Photochem Photobiol B 32:153–157

Merchat M, Spikes JD, Bertoloni G, Jori G (1996b) Studies on the mechanism of bacteria photosensitization by mesosubstituted cationic porphyrins. J Photochem Photobiol B 35:149–157

Metcalf D, Robinson C, Devine D, Wood S (2006) Enhancement of erythrosine-mediated photodynamic therapy of *Streptococcus mutans* biofilms by light fractionation. J Antimicrob Chemother 58:190–192

Minnock A, Vernon DI, Schofield J, Griffiths J, Parish JH, Brown ST (1996) Photoinactivation of bacteria. Use of a cationic water-soluble zinc phthalocyanine to photoinactivate both gram-negative and gram-positive bacteria. J Photochem Photobiol B 32:159–164

Minnock A, Vernon DI, Schofield J, Griffiths J, Parish JH, Brown SB (2000) Mechanism of uptake of a cationic water-soluble pyridinium zinc phthalocyanine across the outer membrane of *Escherichia coli*. Antimicrob Agents Chemother 44:522–527

Muller HP, Holderrieth S, Burkhardt U, Hoffler U (2002) *In vitro* antimicrobial susceptibility of oral strains of *Actinobacillus actinomycetemcomitans* to seven antibiotics. J Clin Periodontol 29:736–742

Nagata JY, Hioka N, Kimura E, Batistela VR, Terada RS, Graciano AX, Baesso ML, Hayacibara MF (2012) Antibacterial photodynamic therapy for dental caries: evaluation of the photosensitizers used and light source properties. Photodiagnosis Photodyn Ther 9:122–131

Nitzan Y, Gutterman M, Malik Z, Ehrenberg B (1992) Inactivation of gram-negative bacteria by photosensitized porphyrins. Photochem Photobiol 55:89–96

Nitzan Y, Dror R, Ladan H, Malik Z, Kimel S, Gottfried V (1995) Structure-activity relationship of porphines for photoinactivation of bacteria. Photochem Photobiol 62:342–347

Ochsner M (1997a) Photodynamic therapy: the clinical perspective. Review on applications for control of diverse tumorous and non-tumorous diseases. Arzneimittelforschung 47:1185–1194

Ochsner M (1997b) Photophysical and photobiological processes in the photodynamic therapy of tumours. J Photochem Photobiol B 39:1–18

Ojetti V, Persiani R, Nista EC, Rausei S, Lecca G, Migneco A, Cananzi FC, Cammarota G, D'Ugo D, Gasbarrini G, Gasbarrini A (2007) A case-control study comparing methylene blue directed biopsies and random biopsies for detecting pre-cancerous lesions in the follow-up of gastric cancer patients. Eur Rev Med Pharmacol Sci 11:291–296

Paardekooper M, Van Gompel AE, Van Steveninck J, Van den Broek PJ (1995) The effect of photodynamic treatment of yeast with the sensitizer chloroaluminumphthalocyanine on various cellular parameters. Photochem Photobiol 62:561–567

Pereira CA, Romeiro RL, Costa AC, Machado AK, Junqueira JC, Jorge AO (2011) Susceptibility of *Candida albicans*, *Staphylococcus aureus* and *Streptococcus mutans* biofilms to photodynamic inactivation: an in vitro study. Lasers Med Sci 26:341–348

Pervaiz S (2001) Reactive oxygen-dependent production of novel photochemotherapeutic agents. FASEB J 15:612–617

Pieslinger A, Plaetzer K, Oberdanner CB, Berlanda J, Mair H, Krammer B, Kiesslich T (2006) Characterization of a simple and homogenous irradiation device based on light-emitting diodes: a possible low-cost supplement to conventional light sources for photodynamic treatment. Med Laser Appl 21:277–283

Prates RA, Yamada AM Jr, Suzuki LC, Eiko Hashimoto MC, Cai S, Gouw-Soares S, Gomes L, Ribeiro MS (2007) Bactericidal effect of malachite green and red laser on *Actinobacillus actinomycetemcomitans*. J Photochem Photobiol B 86:70–76

Qian Z, Sagers RD, Pitt WG (1997) The effect of ultrasonic frequency upon enhanced killing of *P. aeruginosa* biofilms. Ann Biomed Eng 25:69–76

Ramage G, Vandewalle K, Wickes BL, López-Ribot JL (2001) Characteristics of biofilm formation by *Candida albicans*. Rev Iberoam Micol 18:163–170

Ribeiro DG, Pavarina AC, Dovigo LN, Palomari Spolidorio DM, Giampaolo ET, Vergani CE (2009) Denture disinfection by microwave irradiation: a randomized clinical study. J Dent 37:666–672

Romanova NA, Brovko LY, Moore L, Pometun E, Savitsky AP, Ugarova NN, Griffiths MW (2003) Assessment of photodynamic destruction of *Escherichia coli* O157:H7 and *Listeria monocytogenes* by using ATP bioluminescence. Appl Environ Microbiol 69:6393–6398

Ryder MI (2002) An update on HIV and periodontal disease. J Periodontol 73:1071–1078

Saino E, Sbarra MS, Arciola CR, Scavone M, Bloise N, Nikolov P, Ricchelli F, Visai L (2010) Photodynamic action of Tri-meso (N-methylpyridyl), meso (N-tetradecyl-pyridyl) porphine on *Staphylococcus epidermidis* biofilms grown on Ti6Al4V alloy. Int J Artif Organs 33:636–645

Schafer M, Schmitz C, Horneck G (1998) High sensitivity of *Deinococcus radiodurans* to photodynamically-produced singlet oxygen. Int J Radiat Biol 74:249–253

Sharman WM, Allen CM, van Lier JE (1999) Photodynamic therapeutics: basic principles and clinical applications. Drug Discov Today 4:507–517

Simplicio FI, Maionchi F, Hioka N (2002) Terapia fotodinâmica: aspectos farmacológicos, aplicações e avanços recentes no desenvolvimento de medicamentos. Quim Nova 25:801–807

Soukos NS, Goodson JM (2011) Photodynamic therapy in the control of oral biofilms. Periodontology 2000 55:143–166

Soukos NS, Ximenez-Fyvie LA, Hamblin MR, Socransky SS, Hasan T (1998) Targeted antimicrobial photochemotherapy. Antimicrob Agents Chemother 42:2595–2601

Souza RC, Junqueira JC, Rossoni RD, Pereira CA, Munin E, Jorge AO (2010) Comparison of the photodynamic fungicidal efficacy of methylene blue, toluidine blue, malachite green and low-power laser irradiation alone against *Candida albicans*. Lasers Med Sci 25:385–389

Steiner R (2006) New laser technology and future applications. Med Laser Appl 21:131–140

Sternberg ED, Dolphin D (1998) Porphyrin-based photosensitizers for use in photodynamic therapy. Tetrahedron 54:4151–4202

Tardivo JP, Giglio AD, de Oliveira CS, Gabrielli DS, Junqueira HC, Tada DB, Severino D, Turchiello RF, Baptista MS (2005) Methylene blue in photodynamic therapy: from basic mechanisms to clinical applications. Photodiagnosis Photodyn Ther 2:175–191

Turro NJ (1991) Modern molecular photochemistry. University Science, Sausalito, CA

Usacheva MN, Teichert MC, Biel MA (2003) The interaction of lipopolysaccharides with phenothiazine dyes. Lasers Surg Med 33:311–319

Vahabi S, Fekrazad R, Ayremlou S, Taheri S, Zangeneh N (2011) The effect of antimicrobial photodynamic therapy with radachlorin and toluidine blue on *Streptococcus mutans*: an *in vitro* study. J Dent 8:48–54

Vilela SF, Junqueira JC, Barbosa JO, Majewski M, Munin E, Jorge AO (2012) Photodynamic inactivation of *Staphylococcus aureus* and *Escherichia coli* biofilms by malachite green and phenothiazine dyes: an in vitro study. Arch Oral Biol 57:704–710

Vitkov L, Hannig M, Krautgartner WD, Fuchs K (2002) Bacterial adhesion to sulcular epithelium in periodontitis. FEMS Microbiol Lett 211:239–246

Wainwright M, Crossley KB (2004) Photosensitising agents-circumventing resistance and breaking down biofilms: a review. Int Biodeterior Biodegradation 53:119–126

Wainwright M, Phoenix DA, Marland J, Wareing DR, Bolton FJ (1997) A study of photobactericidal activity in the phenothiazinium series. FEMS Immunol Med Microbiol 19:75–80

Wainwright M, Mohr H, Walker WH (2007) Phenothiazinium derivatives for pathogen inactivation in blood products. J Photochem Photobiol B 86:45–58

Walker CB (1996) The acquisition of antibiotic resistance in the periodontal microflora. Periodontol 2000(10):79–88

Wei GX, Campagns AN, Bobek LA (2006) Effect of MUC7 peptides on the growth of bacteria and on *Streptococcus mutans* biofilm. J Antimicrob Chemother 57:1100–1109

Welin J, Wilkins JC, Beighton D, Wrzesinski K, Fey SJ, Mose-Larsen P, Hamilton IR, Svensater G (2003) Effect of acid shock on protein expression by biofilm cells of *Streptococcus mutans*. FEMS Microbiol Lett 227:287–293

White TC, Marr KA, Bowden RA (1998) Clinical, cellular, and molecular factors that contribute to antifungal drug resistance. Clin Microbiol Rev 11:382–402

Williams P (2007) Quorum sensing, communication and cross-kingdom signalling in the bacterial world. Microbiology 153:3923–3938

Wilson M (1993) Photolysis of oral bacteria and its potential use in the treatment of caries and periodontal disease. J Appl Bacteriol 75:299–306

Wilson BC, Patterson MS (2008) The physics, biophysics and technology of photodynamic therapy. Phys Med Biol 53:R61–R109

Wilson M, Burns T, Pratten J, Pearson GJ (1995) Bacteria in supragingival plaque samples can be killed by low-power laser light in the presence of a photosensitizer. J Appl Bacteriol 78:569–574

Wilson M, Burns T, Pratten J (1996) Killing of *Streptococcus sanguis* in biofilms using a light-activated antimicrobial agent. J Antimicrob Chemother 37:377–381

Wong TW, Wang YY, Sheu HM, Chuang YC (2005) Bactericidal effects of toluidine blue-mediated photodynamic action on *Vibrio vulnificus*. Antimicrob Agents Chemother 49:895–902

Wood S, Metcalf D, Devine D, Robinson C (2006) Erythrosine is a potential photosensitizer for the photodynamic therapy of oral plaque biofilms. J Antimicrob Chemother 57:680–684

Yordanov M, Dimitrova P, Patkar S, Saso L, Ivanovska N (2008) Inhibition of *Candida albicans* extracellular enzyme activity by selected natural substances and their application in *Candida* infection. Can J Microbiol 54:435–440

Clinical and Regulatory Development of Antibiofilm Drugs: The Need, the Potential, and the Challenges

Brett Baker, Patricia A. McKernan, and Fred Marsik

Abstract Escalating rates of mortality and morbidity associated with chronic, recurrent, persistent, and increasingly antibiotic-resistant bacterial infections have generated an extremely urgent unmet need for new antibiotics, particularly those with new mechanisms and targets of action. Over 80 % of all infections are associated with biofilms, a growth condition that not only increases their resistance to currently available antibiotics but also enhances their capability for evading many host defenses. Targeting biofilms may therefore be one of the most important new strategies available for the development of novel antibacterials. In spite of this, biofilms remain underappreciated as targets by the current global healthcare system. Deterrents to their clinical development are primarily associated with the antimicrobial regulatory approval process and insufficient financial incentives offered to pharmaceutical developers—issues which are just now becoming recognized by governments and professional health associations around the world. To facilitate the development of antibiofilm drugs, it is critical that infectious disease stakeholders engage in an ongoing dialog with regulatory agencies to develop standardized methods with respect to (a) the rapid, effective diagnostic assessment of biofilm-related infections and (b) the clinical assessment of antibiofilm drug efficacy. It is also imperative that regulatory agencies are willing to exercise maximum flexibility in approving these innovative and much needed new drugs.

B. Baker (✉) • P.A. McKernan
Microbion Corporation, 1102 West Babcock, Suite B, Bozeman, MT 59715, USA

F. Marsik
Microbiology Consultant, formerly Microbiology Team Leader, FDA, Division of Antiinfective Products, New Freedom, PA 17349, USA

1 Microbial Biofilms: Age-Old Life Form, Cutting-Edge Science, Key Component of Antibiotic Resistance

Bacteria living in the highly resistant biofilm phenotype are estimated to account for over 99 % of all bacteria and approximately half of the biomass on Earth. With the fossil record extending billions of years into Earth's past, microbial biofilms are apparently the most successful, persistent form of life that this planet has known. Even so, our knowledge of biofilms, particularly with respect to health-related matters, has only been collected very recently. Over the past two decades, the nascent science of medical biofilms has benefited from an explosion of research activity. Over 98 % of PubMed's several thousand publications on biofilms have been published since 1994 (Bjarnsholt 2013). This impressive body of biofilm research is now growing at a rapid pace, with over 2000 new scientific journal articles being published each year (ibid). Many of these studies have firmly demonstrated the strong relationship between microbial biofilms, antibiotic-resistant infections, and chronic, recurrent, and persistent infections.

Biofilm resistance and tolerance to antibiotics has been described as follows (Ciofu and Tolker-Nielsen 2011):

> One of the most important features of microbial biofilms is their tolerance to antimicrobial agents and components of the host immune system. The difficulty of treating biofilm infections with antibiotics is a major clinical problem. Although antibiotics may decrease the number of bacteria in biofilms, they will not completely eradicate the bacteria in vivo which may have important clinical consequences in the form of relapses of the infection. Therefore, common antibiotic regimes for treating biofilm-associated infections imply the removal of infected tissues or implanted devices (if possible) associated with long-term anti-microbial therapy. In some cases chronic suppressive therapy with antibiotics may be necessary.

> The difficulty of treatment of biofilm infections involves the inefficacy of the immune system to eradicate the biofilm-embedded microorganisms as well as the recalcitrance of biofilms to antimicrobial therapy.

The recalcitrance of biofilms to antimicrobial drugs is related to a range of mechanisms including (a) restricted penetration of antimicrobial drugs, (b) differential physiological activity within different regions of biofilms, (c) development of persister cells and phenotypic variants, (d) genetic tolerance mechanisms conferred by gene products produced specifically in biofilms, and (e) other tolerance mechanisms that are not related specifically to biofilms, such as the production of antibiotic-degrading enzymes (Ciofu and Tolker-Nielsen 2011; Fux et al. 2005). In addition to the combined contributions of these various mechanisms of antibiotic tolerance, biofilms themselves are hypermutable, leading to de novo development of antibiotic resistance within biofilms which may not have initially contained antibiotic-resistant bacteria (Ciofu and Tolker-Nielsen 2011).

Biofilm infections frequently lead to recurrent infections, chronic/persistent infections, amputations, reoperations, implant revisions, disabilities, and particularly in elderly or immune-compromised populations, death. Clearly these infections contribute an extremely heavy burden to overall healthcare costs. The cost in human terms is unacceptable.

2 Lag in Antibiofilm Drug Development Despite Acute Need for New Anti-infectives

Typically there is a lag time between the development of a convincing new body of health-related science—in this case the role of biofilms in infections—and the development of drugs to address the newly identified therapeutic targets. The lack of development of new drugs to deal with biofilm-related infections is taking place concurrently with four serious global conditions and unmet needs, creating a 'perfect storm' that is subjecting public health on a global scale to unnecessary and increasingly life-threatening health consequences.

First, over the past 20 years, a steadily diminishing number of antibacterial drug products have been developed and approved, due to a combination of very demanding regulatory requirements (and associated costs) and insufficient financial incentives for pharmaceutical companies. For many years the Food and Drug Administration (FDA) did not revise its guidances for antibacterial drug development, thus leaving in place many unaddressed barriers to development. Accordingly, the approval of new antibacterial drugs has essentially come to a standstill. With respect to drugs targeting biofilms, the FDA and other regulatory agencies have also lacked access to standardized test methods to assess antibiofilm clinical effects since many of these standardized methods are only now in the early stages of development.

Second, antibiotic resistance is spreading with amazing speed on a global basis. Substantially higher rates of mortality and morbidity are associated with antibiotic-resistant infections—infections that are often very difficult or impossible to treat with current antibiotics. Accordingly, health agencies around the world are voicing urgent requests for the development of new antibiotics, *particularly those with new mechanisms and targets of action*, even as regulatory approvals for new antibacterial agents have essentially dried up. While microbial biofilms and antibiotic resistance are clearly and often causally related, biofilms have not been a focus of the pharmaceutical industry due to unclear regulatory guidelines, lack of diagnostic methods to demonstrate biofilms at the site of infections, and lack of standardized methods to demonstrate in vitro and in vivo correlation of antibiofilm activity with clinical outcomes.

Third, increasing rates of diabetes and obesity as well as aging populations in North American and Europe are leading to an increasing number of chronic and recalcitrant infections. It has been estimated that over 80 % of all infections are related to biofilms and that biofilms have an "almost ubiquitous involvement" in chronic infectious diseases and in many acute infections (Costerton et al. 1999; NIH 1999, 2002). Many long-term, chronic infections involve slow-growing or non-multiplying bacteria, against which current antibiotics are minimally effective. A partial list of these conditions include endocarditis, pulmonary infections associated with cystic fibrosis, recurrent urinary tract infections, chronic otitis media, chronic rhinosinusitis, Crohn's disease, musculoskeletal infections, osteomyelitis, bacterial prostatitis, necrotizing fasciitis, periodontitis chronic wound infections, fistulas, and infections associated with indwelling medical devices including catheters, stents, orthopedic devices, contact lenses, implantable electronic devices, cardiac devices, and breast implants (Costerton et al. 1999; Claret et al. 2007; Lynch and Robertson 2008; Rieger et al. 2013).

And fourth, there is a very great unmet need for rapid, reliable, and affordable biofilm diagnostic technologies. Until rapid, practical diagnostic tools are available to clinicians, the biofilm component of most infections will continue to be unrecognized and untreated, resulting in unacceptable rates of morbidity and all too frequently mortality.

Taken together, these global trends and conditions have generated an extremely urgent unmet need for new antibacterials, particularly those with new mechanisms of action and targets of action such as would be provided by agents with antibiofilm activity. Considering the ubiquity of the biofilm phenotype in nature and in the majority of infections, targeting biofilms may be one of the most important innovations in the history of antibacterial drug development. Standing in the way of the rapid development of biofilm drugs are the four serious global conditions described above, which make the development of this strategic approach an extremely long, risky, expensive, and arduous process. Extended lag time and virtual barriers to entry for antibiofilm drugs are not acceptable if global antibiotic resistance and chronic infectious diseases are to be treated in a timely and clinically relevant manner.

3 Increasingly Urgent Warnings of Global Antibiotic Resistance Issued by Health Agencies, Organizations and Governments

There is currently an urgent and expanding global need to advance the clinical and regulatory development of new anti-infective drugs, as evidenced by the positions taken by leading national and international organizations, government health officials, and infectious disease experts.

In 2011, the World Health Organization (WHO) announced their global campaign to focus attention on combating antimicrobial resistance (WHO 2011). In the keynote address at the 2012 European Union conference on "Combating Antimicrobial Resistance: Time for Action", Dr. Margaret Chan, Director General of the WHO, stated:

> "Hospitals have become hotbeds for highly resistant pathogens, like MRSA, ESBL, and CPE, increasing the risk that hospitalization kills instead of cures." (Chan 2012)

David Livermore, the director of antibiotic resistance monitoring at the UK's Health Protection Agency stated:

> "So much of modern medicine —from gut surgery to cancer treatment to transplants— depends on our ability to treat infection. If resistance destroys that ability then the whole edifice of modern medicine crumbles." (Stovall 2011)

Also in 2011, the European Commission recognized the urgent need for a "new business model" for the development of new antibiotics. This new model would benefit from an unprecedented and collaborative effort between the government and companies developing new antibiotics to combat antibiotic-resistant infections (European Commission 2011).

In 2012, the US Congress reauthorized the Food and Drug Administration Safety and Innovation Act (FDASIA) that contains new legislation entitled "Generating Antibiotic Incentives Now" (GAIN) in its Title VIII (United States Senate Bill S.3187 2012). This legislation provides new pathways to improve the speed of the regulatory review process for new antibiotics with the potential to overcome antibiotic-resistant infections and provides incentives to companies for their development such as extending the period of market exclusivity within the USA. Congress, by establishing the GAIN Act, has clearly demonstrated that development of new antibiotic drugs to address antibiotic resistance is a critical priority.

A letter from the Infectious Disease Society of America (IDSA) written in support of the GAIN Act (Infectious Disease Society of America 2011) stated that:

> Antibiotic-resistant infections significantly increase both health care and societal costs and hospital stays as demonstrated by an analysis of antibiotic-resistant infection data from a study conducted at Chicago Cook County Hospital (Roberts et al. 2009). Extrapolating that analysis nationwide, the authors concluded antibiotic-resistant infections cost the U.S. health care system in excess of $20 billion annually, $35 billion in societal costs, and more than 8 million additional days spent in the hospital. The cost to society of antimicrobial resistant infections in terms of lives lost and the economy will only rise as antimicrobial resistance continues to spread.

The urgent calls to action, as well as the global infectious crisis itself, have continued to grow more desperate during 2013. The United Kingdom's Chief Medical Officer, Dame Sally Davies, recommended that antibiotic-resistant bacteria be "...ranked alongside terrorism and climate change on the list of critical risks to the nation" (McCarthy 2013).

In September of 2013, the CDC announced in a 114-page report entitled *Antibiotic Resistance Threats in the United States, 2013* (CDC 2013) that in the USA alone, over two million people become infected with antibiotic-resistant bacteria each year and that at least 23,000 people die as a direct result of those infections annually. The CDC stated that these estimates are "minimum estimates." While there are many unmet needs in health care today, there are none with more far-reaching, potentially catastrophic consequences on a global level, than antibiotic-resistant infectious disease.

In light of so many urgent calls for (a) new mechanisms of action, (b) new targets, and (c) new strategies to overcome infectious diseases, it is of great concern that the novel, high-potential antibiofilm strategy for treatment of infection faces so many barriers and challenges to clinical implementation.

4 Current Regulatory Guidances: Emphasis on New Treatments for Serious, Antibiotic-Resistant Infections, But No Provision for Antibiofilm Mechanism

While there are a series of FDA clinical/antimicrobial guidances related to the development of antibacterials for the treatment of a variety of infections (FDA website: Guidance Compliance Regulatory Information, Guidances), none of the guidances address the development of antimicrobial drugs targeting biofilms or any other specific microbial targets (e.g., virulence factors). Likewise, the European Medicines Agency document titled "Guideline on the evaluation of medicinal products indicated for the treatment of bacterial infections" (European Medicines Agency website) also does not address the issue of the treatment of infections related to biofilms.

Even so, the July 2013 FDA guidance titled "Guidance for Industry: Antibacterial Therapies for Patients with Unmet Medical Need for the Treatment of Serious Bacterial Diseases" (FDA 2013) provides a discussion that indicates (a) the urgent unmet need for new antibacterial drugs, (b) the difficulty in carrying out antimicrobial regulatory studies, and (c) the effort that the FDA is making to encourage antimicrobial drug development, particularly for life-threatening infections.

> Over the past few decades, efforts to develop new antibacterial drugs have declined substantially. Over this same time period antibacterial drug resistance has become more common even in settings in which attempts were made to slow the rate at which bacterial pathogens become resistant, such as the prudent use of antibacterial drugs and adherence to infection control procedures. As a result, an increasing number of patients are suffering from bacterial diseases that do not respond to currently available antibacterial drugs, with serious consequences, including increased mortality.
>
> Clinical trials for antibacterial drugs can be challenging for a number of reasons, including: (1) for a serious bacterial disease, there is a need to urgently initiate empiric antibacterial drug therapy, which may obscure the effect of the antibacterial drug under study because patients receive effective antibacterial therapy before enrolling in the trial; (2) patients with serious acute bacterial diseases can be acutely ill (e.g. delirium in the setting of acute infection) and obtaining informed consent and performing other trial enrollment procedures in a timely fashion may be difficult; (3) there may be diagnostic uncertainty with respect to the etiology of the patients' underlying disease, including identifying a bacterial etiology; and (4) there may be a need for concomitant antibacterial drug therapy with a spectrum of activity that may overlap with the antibacterial drug being studied.
>
> A decreased rate of antibacterial drug development poses a significant public health concern. As bacteria continue to develop resistance because of selection pressures from empiric and/or inappropriate use of currently available antibacterial therapies, increased

numbers of patients will have unmet medical needs related to effective antibacterial drug therapy. Therefore, it is important for the public health that new antibacterial drugs be developed while also considering how best to ensure appropriate use.

To foster development of new antibacterial therapies for the treatment of serious bacterial diseases, we are exploring approaches that may help streamline development programs for antibacterial drugs, especially for drugs that could address an unmet medical need. As recognized in FDA regulations for the evaluation of drugs intended to treat life-threatening and severely debilitating illnesses.

"The Food and Drug Administration (FDA) has determined that it is appropriate to exercise the broadest flexibility in applying the statutory standards, while preserving appropriate guarantees for safety and effectiveness. These procedures reflect the recognition that physicians and patients are generally willing to accept greater risks or side effects from drugs that treat life-threatening and severely-debilitating illnesses, than they would accept from drugs that treat less serious illnesses. These procedures also reflect the recognition that the benefits of the drug need to be evaluated in the light of the severity of the disease being treated" (ibid).

Despite the FDA's efforts to "exercise the broadest flexibility in applying the statutory standards," a regulatory pathway that specifically enables and facilitates antibiofilm drug development remains to be defined.

5 Current Regulatory Path for Antibiofilm Agents and Challenges for Development of an Antibiofilm-Specific FDA Guidance

There are two types of categories of antibiofilm agents: (1) those with both conventional antibacterial activity against planktonic bacteria and antibiofilm activity and (2) those with antibiofilm activity, but without potent conventional antimicrobial activity, an example of which may be certain biofilm dispersants. Under current FDA guidances the two categories of agents are likely to face very different regulatory hurdles (Figs. 1 and 2). As of October 2013, the FDA has not yet granted a label claim *related specifically to activity against biofilms* for either category of agent, despite the fact that some antibacterial FDA-approved drugs have been demonstrated in nonclinical studies to have antibiofilm activity and also indirectly to be effective against infections related to biofilms (Bjarnsholt 2013).

Importantly, agents in the first category (those that can achieve FDA approval based on conventional antimicrobial activity) have a current, viable path to regulatory approval, as outlined in existing FDA guidances. This path is feasible because in vitro, in vivo, and clinical antimicrobial efficacies can be evaluated using a substantial body of existing standardized methodologies that are acceptable and well known to the FDA. It can therefore be expected to result in the availability of new drugs with antibiofilm activity to treat patients who are in dire need of new treatment strategies. Thus while these new, innovative drugs can be granted label

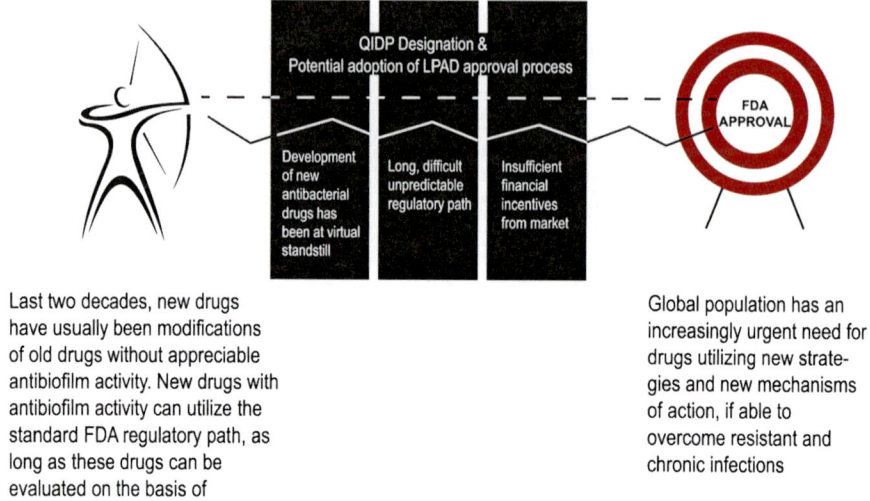

Fig. 1 Challenges to development of new antibacterial drugs (with or without antibiofilm activity)

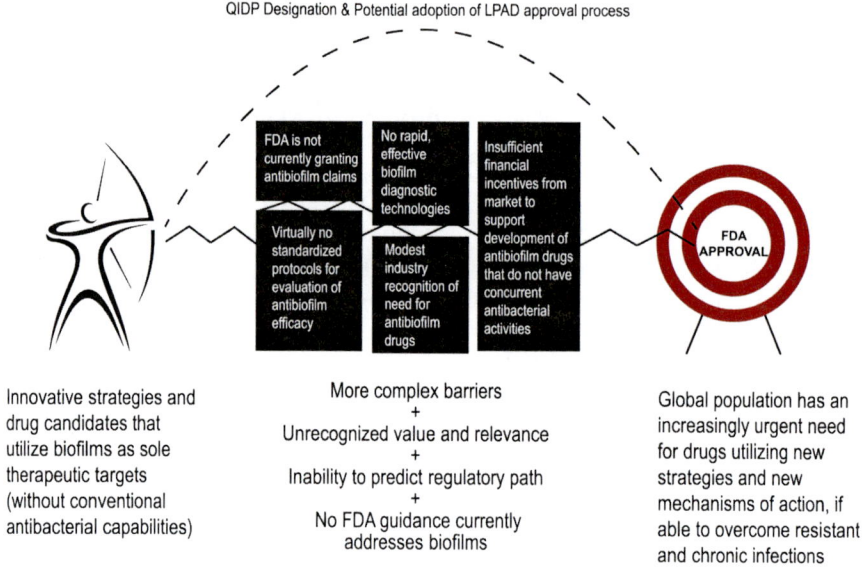

Fig. 2 Challenges to development of new antibiofilm drugs (without conventional antibacterial activity)

claims related to conventional antimicrobial activity and efficacy against specific infectious pathogens or diseases, at this time, their labels will not be able to communicate claims of antibiofilm activity.

Until a substantial body of standardized biofilm testing and diagnostic methodologies is developed for use in the analysis of biofilms and antibiofilm activity, claims related to biofilms are not likely to be awarded. For those agents with antibiofilm effects but without conventional antibacterial activities, an even higher level of regulatory scrutiny is likely to be applied as a direct result of the lack of standardized, clinically relevant methods to evaluate antibiofilm drugs. Drugs seeking antibiofilm claims would likely be required to demonstrate activity against bacteria living in a sessile, non-multiplying or persister (biofilm) clinical state—and *that* is currently far less than straightforward, particularly from a clinical perspective. True, a plethora of knowledge now exists describing the various stages of biofilm development and the role of biofilms in infections (Bjarnsholt 2013), and animal models of biofilm infection are increasingly available. Continual expansion of this body of knowledge will eventually lead us to effective control of microbial biofilms by drugs *with* biofilm claims. However, with only rare exceptions, standardized methods to detect and assess biofilms and antibiofilm efficacy in vitro, in animal models, and in clinical studies have not yet been approved by reputable standardization organizations. To date, a total of five methods developed by the Center for Biofilm Engineering at Montana State University and others have been approved by ASTM (American Society for Testing and Materials) International as standardized biofilm testing methods (ASTM 2011). This is a very important start, but much more expansive work is urgently needed in terms of the development of many more necessary, standardized clinically relevant biofilm tests.

6 Call to Action for the Biofilm Community, Infectious Disease Stakeholders, and Regulatory Agencies

It is incumbent upon those working in the biofilm arena to engage with the FDA, the infectious disease community, the pharmaceutical industry, and healthcare organizations (e.g., Cystic Fibrosis Society) to raise the awareness of the need to develop criteria that could be used to provide satisfactory clinical evidence that antibiofilm agents provide novel and effective treatments for antibiotic-resistant infections. It will be critical for these diverse stakeholders to work together to define and develop a coherent strategy for development of the necessary methodology that will subsequently support a clear and approachable regulatory path for antibiofilm agents.

An example of this type of cooperation occurred in the drug manufacturing arena. The Parenteral Drug Association (PDA) organized a task force composed of European and North American pharmaceutical manufacturers, suppliers of cleaning chemicals, and consultants to formulate and enact a new paradigm to address contamination control in healthcare product manufacturing processes, in part to address the need for prevention and elimination of the microbial biofilms (Madsen and Moldenhauer 2013; PDA 2012).

In 2014, the Center for Biofilm Engineering at Montana State University is taking the initiative to create the first opportunity to bring key industry, academic, and regulatory stakeholders together. Together with FDA, they co-sponsored a conference, "Biofilms, Medical Devices, and Anti-biofilm Technology: Challenges and Opportunities" in the Washington, DC, area to create an opportunity for FDA scientists and administrators to engage in dialog with members of the pharmaceutical, industrial, and academic community. FDA's co-sponsorship of the conference is a clear indication of the agency's increasing awareness of the need for engagement, communication, and collaboration to solve the challenges of developing standardized test methods and a regulatory path for antibiofilm agents.

Moving forward, the challenge for government agencies such as NIH, CDC, and the scientific and medical communities at large will be the commitment of substantial resources to the development of standardized evaluation procedures relevant to (a) the role of biofilms in infections, (b) diagnosis of biofilm infections in patients, and (c) evaluation of antibiofilm drug activity both clinically and in vitro. Such methodology development will allow for a more clinically equivalent and efficient comparison of diagnosis and treatment outcomes. It can be anticipated due to the heterogeneous and obscure nature of biofilms that many new methods and procedures will be required.

7 Potentially Expedited Paths for Antibiofilm Agents That Have Conventional Antimicrobial Activity

Professional organizations such as the Infectious Diseases Society of America (IDSA) are actively lobbying Congress for the development of exceptions to existing regulatory pathways to rapidly facilitate the development of drugs to treat patients with severe and life-threatening infections. The IDSA is encouraging the development of legislation providing for a new Limited Population Antibacterial Drug (LPAD) approval mechanism and is further encouraging the FDA, even in the absence of such legislation, to enact the LPAD mechanism under its current statutory authority. From the perspective of the IDSA,

> FDA has an essential role to play in ensuring that Americans have access to safe and effective drugs. But, in so doing, the agency must ensure that the risks associated with approving new products are appropriately balanced with the need to provide patients in desperate need with access to beneficial products. To date, when it comes to antibiotics, and particularly antibiotics needed to treat patients with the most serious bacterial infections, FDA's benefit-risk equation has been out of balance.

> LPAD will rebalance the benefit/risk equation and provide an important new approval pathway option for companies interested in and able to develop antibacterial drugs that treat the most serious infections where insufficient satisfactory therapeutic options currently exist. At least 15 companies and 24 medical and public health organizations including the American Medical Association (AMA) have lined up with IDSA in support of LPAD's creation.

Why do we need LPAD? It is not feasible for antibacterial drugs that treat serious infections due to highly resistant bacterial pathogens to be developed using traditional, large scale clinical trials due to the limited numbers of patients in which such serious infections occur. Examples of these organisms include *Acinetobacter baumannii* (which is threatening soldiers returning from Afghanistan as well as patients throughout the U.S. and the world), carbapenemase-producing *Klebsiella pneumoniae*, and *Pseudomonas aeruginosa*. Such infections kill an astonishingly high percentage of infected patients (e.g., greater than 50%–60% of patients with infection in the blood, greater than 40%–50% of patients with lung infection, etc.) despite any available treatment. Furthermore, extended-spectrum beta lactamase (ESBL)-producing Enterobacteriaceae (e.g., *Escherichia coli* [*E. coli*] and *Enterobacter* spp.), which often are resistant to all orally administered antibiotics, have spread through health care systems and more recently into communities.

"For serious diseases for which few if any acceptable treatments are available, the tolerable level of uncertainty regarding a potentially life-saving drug's effectiveness and safety profile is much greater. As an example, before the first HIV drug was approved, even highly toxic drugs were appropriately deemed approvable, because the infection itself caused nearly a 100% mortality rate. As more and more new anti-HIV drugs were approved, the death rate from HIV infection plummeted, and there was an increasingly safe group of antiretroviral drugs already on the market. As such, the tolerability for risk for each successively approved new agent became lower and lower, appropriately so. Similar to the early years of HIV drug development, the benefit-risk ratio of approved LPAD drugs will be quite different than for antibiotics approved under traditional development programs where the drug is indicated for use more broadly." (FDA Docket No. 2012-N-1248 2013).

The precedent for regulatory flexibility extended to AIDs drugs is highly relevant to the current need for regulatory facilitation of new treatments for infections caused by antibiotic-resistant bacteria, including antibiofilm agents. As reported by CDC researchers (Klevens et al. 2007), in 2005 there were an estimated 94,000 life-threatening MRSA infections in the USA, resulting in 18,650 deaths, more than the estimated 17,011 Americans who died of AIDs that same year (CDC 2007 revised).

"Under the LPAD approval mechanism, an antibacterial drug's safety and effectiveness would be studied in substantially smaller, more rapid, and less expensive clinical trials—much like the Orphan Drug Program permits for other rare diseases. LPAD products then would be narrowly indicated to be marketed to and used in small, well-defined populations of patients for whom the drugs' benefits have been shown to outweigh their risks. Many bacterial diseases have a broad spectrum of severity. The LPAD mechanism is intended to address the needs of a special population of patients with serious manifestations of such diseases who lack satisfactory treatments. In caring for such severely ill patients with limited treatment options, the patients, health care providers, regulators, and society can tolerate a greater degree of uncertainty about overall risk associated with a drug than can be tolerated in patients with milder manifestations of the disease, or those who have more satisfactory therapeutic options. The LPAD mechanism will not be used to approve antibacterial products that treat more common, less serious infections or infections where sufficient alternative therapeutic options exist" (FDA Docket No. FDA-2012-N-1248 2013).

In the Personal Views section of the March 2013 issue of Lancet Infectious Diseases, a group of pharmaceutical company physicians and researchers proposed a four-tiered regulatory framework for registration of new treatments that address the unmet need for new antibacterial drugs (Rex et al. 2013). This approach would provide expedited regulatory pathways that incorporate concepts from orphan drug registration and other existing registration pathways to balance clinical efficacy requirements with seriousness of the unmet medical need. The authors make the case that the conventional registration requirement for two Phase 3 clinical studies (Tier A in their model) will result in excessive delays in bringing new antibacterials to market, particularly for agents that have narrow-spectrum activity or those that are directed at emerging resistant pathogens, due to the limited number of patients available for enrollment. They propose creation of two new approval tiers that fall between the conventional Tier A requirements and the Animal Rule approval path (their Tier D). Tier C is pathogen-focused, would require only small comparative and descriptive human clinical studies, and would apply to antibacterials with efficacy against uncommon pathogens or mechanisms of resistance. Tier B would be intermediate between Tier A and C and would require only one Phase 3 randomized controlled trial plus Tier C type small comparative and descriptive studies. The label for Tier B and C agents would communicate the limitations, risks, and uncertainties of the clinical data used in support of registration.

The 2012 passage of the GAIN Act as Title VIII of the Food and Drug Administration Safety and Innovation Act and the potential adoption of the LPAD approval process represent avenues of regulatory development that may be utilized to speed the development of antibiofilm drugs, as long as such drugs can also demonstrate the ability to effectively overcome antibiotic-resistant bacteria. This latter attribute may make such drugs eligible for FDA Qualified Infectious Disease Product (QIDP) designation, or, if the LPAD approval process is adopted, demonstrate the ability to effectively treat infections associated with unmet needs. Considering the contribution of microbial biofilms both to antibiotic-resistant infections and to life-threatening persistent infections, these exceptions to the standard FDA regulatory process may indeed be available to aid in the advancement of antibiofilm drugs.

A limited number of drug candidates with likely, purported, or proven antibiofilm activity are currently undergoing clinical development (see Table 1), while many other antibiofilm agents are in pre-clinical or research stages of development. It is indeed notable that three of the eight drug candidates listed in this table have been granted the QIDP designation by the FDA. It is perhaps even more notable that these QIDP designations have been granted to drug candidates with potential antibiofilm activity given that QIDP designation has only been awarded to a total of approximately sixteen drug candidates overall (some candidates have multiple designations for multiple formulations), as of October 2013. The FDA's QIDP designation is anticipated to speed and prioritize the development of these drug candidates, while also extending and fortifying the period of market exclusivity for these drug candidates, if and when approved.

Table 1 Antibiofilm drugs in clinical development[a]

Company	Drug	Type of molecule	Clinical stage	Antibiofilm evidence[a]	Therapeutic indication
C3 Jian http://www.c3-jian.com	C16G2	STAMP—specifically targeted antimicrobial peptide directed at *S. mutans*	Phase 1 completed 8/2013	Company website description with links to publications	Prescription mouth rinse for dental caries
Catholic University of the Sacred Heart, Italy[b,c] http://www.unicatt.it	*N*-acetyl-cysteine (NAC)	*N*-acetyl-cysteine	Post-Phase 1 efficacy study[b]	*H. pylori* clinical trial protocol with biofilm endpoint and peer-reviewed publications ClinicalTrials.gov Identifier: NCT00985608	*Helicobacter pylori* infections
Insmed (licensed from Transave) http://www.insmed.com	Arikace liposomal amikacin for inhalation	Aminoglycoside antibiotic delivered with eFLOW electronic nebulizer using proprietary liposomal technology	Phase 3 *P. aeruginosa* in cystic fibrosis (CF) patients: European and Canadian results announced 7/2013 Phase 2 completed in non-CF *P. aeruginosa* Phase 2 in Non-tubercular mycobacteria (NTM) lung infections enrolling in USA and Canada **FDA granted QIDP and Fast Track designation for NTM**	Multiple peer-reviewed publications when drug was being developed by Transave, however no mention of antibiofilm activity on Insmed website or press releases	Lung infections: *P. aeruginosa* (CF and non-CF) and NTM

(continued)

Table 1 (continued)

Company	Drug	Type of molecule	Clinical stage	Antibiofilm evidence[a]	Therapeutic indication
MerLion Pharmaceuticals (partner with Alcon for ear infections) http://www.merlionpharma.com	Finfloxacin	Fluoroquinolone antibiotic, novel low pH activated form	Phase 3 for acute otitis media initiated 4/2012. Phase 2 for UTI initiated 12/2012. **QIDP and Fast Track Status granted by FDA for oral and IV use for complicated urinary tract infections (UTIs)**	Company website, press releases	Complicated UTI and pyelonephritis. *H. pylori* infections. Topical formulation for acute otitis externa and otitis media being developed with Alcon
Microbion Corp. http://www.microbioncorp.com	BisEDT Bismuth-1,2-ethanedithiol	Bismuth-thiol	Phase 1 completed in the UK. Phase 2 ready for orthopedic infections in the USA. **QIDP Designation awarded by FDA for postsurgical orthopedic infections**	Company website, press releases, multiple peer-reviewed publications	Gel formulation for local administration for postsurgical orthopedic infections. Topical gel for chronic wound infections; skin, burn, and surgical site infections
Nova Bay Pharmaceuticals (partnered with Galderma for dermatology) http://www.novabay.com	NVC-422 Auriclosene: *N,N*-dichloro-2,2-dimethyltaurine	Aganocide	Currently enrolling Phase 2 for urology, Phase 2b for ophthalmology, and Phase 2b with Galderma for skin infections	Early website, publications, press releases but no mention in more recent publicity 2011 and 2012 peer-reviewed publications show efficacy in sheep biofilm infection	Impetigo. Urological infections, viral conjunctivitis (eye)

Company	Drug	Target	Clinical status	Biofilm evidence[a]	Indication
Toltec Pharmaceuticals (No website)	TOL-463	Undisclosed	Phase I completed	Antibiofilm activity demonstrated in vitro employing biofilm model established by the CDC (company communication)	Vaginitis
University of British Columbia http://www.ubc.ca	Esp protein	Esp protein from *S. epidermidis*	Phase 1, not yet recruiting (as per clinicaltrials.gov, accessed October 31, 2013)	Biofilm endpoint in clinical trial protocol on clinicaltrials.gov and peer-reviewed publications ClinicalTrials.gov Identifier: NCT01646502	Chronic wound biofilms

[a] Multiple sources were surveyed for antibiofilm evidence including clinical trials.gov website, company websites, press releases, scientific publications, and general Internet searches. All company websites were most recently accessed on 31 October 2013. Company sponsored agents with minimal or no indication on the company website of intent to develop the drug as an antibiofilm agent, or where information was very dated, generally were not included except in cases where the drug candidate was the subject of recent peer-reviewed publications demonstrating antibiofilm activity

[b] Catholic University of the Sacred Heart clinical study phase and status is unclear. The *H. pylori* study with a biofilm endpoint has not been updated on clinicaltrials.gov since January 2010. Other more recent studies targeting other infectious diseases are listed but without mention of a biofilm endpoint

[c] In addition to Catholic University of the Sacred Heart, other clinical trials of NAC with biofilm endpoints (clinicaltrial.gov, accessed October 31, 2013) are also being sponsored by Buddhist Tsu Chi General Hospital, Taiwan for *H. pylori* infections (NAC in combination with standard triple therapy, ClinicalTrials.gov Identifier: NCT01572597, status: recruiting) and by Minia University, Egypt, for bacterial vaginosis (NAC alone or in combination with metronidazole, ClinicalTrials.gov Identifier: NCT01841411, status: recruiting)

8 Summary of the Challenge of Antibiofilm Drug Development

In summary, the challenges facing clinical medicine and the regulatory arena are related to the comparatively recent scientific recognition of the role that biofilms play in the majority of human infections and the urgent need to facilitate methods for the development of antibacterials to address these ubiquitous microbial phenomena. The challenges are primarily associated with the antimicrobial regulatory approval process and insufficient financial incentives offered to pharmaceutical developers. These substantial barriers to entry have been recognized by governments around the world and initial steps are being taken to both streamline the development of antimicrobial drugs and to improve the financial incentives associated with new antimicrobial development.

As a result of the expanding global crises relating to antimicrobial resistance and chronic infectious diseases, there is now an extremely urgent need to rapidly develop new drugs that have antibiofilm capabilities. To this end biofilm drug developers must develop standardized protocols for evaluation of biofilm activity, as well as rapid, effective biofilm diagnostic technologies. It is also imperative that regulatory agencies exercise maximum flexibility in approving these innovative new drugs. As was done in the early development of drugs to treat AIDS, regulatory agencies need to encourage the development of new drugs targeting microbial biofilms in recognition of the role the biofilms play in the majority of infectious diseases. It is time for the biofilm research and product development communities to make our voices heard, to connect the dots between antibiotic-resistant infections and biofilms, and to bring attention to the fact that biofilm strategies may offer one of the most timely and critically important answers to the global, expanding infectious disease crisis.

References

ASTM International (2011) Standardization news. http://www.astm.org/standardization-news/update/biofilm-ja11.html

Bjarnsholt T (2013) The role of bacterial biofilms in chronic infections. APMIS 121:1–58. doi:10.1111/apm.12099

CDC (2007 revised) HIV/AIDS surveillance report, cases of HIV infection and AIDS in the United States and dependent areas, 2005. http://www.cdc.gov/hiv/pdf/statistics_2005_HIV_Surveillance_Report_vol_17.pdf. Accessed 29 Oct 2013

CDC (2013) Antibiotic resistance threats in the United States, 2013. http://www.cdc.gov/drugresistance/threat-report-2013/pdf/ar-threats-2013-508.pdf. Accessed 28 Oct 2013

Chan DM (2012) Antimicrobial resistance in the European Union and the world. In: Keynote address at the conference on Combating antimicrobial resistance: time for action, Copenhagen, Denmark, 14 March 2012

Ciofu O, Tolker-Nielsen T (2011) Antibiotic tolerance and resistance in biofilms. In: Bjarnsholt T et al (eds) Biofilm infections. Springer Science + Business Media, LLC, New York, NY, pp 215–230

Claret L, Miquel S, Vieille N, Ryjenkov DA, Gomelsky M, Darfeuille-Michaud A (2007) The flagellar sigma factor FliA regulates adhesion and invasion of Crohn disease-associated Escherichia coli via a cyclic dimeric GMP-dependent pathway. J Biol Chem 282(46): 33275–33283. doi:10.1074/jbc.M702800200

Costerton JW, Stewart PS, Greenberg EP (1999) Bacterial biofilms: a common cause of persistent infections. Science 284(5418):1318–1322. doi:10.1126/science.284.5418.1318

European Commission (2011) Communication from the Commission to the European Parliament and the Council: action plan against the rising threats from antimicrobial resistance. http://ec.europa.eu/dgs/health_consumer/docs/communication_amr_2011_748_en.pdf. Accessed 29 Sept 2013

European Medicines Agency (2011) Guideline on the evaluation of medicinal products indicated for the treatment of bacterial infections. http://www.ema.europa.eu/docs/en_GB/document_library/Scientific_guideline/2009/09/WC500003417.pdf. Accessed 15 Oct 2013

FDA (2013) Guidance for industry: antibacterial therapies for patients with unmet medical need for the treatment of serious bacterial diseases. https://www.federalregister.gov/articles/2013/07/02/2013-15783/draft-guidance-for-industry-on-antibacterial-therapies-for-patients-with-unmet-medical-need-for-the. Accessed 28 Sept 2013

FDA Docket No. FDA-2012-N-1248 Statement of the Infectious Diseases Society of America modified from comments delivered at the Food and Drug Administration's hearing by R. J. Guidos, Creating an alternative approval pathway for certain drugs intended to address unmet medical need, 4 February 2013

FDA Guidance compliance regulatory information, guidances (2013) http://www.fda.gov/drugs/guidancecomplianceregulatoryinformation/guidances/ucm064980.htm. Accessed 29 Oct 2013

Fux CA, Costerton JW, Stewart PS, Stoodley P (2005) Survival strategies of infectious biofilms. Trends Microbiol 13(1):34–40. doi:10.1016/j.tim.2004.11.010, S0966-842X(04)00264-1 [pii]

Infectious Disease Society of America (2011) Letter from the Infectious Disease Society of America dated November 22, 2011, to Senators Richard Blumenthal and Bob Corker in support of the Generating Antibiotic Incentives Now (GAIN) Act. http://www.idsociety.org/uploadedFiles/IDSA/Policy_and_Advocacy/Current_Topics_and_Issues/Antimicrobial_Resistance/10x20/Letters/To_Congress/IDSA%20Letter%20to%20Blumenthal%20and%20Corker%20re%20GAIN%20with%20Attachment%20112211.pdf. Accessed 20 Oct 2013

Klevens RM, Morrison MA, Nadle J, Petit S, Gershman K, Ray S, Harrison LH, Lynfield R, Dumyati G, Townes JM, Craig AS, Zell ER, Fosheim GE, McDougal LK, Carey RB, Fridkin SK (2007) Invasive methicillin-resistant Staphylococcus aureus infections in the United States. JAMA 298(15):1763–1771. doi:10.1001/jama.298.15.1763, 298/15/1763 [pii]

Lynch AS, Robertson GT (2008) Bacterial and fungal biofilm infections. Annu Rev Med 59: 415–428. doi:10.1146/annurev.med.59.110106.132000

Madsen R, Moldenhauer J (eds) (2013) Contamination control in healthcare product manufacturing, vol 1. PDA DHI, Bethesda, MD

McCarthy M. Chief Medical Officer Dame Sally Davies: resistance to antibiotics risks health 'catastrophe' to rank with terrorism and climate change. The Independent, 11 March 2013. http://www.independent.co.uk/news/science/chief-medical-officer-dame-sally-davies-resistance-to-antibiotics-risks-health-catastrophe-to-rank-with-terrorism-and-climate-change-8528442.html. Accessed 29 Sept 2013

NIH (1999) SBIR/STTR Study and control of microbial biofilms, PA number: PA-99-084. http://grants.nih.gov/grants/guide/pa-files/PA-99-084.html. Accessed 28 Oct 2013

NIH (2002) NIH Guide: research on microbial biofilms, PA number: PA 03-047. http://grants.nih.gov/grants/guide/pa-files/PA-03-047.html. Accessed 4 Oct 2013

PDA (2012) PDA Technical Report No. 29, Revised 2012 (TR 29) Points to consider for cleaning validation, Parenteral Drug Association. http://www.pda.org. Accessed 15 Oct 2013

Rex JH, Eisenstein BI, Alder J, Goldberger M, Meyer R, Dane A, Friedland I, Knirsch C, Sanhai WR, Tomayko J, Lancaster C, Jackson J (2013) A comprehensive regulatory framework to address the unmet need for new antibacterial treatments. Lancet Infect Dis 13(3):269–275. doi:10.1016/s1473-3099(12)70293-1

Rieger UM, Mesina J, Kalbermatten DF, Haug M, Frey HP, Pico R, Frei R, Pierer G, Luescher NJ, Trampuz A (2013) Bacterial biofilms and capsular contracture in patients with breast implants. Br J Surg 100(6):768–774. doi:10.1002/bjs.9084

Roberts R, Hota B, Ahmad I, Scott RN, Foster S, Abbasi F, Schabowski S, Kampe L, Ciavarella G, Supino M, Naples J, Cordell R, Levy S, Weinstein R (2009) Hospital and societal costs of antimicrobial-resistant infections in a Chicago teaching hospital: implication for antibiotic stewardship. Clin Infect Dis 49(8):1175–1184. doi:10.1086/605630

Stovall S (2011) WHO calls for action on superbugs. Wall Street J, 8 Apr 2011. http://online.wsj.com/news/articles/SB10001424052748704013604576248182661678522. Accessed 26 Oct 2013

United States Senate Bill 3187 (2012) Food and Drug Administration Safety and Innovation Act, Title VIII, Generating Antibiotic Incentives Now (GAIN Act). http://www.gpo.gov/fdsys/pkg/BILLS-112s3187enr/pdf/BILLS-112s3187enr.pdf. Accessed 22 Oct 2013

WHO (2011) World Health Day 2011: policy briefs. http://www.who.int/world-health-day/2011/policybriefs/en/index.html. Accessed 31 Oct 2013

Index

A

Acetyl-11-keto-b-boswellic acid (AKBA), 209
Achyranthes aspera, 206
Acinetobacter baumannii, 7, 173, 343
N-Acyl homoserine lactones (AHLs), 96, 97, 137–138, 333
AI-2. *See* Autoinducer-2 (AI-2)
American Society for Testing and Materials (ASTM), 471
Amoxicillin, 21, 304
Amphotericin B, 272–273, 275, 276
AMPs. *See* Antimicrobial peptides (AMPs)
Antibiotics, oral biofilm
    amoxicillin, 304
    antibiotic concentration, 303
    antimicrobial peptides, 305–306
    azithromycin and tetracycline, 304
    ciprofloxacin–metronidazole, 304
    drug delivery system, 304
    penicillin, 305
Antibiotic therapy, 24, 117, 428, 453
Antimicrobial coatings
    central venous catheters
        chlorhexidine silver sulfadiazine, 174–175
        minocycline/rifampicin, 175–176
        organoselenium coating, hemodialysis catheters, 176–178
        silver, 174
    contact lenses (*see* Contact lenses (CLs))
    urinary tract catheters
        antibiotics, 181–182
        biological coatings, 182–183
        gendine, 181
        hydrogels, 179
        nitric oxide, 181
        quorum-sensing inhibitors, 182
        silver, 179–180
        triclosan, 180–181
Antimicrobial peptides (AMPs), 305–306
Antimicrobial photodynamic therapy (APDT), 127, 444–445
Antiseptics
    amine fluoride, 303
    cetylpyridinium chloride, 303
    chlorhexidine mouth rinses, 300–302
    listerine amd hexetidine, 303
    triclosan, 302
ATP binding cassette (ABC) transporters, 48–49
Autoinducer-2 (AI-2), 96, 99, 104, 138, 308–309, 329, 333
Autoinducing peptides (AIPs), 138, 234, 235
Azithromycin, 126, 263, 304, 329
Azole responsive cis-acting elements (ARE), 46

B

Bacteraemia, 453
Bacterial biofilms
    HVPC, 431–432
    LIDC
        human studies, 428
        in vitro studies, 427
        in vivo studies, 427–428
        medicinal plants and phytocompounds (*see* Medicinal plants and phytocompounds)
    RFAC, 434–435
Bacteriophages, 131–132, 369–370
Benzyldimethyldodecylammonium chloride (BDMDAC), 158–159, 164, 165

Biofilm-associated infections
  biofilm inhibitors
    microbial metabolites (*see* Microbial metabolites)
    plant products, QS system (*see* Quorum sensing (QS) systems)
  combination therapy, 118
  CRBSI (*see* Catheter-related blood stream infections (CRBSIs))
  disruption of
    biofilm-disrupting enzymes, 126–127
    detachment-promoting agents, 125
    kinetic model, 125
    PDT, 127–130
  microbial interactions/interference, 127–130
    biofilm formation and development, 130
    commensal bacteria, 130
    phages, 131–133
    probiotics, 130–131
  small molecule control
    chemical library screening, 142–144
    HTS, 140
    inhibit/disperse bacterial biofilms, 140
    natural products, 141
    principle of, 141, 142
Biofilms
  antimicrobial coatings (*see* Antimicrobial coatings)
  conjugative plasmid transfer
    gram-negative bacteria, 73–74
    gram-positive bacteria, 74–76
  in disease
    acute respiratory infections, 1
    bacterial infections, 2
    biofilm formation, 4–6
    biofilm resistance, 4
    Candida albicans, 3
    coagulase-negative staphylococci, 2
    cystic fibrosis, 3
    dental plaque biofilm, 3
    drug development, 8
    Fusarium ssp., 3
    keratitis, 3
    life-threatening secondary infections, 3
    medical biofilm, 2
    middle-ear infection, 3
    novel antibiotics, 8
    pathogens, 6–7
    Staphylococcus aureus, 2
    urinary tract infections, 3
  in vitro and in vivo analysis, 8
  environmental factors, 81–82
  fungal biofilm (*see* Fungal biofilm)
  infection (*see* DNA methods)
  inhibition
    microbial extracts, 380–381
    nanoparticles (*see* Nanoparticles)
    plant products, 381
    synthetic chemicals, 380
  MDR, fungal cells (*see* Multidrug resistance (MDR))
  medical (*see* Medical biofilms)
  mobile genetic element, 82
  oral biofilm (*see* Oral biofilm)
  planktonic and biofilm modes, HGT in (*see* Horizontal gene transfer (HGT))
  polymicrobial biofilm (*see* Polymicrobial biofilm)
  prevention (*see* Drug delivery systems)
  Pseudomonas aeruginosa treatment (*see Pseudomonas aeruginosa* biofilms)
  quorum sensing in (*see* Quorum sensing (QS) systems)
  Staphylococcus aureus (*see Staphylococcus aureus* biofilms)
  study model organisms, 158
  tissue samples (*see* Tissue samples)
  transduction, 78–80
  transformation, 76–78
  wound biofilm (*see* Wound biofilm)
Bis-(3'5')-cyclic di-guanylic acid (c-di-GMP), 138
Bismuth-1,2-ethanedithiol, 476
Blue/white screening, 160
Brown and Brenn Gram staining, 30

C

Candida albicans biofilm chip (CaBChip), 278
Caspofungin, 275–277
Catheter-related blood stream infections (CRBSIs), 173–176
  healthcare-associated infections, 119
  implant-associated infection, 119
  implanted materials, 120
  lock therapy, 119–120
  surface coatings, 121–123
  surface properties, 121
  use of nanoparticles, 123–125

Index

Catheter-related candidiasis, 270
Central venous catheters (CVCs)
    antimicrobial coatings
        chlorhexidine silver sulfadiazine, 174–175
        minocycline/rifampicin, 175–176
        organoselenium coating, hemodialysis catheters, 176–178
        silver, 174
    CRBSIs, 173
Cetylpyridinium chloride (CPC), 302, 303
Chelerythrine (CH), 210
Chitosan, 123, 381, 411–412
Chlorhexidine silver sulfadiazine (CH-SS), 174–175
Chronic wounds
    biofilm pathogenesis, 420–421
    electrical stimulation (see Electrical stimulation)
    ulcer wounds, 420
CL acute red eye (CLARE), 184
Clotrimazole, 276
CL peripheral ulcer (CLPU), 184
CLs. See Contact lenses (CLs)
Competence stimulating peptide (CSP), 99, 291, 309–310
Confocal laser scanning microscopy (CLSM), 8, 221
Contact lenses (CLs)
    advantages, 185
    contact lens cases
        bacterial contamination, 190
        organoselenium, 191
        silver, 190
    contact lens-related bacterial infection, 183–184
    covalent attachment
        cationic peptides (melimine), 185, 187–188
        fimbrolides (furanones), 185, 187
        organoselenium, 188
        polyquaternium compounds and polymeric pyridinium compounds, 185, 187
    disadvantages, 185
    non-covalent attachment
        furanones, 185, 186
        polyquaternium compounds, 185, 186
        silver, 185–187
    selenium, antimicrobial mechanism of, 188–189
CPC. See Cetylpyridinium chloride (CPC)
Cranberry, 207

CRBSIs. See Catheter-related blood stream infections (CRBSIs)
Cryoembedding, 33
Cryosectioning, 33
CSP. See Competence stimulating peptide (CSP)
CVCs. See Central venous catheters (CVCs)
Cystic fibrosis transmembrane conductance regulator (CFTR), 3

D
DCU. See Dental chair units (DCU)
Decubitus direct current treatment (DDCT), 426
Delisea pulchra, 106, 133, 135
Dendrimer
    architectural region, 410
    convergent approach, 410–411
    divergent approach, 410
    hydrophobic drugs, 411
Dental caries, 290–291, 296, 310, 313
Dental chair units (DCU), 292
Dental plaque, 287–289, 291, 296. See also Oral biofilm
Dental Unit Water Systems (DUWS), 292–293, 300, 301
Deoxyribonuclease (DNase)
    Pseudomonas aeruginosa biofilm, 259
    Staphylococcus aureus, 237
Deoxyribonuclease I (DNase I), 126
Dihydroxybenzofuran (DHBF), 210, 211
Dispersin B, 126, 237–238, 331
DNA methods
    advantages of, 20–22
    bioinformatics, 17
    clinical cultures, 17–19
    clinical use, 22–24
    data analysis process, 16
    HMP, 14
    PCR, 14–15
    sequencing instruments, 16
    16S ribosomal DNA, 16
    16S/18S rDNA gene, 15
DNA transfer and replication (Dtr) proteins, 68
Drug delivery systems, 401
    chitosan, 411–412
    dendrimer
        architectural region, 410
        convergent approach, 410–411
        divergent approach, 410
        hydrophobic drugs, 411
    drug carriers, 402

Drug delivery systems (cont.)
  magic bullet, 402
  polymer-based drug delivery system
    biocompatible/biodegradable polymers, 408–409
    hydrophilic/hydrophobic region, 407
    poly(lactide-co-glycolide), 409, 413
    poly(2-hydroxyethyl methacrylate) hydrogel, 410
  vesicular systems
    liposomes (see Liposomes)
    niosomes, 406–407
    physicochemical characteristics, 403
Drug development and regulation
  antibiotic resistance, 465
  arikace liposomal amikacin, 475
  biofilm resistance and tolerance, 464
  bismuth-1,2-ethanedithiol, 476
  CDC, 467, 473
  Center for Biofilm Engineering, 472
  C16G2, 475
  challenges, 469–471
  costs, 465
  diabetes and obesity rates, 465–466
  diagnosis, 466
  esp protein, 477
  FDA guidance, 468–469
  FDASIA, 467
  finfloxacin, 476
  GAIN Act, 467, 474
  government agencies, 472
  Lancet Infectious Diseases, 473–474
  LPAD, 472–473
  mechanisms, 464–465
  N-acetyl-cysteine, 475
  NVC-422, 476
  Parenteral Drug Association, 471
  QIDP designation, 474
  TOL-463, 477
  World Health Organization, 466–467
DUWS. See Dental Unit Water Systems (DUWS)

E

Echinocandins, 273–275
EGCG. See Epigallocatechin 3 gallate (EGCG)
Electrical stimulation
  alternating current, 433
  current of injury
    porcine model system, 423
    trans-epithelial potentials, 422
  ESTHR, 421–422

HVPC
  antibacterial mechanism, 432–433
  bacterial biofilm, 431–432
  clinical studies, 430–431
  twin-peaked wave form, 422, 430
LIDC
  antibacterial mechanisms, 429–430
  bacterial biofilms, 427–428
  clinical studies, 426
  contraindications, 430
  wound care adjunctive therapy, 425
RFAC
  bacterial biofilms, 434–435
  tumor treating fields, 433
wound healing
  overlapping phases, 423
  in vitro studies, 424
  in vivo studies, 425
Electrical stimulation for tissue healing and repair (ESTHR), 421, 425, 430, 435
Electrolyzed water (EW), 362–363
Ellagic acid (EA), 136, 208, 215
Enterococcus faecium, 75–76, 131
Epigallocatechin 3 gallate (EGCG), 307, 308
EPS. See Extracellular polymeric substances (EPS)
ESTHR. See Electrical stimulation for tissue healing and repair (ESTHR)
Ethylenediaminetetraacetic acid (EDTA), 242
EW. See Electrolyzed water (EW)
Extracellular DNA (eDNA)
  QSIs, 258
  Staphylococcus aureus biofilm, 232–234
Extracellular matrix (ECM), 55–57
Extracellular polymeric substances (EPS), 72, 78
  carbohydrates, proteins, 34
  detachment-promoting agents, 125
  DNA, 34
  fluorescently labeled lectin, 35
  host tissue, extracellular matrix, 39–40
  microbial cells, 35
  nanoparticles, 387
  pO157, 162
  polysaccharides, 325, 326, 331, 380
  RFAC, 435
  stained HECM, 35
  staining bacteria, 34
  WGA-TR, 35

F

Fecal microbiota transplantation (FMT), 312
Finfloxacin, 476

Fluconazole (FLC), 45–50
Fluorescent in situ hybridization (FISH), 36–38
FMT. *See* Fecal microbiota transplantation (FMT)
Food and Drug Administration Safety and Innovation Act (FDASIA), 467
Food processing
  antibiofilm strategies
    Ag-PTFE-coated stainless steel, 365
    bacteriophages, 369–370
    benzyldimethyldodecylammonium chloride, 365–366
    biocides, 364–365
    biosurfactants, 367
    cell-to-cell communication, 368–369
    chemical disinfection, 360–361
    cleaning-in-place, 359
    electrical methods, 362
    electrolyzed water, 362–363
    enzymes, 367–368
    essential oils, 366–367
    Hazard Analysis and Critical Control Points, 359–360
    nitrogen-doped titanium dioxide, 365
    polyethylene glycol, 366
    pressure washing, 361–362
    quorum sensing, 368–369
    silver-montmorillonite, 384
    ultrasonication, 362
  drinking water systems, 356
  exopolymeric substances, 355
  foodborne pathogen
    Bacillus spp., 358–359
    Campylobacter spp., 358
    Escherichia coli, 358
    Listeria monocytogenes, 357
    Salmonella spp., 357–358
Fresh garlic extract (FGE), 134, 221–222
Fungal biofilm
  antifungal lock therapy, 276
  clinical implication, 270–271
  cocktail solution, 277–278
  drug resistance
    amphotericin B, 272–273
    echinocandins, 273–274
    potential contributory mechanisms, 274
  farnesol, 277
  formation stages, 271–273
  inhibition, 274
    combination therapy, 275–276
    conventional antifungal agents, 275
  nano-biofilm chip, 278–279
  novel anti-biofilm agents, 276–277
  photodynamic therapy, 278
  plastic impregnation, 277
  structural characteristics, 272, 273

G

GAIN. *See* Generating Antibiotic Incentives Now (GAIN)
Gain of function (GOF), 46, 51
Gendine, 181
Generating Antibiotic Incentives Now (GAIN), 467, 474
Genomic islands (GEIs)
  coding capacity of, 71
  Enterococcus faecalis pathogenicity island, 80
  SGI1, 71
  T4SS, 80
  Yersinia pseudotuberculosis, 81
Green fluorescent protein (GFP), 84, 103, 104
Group B Streptococci (GBS), 7

H

Hazard Analysis and Critical Control Points (HACCP), 359–360
Hematoxylin and eosin (H&E), 33
Hexetidine, 302, 303
High throughput screening (HTS), 140, 142, 278
High-voltage pulsed current (HVPC)
  antibacterial mechanism, 432–433
  bacterial biofilm, 431–432
  clinical studies, 430–431
  twin-peaked wave form, 422, 430
Horizontal gene transfer (HGT)
  biofilm formation, 72–73
    environmental factors, 81–82
    mobile genetic element, 82
  liquid, surfaces, biofilm mode
    conjugative transfer, 73–76
    genomic islands, 80–81
    transduction, 78–80
    transformation, 76–78
  modes of
    conjugative transfer, 68–69
    DNA transduction, 70
    DNA uptake, transformation, 69–70
    genomic islands, 70–71
  monitoring of
    antibiotic resistance genes detection, 83
    DsRed, 84
    fluorescent microscopy, 85
    GFP, 84, 85

high-resolution in situ analysis, 84
RFP, 85
socio-microbiology and bacterial evolution, 66, 67
Host extracellular matrix (HECM), 34
Human microbiome project (HMP), 14
HVPC. *See* High-voltage pulsed current (HVPC)
Hyaluronate lyase, 237

I

3-Indolylacetonitrile (IAN), 209
Infectious Disease Society of America (IDSA), 467, 472
Institute of Chemistry and Cell Biology-Longwood (ICCB-L), 143
Integrative conjugative elements (ICEs), 68–69, 71, 74
Invasive aspergillosis, 271

K

Kaurenoic acid (KA), 213

L

LIDC. *See* Low-intensity direct current (LIDC)
Life-threatening meningitis, 271
Limited Population Antibacterial Drug (LPAD), 472–474
Liposomes
  characteristics, 405
  classification, 404
  liposomal formulation, 405–406
  preparation, 403–404
Listerine, 302, 303
Low-intensity direct current (LIDC)
  antibacterial mechanisms, 429–430
  bacterial biofilms
    in vitro studies, 427
    in vivo studies, 427–428
  clinical studies, 426
  contraindications, 430
  wound care adjunctive therapy, 425
LPAD. *See* Limited Population Antibacterial Drug (LPAD)
Lyase, 127, 237
Lysostaphin, 126–127, 238

M

Major facilitator superfamily (MFS) transporters, 49–50
MDR. *See* Multidrug resistance (MDR)

Medical biofilms
  antibiotic therapy, 117
  biofilm-associated infections (*see* Biofilm-associated infections)
  biofilm growth, 115, 116
  biotic/abiotic surfaces, 114
  Candida albicans, 116, 117
  Staphylococcus epidermidis, 116, 117
Medicinal plants and phytocompounds
  antibiofilm compounds, 204, 205
  bacterial biofilm inhibitors
    Achyranthes aspera, 206
    Aesculus hippocastanum, 208
    AKBA, 209
    Andrographis paniculata, 204
    anti-*Staphylococcus epidermidis*, 205
    baicalin, 207
    Carex pumila, 211
    chelerythrine, 210
    Curcuma longa, 206
    DHBF, 210, 211
    ellagic acid, 208
    Epimedium brevicornum, 209, 210
    IAN, 209
    kaurenoic acid, 213
    Listeria monocytogenes, 213
    Polygonum cuspidatum, 209, 210
    proAc, 210, 211
    proanthocyanidins, 207–208
    Propionibacterium acnes, 209, 210
    Rhodiola crenulata, 209, 210
    Rubus ulmifolius, 205–206
    Salvia triloba, 204
    sanguinarine, 210
    Streptococcus mutans, 204, 205
    tannic acid, 208, 209
    trans-resveratrol, 211, 212
    Vaccinium macrocarpon, 207
    ε-viniferin, 211, 212
    zingerone, 212
  broad spectrum biofilm inhibitor
    allicin, 215, 216
    casbane diterpene, 216
    epigallocatechin, 215
  fungal biofilm inhibitors
    Candida albicans, 213–215
    Cymbopogon citratus, 215
    epigallocatechin-3-gallate, 213, 214
    Muscari comosum, 214
    purpurin, 214
    Syzygium aromaticum, 215
  multidrug resistance, 203
  quorum-sensing inhibitors
    apigenin, 219, 220
    bergamottin, 218, 219

Index

Cuminum cyminum, 217
curcumin, 221
dihydroxybergamottin, 218, 219
hamamelitannin, 217, 218
kaempferol, 219, 220
Melia dubia, 217
naringenin, 219, 220
quercetin, 219, 220
Terminalia catappa, 217
Terminalia chebula, 221
ursolic acid, 217, 218
vanillin, 218, 219
use of, 203
in vivo efficacy, 221–222
Methicillin-resistant *Staphylococcus aureus* (MRSA), 23, 24, 129, 204, 238
Microban®, 364–365
Microbial metabolites
biosurfactants, 139–141
E. coli group II capsular polysaccharides, 137
furanone to exo-polysaccharides, 137
marine *Vibrio* sp., 137
QS molecules, 137–138
Microbial Surface Components Recognizing Adhesive Matrix Molecules (MSCRAMMs), 232, 233, 235
Minimum biofilm eradication concentration (MBEC), 144
MLVs. *See* Multilamellar vesicles (MLVs)
Mobile genetic elements (MGE), 65–66, 73, 82
Multidrug resistance (MDR)
and biofilms
ECM, 56, 57
NACS, 56
Trichosporon asahii, 56
in vitro model, 55
drug efflux
ABC transporters, 48–49
MFS transporters, 49–50
drug import, 47–48
ergosterol biosynthetic pathway, 46–47
MDR genes, regulation of, 51–52
mechanisms of
iron levels, 55
lipids in, 54
mitochondria and cell wall integrity, 52–54
overexpression of P45014DM, 46
pathogens, 379–380
target alteration, 45–46
Multilamellar vesicles (MLVs), 404, 407

N

Nanoemulsions chloro-aluminum phthalocyanine (NE-ClAlPc), 128–129
Nanoparticles
application
antimicrobial activity, 386, 388–389
biotechnology and medicine, 385–386
industry, 384
silver nanoparticles, 387
titanium oxide, 387
disadvantages
adverse effects, 387
inhalation, 387, 389
podoconiosis, 390
reactive oxygen species, 389
oral biofilm, 298–299
properties and synthesis
biological synthesis, 383
gas/vapor phase nanoparticle fabrication, 382
liquid phase fabrication, 383
size, 381
QPEI, 124
silver nanoparticles, 124
zinc oxide (ZnO), 124
National Institutes of Health, 2, 202, 336
Niosomes, 406–407
Nitric oxide (NO), 181
Non-albican *Candida species* (NACS), 56
Norfloxacin, 181
Nucleotide binding domains (NBDs), 49, 50

O

Oligon vantex silver catheter (OVSC), 174
Oral biofilm
biofilm-associated oral diseases
dental caries, 290–291
periodontitis, 291–292
clinical history, 287–288
composition and structure, 290
dental plaque, 287–289
oral microbiome, 289–290
pellicle formation, 288
prevention, 293–295
antibiotics (*See* Antibiotics, oral biofilm)
antiseptics (*see* Antiseptics)
DUWS, 292–293, 300, 301
mechanical plaque control, 295–296
nanoparticles, 298–299, 314

natural products, 306–308
photodynamic therapy, 297–298, 314
probiotics, 312–314
quorum sensing (*see* Quorum sensing (QS) systems)
vaccination, 310–312, 314

P

Parenteral Drug Association (PDA), 471
PDT. *See* Photodynamic therapy (PDT)
Peptide nucleic acid-based fluorescence in situ hybridization (PNA-FISH), 37–38
Periodontitis, 291–292, 311
Phage therapy, 131–133, 243
Phenol-soluble modulins (PSMs), 6, 101, 232, 234, 235
Photoactivated disinfection (PAD), 445, 455
Photodynamic therapy (PDT), 127–130, 314
 advantages, 297
 antimicrobial photodynamic therapy, 444–445
 dentistry, 444
 fungal biofilm, 278
 gram-positive/gram-negative bacteria, 445
 mechanism of action
  lethal damage, 446
  light source, 449–450
  photon absorption, 446
  photooxidation, 446
  photosensitiser, 448–449
  singlet oxygen, 446–447
 medical history, 443–444
 methylene blue, 298
 natural photosensitizers, 298
 oral biofilm
  advantages, 456
  bacteraemia, 453
  bacterial plaque, 450
  Candida spp., 451–452
  periodontal disease, 453
  Staphylococcus aureus, 451, 452
  treatment, 454–456
  in vitro susceptibility tests, 453
 photodestruction, 445
 polymicrobial biofilm, 336–338
 toluidine blue O/disulfonated aluminum phthalocyanine, 297–298
Photosensitizers (PS), 127–128, 278, 297, 298, 337
Plasmids
 biofilm resistance
  BDMDAC, 164–165

flow cell system, 163–164
QAC, 164–166
interfering substances, 162–163
pET28 vector, 159, 160
physiological effect
 cellular viability, 161
 conjugative plasmids, 161–162
 metabolic burden, 160
 non-conjugative plasmids, 162
 size, 161
pUC8 vector, 159–160
Poly(D,L-lactide-co-glycolide) (PLGA), 298–299, 409
Polymerase chain reaction (PCR), 14, 15, 18, 23
Polymer-based drug delivery system
 biocompatible/biodegradable polymers, 408–409
 hydrophilic/hydrophobic region, 407
 poly(lactide-co-glycolide) (PLGA), 409, 413
 poly(2-hydroxyethyl methacrylate) hydrogel, 410
Polymicrobial biofilm
 antimicrobial coating surface
  antibiotics, 340–341
  liquid coating, 341
  polyethylene glycol, 341
 biofilm dispersal/de-adhesion
  dispersin B, 331
  DNase, 331–332
  erosion, sloughing and seeding, 330–331
  nonenzyme-based approach, 332
 cell–cell signaling inhibitor
  acyl homoserine lactones, 333
  autoinducer-2, 329, 333
  sociomicrobiology, 332
  STAMPs, 333–334
 coaggregation
  definition, 334
  EDTA, 335–336
  guerrilla warfare approach, 335
  requirements, 334–335
 debridement, 325, 327
 definition, 324
 development of, 324–326, 342–343
 efflux pumps, 329
 extracellular polymeric substances, 325
 growth rate alteration, 328
 incidence, 324
 nanotechnology, 340–341
 nonthermal plasma technology

Index 495

cool plasma, 338
etching, 339
inactivating bacteria, 338–339
titanium, 340
photodynamic therapy, 336–338
reaction diffusion limitation, 327–328
Poly-morphonuclear neutrophils (PMN), 256, 257, 259
Poly N-acetylglucosamine (PNAG). *See* Polysaccharide intercellular adhesin (PIA)
Polyphenol, 307–308
Polysaccharide intercellular adhesin (PIA), 232, 238
Polytetrafluoroethylene (PTFE), 365
ProAntho-cyanidin A2-phosphatidylCholine (proAc), 210, 211
Probiotics, 130–131, 312–313
Protein kinase C (PKC), 52, 53
Proteus mirabilis, 106, 132, 180, 181
Pseudomonas aeruginosa biofilms, 254–255
cystic fibrosis, 254
tolerance of, 255
treatment strategies, 255–256
DNase, 259
IgG1 monoclonal antibodies, 262
IgY egg yolk antibodies, 261–262
perspectives for, 262–263
QSIs, 256–258
silver, 260–261

Q

QSIs. *See* Quorum sensing inhibitors (QSIs)
QS system. *See* Quorum sensing (QS) systems
Qualified Infectious Disease Product (QIDP) designation, 474
Quaternary ammonium compounds (QACs), 158–159, 164–166, 360–361
Quaternary ammonium poly(ethylene imine) (QA-PEI), 299
Quorum sensing inhibitors (QSIs)
medicinal plants and phytocompounds
apigenin, 219, 220
bergamottin, 218, 219
Cuminum cyminum, 217
curcumin, 221
dihydroxybergamottin, 218, 219
hamamelitannin, 217, 218
kaempferol, 219, 220
Melia dubia, 217
naringenin, 219, 220
quercetin, 219, 220
Terminalia catappa, 217
Terminalia chebula, 221
ursolic acid, 217, 218
vanillin, 218, 219
Pseudomonas aeruginosa
control biofilm tolerance, 257
eDNA, 258
prophylactic administration, 257
rhamnolipids, 256–257
and tobramycin, 257–258
in vitro and in vivo studies, 257
urinary tract catheters, biofilm development, 182
Quorum-sensing modulators (QSMs), 105–107
Quorum sensing (QS) systems
AHL-mediated QS, 96, 97
AI synthase, 96, 97
biofilm pathogens, 6–7
environmental conditions, 98
food processing, 368–369
methods for
biofabricated microenvironments, signaling dynamics, 104–105
monitor strains, 103
microbial metabolites, 137–138
oral biofilm
AHLs and AIPs, 308
autoinducer-2, 308–309
competence stimulating peptide, 309–310
definition, 289
plant products
Allium sativum, 134
antimicrobial agents and immune modulation, 135
biocide-free antibiofilm agents, 136
Boesenbergia pandurata, 133
Candida albicans, 134, 135
Capparis spinosa, 134
Curcuma xanthorrhiza, 133
drug target, 134
marine plants, 133
Ocimum americanum, 133
Pseudomonas aeruginosa, 136
TCMs, 136
QS-dependent biofilm processes
agr QS systems, 101
HSL-based QS systems, 101–102
LuxS/AI-2 QS system, 102
microorganisms, 98–100
stages of, 98, 100, 101

Quorum sensing (QS) systems (cont.)
    QS modulators, efficacy of, 105–107
    traditional microbiological investigations, 95–96

R
Radio frequency alternating electric current (RFAC), 434–435
Recurrent meningitis, 271
Red fluorescence protein (RFP), 85
RNAIII activating protein (RAP), 182
RNAIII-inhibiting peptide (RIP), 182
Rubus ulmifolius, 205–206, 240

S
Saccharomyces cerevisiae, 45, 47, 50–54
Salmonella genomic island 1 (SGI1), 71
Sanguinarine (SA), 210
Scanning electron microscopy (SEM), 38, 204, 212, 214
Selectively targeted antimicrobial peptides (STAMPs), 333–334
Selenocyanato diacetic acid (SCAA), 176–177
Sesquiterpene lactones (SLs), 221
Silver
    antimicrobial coatings
        contact lenses, 185–187
        CVCs, 174
        urinary tract catheters, 179–180
    Pseudomonas aeruginosa, 260–261
Silver-montmorillonite (Ag-MMT), 384
Silver sulfadiazine (SSD), 260, 261
Singlet oxygen, 446–447
Slippery Liquid-Infused Porous Surfaces (SLIPS), 243
Sociomicrobiology, 332
STAMPs. *See* Selectively targeted antimicrobial peptides (STAMPs)
Standard wound care (SWC), 421, 426, 431
Staphylococcus aureus biofilms
    biofilm development, 231, 232, 242
    enzymatic mechanisms, 232, 233
        dispersin B, 237–238
        hyaluronate lyases, 237
        lysostaphin, 238
        nuclease and DNases, 237
        proteases, 236
    healthcare systems, 230
    matrix components, 242
        eDNA, 232–234
        MSCRAMMs, 232, 233
        natural signaling, 232, 233
        PIA, 232
        β-toxin, 234
    quorum sensing system
        agr system, 234, 235, 242
        AIP signal, 234, 235
        Aur metalloprotease, 235
        PSMs, 234, 235
        RNAIII, 234
        SspA, 235
        Staphopains, 235, 236
    small-molecule inhibitors
        natural product inhibitors, 239–241
        synthetic inhibitors, 239, 241–242
Structure-based virtual screen (SB-VS), 143
SWC. *See* Standard wound care (SWC)

T
Tannic acid (TA), 208, 209
TENS. *See* Transcutaneous electrical nerve stimulation (TENS)
Tissue samples
    Brown and Brenn Gram staining, 30
    conventional and fluorescent stains, 33–34
    CSLM, 38–39
    definition, 39
    embedding and sectioning, 32–33
    fixation, 31
    fluorescent stain selection
        FISH, 36–38
        immunofluorescence, 35–37
        LIVE/DEAD® kits, 34
        SYTO® 9, 34
    Gram staining, 33–34
    H&E, 33
    HECM, 34
    intrinsic fluorescence, 31
    light microscopy, 38
    nonspecific binding, 31–32
    quantifying biofilm, 40
    specimen collection, 30
    transmission electron microscopy, 38
    tridimensional microbial communities, 30
Tobramycin (TOB)
    and QSIs, 257–258
    and silver sulfadiazine, 261
Toluidine blue O-induced photodynamic therapy, 298
Traditional Chinese medicines (TCMs), 136
Transcription factor (TF), 51–53
Transcutaneous electrical nerve stimulation (TENS), 433

Index

Transmembrane domains (TMDs), 49
Triclosan, 180–181, 302
Tumor treating felds (TTFields), 434
Type IV secretion systems (T4SS),), 68, 69, 80

U
Unilamellar vesicles (ULVs), 404
Urinary tract catheters
    antimicrobial coatings
        antibiotics, 181–182
        biological coatings, 182–183
        gendine, 181
        hydrogels, 179
        nitric oxide, 181
        quorum-sensing inhibitors, 182
        silver, 179–180

        triclosan, 180–181
    catheter-associated urinary tract infections, 177–179
U.S. National Institutes of Health, 202

V
Ventilator Associated Pneumonia (VAP), 257

W
96-Well microtiter plate model, 272
Wound biofilm
    electrical stimulation (*see* Electrical stimulation)
    pathogenesis, 420–421
    ulcer wounds, 420

Printed by Publishers' Graphics LLC
LMO140522.23.33.40